NanoScience and Technology

For further volumes:
http://www.springer.com/series/3705

The series NanoScience and Technology is focused on the fascinating nano-world, mesoscopic physics, analysis with atomic resolution, nano and quantum-effect devices, nanomechanics and atomic-scale processes. All the basic aspects and technology-oriented developments in this emerging discipline are covered by comprehensive and timely books. The series constitutes a survey of the relevant special topics, which are presented by leading experts in the field. These books will appeal to researchers, engineers, and advanced students.

Quan Li
Editor

Nanoscience with Liquid Crystals

From Self-Organized Nanostructures to Applications

Springer

Editor
Quan Li
Liquid Crystal Institute
Kent State University
Kent, OH
USA

ISSN 1434-4904 ISSN 2197-7127 (electronic)
ISBN 978-3-319-04866-6 ISBN 978-3-319-04867-3 (eBook)
DOI 10.1007/978-3-319-04867-3
Springer Cham Heidelberg New York Dordrecht London

Library of Congress Control Number: 2014935223

Printed on acid-free paper

Springer is part of Springer Science+Business Media (www.springer.com)

Preface

Liquid crystals (LCs) represent a fascinating state of matter which combines order and mobility at different hierarchical levels, i.e. from nanoscale to macroscale or in other words from molecular level to macroscopic levels. This unique combination enables such system to respond to different external stimuli such as temperature, magnetic field, electric field, mechanical stress, light, chemical reaction, and electrochemical reaction by finding a new configuration of minimum energy. Due to their dynamic nature, liquid crystalline nanostructures efficiently transmit and amplify information and properties over macroscopic scales. The responsive nature and diversity of LCs provide tremendous opportunities as well as challenges for insights in fundamental science, and open the door to numerous practical applications. Conventional nematic LCs have become the quintessential materials of the LC displays (LCDs) with annual more than $100 billion market and have drastically revolutionalized the way that we present information. With the LC displays ubiquitous in our daily life, the research and development of LCs are moving rapidly into the forefront of nanoscience and nanotechnology. With the advent of highly sophisticated instruments and techniques, the self-organized nanostructures of LCs can be probed in multiple dimensions and over different length scales and help in unraveling the structure-property relationships to an unprecedented level. Similarly, the development of devices with nanoscale control of morphologies has provided useful guidelines to design new materials with targeted properties and functionality. Recently, the nanoconfinement and nanoparticles of LCs exhibit new and exciting phenomena compared to bulk liquid crystalline phases owing to the presence of unusually large number of molecules on the surface. Promising new techniques have been developed to controllably orient liquid crystalline nanostructures in the desired direction in device configurations.

It is understood that nanotechnology is hunting for simple and versatile bottom-up self-assembly-based processes to assemble and (re)organize molecules and/or nanoparticles into well-defined functional superstructures in multiple dimensions over multiple length scales for many advanced technological applications. LCs are emerging as the viable systems for the dynamic self-assembly of nanomaterials in LC media as well as molecular dynamic self-assembly of LC molecules themselves. For example, it has been demonstrated that LCs enable reconfigurable and switchable self-assembly of dispersed nanoparticles directed by the nanoscale molecular ordering in these anisotropic fluids with a broad range of mesomorphic states. Furthermore, various nanostructured LC phases have been used as "templating" media for the preparation of anisotropic nanoparticles with tunable shape, size, and controllable polydispersity. Liquid crystalline nanostructures have also been employed for encapsulation and delivery of drugs. Processing nanomaterials through liquid crystalline phase yields high performance materials and devices, and LC noble metal nanoparticle superstructures are envisaged to furnish optical metamaterials. Thus, the focus of this book is on nanoscale phenomena of LCs and their manipulation, mutually beneficial properties of LCs and nanoparticles, and nanoscale control of self-assembly and self-organization of LCs toward efficient and enhanced device performances.

This book does not intend to exhaustively cover all the topics on nanoscience with LCs as it is extremely difficult to do so within a single book. Instead, the book focuses on the recent development of the most fascinating theme about nanoscience with LCs: from self-organized nanostructures to applications. The chapters span the following topics: holographic LCs for nanophotonics (Chap. 1), directing 3D topological defects in smectic LCs and their applications as an emerging class of building blocks (Chap. 2), liquid crystalline 1D and 2D carbon materials (Chap. 3), LC-gold nanoparticle hybrid materials (Chap. 4), photoresponsive chiral LC materials: from 1D helical superstructures to 3D periodic cubic lattices and beyond (Chap. 5), glassy LCs as self-organized films for robust optoelectronic devices (Chap. 6), directing self-organized columnar nanostructures of discotic LCs for device applications (Chap. 7), discotic liquid crystalline blends for nanostructure formation toward bulk heterojunction active layer in organic photovoltaics (Chap. 8), ion-based LCs: from well-defined self-organized nanostructures to applications (Chap. 9), nanotechnology and nanomaterials in photodeformable liquid crystalline polymers (Chap. 10), self-assembled liquid crystalline conjugated polymers: synthesis, development, and their advanced electro-optical properties (Chap. 11), and solubilization and delivery of drugs from GMO-based lyotropic liquid crystals (Chap. 12). In each chapter, the state of the art, along with future potentials in the respective fields is discussed and highlighted by the leading experts.

This book offers up-to-date and accessible coverage of nanoscience with LCs, with emphasis on self-organized nanostructures and their applications for undergraduate students and graduate students, as well as the researchers in both academia and industries in the fields of organic chemistry, polymer science, liquid crystals, materials science, material engineering, electrical engineering, chemical engineering, photonics, optoelectronics, nanotechnology, medicine, pharmacy, and renewable energy. I hope that readers will find this book professionally valuable and intellectually stimulating in the rapidly emerging area of nanoscience with liquid crystals.

Finally, I would like to express my gratitude to Claus Ascheron at Springer for inviting us to bring this exciting field of research to a wider audience, and to all our distinguished contributors for their dedicated efforts. I am indebted to my wife Changshu, my sons Daniel and Songqiao for their great support and affectionate encouragement.

Kent, December 2013 Quan Li

Contents

Contributors

Kazuo Akagi Department of Polymer Chemistry, Kyoto University, Katsura, Kyoto 615-8510, Japan, e-mail: akagi@fps.polym.kyoto-u.ac.jp

Abraham Aserin Casali Institute of Applied Chemistry, The Institute of Chemistry, The Hebrew University of Jerusalem, 91904 Jerusalem, Israel

Hari Krishna Bisoyi Liquid Crystal Institute, Kent State University, Kent, OH 44242, USA

Haider Butt School of Mechanical Engineering, University of Birmingham, Birmingham B15 2TT, Edgbaston, UK, e-mail: h.butt@bham.ac.uk

Shaw H. Chen Department of Chemical Engineering and Laboratory for Laser Energetics, University of Rochester, Rochester, NY 14623-1212, USA, e-mail: shch@LLE.rochester.edu

Nissim Garti Casali Institute of Applied Chemistry, The Institute of Chemistry, The Hebrew University of Jerusalem, 91904 Jerusalem, Israel, e-mail: garti@vms.huji.ac.il

Apiradee Honglawan Department of Chemical and Biomolecular Engineering and Materials Science and Engineering, University of Pennsylvania, 3231 Walnut Street, Philadelphia, PA 19104, USA

Sandeep Kumar Raman Research Institute, Bangalore, India

Quan Li Liquid Crystal Institute, Kent State University, Kent, OH 44242, USA, e-mail: qli1@kent.edu

Yannian Li Liquid Crystal Institute, Kent State University, Kent, OH, USA

Dima Libster Casali Institute of Applied Chemistry, The Institute of Chemistry, The Hebrew University of Jerusalem, 91904 Jerusalem, Israel

Hiromitsu Maeda College of Pharmaceutical Sciences, Ritsumeikan University, Kusatsu 525-8577, Japan, e-mail: maedahir@ph.ritsumei.ac.jp

Yunuen Montelongo Electrical Engineering Division, University of Cambridge, Cambridge, UK

Jane J. Ou Department of Chemical Engineering, University of Rochester, Rochester, NY 14623-1212, USA

H.-M. Philp Chen Department of Photonics and Institute of Display, National Chiao-Tung University, Hsinchu 30010, Taiwan

Benedict A. San Jose Department of Polymer Chemistry, Kyoto University, Katsura, Kyoto 615-8510, Japan

Yo Shimizu Research Institute for Ubiquitous Energy Devices, National Institute of Advanced Industrial Science and Technology, Kansai Center (AIST-Kansai), Osaka 563-8577, Japan, e-mail: yo-shimizu@aist.go.jp

Timothy D. Wilkinson Electrical Engineering Division, University of Cambridge, Cambridge, UK, e-mail: tdw13@cam.ac.uk

Chenming Xue Liquid Crystal Institute, Kent State University, Kent, OH, USA

Shu Yang Department of Chemical and Biomolecular Engineering and Materials Science and Engineering, University of Pennsylvania, 3231 Walnut Street, Philadelphia, PA 19104, USA, e-mail: shuyang@seas.upenn.edu

Haifeng Yu Department of Materials Science and Engineering, College of Engineering, Peking University, Beijing, China, e-mail: yuhaifeng@pku.edu.cn

Li Yu Department of Materials Science and Engineering, College of Engineering, Peking University, Beijing, China

Chapter 1
Holographic Liquid Crystals for Nanophotonics

Timothy D. Wilkinson, Haider Butt and Yunuen Montelongo

Abstract Nanotechnology offers a new paradigm in ways of controlling light in optical systems. Optically enhanced effects such plasmonic resonances and nano-antennas combined with diffraction and photonic bandgap effects can create new mechanisms to enhance the performance of modulation technologies in applications such as three dimensional displays. The power of these optical effects can then be made even more effective by adding in a variable refractive index material such a liquid crystal. This allows the optical properties to be tuned or modulated and creates a new class of optical devices which utilise features on the nano-scale. This chapter pulls together the various strands that have been developed in this area to make an initial investigation into these types of devices. The power of diffraction is introduced to propagate light in a manner which ideally suits nanotechnology. This is then combined with the algorithms used to create computer generated holograms to demonstrate that the diffraction process is indeed the key to the optical control mechanisms at length scales of the order of the wavelength of the light. The key properties of carbon nanotubes and liquid crystals are then introduced to provide the means to create enhanced diffraction through resonant effects which can then be tuned through the variable refractive index properties of the liquid crystals. The most important property of the nanotechnology is the ability to have electrically conductive structures on the nanometre length scale, which allows the rules of electric field interaction to ne manipulated by plasmonics. These effects are demonstrated using both conducting multiwall carbon nanotubes as well as silver nano-antennas. Plasmonic resonance in arrays of nanotubes show the predicted wavelength cut off due to a negative dielectric constant. The same effects are then linked with diffraction to create quasi-crystalline diffraction patterns and fully

T. D. Wilkinson (✉) · Y. Montelongo
Electrical Engineering Division, University of Cambridge, Cambridge, UK
e-mail: tdw13@cam.ac.uk

H. Butt
School of Mechanical Engineering, University of Birmingham, Birmingham B15 2TT, Edgbaston, UK
e-mail: h.butt@bham.ac.uk

Q. Li (ed.), *Nanoscience with Liquid Crystals*, NanoScience and Technology,
DOI: 10.1007/978-3-319-04867-3_1,

synthetic computer generated holograms. These effects are expanded further with the silver nano-antennas, where the enhanced resonance effects allow the control of polarisation as well as the wavefront through diffraction. Finally the liquid crystal element of variable refractive index is added to the devices to control the resonance and tune its performance. While this is still at a very early stage of research, it already demonstrates the power and versatility created by the combination of these different optical effects.

1.1 Introduction

Computer generated holography has a very long and distinguished career in the world of photonic research [1–5], however it has never quite lived up to its potential due to the limitations of the technology chosen to display the holograms. Classical photographic film holograms as pioneered by Gabor [6, 7] and later Leith [8], demonstrate the true potential of holograms for the display of three dimensional (3D) images, however research has yet to come up with a truly reconfigurable version that would allow computer generated holograms (CGHs) to live up to their true potential. Recent work has demonstrated reconfigurable holograms for 3D display based on photo-addressed polymer materials [9], but the technique is still limited by the need for an optical assembly to allow the interference patterns to be recorded. The sort of modulation that is required to replicate these holograms is highly complex in nature, very high resolution and uses sub-wavelength pitch pixels. The challenge is to find a dynamically addressed technology that can in some way come close to the sort of modulation that has been achieved with optically recorded interference patterns.

The power of the CGH is that the hologram can be calculated offline and then stored for later display, or can be calculated in real time using an appropriate algorithm. Once the CGH has been calculated, the next stage is then to match the hologram to the desired modulation technology. Liquid crystals (LCs) offer a tantalizingly close approximation to the sort of refractive index change seen in photographically reordered holograms, only with a much reduced resolution and larger pixel size. The holograms written in [4] show that LCs can get close to the sort of resolutions desired in these holograms and recent advances in ultra high-definition displays show that it is possible to electrically address the huge amount of information contained within these holograms. The final hurdle is to reduce the pixel pitch in these LC devices to get to the order or even better below the wavelength of the light used to illuminate the hologram. This is still a very significant challenge, however nano-photonic devices are beginning to hint at the potential for this technology. Materials such as multi-walled carbon nanotubes (MWCNTs) off the possibility of integrating sub-wavelength structures which can be sued to electrically control the LC director [10]. These materials are also demonstrating enhanced electronic properties such as plasmonic resonances [11] which further enhance their optical properties and expand the potential of their use in holography. In this chapter we explore the potential of these structures and where applicable their interaction with LCs in order to make the first steps towards a truly electrically reconfigurable holographic structure.

1.2 Computer Generated Holography

1.2.1 Diffraction Through an Aperture

In order to understand how nano technology can be used in computer generated holography, we need understand the basics of diffraction theory. This is covered in much better detail in several fundamental texts [12, 13], so we will only summarise the key aspects in this chapter. Let us assume we have an arbitrary aperture (hole) function, $A(x, y)$ in the plane S as shown in Fig. 1.1, with coordinates $[x, y]$. The light passing through this aperture will be diffracted at its edges and the exact form of this pattern can be calculated. We want to calculate the field distribution at an arbitrary position away at the point P, which is a distance R from the aperture.

If we consider an infinitely small differential of the aperture, dS, we can model this as a point source of light emitting spherical 'Huygens' wavelets with an amplitude of $A(x, y)dS$. The wavelet acts as a radiating point source, so we can calculate its field at the point P, a distance r from dS. The point source dS can be considered to radiate a spherical wave front of frequency ω. The total field distribution at P is evaluated by superposition (summation) of all the wavelets across the aperture. The process of interference of these spherical wavelets is called diffraction and is based on the Huygens-Fresnel approximation. In order to understand and analyse the propagating wavelets, a series of approximations and assumptions must be made. If we consider only the part of the wavelets which are propagating in the forward ($+z$) direction and are contained in a cone of small angles away from the z axis, then we can evaluate the change in field dE at the point P, due to dS. As the wavelet dS acts as a point source, we can say that the power radiated is proportional to $1/r^2$ (spherical wavefront), hence the field dE will be proportional to $1/r$. We can see that for a real propagating wave of frequency ω and wave number k, ($k = 2\pi/\lambda$) we have at point P,

$$dE = \frac{A(x,y)}{r} e^{jwt} e^{-jkr} dS. \tag{1.1}$$

Now, we need to change coordinates to the plane containing the point P, which are defined as $[\alpha, \beta]$ to give the full expression for each wavelet in terms of x and y ($dS = dxdy$) for dE will now be.

$$dE = \frac{A(x,y)e^{j\omega t} e^{-jkR\sqrt{1-\frac{2\alpha x+2\beta y}{R^2}+\frac{x^2+y^2}{R^2}}}}{R\sqrt{1-\frac{2\alpha x+2\beta y}{R^2}+\frac{x^2+y^2}{R^2}}} dxdy. \tag{1.2}$$

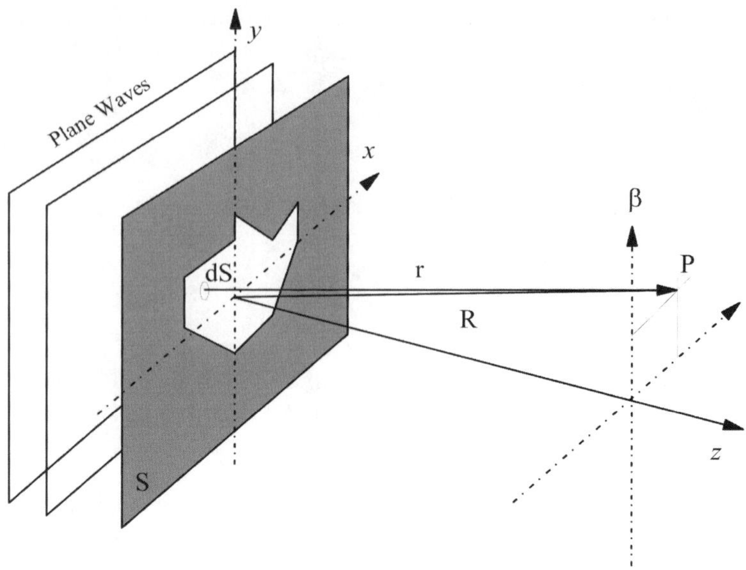

Fig. 1.1 Aperture and diffraction coordinate system

Such an expression can only be solved directly for a few specific aperture functions. To account for an arbitrary aperture, we must approximate, simplify and restrict the regions in which we evaluate the diffracted pattern. If the point P is reasonably coaxial (close to the z axis, relative to the distance R) and the aperture $A(x, y)$ is small compared to the distance R, then the lower section of (1.2) for dE can be assumed to be almost constant and that for all intents and purposes, $r = R$. The similar expression in the exponential term in the top line of (1.2) is not so simple. It cannot be considered constant as small variations are amplified through the exponential. To simplify this section we must consider only the far field or Fraunhofer region where,

$$R^2 \gg x^2 + y^2. \tag{1.3}$$

In this case, the final term in the exponential can be considered negligible. To further simplify, we use the binomial expansion, and keep the first two terms only to further simplify the exponential expression. Hence the simplified version of the field dE, can be expressed as,

$$dE = \frac{A(x, y)}{R} e^{j(\omega t - kr)} e^{jk\left(\frac{\alpha x + \beta y}{R}\right)} dxdy. \tag{1.4}$$

The regions of the approximation are defined such that in the far field or Fraunhofer region, the approximations are accurate, hence the field distribution $E(x, y)$ only changes in size with increasing z, rather than changes in structure as demonstrated in Fig. 1.2. In the case where the approximation is bearably accurate,

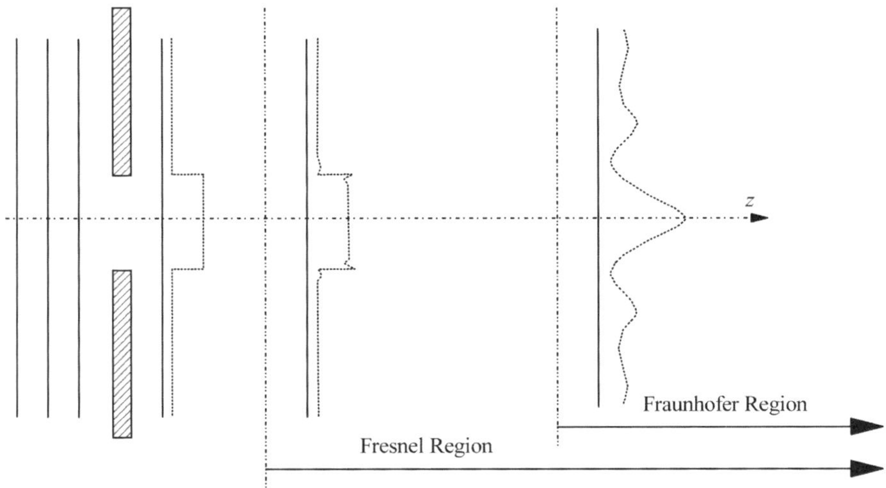

Fig. 1.2 Diffraction regions for a square aperture

we are in the Fresnel region. Before the Fresnel region, the evaluation of E is extremely difficult and is defined as the near field diffraction pattern. The exact boundary of the Fresnel region will depend on the acceptable accuracy and can be found in Goodman [12].

The total effect of the dS wavelets can be integrated across dE to get an expression for the far field or Fraunhofer diffraction pattern. The initial exponential term $e^{j(\omega t - kR)}$ in (1.4) refers the wave to an origin at $t = 0$, but we are only interested in the scaling of relative points at P with respect to each other, so it is safe to normalise this term to unity. Thus, our final expression for the far field diffraction pattern becomes,

$$E(\alpha, \beta) = \iint_A A(x, y) e^{jk(\alpha x + \beta y)/R} dxdy. \tag{1.5}$$

Hence the far field diffraction pattern at the point P is related to the aperture function $A(x, y)$, by the Fourier transform. The final step is to remove the scaling effect of R in the equation, as it does not affect its structure, just its size. The coordinates $[\alpha, \beta]$ are absolute and are scaled by the factor R. For this reason, we normalise the coordinates and define the Fourier transform of the aperture in terms of its spatial frequency components $[u, v]$,

$$u = \frac{k\alpha}{2\pi R} \quad \text{and} \quad v = \frac{k\beta}{2\pi R}, \tag{1.6}$$

and the final relationship is,

$$E(u, v) = \iint_A A(x, y)e^{2\pi j(ux+vy)}dxdy. \tag{1.7}$$

And inversely we can calculate the aperture from the far field pattern.

$$A(x, y) = \iint_A E(u, v)e^{-2\pi j(ux+vy)}dudv. \tag{1.8}$$

This relationship also gives us the scaling law for computer generated holograms. As an example, the far field pattern of a square aperture of width 1 mm, at a wavelength of 633 nm can only be accurately measured 10 m away from the aperture. Such a far field distance is clearly difficult to achieve in practical terms, hence either smaller apertures are required, or a positive focal length lens can be included directly after the aperture to create the far field of the aperture in the focal plane of the lens [12]. A positive lens performs a Fourier transform of the aperture placed behind it. The final result for the diffracted aperture $A(x, y)$ a distance d behind a positive focal length lens of focal length f, given by,

$$E(\alpha, \beta) = e^{\frac{jk}{2f}\left(1-\frac{d}{f}\right)\left(\alpha^2+\beta^2\right)} \iint_A A(x, y)e^{\frac{jk}{f}(\alpha x+\beta y)}dxdy. \tag{1.9}$$

Once again we can translate the equation into spatial coordinates such that now the scaling coordinates are,

$$u = \frac{k\alpha}{2\pi f} \quad \text{and} \quad v = \frac{k\beta}{2\pi f}. \tag{1.10}$$

We want to calculate the far field or Fraunhofer region for a square aperture. This aperture can be represented in one dimension as a rectangular or 'Rect' function. Hence the Fourier transform of this will be a one dimensional sinc function, where $\text{sinc}(x) = \sin(x)/x$. The sinc finction is shown in Fig. 1.3 and is one of the fundamental structures that dominates the information generated by computer generated holograms

It is simple to repeat this calculation in the orthogonal direction to create the far field diffraction pattern in two dimensions. The far field of a square aperture with transmission of A and width a therefore its Fourier transform,

$$F(u, v) = Aa^2\text{sinc}(\pi au)\text{sinc}(\pi av). \tag{1.11}$$

The resulting two dimensional sinc function has a very strong effect on the structure of the replay field of a hologram as it forms the overall envelope that will contain the desired information as well as having repeating sidelobes that represent the higher order diffraction terms as shown in Fig. 1.3. Now we will look at what

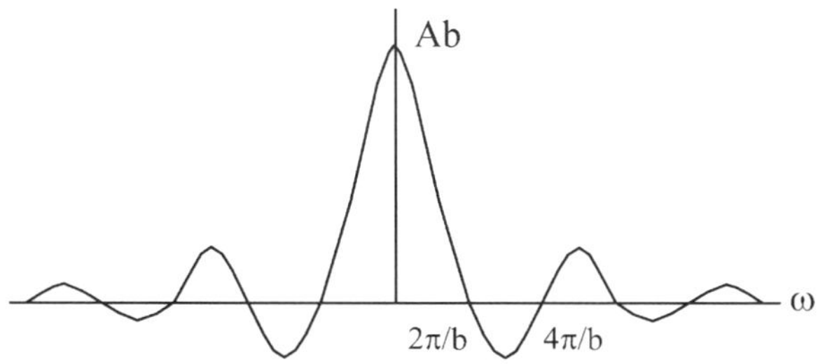

Fig. 1.3 The one dimensional sinc function (sinc envelope)

happens when we shift the aperture by a distance *a*/2 from the origin of the plane using the shift theorem for Fourier transforms.

$$F(u,v) = Aa^2 e^{j\pi au}\mathrm{sinc}(\pi au)\mathrm{sinc}(\pi av). \tag{1.12}$$

The result in (1.12) is same sinc function as (1.11), but with an additional phase term due to the shift. By combining an array of these apertures (now referred to as pixels) at various positions on a regular sampled grid, it is possible to generate a complex amplitude function in the far field. Such a two dimensional (2D) combination of these pixels in various positions is defined as a Hologram and the pattern generated by the hologram if the far field is the Replay Field. Hence a hologram can be designed by selecting the positions of the apertures that will combine together through diffraction to generate the desired optical intensity in the replay field. By altering the value of the transmitted amplitude *A* of each pixel, centred on a grid of interval *a* (note that the sampling grid does not have to equal the size of the aperture), it is possible to add up the 2D sinc functions and create an arbitrary 2D distribution in the far field region. By superimposing all the exponential phase terms due to the shift and varying the amplitude A, it is possible to create useful patterns in the far field. In general terms, the broader the feature or combination of pixels, the smaller or more delta function-like the replay object. Also, repetitive pixel patterns in the hologram leads to repetitive features in the replay field.

The exact structure of the replay field distribution depends on the shape of the "fundamental" pixel and the number and distribution of these pixels in the hologram. The pattern we generate with this distribution of pixels is repeated in each lobe of the sinc function from the fundamental pixel. The lobes can be considered as spatial harmonics of the central lobe, which contains the desired 2D pattern. For example, a line of square pixels with alternate pixels being one or zero (i.e. a square wave) would have the basic replay structure seen in Fig. 1.4.

There is a direct analogy between the one dimensional (1D) and 2D examples. The repetition of a square wave leads to discrete sampling in the frequency domain. In the case of the square wave, there is a series of odd harmonics generated. In 2D, these harmonics appear as orders radiating out in the lobes of the sinc function from the dimensions of the fundamental aperture or pixel. The more pixels we have in the hologram, the closer we get to the infinite case and spots generated become more like delta functions.

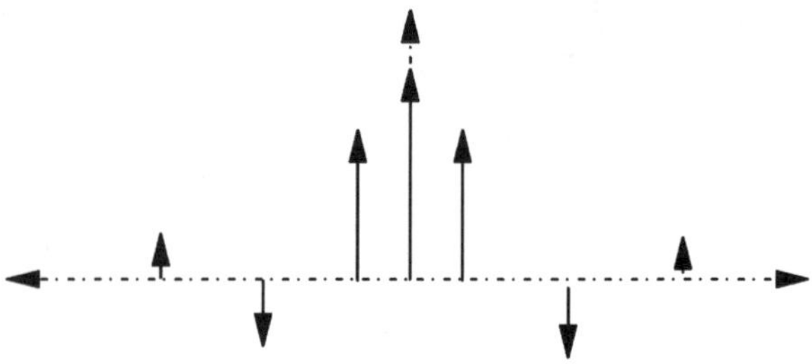

Fig. 1.4 Fourier transform of a square wave (50/50 mark space ratio)

So far we have assumed that the aperture A can only modulate intensity or amplitude between pixels. There is a limitation if we have binary amplitude modulation $A \in [0, 1]$ which is that the central point at the origin (defined as the zero order) of the replay field can only be zero if all values of A are zero. This is because the point at the origin of the replay field is proportional to the average of the pixels in the grating or hologram. A better modulation scheme would be binary phase modulation where $A \in [+1, -1]$. If there are the same number of pixels set to $+1$ as set to -1, then the average will be zero and the zero order will be zero too. A simple proof of this is as follows. Hence, with binary phase modulation ($A \in [+1, -1]$), the pixel in the centre of the replay field can be defined by the structure of the hologram. A drawback of both these binary modulation schemes is that the hologram will always be a real function.

$$F_T[h(x, y)^*] = H(-u, -v). \tag{1.13}$$

If function $h(x, y)$ is real then $h(x, y)^* = h(x, y)$, hence we cannot differentiate between $H(u, v)$ and $H(-u, -v)$ which means that both must appear in the replay field. Hence any replay field generated by a binary phase or amplitude hologram will always have 180° rotational symmetry. This symmetry restricts the useful area of the replay field to the upper half plane of the sinc envelope, as any pattern generated by the hologram will automatically appear as desired as well as rotated about the origin by 180°.

The next step is to look at 2D patterns such as the chequerboard pattern in Fig. 1.5, of pixels on an equally spaced grid (for simplicity A is restricted to binary values (such as 0 or 1). The chequerboard can be generated by the XOR of a 2D grating with itself rotated by 180°. Hence, the replay field will be made from the convolution of the FT of the two gratings.

1.2.2 Calculating Computer Generated Holograms

There is no simple way of generating a hologram analytically except for general cases such as gratings [14] and chequerboards. In order to create a hologram which generates an arbitrary replay field we need a more sophisticated algorithm. In order

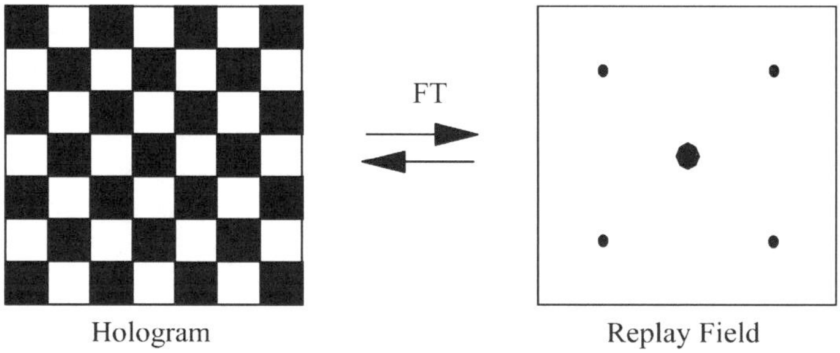

Fig. 1.5 Chequerboard hologram and its far field diffraction pattern

Fig. 1.6 Target function of 4 × 4 spots (on a 64 × 64 grid)

to achieve this we must use optimisation techniques such as simulated annealing [15], Gerchberg-Saxon [16] or the genetic algorithm [17]. To illustrate the basic concept of these algorithms we shall look at a simple binary phase only computer generated hologram (CGH) example based on direct binary search, which is a simplification of simulated annealing. Say we want an array of 4 × 4 delta functions in a square grid spaced as seen in Fig. 1.6.

If we take the FT of the target replay field of Fig. 1.7, then take the phase and threshold it about $\pi/2$ to create a binary hologram as shown in the left side of Fig. 1.7.

The "hologram" and replay field in Fig. 1.8 are clearly not what we desire. Hence we need a better way of finding the optimum combination of pixels to give us the target replay field we desire. The simplest method of generating a CGH, avoiding the problems of Fig. 1.8, is direct binary search (DBS). In this technique we take a hologram of random pixel values and then calculate its replay field. We then flip the binary value of a randomly positioned pixel and calculate the new replay field. We then subtract the two replay fields from the target replay field, sum up the differences to form a cost function for the hologram before and after the pixel change. If the cost function after the pixel has been flipped is less than the

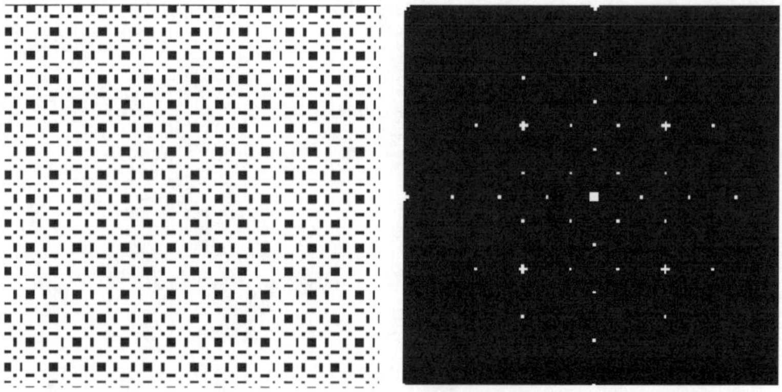

Fig. 1.7 CGH generated by thresholding the phase of FT (target). *Left*, binary phase hologram, right is the replay field generated from it

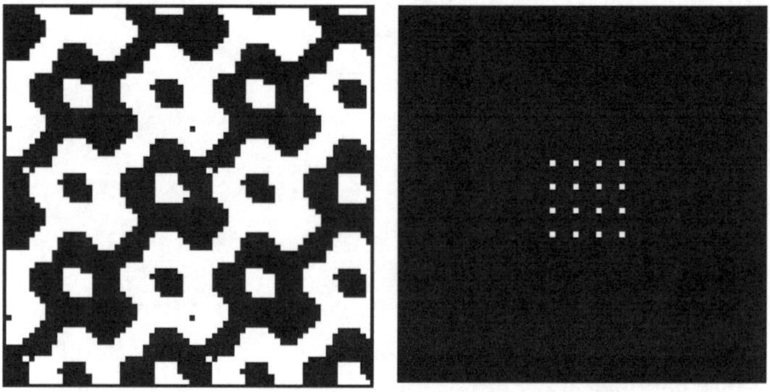

Fig. 1.8 Hologram generated by DBS (*left*) and its replay field (*right*)

cost function before the pixel was flipped, then the pixel change is considered to be advantageous and is accepted. The new cost function is then used in comparison to another randomly chosen flipped pixel. The process repeats until no further pixels can be flipped to give an improvement in the cost function.

Direct Binary Search

(1) Define an ideal target replay field, $\boldsymbol{T}$ (desired pattern as in Fig. 1.6).
(2) Start with a random array of binary phase pixels.
(3) Calculate its replay field (FT), $\boldsymbol{H_0}$.
(4) Take the difference between $\boldsymbol{T}$ and $\boldsymbol{H_0}$ and then sum up to make the first cost, $\boldsymbol{C_0}$.
(5) Flip a pixel state in a random position.
(6) Calculate the new replay field, $\boldsymbol{H_1}$.

(7) Take the difference between $\boldsymbol{T}$ and $\boldsymbol{H_1}$ then sum up to make the second cost, $\boldsymbol{C_1}$.
(8) If $\boldsymbol{C_0} < \boldsymbol{C_1}$ then reject the pixel flip and flip it back.
(9) If $\boldsymbol{C_0} > \boldsymbol{C_1}$ then accept the pixel flip and update the cost $\boldsymbol{C_0}$ with the new cost $\boldsymbol{C_1}$.
(10) Repeat steps 4–9 until $|\boldsymbol{C_0} - \boldsymbol{C_1}|$ reaches a minimum value.

This is not a fully optimum means of generating a hologram, but it gives a very good approximation as demonstrated by Fig. 1.8. More sophisticated techniques are required to fully exploit the possible combinations of pixel values in the hologram. One such algorithm is simulated annealing, which uses DBS, but also includes a probabilistic evaluation of the cost function, which changes as the number of iterations increases. The idea is to allow the hologram to "float" during the initial iteration, with good and bad pixel flips being accepted. This lets the optimisation float into a more global minima rather than getting stuck in a local minima as is the case with DBS.

1.3 Carbon Nanotubes

Carbon is the most versatile element in the periodic table because of the different arrangements of electrons around the nucleus of the atom and hence the number of bonds it can form with other elements. There exists three allotropic forms of carbon; graphite, diamond and buckminsterfullerene. Graphite consists of layered planar sheets of sp^2 hybridized carbon atoms bonded together in a hexagonal network. Diamond has a crystalline structure where each sp^3 hybridized carbon atom is bonded to four others in a tetrahedral arrangement. The third allotrope buckminsterfullerene or fullerene (C_{60}) is made up of spheroidal or cylindrical molecules with all the carbon atoms sp^2 hybridize. Smalley [18] proposed the existence of a tubular form of fullerene termed a carbon nanotube (CNT). The experimental evidence of the existence of carbon nanotubes was discovered by Iijima [19] by imaging multi-walled CNTs using transmission electron microscopy. Iijima discovered single walled carbon nanotubes as well 2 years later. A single walled carbon nanotube can be considered as a rolled graphene sheet. CNTs are one of the more promising materials for device fabrication because of their excellent electrical, thermal and mechanical properties in addition to high aspect ratio and high resistance to chemical and physical attack.

CNTs are metallic or semiconducting based on the exact way the CNTs are wrapped. Multi-walled CNTs are metallic in most cases as there are so many layers and the probability is very high of having one wrapped layer in the group to being metallic.

CNTs find applications in the areas such as micro electronics, field emission displays, X-ray sources and gas sensors. Single walled and multi-walled CNTs can be grown using high pressure arcs, laser ablation and chemical vapour deposition.

It is essential to develop a process which enables high yield, high uniformly perfectly aligned nanotubes at precise locations on the substrate in order to utilize their unique properties. The plasma enhanced chemical vapor deposition (PECVD) technique was used to grow uniform, perfectly aligned nanotubes on a Si substrate [20]. The process of using PECVD along with lithographic patterning without the deposition of amorphous carbon (a-C) has been largely ignored in the literature but this work showed the impact of the C_2H_2: NH_3 ratio in the growth of the carbon nanotubes and a-C deposition in the conventional PECVD technique. The plasma is not necessary for the nucleation and growth of nanotubes, but the electric field induced by the plasma is required for the alignment of the nanotubes. The C_2H_2 provides carbon for the growth of the nanotubes, while the NH_3 etches the carbon and hence a balance between the growth of the tubes and removal of a-C is achieved. The C_2H_2: NH_3 ratio was varied from 15 to 75 % in order to investigate the influence of the gas composition on the formation of a-C. The chemical composition of the unpatterned Si areas was investigated by Auger electron spectroscopy (using an in situ 2 keV Ar-ion gun). When the C_2H_2 ratio increased beyond 30 %, a-C deposition also increased and they found a peak when it crossed 50 %. An interface region of Si, C, N_2 and O_2 formed for all C_2H_2 concentrations. It was found that the optimal C_2H_2 ratio for clean nanotubes deposition lies between 15 and 30 %. The thickness of the Ni thin film catalyst controls the diameter, height and density of the nanotubes. The spacing between the nanotubes can be controlled via lithography and the height of the nanotubes by the deposition time of the tip growth mechanism. This pattern and growth method was used for growing uniform patterned multi-walled nanotubes in the nanophotonic devices.

Where V the voltage is applied across the electrodes and d is the gap spacing between the electrodes. If one of the electrodes is replaced by a sharp protrusion or a carbon nanotube, then $d$is interpreted as the minimum distance between the electrodes. So the local field (within 1–2 nm of the surface atoms) is much higher than the applied field. So the field is very high close to the tip and causes field emission of electrons above a threshold value. The shape of the electric field that spawns from the carbon nanotube in vacuum is also found to be near Gaussian [21].

1.4 The Optics of Nematic Liquid Crystals

The nematic mesophase is the one of the most common calimatic (rod-like) LC mesophases [22]. A large birefringence and a low control voltage distinguish the nematic phase from other electro optical materials. The molecules in the nematic phase only have long range order and no longitudinal order (i.e. do not form layers). This is the least ordered mesophase before the isotropic. The calimatic molecular shape gives order in the nematic phase, which means that on the average the molecules spend slightly less time spinning about their long axis than they do about their light short axis.

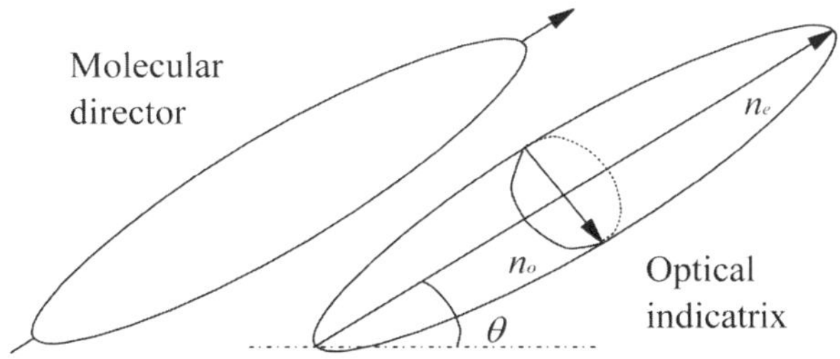

Fig. 1.9 Refractive indicatrix of a nematic liquid crystal molecule

The optical indicatrix or index ellipsoid of a liquid crystal is a 3D structure to represent the variation of the velocity of light passing through it with respect to the molecular director. When light enters a birefringent material like a liquid crystal, its electric field is split into two orthogonal components termed ordinary and extra ordinary rays corresponding to the ordinary refractive index (n_o) and extra ordinary refractive index (n_e) as shown in Fig. 1.9. In the case of uniaxial samples, one of the components lies on the equatorial plane and always has the same value, called ordinary component. The other component varies with the angle of incidence and is known as the extra ordinary component. All these parameters are bulk parameters, which are estimated by taking the statistical average across billions of individual molecules.

If the optical indicatrix is oriented at an angle θ to the plane of the cell such as the plane of the glass walls and ITO electrodes, then the refractive index seen by light passing perpendicular to the cell wall is given by

$$n(\theta) = \frac{n_e n_o}{\left[n_e^2 \sin^2\theta + n_o^2 \cos^2\theta\right]^{\frac{1}{2}}}. \tag{1.14}$$

From the above equation, the optical retardance Γ can be calculated for a given sample of thickness d and at a wavelength λ when the material is oriented at an angle θ to the light polarization direction.

$$\Gamma = 2\pi d(n(\theta) - n_o)/\lambda. \tag{1.15}$$

The main properties that are used in our holographic and nanophotonic device developments are birefringence, dielectric anisotropy and fluid viscosity.

- Microscopic Anisotropy + ordering → Macroscopic Anisotropy
- Fluidity → Permits reorganization.

The speed of light in a medium is a direct function of the refractive index of that medium. Birefringent materials have refractive indices that are direction dependent. If the refractive indices for parallel and perpendicular polarised light are different within a material, then the light will travel at different velocities depending on its polarisation (the orientation of the electric field) through the birefringent medium. Interaction between the light waves polarised parallel and perpendicular to the molecular axis, depend on the wavelength and give rise to different colours when viewed through an optical microscope with white light. Dielectric anisotropy of liquid crystals may be defined as the difference between the dielectric permittivities parallel and perpendicular to the director. Interaction of a liquid crystal with the external electric field is very much dependent on their dielectric properties.

1.5 Carbon Nanotube Plasmonic Devices

The optical properties of individual multi-wall CNTs (MWCNTs) are defined by their dielectric function, which is anisotropic in nature and matches very closely with that of bulk graphite [23]. However, the highly dense periodic arrays of MWCNTs display an artificial dielectric function, with a lower effective plasma frequency in a few hundreds of terahertz. Pendry [24] demonstrated that the electromagnetic response of a metallic array composed of thin metallic wires, excited by an electric field parallel to the wires is similar to that of a low-density plasma of very heavy charged particles, with a plasma frequency $\omega_{p:}$

$$\omega_p^2 = \frac{2\pi c_0^2}{a^2 \ln(a/r)} \tag{1.16}$$

where c_0 is the velocity of light in vacuum, a is the lattice constant of the 2D wire array, r is the radius of the wires. This concept can be used for lowering the plasma frequency in the nanotube based applications and achieving negative dielectric constant for producing metamaterials. The lowering of the plasma frequency is due to the increase in the effective electronic mass on the nanotubes due to the induced current and corresponding magnetic field around them. According to (1.16), the effective plasma frequency strongly depends on the nanotube radius and lattice constant. Their values can discretely be chosen to engineer the MWCNT arrays of a desired frequency. The resultant frequency dependent permittivity can be calculated using the Drude model for metals [25, 26] described as

$$\varepsilon(\omega) = 1 - \frac{\omega_p^2}{\omega^2} \tag{1.17}$$

The effective permittivity $\varepsilon(\omega)$ is negative for frequencies less than ω_p, therefore no wave propagation will take place inside the material. Electromagnetic waves propagation only occurs above ω_p, due to which the structure acts as a nanophotonic high-pass filter. The structure was realized with square lattice array of MWCNTs having radius of 50 nm and lattice constant of 400 nm. The growth of high *a/r* aspect ratio arrays of MWCNTs was a significantly difficult task overcome through e-beam lithography and optimized growth recipe. To the best of our knowledge such highly dense periodic arrays of nanotubes have not been accomplished by any other research groups. Such small lattice constants are only reported in non-periodic patterned CNT growth or forests form growth of CNT arrays.

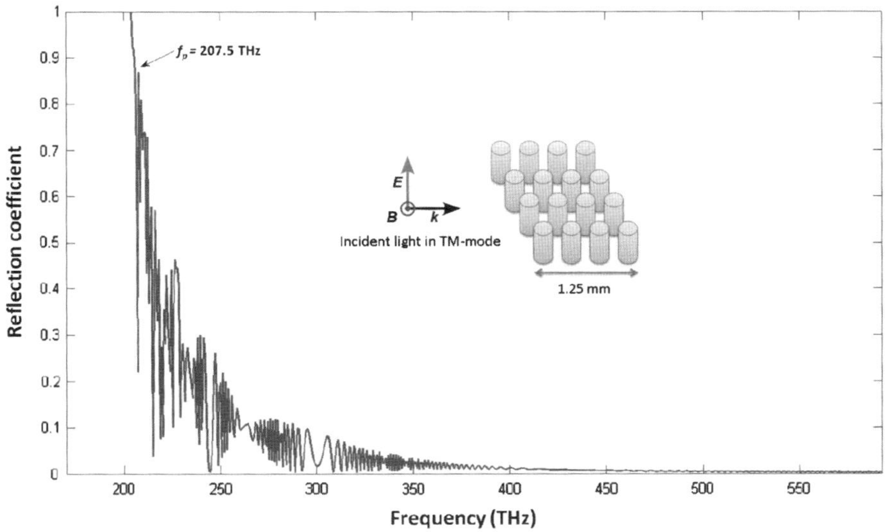

Fig. 1.10 The calculated reflection coefficient for a CNT based plasmonic high pass filter. At plasma frequency of 207.5 THz a sharp drop in reflectivity is calculated. The incident light is polarized parallel to the wire structure

The frequency dependent reflection coefficient was calculated as shown in Fig. 1.10. A sharp drop in the reflection is calculated at the plasma frequency, displaying a high transmission of the electromagnetic waves with larger frequencies [11]. The small peaks in the plot can be attributed to the multiple reflections at the interfaces. The result significantly matches the band gaps simulations of a 2D square lattice array of MWCNTs. For an array of the same lattice aspect ratio and incident light polarized parallel to the nanotubes (transverse magnetic (TM) mode), a band gap is observed close to the plasma frequency. However, in transverse electric mode (TE) (light polarized perpendicular to the nanotubes) no significant reflection band was observed in this regime, suggesting the plasmonic response of the structure in strongly dependent on polarization of the incident light.

To achieve this plasmonic high pass filter in the optical domain, metallic cylinders of nano-scale dimensions and interspacing are required. MWCNTs are promising materials to establish such metamaterials structures and the advancement in nano-technology facilitates the fabrication of high *a/r* aspect ratio nanotube arrays. Square lattice arrays of vertically aligned MWCNTs were grown on silicon substrates. An array of individual nanotubes is shown in the electron microscopy image of Fig. 1.11a and with a higher magnification view in Fig. 1.11b.

Due to the small lattice constant the nanotubes were tangled at various regions of the substrate. Additionally, some array defects were produced due to inhomogeneous depth of the nickel catalyst layer, affecting the decomposition of the carbon source, and hence, producing shorter MWCNTs. However, a steady periodicity required for a 2D square lattice was common throughout. Achieving longer

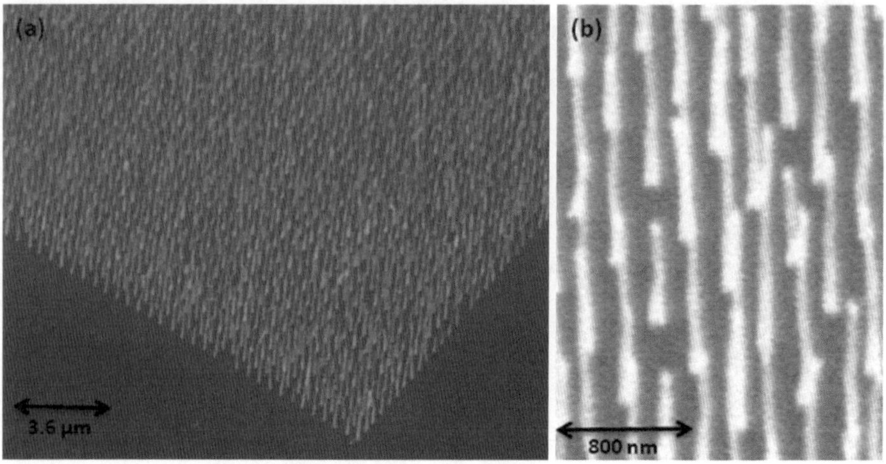

Fig. 1.11 **a** Electron microscopy image of a 2D square lattice array of MWCNTs having radius of 50 nm and lattice constant of 400 nm, grown on silicon substrate using PECVD. **b** The same in higher resolution

MWNCTs also remains a difficult task. Beyond the height of 4 μm the nanotubes become fairly thin leading to their tips collapsing with the neighboring tubes. Growth of longer tubes will require further optimization of the plasma generation technique and stronger electric fields for the vertical alignment of the tubes. According to the theory [24] the height of the wire structures must be longer than the operating wavelengths. Therefore, a MWCNT height of 2 μm was sufficient for this study as it established an aspect ratio (height to radius) of 40 and is longer than the plasma wavelength $\lambda_p = 1.45$ μm ($\lambda_p = c_0/f_p$) of the material.

The measured reflection spectrum at near normal incidence for light polarized parallel (TM) and perpendicular (TE) to the nanotubes is shown in Fig. 1.12. It shows a rapid drop in reflection after a frequency of 207.3 THz and significantly matches the calculated plasma frequency of the sample. Some small peaks were observed which can be explained by the defects in the periodic array of MWCNTs. The spectrum with light polarized perpendicular to the nanotubes did not show any significant cut-off effect, showing that most of the light was transmitted. However, a rapid drop in reflection at f_p for parallel polarized light shows that the periodic array of MWCNTs acts as high pass filter for near optical frequencies. The plasma frequency for the metamaterial can be further reduced into the optical regime by increasing the material density of the sample, i.e. by increasing the radius of the tubes and decreasing the lattice constant as presented in (1.15). Growth of well aligned MWCNTs at interspacing of less than 500 nm was a considerable achievement. Smaller lattice constant can be achieved by optimizing the catalyst layer and growth time in the PECVD process [27].

The results in Fig. 1.12 show how plasmonic effects in MWCNT arrays can be harnessed to control light in an effective manner. Ideally we could now add a

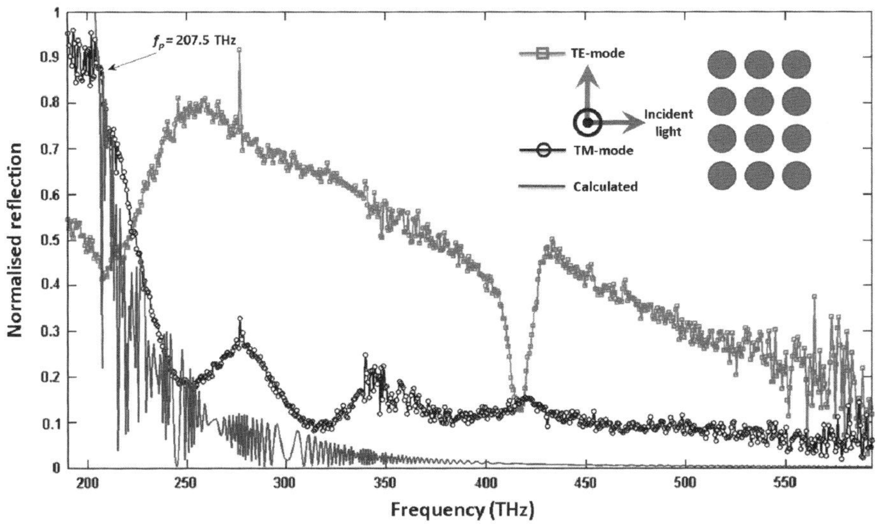

Fig. 1.12 The reflection measurement from the sample for light polarized parallel (*TM*) and perpendicular (*TE*) to the multi-walled carbon nanotubes. A sharp drop in the reflected TE-mode is observed at plasma frequency (207.5 THz) which significantly matches the theoretical result. No such cut-off effect is observed for the TE-mode

liquid crystal material to the arrays of MWCNTs, allowing us to vary the refractive index about the CNTs with an applied electric field and therefore allow us to vary the plasmonic resonant cut off. Unfortunately adding a higher refractive index material to (1.15) pushes the plasmonic resonance further into the longer wavelengths (near infrared) which does not simplify the processing of characterising these devices. In order to reduce the plasmonic resonant frequency into the visible regime, MWCNT spacings of around 100 nm would be required which has not yet been possible using PECVD techniques. As will be seen on the following sections, the light does not have to illuminate the MWCNTs parallel to their length in order to excite plasmonic effects. Light can be incident along their length and still see plasmonic effects due to the excitement of the electrons across the lateral diameter of the MWCNTs [28]. The physics of this process is not yet fully understood, however there have been successful observations of this effect which can be used to create MWCNT based diffractive holograms.

1.6 Quasi Crystalline Diffraction from Nanotube Arrays

A very effective demonstration of enhanced plasmonic effects from laterally illuminated MWCNTs has been demonstrated by the diffraction patterns observed from a 2D Penrose tiled quasicrystal structure [29]. This structure has been seen previously by etching air cylinders into a quartz substrate [30], where laser light at 633 nm was transmitted through the dielectric quasi-crystal formed by the air cylinders in the quartz and relatively weak diffraction spots were observed. The air cavities were 3 μm in diameter, 700 nm in depth and placed 10 μm apart on a tile side (the pattern area was 0.013 mm^2). An alternative approach is to use a finer 'inverse' structure of quasi-periodic pentagonal arrays of MWCNTs on a Si substrate. The MWCNT diameter, related to patterned catalyst diameter, was 80–120 nm, and length 700 nm. The spacing between MWCNTs on a tile side was 1400 nm and the pattern area 2.5 mm^2. In contrast to the previous work, the MWCNTs act as an array of metallic nanometre scale rods forming a hologram for reflected light. This allows the reversibility of the optical waves from the light source through reflection to be exploited to get a "beaming" effect due to the plasmonic resonance and is often compared to that emanating from a phased array antenna. Since the MWCNTs are placed in a 2D quasi-crystalline array (i.e. having periodic breaking of symmetries), a rich diffraction pattern, or many spots are seen. The MWCNT arrays were excited with 532 nm laser light and the results compared with theoretical predictions of diffraction for crystals with tenfold symmetry. The flexibility and accuracy of the growth system used allows the production of array patterns over large areas on wafers up to 8″ diameter, so that the exciting beam of laser light falls entirely on the array without the need for complicated alignment procedures.

The calculation of the diffraction pattern for a periodic system revolves around the construction of the reciprocal lattice and subsequent placement of the first Brillouin zone; however, in this case the aperiodicity of the pentagonal array requires a different approach due to the lack of translational symmetry. The reciprocal lattice of such an array is densely filled with reciprocal lattice vectors, with the consequence that the wave vector of a transmitted/reflected light beam encounters many diffraction paths. The resultant replay fields can be accurately calculated by taking the FT of the holograms. To perform the 2D fast Fourier transform (FFT) of the quasi-crystalline nanotube array, a normal scanning electron micrograph was taken, as shown in Fig. 1.13.

The Si wafer bearing the pentagonal array was mounted onto a post with a flat screen set above it. The plane of the screen is parallel to the plane of the sample and the screen has a small aperture in its centre. A 532 nm (green) laser was mounted above the screen, arranged so that the beam was normally incident at the sample. The reflected diffraction pattern produced when the sample was illuminated was captured by a camera and corrected for distortion due to the off-axis camera position. The observed diffraction pattern is in very good agreement with the calculated results. A comparison of the two is presented in Fig. 1.14a, b. The pattern order also shows good agreement when compared with the Penrose reciprocal lattice calculated using the five lattice vectors, Fig. 1.14c. The diffraction pattern in Fig. 1.14b shows bright outer spots which correspond to the superposition of two pentagrams rotated by angle $\pi/5$ with respect to each other. The Penrose tiling as seen in the SEM image of Fig. 1.13a is generated from "fat" and "skinny" rhombi (P3 group), whose base angle is $\pi/5$.

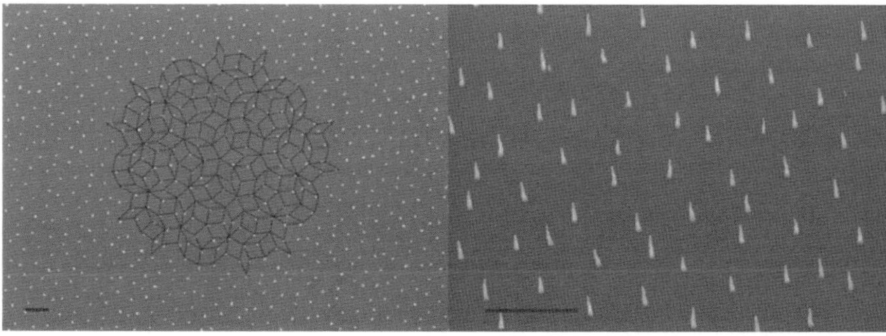

Fig. 1.13 SEM images of a quasi-periodic array of carbon nanotubes. *Left panel* Normal view, with Penrose tile scheme overlaid. *Right panel* Array tilted by 45°. Scale bars 2 μm. CNT heights are ~700 nm

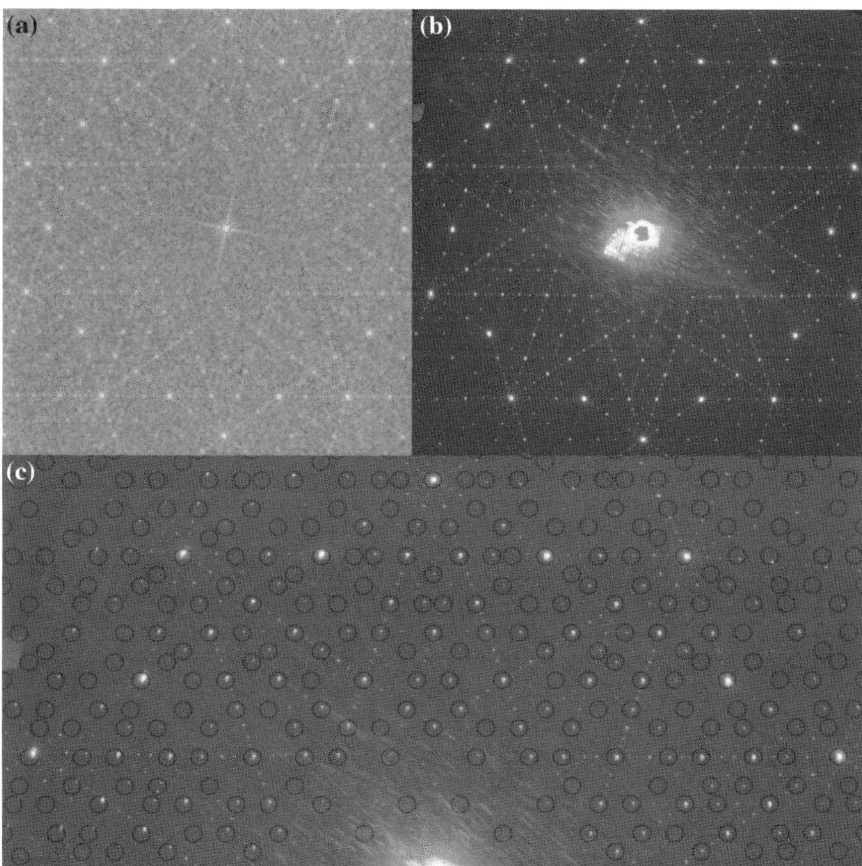

Fig. 1.14 Comparison of **a** FFT of SEM image of pentagonal array nanotube tips, **b** diffraction of normally-incident *green* laser light (the *dark central* spot is the aperture in the screen through which the laser enters) and **c** shows an overlay of the reciprocal lattice points as calculated onto half of the laser diffraction pattern

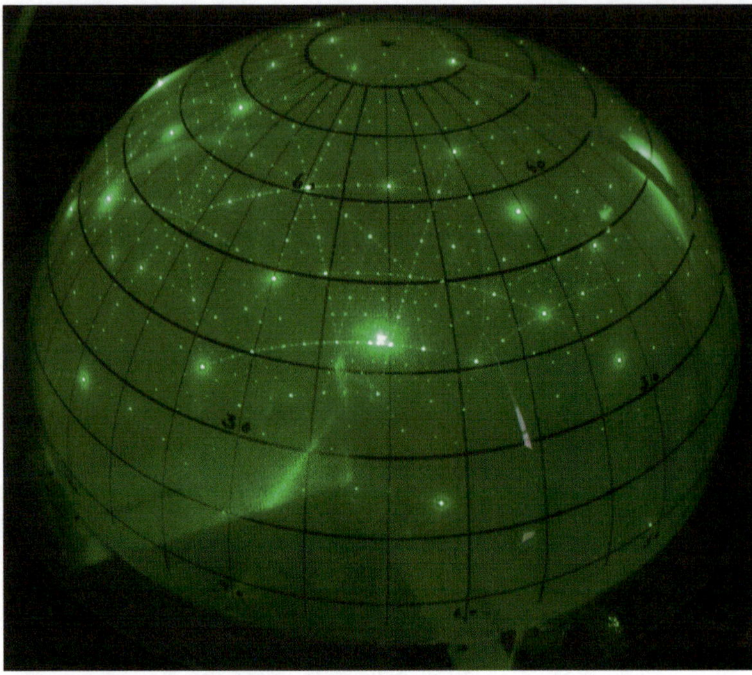

Fig. 1.15 Penrose diffraction of 532 nm laser, incident at an angle of 50̊ from the normal, captured with a semi-transparent dome

The remarkable “richness” and intricacy of the optical replay field emanating from the quasi-crystalline MWCNT array is best seen in the spherical diffraction pattern shown in Fig. 1.15. The results clearly show that the quasiperiodic array of nanotube antennas act as holograms (apertures) for the reflected light, producing remarkable and striking diffraction patterns (replay fields).

These patterns are complimentary to the materials that have shown patterned MWCNT arrays which form a regular 2D crystal can also be used to obtain metamaterial properties associated with nanoscale metal wires [11]. In this mode the reduced plasma frequency in the MWCNT due to induced current flow in the axial direction and the resulting circumferential magnetic field is exploited to obtain a negative permittivity in the optical frequencies In the quasi-periodic array used here r is nominally 100 nm and the spacing a between them is 1400 nm. Given that the permittivity function as a function of plasma frequency given by (1.17). The negative permittivity regime, when no light can be transmitted into the array, will only result at wavelengths larger than $\lambda_p = 5700$ nm. The optical wavelengths at which the diffraction is measured are well below this. Hence light transmission into the MWCNT array is not prohibited. On the other hand the orientation of the axes of the MWCNTs is normal to the plane of incidence of the

EM. This makes both the electric and magnetic field components normal to the MWCNT axis.

Micro-scaled patterns of light resulting from the quasi-periodic MWCNT arrays as shown here are difficult to replicate/forge. A ready application for them would therefore be for security purposes such as to validate the originality of brands, passports, cheques, important documents etc. They can also be used, in either periodic or quasi-periodic form, for directing a single beam of coherent light to many spatial locations simultaneously. There could of course be many other applications hitherto not foreseen, such as optical coding, where the rich and layered diffraction patterns resulting from these 2 dimensional quasi-crystals can utilised.

1.7 CNT Based CGH Holograms

As seen in the case of the quasi-crystalline structures, the far field diffraction patterns (and the field of view) are dictated by the pattern of the carbon nanotube arrays. The next objective is to control the angular diffraction from CNT array by fabricating these arrays in the form of CGH patterns. The array of nanotubes acts like a hologram (grid of apertures) towards the incident light, producing a diffraction pattern (in this case the word CAMBRIDGE) in the far field [31]. This replicates the mechanism of an optical projector or spatial light modulator while using the world's smallest pixel defined by a nanoscaled carbon nanotube. Fourier optics was utilised to calculate the MWCNT array (hologram) that would produce a CAMBRIDGE image in the replay field. The methodology used was to optimize a binary amplitude intensity mask using the Gerchberg–Saxton algorithm [16]. The CNT hologram was modeled as a square array of 300 $\times$ 300 pixels spaced by a lattice constant of 400 nm and a tube radius of 50 nm. Through an iterative process the optimized binary intensity hologram (carbon nanotube array) was calculated as shown in Fig. 1.16 with its corresponding replay field. The several orders of symmetrical conjugate images were as predicted for the hologram. The calculations were performed based on the principle of diffraction optics assuming the carbon nanotubes acted as diffracting elements. The calculations did not directly take into account near field optical effects like surface plasmons, usually attributed to nanophotonic structures like carbon nanotubes.

Based on the calculated CGH pattern in Fig. 1.16 the fabrication of carbon nanotube array was fabricated on a 10 $\times$ 10 mm silicon (Si) substrate. A growth time of 15 min at a pressure of $\sim$3 mBar yielded a MWCNT array with a tube length of approximately 1500 nm. Scanning electron microscrograph (SEM) of the fabricated nanotube array is shown in Fig. 1.17. A highly ordered array of vertically aligned MWCNTs was obtained with inter-nanotube spacings of $\sim$400 nm. Figure 1.17c shows a cropped SEM image of the nanotubes CGH hologram taken from vertically above the substrate. It can be observed that the square lattice consist of less than 4 % of the nanotube area. The FT of this image is shown in Fig. 1.17d. The highly intense zero order due to the un-diffracted light has been

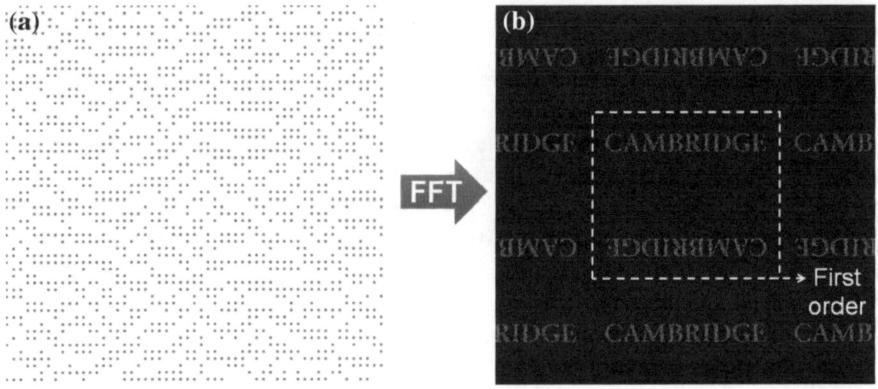

Fig. 1.16 The CGH designed. **a** 2D hologram based on 300 × 300 *square* array of multi-walled carbon nanotubes (lattice constant 400 nm). **b** FT of the CGH

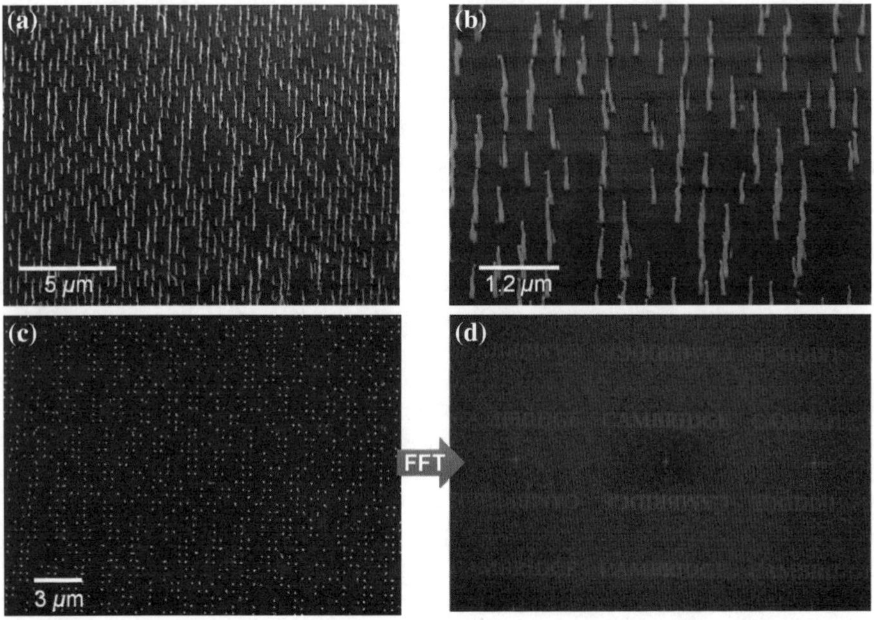

Fig. 1.17 Fabricated images of the carbon nanotube array. **a** SEM image of the fabricated MWCNT array (30° from vertical). **b** Higher magnification. **c** SEM image of the nanotube array taken at a normal angle. **d** 2D FT of the SEM image

removed from the image to make the image clearer, and the efficiency of the hologram was estimated at 4 %. The results showed that the nanotube array was accurately fabricated and could perform as a binary amplitude intensity hologram.

An optical characterization of the nanotubes CGH was undertaken next. The CGH was mounted onto a post with a semi-transparent hemispherical screen set above it. The hemispherical screen had a radius of 20 cm which allowed sufficient distance to capture diffraction patterns in the far field. The base of the screen was placed parallel to the plane of the sample and the screen had a small aperture in its centre to allow the passage of an incident laser as shown in Fig. 1.18a. A 532 nm laser was shone through the aperture, arranged so that the beam was normally incident to the sample. The reflected diffraction pattern produced when the CNT hologram was illuminated was captured using a digital camera.

Figure 1.18b shows the observed diffraction pattern which was in excellent agreement with the calculated results. The first order diffraction pattern (replay field) consisted of two symmetrically conjugate CAMBRIDGE images placed on opposing sides of the screen. The higher orders diffraction patterns were not observed as they were projected large angles (>90° to the normal) due to the nanoscale periodicity of the MWCNT array. A high intensity zero order was observed consisting of the un-diffracted light which did not interact with the CNTs and was mostly reflected from the Si substrate. According to calculations around 94 % of the incident light should be present in the zero order as CNTs occupy only about 4 % of the area in the binary hologram. Under these circumstances the zero order should dominate the replay field and be bright enough to saturate the camera resulting in the first order not being visible. However, as observed in Fig. 1.18, the CAMBRIDGE pattern was clearly visible even in the presence of the zero order suggesting the CNT based hologram presents higher diffraction efficiency than expected. Later testing indicated that the diffraction efficiency was more like 25 %, which indicates that the hologram is performing more than just amplitude modulation This is related to the highly metallic character and large effective scattering cross-section of the MWCNTs [28] which allows them to act as exceptional optical scattering elements.

Also the nanoscale dimensions of the CNT array ($a = 400$ nm) causes the diffraction of light at large angles increasing the field of view. According to the principle of Bragg diffraction the diffraction angle is dependent on wavelength and lattice constant. The diffraction patterns from the MWCNT hologram were studied under blue (454 nm), green (532 nm) and red (635 nm) lasers. With the increase in laser wavelength the image was diffracted at larger angles producing larger images. The effect was consistent with Bragg's law. The red laser image was diffracted to the largest angle of around 35° from the vertical. Furthermore, in contrast to the conventional 2D holograms, CNT arrays have a 3D structure that causes the optical diffraction in an anisotropic manner. The diffraction from nanotube arrays was expected to be dependent on the angle of incidence and the polarization of light. The diffraction efficiency measurements were performed by measuring the intensity of light in the zero and the first order. The first order image was measured by placing a high numerical aperture lens near the sample to focus all the diffracted light into the photodiode. Measurements were performed with red laser light polarized parallel and perpendicular to the axis of the carbon nanotubes.

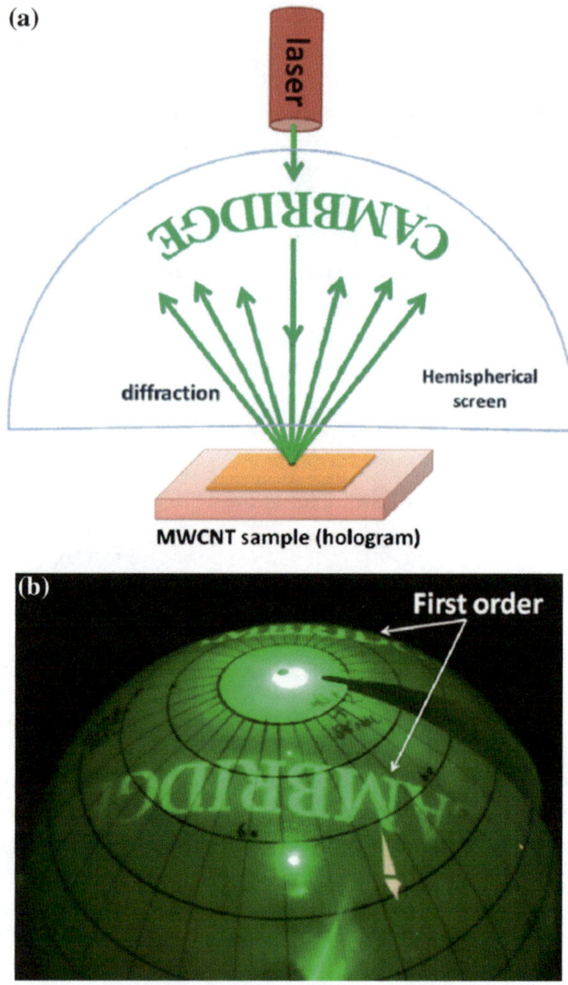

Fig. 1.18 Experimental setup and measured diffraction pattern results. **a** The experimental setup employed to capture the diffraction pattern. **b** The pattern was obtained on a semi-transparent hemispherical screen

The role of the MWCNT plasmonic effect is still not completely understood in this work, however there is a clear potential for these devices to increase the capabilities of future optical systems. The ability to control light using these structures can be further enhanced using liquid crystal materials and several devices have been made which combine the ability to control the light through plasmonic resonance with the variable refractive index of a LC material. Figure 1.19 shows a 400 nm spacing MWCNT similar to the one shown in Fig. 1.12. BLO006, a nematic LC mixture from Merck was added above the array in a 10 μm thick layer using an ITO coated glass top substrate with rubbed polyimide alignment. As the Si is conducting it is possible to place an electric field across the cell and switch the LC material.

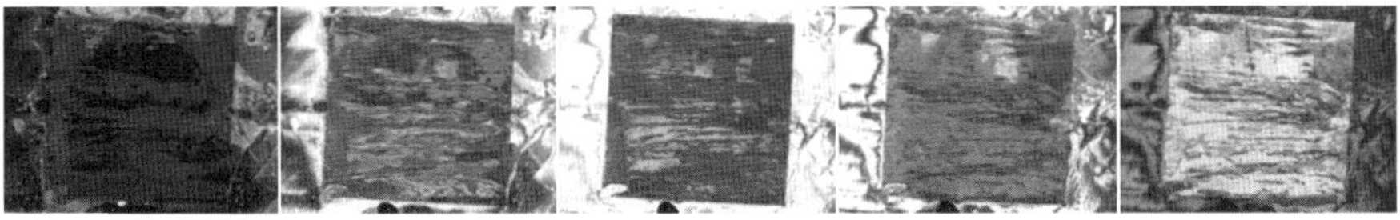

Fig. 1.19 A 400 nm pitch MCWNT array with 10um of nematic LC seen between crossed polarisors at different applied electric fields

The images shown in Fig. 1.19 show the colour changes induced by applying an electric field. These colours are much more vivid than the ones seen in normal nematic LC devices and are indicative of a shift in the Bragg grating wavelength of the MWCNT array. Further investigation of this effect has been hampered by the geometry of the devices and the difficulty in getting through cover glass (0.7 mm thick) in order to interrogate the LC/MWCNT combination. This limitation is both tantalizing but also shows that this type of structure is inherently limited by the growth direction of the MWCNTs. In order to see what sort of effects might be possible using these sorts of materials, a better resonant structure is needed which allows much easier access to the resonant features.

1.8 Nanophotonic Antennas

An alternative geometry to the vertically grown MWCNT arrays is to create nanophotonic structures that are parallel to the substrate (i.e. laying on the surface). Such structures in this geometry are often referred to as antennas as they have similar field properties to the larger scale versions used in communications technology. This form of plasmonic nano-antennas has been of interest due to their unique optical properties [32]. For instance, a perfect conducting rod with a negligible skin depth will have a half-wave dipolar resonance when the length of the long axis is equal to any multiple of half the wavelength. In contrast, when the polarisation is rotated towards the short axis the resonant conditions change. A critical criterion to consider is that antennas do not scale linearly across the whole spectra. The reason for this is that at short wavelengths the skin depth has a comparable size with the antenna. Hence, radiation penetrates into the metal and gives rise to oscillations of the free-electron gas within. Consequently, the effective resonant wavelength inside the metal can be reduced to a fraction of the size [33, 34]. Although this scalability disproportion might seem a problem from the fabrication point of view, a nano-antenna with sub-wavelength sizes allows one to widen the traditional concept of diffractive optics.

It has been proposed that sub-wavelength dielectric structures alter the effective medium without affecting the diffraction pattern in an optical system [35]. An expansion of the effective medium approach was also suggested for metals [36].

The effective medium theory can simplify the design of a hologram, but the effect of an individual nano-structure is not considered. A single carbon nanotube as a diffractive unit has been demonstrated in the previous section. In contrast with the effective medium theory, the scattering of a sub-wavelength structure was used as a diffractive unit. Interestingly, when these nano-structures have large scattering cross section, the diffraction effect is highly enhanced [28, 31]. It is not uncommon that a metallic nano-structure scatters above five times more light than that calculated purely from its geometrical cross-section. Another completely different approach is selectively and simultaneously illuminating different nano-structures with a predefined field to superpose their radiation as eigenmodes, so patterns of only a few pixels can be reconstructed.

In order to produce a diffractive effect, it is necessary that electrons in all of the nano-antennas resonate in phase producing a coherent source. When a laser beam is applied normal to the hologram plane, it is possible to induce emissions in phase. However, only those nano-antennas oriented in the direction of the electric field will emit. This means that by changing the incident light polarisation the nano-antennas can be "switched on" or "switched off". To take advantage of this we have superposed two transversal polarisations at a sub-wavelength distance. This is based on the assumption that two off-axes dipolar antennas with transversal polarisations have minimal coupling. In the case of particles, it has been suggested that dipolar coupling is negligible when the distance between them is larger than ~ 3 times the radii [37]. To avoid the interaction between transversal polarisations we have chosen to place two nano-antennas in an "L" shape separated by a fixed distance. This task would be impossible with typical antennas if the pitch of the sampled hologram was the same as the antenna length. However, in the case of optical nano-antennas, this is easily achievable with a wide separation between them. In contrast, if coupling is required, a "T" shape can produce an anisotropic interaction between the nano-antennas in one of the polarisations.

The steps to produce a hologram start with two arbitrary and independent designs either in two-dimensions or three-dimensions. Each of those designs represents the reconstructions for the relevant transversal polarisation. The diffraction pattern is then obtained by using a retrieving algorithm for CGHs [16]. Once the diffractive pattern is obtained, the sampling process should be performed. In this step, the two independent holograms have to be sampled giving the two transversal nano-antennas. A square grid is optimum due to its geometrical match with the two transversal polarisations. The last step in the hologram design consists of merging the two holograms. To maximise the inter-antenna distance, each hologram has to be shifted by half the sampling pitch in both directions. The displacement between antennas shifts the phase of the reconstruction without altering the intensity pattern. Figure 1.20 shows the design process and the final "L" shape formed by both transversal nano-antennas.

In order to prove the concept, the nano-antenna characterization had to be performed on a suitable device. The metal chosen was silver because of its favourable plasmonic properties in the visible spectrum [39]. The optimum parameters for the nano-antennas were calculated according to literature [34],

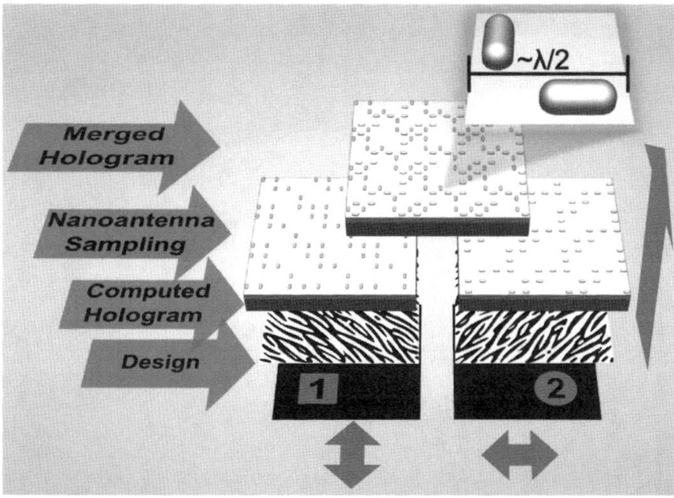

Fig. 1.20 The hologram design consist of four steps: the digital creation of two independent images for each polarisation, the retrieval of independent diffraction patterns, the nano-antenna sampling of the fringes and finally the blending of both sampled designs

simulations, and experiments. The nano-structure chosen was a nano-rod with dimensions of about 60 nm in diameter and 170 nm in length. This nano-antenna showed a strong emission for the red wavelength (650 nm) with a polarisation in the direction of its long axis and no interaction with the transversal polarisation. We tested experimentally our approach by evaluating each hologram in the far field. The target of the two transversal polarisations consisted of two independent images, one containing a square with an inscribed number "1" and the other containing a circle with an inscribed number "2". The independent CGHs that reconstruct these images were retrieved using the Gerchberg-Saxon algorithm. A computer program was developed to generate two arrays of 250 $\times$ 250 nano-antennas with a pitch of 390 nm. Finally the two sampled CGHs were merged into a single overall hologram pattern.

The nano-antennas where fabricated on top of a 200 nm silicon dioxide layer on a standard silicon wafer. The substrate was spin coated with high resolution positive PMMA 950 K resist and electron beam lithography was used to define the nano-antenna structures. After exposure, the samples where developed in a MIBK:IPA solution with a 1:3 composition. Finally, silver was thermally evaporated and lifted-off to remove of the unwanted residual metal areas. The total number of nano-antennas created was 6.25 $\times$ 10^6 distributed over an area of 2.13 mm^2. Although the final array of nano-antennas was laid on top of a dielectric layer of silicon dioxide, the silicon substrate obstructed the transmitted light. In Fig. 1.21, SEM pictures at different magnifications show the obtained results through this process. The two transversal nano-antennas in an "L" shape form a

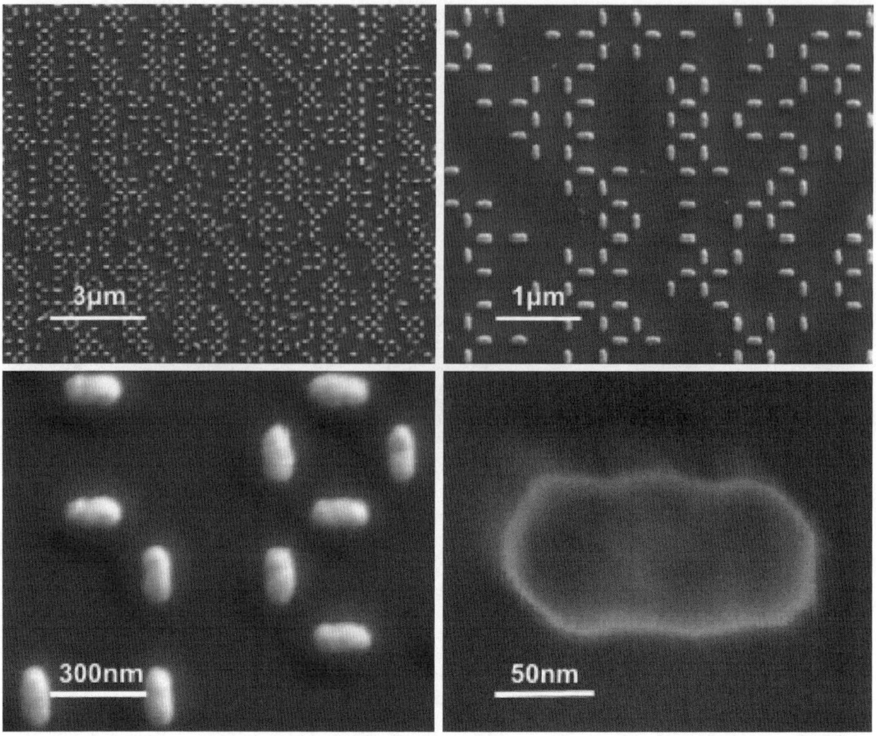

Fig. 1.21 *Four* different magnifications of a section of the fabricated CGH

large array with a profile similar to a weave. The dimensions of the nano-rods are uniform but some roughness can be observed on the surface of the nano-antennas.

The upper part of Fig. 1.22 shows the simulation of the projected waves for polarisations at 0°, 45° and 90°. The simulation shows the intensity of the field at the boundaries of the two transversal nano-antennas from a single diffractive unit of 390 nm × 390 nm. Additionally, it shows the field intensity in two different planes, one at 20 nm and the other at 120 nm from the top of the nano-antenna. The intensity at the boundaries (and the first plane) shows the dipolar behaviour while the second plane shows the radiated far-field. It can be observed, that the interaction occurs mainly when the electric field of the light is oriented in the direction of the long axis, so the radiated pattern in transversal polarisation is negligible. It is interesting to note that the polarisation oriented at 45° induces a resonance in both antennas equally. Simulations were performed with the boundary element method as shown in [38]. The real reconstruction was observed experimentally by using a low intensity polarised laser diode (650 nm) in a direction normal to the hologram plane. The image is switched just by rotating the laser along the normal axis. The images in the lower part of Fig. 1.22 show the projected image on a semi-transparent sphere (half a ping pong ball) for the

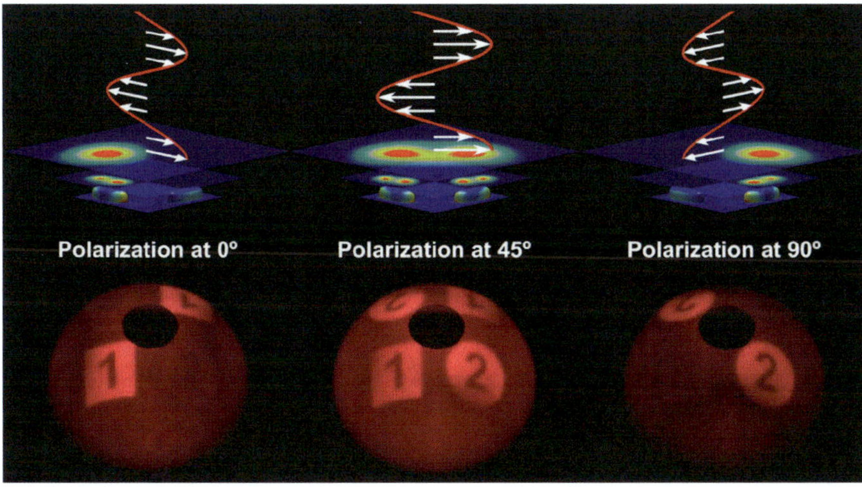

Fig. 1.22 Simulations (*upper half*) and experimental results (*lower half*) for two independent holograms at 0°, 45° and 90°

excitation of the two independent nano-antenna CGHs. In addition, the 45° image shows the excitation of both polarisations producing the two images simultaneously. The reconstruction background noise covers the antenna radiation pattern as an envelope across the whole sphere. This pattern represents an important characteristic of a diffractive antenna array, the two opposite lobes are imaged in the direction of the axis for each excited antenna. Also, the intensity does not decrease at large diffractive angles, in contrast with diffractive optical elements based on embossing or relief, nano-antennas can be highly omnidirectional [39] and holograms can achieve a wide field of view with a uniform angular intensity distribution. This CHG produced a field of view of 112.8°.

The relation between diffracted intensities in the far field produced by both polarisations can be found by projecting the component of the electric field vector in the nano-antenna resonant axis. The solution is given by $I_i cos^2(\theta)$, where I_i is the intensity of the incident beam and θ is the angle between the resonant axis and the electric field vector. This solution is similar to a pair of polarisers, but in this case the solution represents the diffraction produced by radiated light. We have proved this by measuring simultaneously the intensities of both diffracted images (see Fig. 1.23a). The polarisation extinction ratio in our experiments is minimal. The polariser used had an intrinsic leakage of about 1 %, but according to simulations, the transversal resonance of the silver nano-antenna used is about 5 % with respect to the parallel resonance. However, the transversal resonance does not necessarily radiate in an omnidirectional way, and therefore its intensity might be even dimmer in the far-field (this transversal resonance can be decreased even further with optimization).

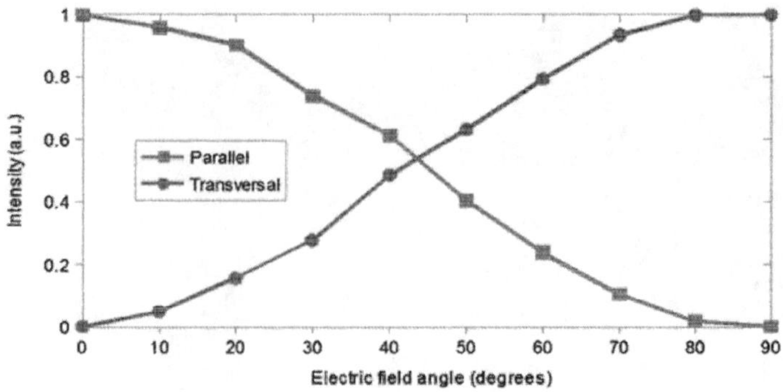

Fig. 1.23 Intensity relation of the two diffracted emissions for an electric field oriented at different angles

The total back-scattered efficiency measured at the far-field with a 650 nm beam was around 3 % with respect to the zero order. However, this is not a clear picture of the real diffraction efficiency because our samples couple most of the light inside the silicon dioxide layer which then gets trapped or absorbed by the silicon wafer. Catchpole showed that hemispherical shapes couple around 90 % of the light into the substrate [40]. Furthermore, silver nanoparticles in these dimensions have a negligible absorption comparing with their scattering. Through an alternative measurement of the zero order, it was found that about 17 % of the light incident on our samples was scattered (considering minimal absorption). This figure seems to be more reasonable if the total geometrical fill factor of the nano-antennas is about 3 % and the scattering cross-section enhances the diffraction by a factor of 5. Hence, we can predict that the efficiency can be highly enhanced by using forward scattering with a transmissive substrate and by matching the refractive index of the substrate.

From this work we see theoretical approach to control the emission of nano-structures by exploiting the plasmonic properties of anisotropic nano-structures. Hence, we have fabricated a switchable CGH capable of reconstructing high resolution images along a wide field of view. The next stage is then to create dynamic switching of the CGH by using a nematic LC material to control the incident polarisation state onto the antennas. Equation 1.15 states that if the LC material is oriented to give a half wave retardance ($\Gamma = \pi$), then the incident polarisation state can be rotated by 90°, flipping from horizontal to vertical (or vice versa). To test this, a nano-antenna substrate was assembled into a LC device using the same structure as that used with the MWCNT array. 10 μm of the nematic LC mixture BLO006 was placed between the array and a glass top substrate with both an ITO electrode and also a rubbed polyimide alignment later. In this device, the rubbing direction was set to as to create planar alignment of the LC at 45° to both directions of the nano-antennas (i.e. it split the "L" of the structure designed). The means that when the LC is set to be a half wave plate, then the polarisation of

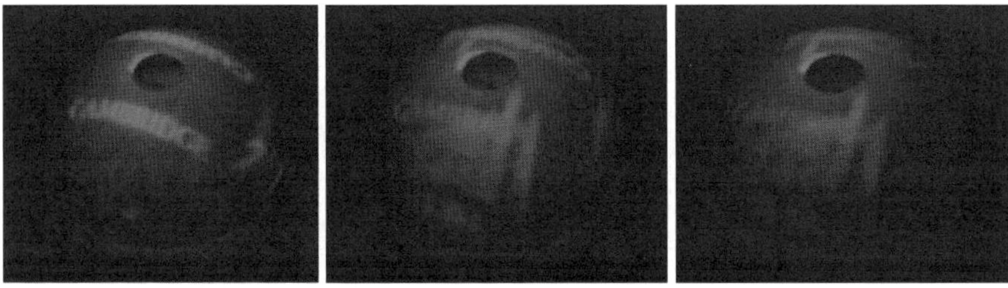

Fig. 1.24 Experimental images from the LC/antenna device. *Left*, 0 V applied (*vertical*). *Centre*, 0.44 Vum^{-1} applied (45°). *Right*, 0.65 Vum^{-1} applied (*horizontal*)

the incident light will be flipped from one set of antennas to that of the orthogonal set.

Figure 1.24 shows the experimental testing of this device with an applied electric field between the Si and the glass substrate (i.e. across the nematic LC). The first image shows the replay filed from the CGH measures using a 650 nm laser, with its polarisation axis aligned to be parallel to the vertical array of nano-antennas. As a result, the word "CAMBRIDGE" can be seen produced by the diffraction from this array. As the applied electric field across the LC is increased, the polarisation of the incident light is rotated and both replay fields can bee seen in the centre image of Fig. 1.24. The image now shows both replay images (the word and also an image of Kings College Chapel) together, as both CGHs are being excited by the incident light. The final image on the right of Fig. 1.24 shows a further increase of the applied electric field to 0.65 Vum^{-1}, and the work CAMBRIDGE is extinguished. This image is very difficult to obtain with this setup, as there is a degree of polarisation wander from the laser, and the rotation of the laser must be made to keep the system stable. The array is also very sensitive to the applied electric field, as the response of the LC is quite steep and difficult to maintain with the voltage source used in this experiment. All three images in Fig. 1.24 were obtained for the same orientation of the laser and purely by variation of the applied electric field.

The results shown in Fig. 1.24 show preliminary switching of the polarisation state for an applied electric field across the nematic LC. Better results could be obtained using a more stable setup, but these results are still very important to the understanding on LN nano-photonic devices. The change in orientation of the nano-structures has allowed polarisation to be interrogated in a much more rigorous way than was possible with the LC/MWCNT devices. Hence it is possible to conclude that the plasmonic resonance of these structures is maintained within the LC and is not affected by the interface with the LC. It also shows that it is possible to engineer suitable device structures based on hybrid combinations of these technologies using solid physical and engineering principles, to give predictable device performance.

1.9 Conclusions and Discussion

In this chapter, we have presented a nanophotonic device based on a hybrid combination of multi-walled carbon nanotubes, silver nano-antennas and LCs. The carbon nanotube electrode arrays were grown on silicon substrate by plasma enhanced chemical vapour deposition after employing e-beam lithography and covered with nematic LC. The multi-walled carbon nanotubes act as individual electrode sites that spawn an electric-field profile, dictating the refractive index profile within the liquid crystal and hence creating a series of graded index profiles, which form various photonic elements. The device was analyzed under an optical microscope and it was found that photonic element was formed at each nanotube site and the photonic elements switched with respect to the applied voltage. This has lead to the conclusion that micro/nano regions in the device interact with light. The silver nano-antennas were fabricated using a standard lift off procedure and created structures that exhibited plasmonic resonances at visible wavelengths.

Both of these nano-photonic structures were combined with classical CGH design techniques to control light by diffraction. Both proved very successful and both also exhibited a higher degree of efficiency than was expected from a purely amplitude based hologram. This indicates that the plasmonic effects seen in the nano-structures are interacting with the light in a complex manner and creating a diffractive element with pixels that appear larger than have been fabricated. Both structures have also been combined with LC materials to create switchable nanophotonic devices. The MWCNT device showed a voltage dependent shift in the Bragg diffraction due to the periodic array of MWCNT. Further characterisation of the MECNT device is on-going to confirm the effect seen is plasmonic in origin; however the geometry of the device makes this very difficult to interrogate. The silver nano-antennas allow us to confirm the effects of the LC material directly through the observed switching in the polarisation state above the hologram. This demonstrates the ability to tune the plasmonic resonances to a particular polarisation state and then switch between the states using as LC material. Further devices are being fabricated which will further exploit this effect across both the polarisation states as well as the visible wavelength range.

The potential applications of this type of nanophotonic device range from 3D displays through to adaptive optics such as tweezers and wavefront sensors. Even more importantly these types of devices open up several new optical and electro-optical effects that can be used to control the light. If such effects were integrated into an existing display technology, then the possibilities are endless with both polarisation, wavelengths and spatial control selectable at the pixel level of the LC devices. Such innovations push the technology ever closer to replicating the sort of optical effects seen in classical photographic holograms.

References

1. B. Brown, A. Lohmann, Complex spatial filtering with binary masks. Appl. Opt. **5**, 967–969 (1966)
2. W. Lee, Sampled Fourier transform hologram generated by computer. Appl. Opt. **9**, 639–643 (1970)
3. A. Jendral, R. Brouer, O. Bryngdahl, Synthetic image holograms: computation and properties. Opt. Comm. **109**, 47–53 (1994)
4. M. Stanley et al., 100-Megapixel computer-generated holographic images from active tiling: a dynamic and scalable electro-optic modulator system. Proc. SPIE **5005**, 247–258 (2003)
5. R.H.-Y. Chen, T.D. Wilkinson, Computer generated hologram with geometric occlusion using GPU-accelerated depth buffer rasterisation for 3D display. Appl. Opt. **48**, 4246–4255 (2009)
6. D. Gabor, A new microscopic principle. Nature **161**, 777 (1948)
7. D. Gabor, Microscopy by reconstructed wave-fronts. Proc. Roy. Soc. (London) A **197**, 454 (1949)
8. E.N. Leith, J. Upatnieks, Wavefront reconstruction with diffused illumination and three-dimensional objects. J. Opt. Soc. Am. **54**, 1295–1301 (1964)
9. S. Tay, P.-A. Blanche, R. Voorakaranam, A.V. Tunc, W. Lin, S. Rokutanda, T. Gu, D. Flores, P. Wang, G. Li, P. St Hilaire, J. Thomas, R.A. Norwood, M. Yamamoto, N. Peyghambarian, An updatable holographic three-dimensional display. Nature **451**, 694–698 (2008)
10. T.D. Wilkinson, X. Wang, K.B.K. Teo, W.I. Milne, Sparse multiwall carbon nanotube electrode arrays for liquid-crystal photonic devices. Adv. Mater. **20**, 363–366 (2008)
11. H. Butt, Q. Dai, P. Farah, T. Butler, T.D. Wilkinson, J.J. Baumberg, G.A.J. Amaratunga, Metamaterial high pass filter based on periodic wire arrays of multiwalled carbon nanotubes. Appl. Phys. Lett. **97**, 163102 (2010)
12. J.W. Goodman, *Introduction to Fourier Optics*, 2nd edn. (McGraw-Hill Companies, New York, 2005), pp. 55–58
13. R.G. Wilson, *Fourier Series and Optical Transform Techniques in Contemporary Optics* (Wiley, New York, 1995)
14. H. Dammann, K. Görtler, High-efficiency in-line multiple imaging by means of multiple phase holograms. Opt. Commun. **3**, 312–315 (1971)
15. M.A. Seldowitz, J.P. Allebach, D.W. Sweeney, Synthesis of digital holograms by direct binary search. Appl. Opt. **26**, 2788–2798 (1987)
16. R.W. Gerchberg, W.O. Saxon, A practical algorithm for the determination of phase from image and diffraction plane pictures. Optik **35**, 237–246 (1972)
17. J.H. Holland, Genetic algorithms. Sci. Am. **267**, 66–72 (1992)
18. A. Thess, R. Lee, P. Nikolaev, H. Dai, P. Petit, J. Robert, R.E. Smalley, Crystalline ropes of metallic carbon nanotubes. Science **273**, 483–487 (1996)
19. S. Iijima, Helical microtubules of graphitic carbon. Nature **354**, 56–58 (1991)
20. K.B.K. Teo, M. Chhowalla, G.A.J. Amaratunga, W.I. Milne, D.G. Hasko, G. Pirio, P. Legagneux, F. Wyczisk, D. Pribat, Uniform patterned growth of carbon nanotubes without surface carbon. Appl. Phys. Lett. **79**, 1534–1536 (2001)
21. W.I. Milne, K.B.K. Teo, M. Chhowalla, G.A.J. Amaratunga, S.B. Lee, D.G. Hasko, H. Ahmed, O. Groening, P. Legagneux, L. Gangloff, J.P. Schnell, G. Pirio, D. Pribat, M. Castignolles, A. Loiseau, V. Semet, V.T. Binh, Electrical and field emission investigation of individual carbon nanotubes from plasma enhanced chemical vapour deposition. Diam. Relat. Mater. **12**, 422–428 (2003)
22. P.J. Collings, M. Hird, *Introduction to Liquid Crystals, Chemisrty and Physics* (Taylor & Francis Group, London, 1998)
23. M.F. Lin, F.L. Shyu, R.B. Chen, Optical properties of well-aligned multiwalled carbon nanotube bundles. Phys. Rev. B **61**, 14114 (2000)

24. J.B. Pendry, A.J. Holden, D.J. Robbins, W.J. Stewart, Low frequency plasmons in thin-wire structures. J. Phys.: Condens. Matter **10**, 4785–4809 (1998)
25. P. Drude, Zur Elektronentheorie der metalle. Ann. Phys. **306**, 566 (1900)
26. P.G. Etchegoin, E.C. Le Ru, M. Meyer, An analytic model for the optical properties of gold. J. Chem. Phys. **125**, 164705 (2006)
27. H. Butt, Q. Dai, R. Rajesekharan, T.D. Wilkinson, G.A.J. Amaratunga, Plasmonic band gaps and waveguide effects in carbon nanotube arrays based metamaterials. ACS Nano **5**, 9138–9143 (2011)
28. Y. Montelongo, H. Butt, T. Butler, G.A.J. Amaratunga, T.D. Wilkinson, Computer generated holograms for carbon nanotube arrays. Nanoscale **5**, 4217–4222 (2013)
29. H. Butt, T. Butler, Y. Montelongo, R. Ranjith, G.A.J. Amaratunga, T.D. Wilkinson, Continuous diffraction patterns from circular arrays of carbon nanotubes'. Appl. Phys. Lett. **101**, 251102 (2012)
30. M.A. Kaliteevski, S. Brand, R.A. Abram, T.F. Krauss, R. De La Rue, P. Millar, Two-dimensional Penrose-tiled photonic quasicrystals: from diffraction pattern to band structure. Nanotechnology **11**, 274 (2000)
31. H. Butt, Y. Montelongo, T. Butler, R. Rajesekharan, Q. Dai, S.G. Shiva-Reddy, G.A. Amaratunga, T.D. Wilkinson, Carbon nanotube based high resolution holograms. Adv. Mater. (2012). doi:10.1002/adma.201202593
32. S.A. Maier, *Plasmonics: Fundamentals and Applications* (Springer, New York, 2007)
33. G.W. Bryant, F.J. Garcia de Abajo, J. Aizpurua, Mapping the plasmon resonances of metallic nanoantennas. Nano Lett. **8**, 631–636 (2008)
34. L. Novotny, Effective wavelength scaling for optical antennas. Phys. Rev. Lett. **98**, 266802 (2007)
35. W. Yu, K. Takahara, T. Konishi, T. Yotsuya, Y. Ichioka, Fabrication of multilevel phase computer-generated hologram elements based on effective medium theory. Appl. Opt. **39**, 3531–3536 (2000)
36. S. Larouche, Y.-J. Tsai, T. Tyler, N.M. Jokerst, D.R. Smith, Infrared metamaterial phase holograms. Nat. Mater. **11**, 450–454 (2012)
37. W. Khunsin, B. Brian, J. Dorfmüller, M. Esslinger, R. Vogelgesang, C. Etrich, C. Rockstuhl, A. Dmitriev, K. Kern, Long-distance indirect excitation of nanoplasmonic resonances. Nano Lett. **11**, 2765–2769 (2011)
38. P. West, S. Ishii, G.V. Naik, N.K. Emani, V.M. Shalaev, A. Boltasseva, Searching for better plasmonic materials. Laser Photonics Rev. **4**, 795–808 (2010)
39. T. Hessler, M. Rossi, R.E. Kunz, M.T. Gale, Analysis and optimization of fabrication of continuous-relief diffractive optical elements. Appl. Opt. **37**, 4069–4079 (1998)
40. K.R. Catchpole, A. Polman, Design principles for particle plasmon enhanced solar cells. Appl. Phys. Lett. **93**, 191113 (2008)

Chapter 2
Directing 3D Topological Defects in Smectic Liquid Crystals and Their Applications as an Emerging Class of Building Blocks

Apiradee Honglawan and Shu Yang

Abstract Controlling topological defects in three-dimensional (3D) liquid crystals (LCs) with complex geometries is not only a scientific curiosity but also crucial to the development of advanced technologies, including displays, sensors and self-assembly of nanostructures. It was not until recently that the smectic phase of LCs, characterized by arrangement of anisotropic molecules into layers with the long molecular axis parallel to the layer normal, has drawn significant interests because in thin film geometry topological defects known as focal conic domains (FCDs) can self-organize into a variety of highly regular micro- and nanostructures over a large area. Depending on surface anchoring at interfaces and elastic properties of smectic LCs, FCDs of different size and symmetry can be obtained by tailoring surface chemistry and physical confinement from chemically or topographically patterned substrates. This chapter will review the recent advances in design of geometric confinement of smectic LCs with uniform and mixed surface anchoring to control FCD formation, its size, size distribution and packing symmetries. We then discuss the applications of these highly ordered structures to various technologies, including patterning, smart surfaces, microlen arrays and directed self-assembly of functional materials (e.g. nanoparticles). The review concludes with perspectives on future directions and potential technological impacts.

2.1 Introduction

Liquid crystals (LCs) are a fascinating class of soft matters, which exhibit unique physical properties between liquid and crystalline phases. Comprised of rod-like or disk-like molecules, LC manifests its anisotropy through a variety of remarkable

A. Honglawan · S. Yang (✉)
Department of Chemical and Biomolecular Engineering and Materials Science and Engineering, University of Pennsylvania, 3231 Walnut Street, Philadelphia, PA 19104, USA
e-mail: shuyang@seas.upenn.edu

Q. Li (ed.), *Nanoscience with Liquid Crystals*, NanoScience and Technology,
DOI: 10.1007/978-3-319-04867-3_2,

optical, electrical and magnetic properties, such as birefringence, polarization, dielectric and diamagnetic phenomena, and formation of distinct patterns and structures (e.g. viscous fingering) based on their molecular alignments that are highly sensitive to external stimuli (e.g. electric and magnetic fields), surface chemistry and confined geometry [1–7]. Because of these unique characteristics of LCs, they have played important roles in a number of technologies today, including LC displays, chemical and biological sensors, actuators, imaging devices, tunable optical components and optical switching devices [8, 9] (e.g. polarizers, filters, diffraction grating and Bragg switches), and energy-absorbing-bullet proof materials [4, 10].

LC molecules also offer excellent prototypes to study self-assembly of soft materials as (1) their structures resemble a number of biological systems (e.g. proteins, viruses, lipids, carbohydrates and nucleic acids), and (2) their self-organization principals based on molecular interactions, such as van der Waals interaction, dipolar and quadruple interactions, charge transfer and π–π interaction, metal coordination and hydrogen bonding, are common in living systems [11, 12]. Depending on the level of ordering of the LC molecules, they can exhibit different phases: (1) nematic phase having only orientational order where molecules self-align along their common long axis known as director, (2) smectic phase showing a degree of both orientational and translational long range ordering, and thus resulting in molecular alignments into layers or planes, and (3) other complex LC phases, including cholesteric or chiral nematic exhibiting chirality, blue phases having a regular 3D cubic structure of defects and possessing icosahedral symmetry similar to quasicrystals, discotic phases assembled from disc-like molecules.

Over the past decades, there have been many technological interests to control LCs in complex geometries. For example, in modern cosmology formation of topological defects in LCs serves as theoretical and optical topological models found in a number of intriguing phenomena, including condense matter experiments [13], ultra high energy cosmic rays [14], a class of elementary particle models [15], phase transition in the early universe [16] and its succeeding evolution [17]. Like in cosmology, defects in any LC systems arises as a solution to enigma of broken continuous symmetry in response to external forces such as stresses [18, 19], local and global boundary conditions [20] and phase transitions into an ordered phase [21]. Importantly, they play a crucial role in generation of various phases of matter such as the twist-grain boundary [22–24], blue phases [25, 26] of liquid crystals and the Abrikosov flux-lattice phase of superconductors [27]. From a technological aspect, the manufacturing of advanced materials has progressively transformed from defect-free systems (i.e. the first geometries of displays) to control defects as a means to create innovative materials [28, 29]. For this reason, the ability to tailor LCs with complex, topological defect arrays will offer new clues for the design and fabrication of the next generation switchable devices.

Much effort has been devoted to discovering and understanding topological defects in nematic phases, for instance, by controlling the stability of emulsions and interaction between colloidal particles in the elastic ocean of nematic LCs [30, 31]. In comparison, little attention has been paid to smectic LCs primarily because

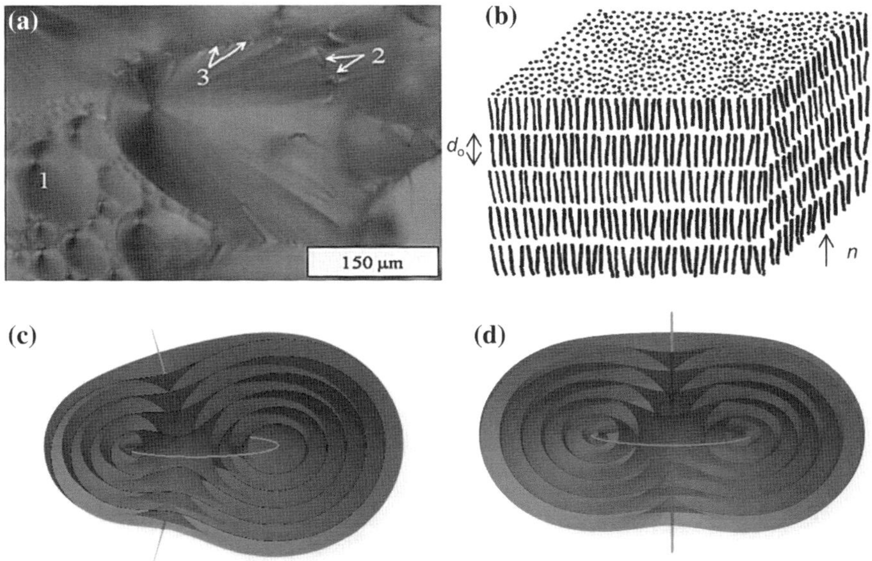

Fig. 2.1 **a** Polarized optical microscopy (POM) image of FCDs in a bulk thick (120 mm) cell of dodecylcyanobiphenyl (12CB) with different orientation: (*1*) ellipses parallel to the plane of view, (*2*, *3*) ellipses perpendicular to the plane of view [33]. Reproduced with permission [33]. Copyright 2009, Taylor & Francis. **b** A schematic illustration of "ideal" SmA phase [34]. Each rod represents LC molecules. Reproduced with permission [34]. Copyright 1991, EDP Sciences Schematic illustrations of layer structure of a negative Gaussian curvature focal conic domain with (**c**) non-zero and (**d**) zero eccentricity

formation of topological defects known as focal conic domains (FCDs) is often irregular in bulk (see Fig. 2.1a). The FCD texture was first recognized in 1922 by G. Friedel [32]. This FCD structure is the most commonly observed texture in the smectic-A (SmA) phase. An "ideal" SmA phase without any defect is characterized as a one dimensional (1D) crystal structure of planar and parallel stacking two-dimensional (2D) liquid layers with an equidistant separation (d_o), see Fig. 2.1b. For most thermotropic SmA, each LC layer, where the rod-like LC molecules are oriented normal to the layer, has a thickness of approximately 30 Å. In fact, the "true" SmA LCs are low dimensional crystals, including large bulk distortions of pure curvature, resulting in undilated 1D crystals. This aspect shows a large contrast to 3D crystals where bulk distortions dilate 3D lattice; each domain of uniform orientation is separated by grain boundaries rather than curvature.

Not until recently that FCDs have attracted significant interests of how to control their arrangement to form ordered FCDs on a flat or patterned substrate [33, 35–39]. The smectic layers in each FCD form concentric sections of Dupin cyclides with two mutually perpendicular linear focal sets (centers of curvature, red lines in Fig. 2.1c and d), an ellipse and a confocal hyperbola [40]. Experimentally, one can observe FCDs with all possible values of eccentricity of the ellipse

($0 \leq e < 1$, $e = \sqrt{\frac{(a^2-b^2)}{a^2}}$, where a and b are the semi-major and semi-minor axis lengths of the ellipse, respectively) that describes symmetry of the structure. One unique characteristic of FCD is that the family of curved surfaces of smectic layers can be enfolded around these two focal lines at the vanishing energy while preserving their equidistance everywhere except at the very defect cores. The 3D configuration of the directors inside the domain can be complex, but the local optic axis can be readily identified by any line that connects a point on the ellipse to a point on the hyperbola.

While FCDs arise as the prototypical, kinetically-trapped texture in bulk, a 2D hexagonal lattice of axially symmetric toric FCDs (TFCDs) with negative Gaussian curvature can be robustly produced in thin smectic films with antagonistic boundary conditions of tangential anchoring at the surface of substrate and homeotropic anchoring toward the air (see schematic in Fig. 2.2a). In this case, the ellipse lying in the plane of the substrate is degenerated to a circle and the hyperbola to a straight line passing through the center of the circle owing to the symmetrically defined boundary conditions at the LC/substrate and LC/air interfaces as shown in Fig. 2.1d. The formed defects with circular focal curves have zero eccentricity. At the LC/air interface, a defect domain appears as a circular, cone-shaped dimple (Fig. 2.2b), forming a characteristic Maltese cross pattern under the crossed polarizer (Fig. 2.2c). The observed polarized texture represents the projected deformation of the director field across the thin film—the director field with respect to the plane of the substrate is radial bounded by the concentric basis of a domain ellipse. The dark background between each domain corresponding to zero birefringence indicates that the smectic layers are parallel to the substrate and air interfaces. In the standard smectic ground state, the smectic layers are flat and parallel to the substrate and thus the molecular orientation points normal to both the LC/air and LC/substrate interfaces. The TFCDs form spontaneously when the decrease in surface energy obtained by tangential anchoring on the substrate outweighs the elastic energy cost of bending the layers and the increase in surface energy due to the dimple-like deformation of the LC/air interface.

While the formation of FCD structures has long been recognized since the 1900s, it was not until the last decade significant progress has been made toward the understanding and controlling of the type, feature size, and spatial distribution of FCD structures in 2D and 3D. This review chapter discusses the current knowledge of FCDs and its organization in thin films with the emphasis on recent advances in experimental designs of chemical and geometric confinement of LCs with uniform and mixed surface anchoring conditions to control LC alignment. A number of interesting applications of these highly ordered structures of defects in various nanotechnologies are embodied in this review, which will be concluded with perspectives in this field of research and its technological impacts.

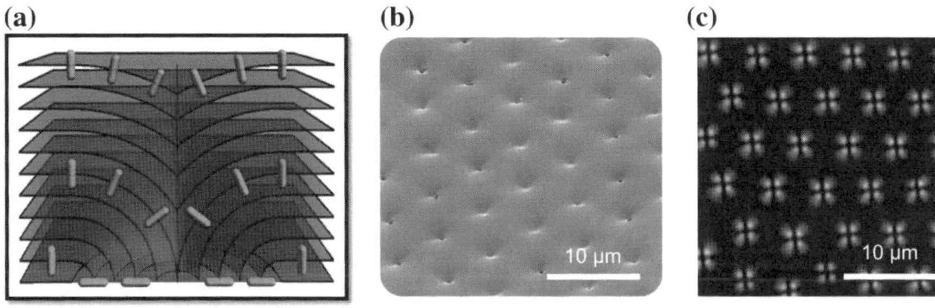

Fig. 2.2 **a** Schematic illustration of smectic layer (*red plane*) construction of a tori focal conic domain on randomly planar anchoring induced surface. The *green rods* represent liquid crystal molecules. **b** SEM image of array of tori focal conic domains and (**c**) its corresponding cross-polarized optical microscopy image

2.2 Engineering Focal Conic Domain Structure Through Confinement

Advancement in chemistry and material sciences has empowered a number of patterning techniques en route to control the growth and arrangement of FCD arrays in thin films through chemical or physical confinement. The effect of confinement, where LC is accommodated by a substrate, imposing a restrictive area for growth of domains, can be realized by utilizing either chemically or topologically patterned substrates [41, 42]. Both physical confinement and surface chemistry of a substrate—closely associated with a wetting phenomenon and dipolar ordering between LC molecules and the interface—impose the effect of confinement by inducing anchoring orientation of LC molecules at the interface in a fashion analogous to epitaxy observed in the growth of solid crystalline. The main parameters controlling the macroscopic structure of the LC therefore are the anchoring condition at interfaces and the global constraint from the bulk elasticity and the positional order. The most commonly obtained anchoring conditions are homeotropic, tangential (or degenerate planar) and planar anchoring where the molecules are oriented respectively perpendicular, randomly parallel or uniformly parallel to a plane of an interface. It is important to note that the interface induced anchoring is not necessarily permanent. The anchoring can be either monostable, multistable, or degenerate, depending on the type of LCs and its confinement. While multistable anchoring having minima at two or more anchoring states—which exists as another immense topic for a review elsewhere—is difficult to obtain, it is the key ingredient to creating switchable devices [9, 10, 43].

2.2.1 Confinement: Chemically Patterned Surfaces

Surface chemistry impacts LC anchoring in a complex manner that is yet to be fully elucidated. Most studies infer that the selection of LC anchoring direction through a chemical means is related to a number of parameters including interfacial energy at a substrate, wettability of LC, history of LC phases and direction of LC flow, most of which can be easily controlled through surface modification on a substrate with surfactants, silanes or evaporated metals [44]. In general, degenerate planar anchoring and planar anchoring (having an interfacial energy in the order of 10^{-5} J/m^2) can be obtained from a substrate with a surface chemistry similar to that of a mean chemical nature of LC molecules whereas homoeotropic anchoring (having a much larger value of interfacial energy in the order of 10^{-3}–10^{-2} J/m^2) is found on a less wettable surface of LC or at a free surface of a LC/air interface [45]. For instance, Bramble and colleagues demonstrated application of self-assembled monolayers (SAMs) technique of functionalized organothiols and microcontact printing to organize FCDs by generating two chemically patterned regions on gold substrates promoting either homeotropic or degenerate planar alignment of LC through a choice of assembled molecules [36]. Topped with a homeotropic anchoring surface from CF_3-terminated SAM, 9CB (4-n-nonyl-4'-cyanobiphenyl) -with a long hydrocarbon tail that dislikes the low energy fluorinated compound of CF_3-terminated SAM—formed FCDs organized in lines and square patterns above the pre-defined degenerate planar anchoring regions from COOH-terminated SAM (see Fig. 2.3a and b) [36]. The study beautifully illustrated a 1D control of FCD patterns sandwiched by two solid substrates maintaining the LC film thickness constant; however, a number of interesting aspects of smectic FCDs such as a relationship between the dimension of the confinement and FCDs were left unexplored. Similarly, Guo et al. presented a simple method to pattern Si-wafers by thermal evaporation of gold through standard TEM grids as lithographic masks to obtain alternating molecular anchoring surfaces such that the gold coated surface promoted homeotropic alignment of LC molecules while the uncoated regions with bare silicon surface provided random planar anchoring [39]. On a square coated pattern separated by 12 μm wide stripes of uncoated regions, FCDs of 8CB (4-*n*-octyl-4'-cyanobiphenyl) were found above the stripe regions, of which the width restricted the maximum size of the defect, shown by the AFM image in Fig. 2.3c. The corresponding height profile along the dash blue line (in Fig. 2.3d) revealing a FCD at the crossing of the two stripes with the equilibrium size approximately $\sqrt{2}$ larger than the others due to larger available surface area substantiates the observed effect of confinement for smaller FCDs on the strip regions [39].

In thin film, the relation between film thickness (h) and the size of TFCD, specifically its lateral diameter ($2r$) is well established by Fournier et al. [46] and The total energy of FCD results from a balance between the elastic energy density of LC, ΔF_{el}, and the surface energy at both the air and substrate interfaces $\Delta F_{air}, \Delta F_{subs}$ as [5]

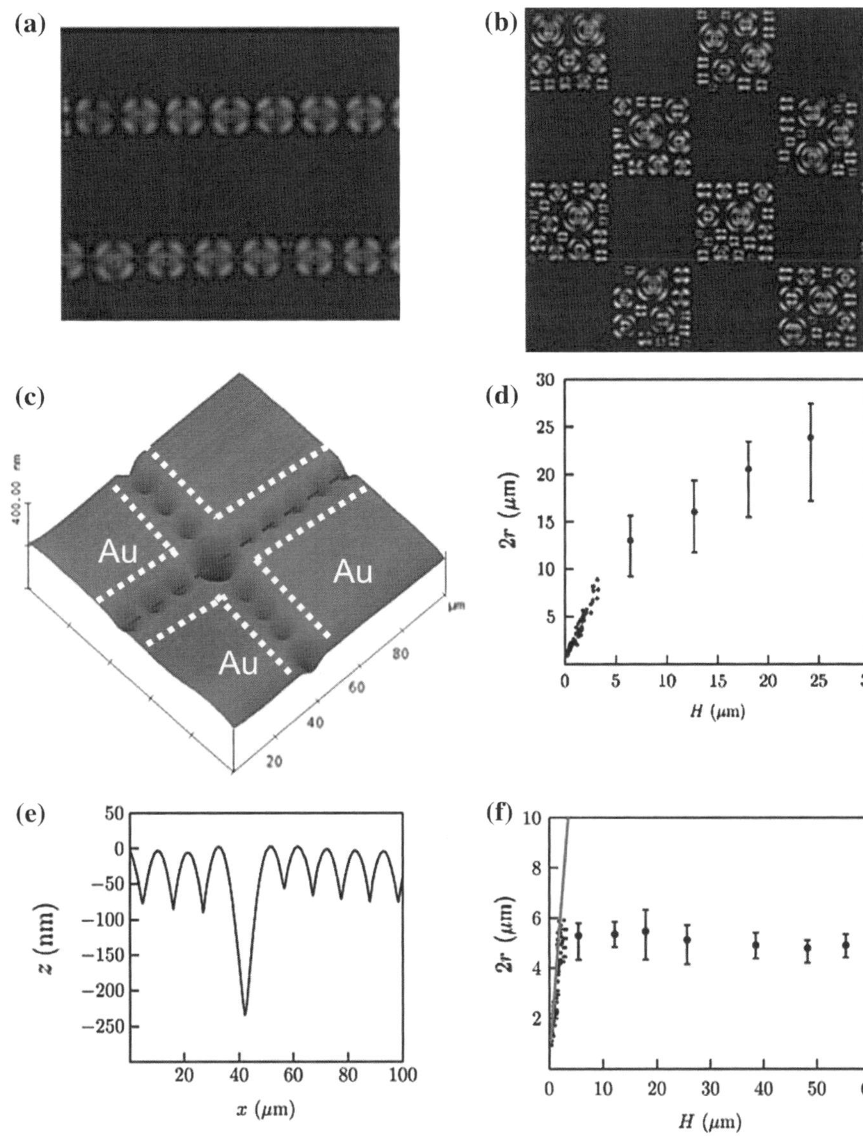

Fig. 2.3 POM images of 9CBs confined in 23 μm thick LC cells with pre-patterned surfaces encompassing degenerate planar anchoring regions with COOH-terminated SAM (**a** for line patterns; **b** for square patterns) and homeotropic anchoring regions with CF_3-terminated SAM. **a** FCDs formed linear arrays with 10 mm line-width and 30 mm periodicity; a checkerboard pattern with a square size of length (**b**) 16 mm. The cross polarizers are aligned parallel to the edges of the images [36]. Reproduced with permission [36]. Copyright 2007, Taylor & Francis Group. **c** AFM image of 100 × 100 μm^2 8CB film forming FCD structure on a silicon substrate patterned with gold-coated squares with 12 μm wide stripes at crossing separating between gold coated regions [39]. **d** The height profile of FCD arrays along the *dash blue line* in (**c**) [39]. **e** A plot of FCD diameter, $2r$ as a function of LC thickness, H of smectic 8CB on a bare silicon substrate [39]. **f** A plot of FCD diameter as a function of LC thickness of 8CB on a circularly patterned silicon substrate with diameter of 6 μm [39]. Reprinted with permission [39]. Copyright 2008, American Chemical Society

$$\Delta F = \Delta F_{el} + \Delta F_{air} + \Delta F_{subs} \tag{2.1}$$

$$F = 2\pi\beta Kh + 2\pi\alpha Kr - \pi r^2 \Delta\sigma_{sub} + \frac{\pi\sigma_{air}}{12h^2} r^4 \tag{2.2}$$

where K is the mean elastic constant, σ_{air} is the surface tension of air, $\Delta\sigma_{sub}$ the surface tension difference between hometotropic and planar anchoring on the substrate, α and β are dimensionless unknown constants.

By minimizing F with respect to r, h is found to be linearly proportionate to r as shown

$$h = \left[\frac{\sigma_{air} r^3}{6\sigma_{sub}\left(r - \frac{\alpha K}{\sigma_{sub}}\right)}\right]^{1/2} \tag{2.3}$$

which is in agreement with most smectic systems: the equilibrium size of the domain increases with LC thickness on an unconfined surface (see Fig. 2.3e for unconfined FCD system from 8CB) [38, 39]. However, this linear relationship can be undermined with the phenomenon of confinement under a substrate that is patterned chemically. As Guo et al. illustrated that a FCD restricted in a circularly patterned region of 6 μm in diameter silicon substrate where the region outside the pattern induced homeotropic anchoring did not grow larger than 6 μm with increasing LC thickness up to 150 μm (see Fig. 2.3f) [39]. This work thus provided a major leap toward controlling the positioning and size of FCD structures through a chemical confinement method.

2.2.2 Confinement: Topographical Surfaces

Development in lithographic techniques such as photolithography and soft lithography—that offers a simple route to create a variety of complex geometries and patterns in more than one dimensions—enables a whole new platform to study formation of FCDs and its arrangement through a topographic confinement as a guiding template for LC assembly. The initial effort to apply lithographic templates in LC study was mainly driven by interests to create a highly ordered array of the LC defect domains due to its unique nonlinear optical property. Choi and colleagues first applied the concept of geometric confinement to effectively control the size and spatial patterning of defect domains [5]. The study involved assembly of SmA 8CBs in 1D silicon based microchannels. The channels were fabricated by photolithography and reactive ion etching techniques. The authors emphasized the importance of anchoring conditions to uniformity of FCDs by coating a layer of polyethyleneimine (PEI) on the channel walls and bottom to promote strong degenerate planar anchoring for 8CBs. Together with homeotropic anchoring condition at LC/air interface, the physical and anchoring boundary conditions in the channels energetically favored the formation of uniform toroidal defects, which were effectively terminated at the side walls of the channels. In brief, the study showed that the defects behaved like colloidal objects: confining 8CBs in the surface-modified microchannels yielded uniform FCDs that arranged themselves in quasi-2D ordered patterns with triangular symmetry as shown in Fig. 2.4 [5]. In addition, Kim et al. reported a systemic study of assembly of high density TFCDs in 1D microchannels, and concluded a key finding that domain formation was strongly influenced by both the channel width (*W*) and, even more dramatically, by the channel depth (*H*) [38]. They found that in order to form an energetically stable, hexagonal array of TFCDs shown in Fig. 2.4b and c, $W > W_c \sim 4$ μm and $H > H_c \sim 2$ μm for the synthesized rod-like biphenyl LC molecules with fluorinated tail (here denoted as F-LC, see Fig. 2.5a for the chemical structure and results in Fig. 2.5b and c) with smectic phase at 114 °C [38].

2.2.3 3D Confinement of Focal Conic Domains

For many years, most attention had been devoted to a precise manipulation of the size and arrangement of FCD arrays in 2D lattices by confining defect domains within small regions through patterning of the substrate. Little was known, however, about a higher level of control of FCDs in 3D where new phenomena could be anticipated. Thus, more recent research has shifted focus to engineering more and more complex templates to study topological defect structures in 3D. We first reported systematic studies of TFCDs from controlled assembly of FCDs from F-LC, using SU-8 micropillars as a 3D topographic confinement [37]. Pillar arrays with variable size, shape, and symmetry (e.g. square and hexagonal lattices) were fabricated by photolithography and soft lithography techniques. SU-8 is a multi-functional epoxy photoresist, which naturally promotes degenerate planar anchoring to the F-LC molecules. By design, the LCs cast on the SU-8 pillar array were confined by a 3D hybrid cell of mixed boundary conditions for local FCDs: degenerate planar anchoring at the flat bottom of the arrays and vertical side walls of the pillars and homeotropic anchoring at air/film interface. As a result, an individual FCD could be directly grown either in between the neighboring pillars or on top of each pillar, yielding highly ordered arrays (square or hexagonal lattice) of TFCDs that were in registry with the guiding templates (see Fig. 2.6a and b). The study revealed that the most important basis for the directed assembly and control of TFCD formation was the appropriate design of the 3D template, specifically its dimension, including height, diameter, *D*, center-to-center spacing of the nearest two diagonal pillars (for a square lattice), *S*. From a series of experiments, the critical values, above which the system energetically favored the formation of stable TFCDs, for each parameter defining the pillar array was found. Based on the free elastic energy model, the critical dimension for both *D* and *S* of the pillar was approximately 0.5 μm for all LC thicknesses, $h \geq 2$ μm, where calculated *F* changed from positive to negative (see Fig. 2.7a). The calculations were in well agreement with the experiments. Furthermore, the results verified the

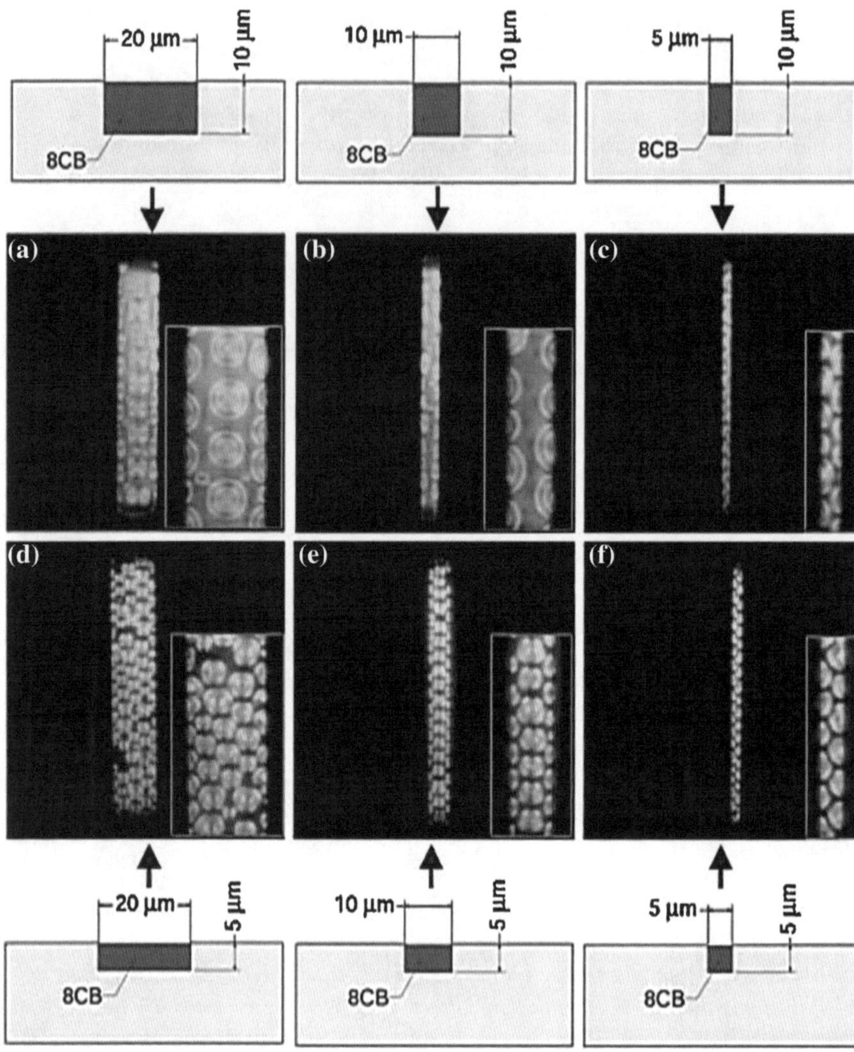

Fig. 2.4 Polarized light microscopy images of FCDs of 8CBs in PEI-coated microchannels with corresponding illustrative schematics shown *above* and *below* the images. The channels had the width of 20, 10, 5 µm from *left* to *right* and the depth of 10 µm for *top images* and 5 µm for *bottom images* [5]. Reproduced with permission [5]. Copyright 2004, National Academy of Sciences, USA

dependence of the domain size to the pillar dimensions, signifying the effect of confinement on local FCDs found at the minimum of the totally free energy (see Fig. 2.7b). The designed pillars allowed a variety of new TFCD arrays beyond the close-packed hexagonal arrangement to be produced uniformly where both dimension and symmetry of the pillar arrays propagated through the TFCDs for

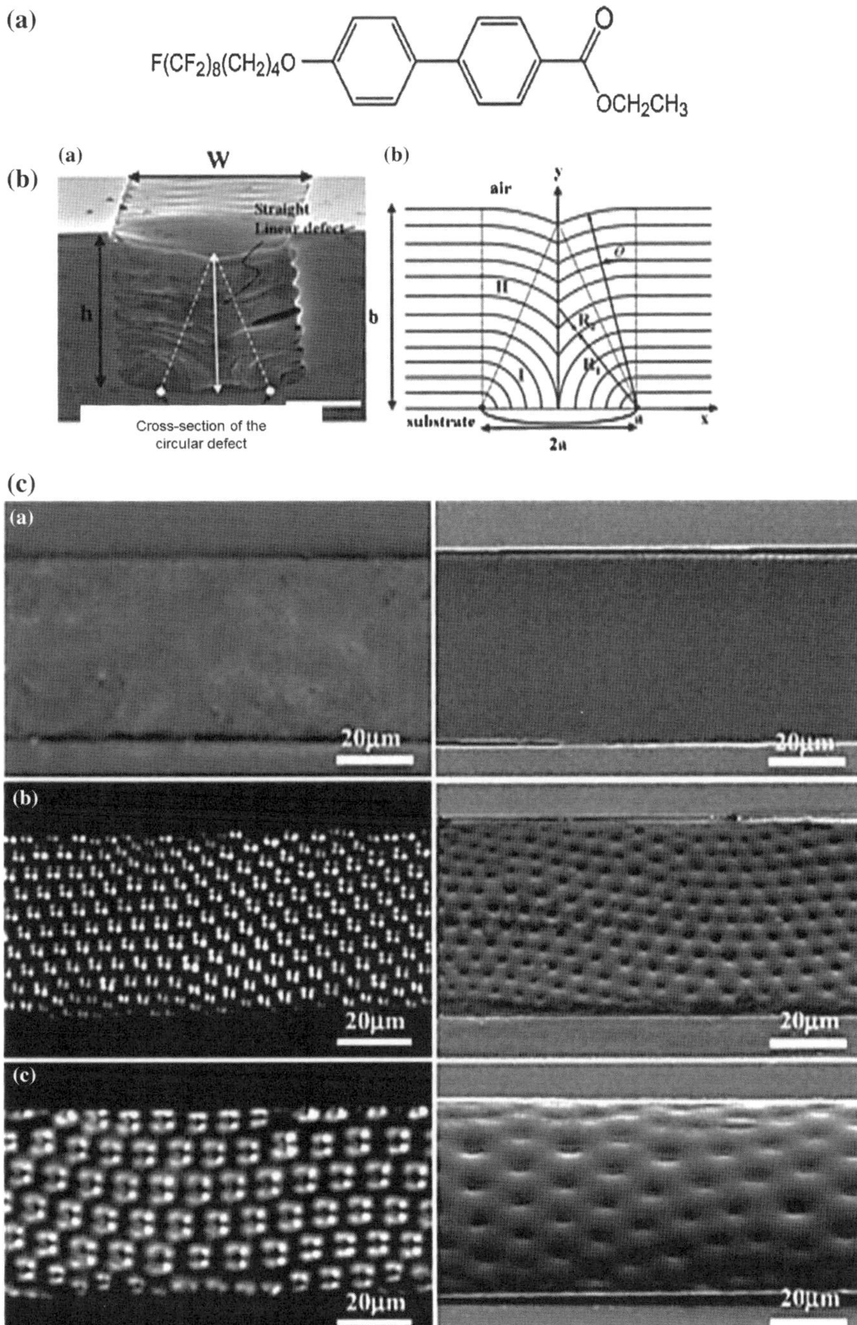

Fig. 2.5 **a** A chemical structure of F-LC [38]. **b** A cross-sectional SEM image of FCDs from fluorinated LC polymers confined in fluorinated polymer coated microchannels (*left*) and the corresponding illustration of the LC geometry in the channel [38]. **c** Polarized optical microscope (POM) images (*left*) and the corresponding SEM images (*right*) of the FCD arrays in fluorinated polymer coated microchannels of various depths: 2, 5 and 10 μm (*top* to *bottom*) and fixed width of 50 μm [38]. Reprinted with permission [38]. Copyright, 2009. American Chemical Society

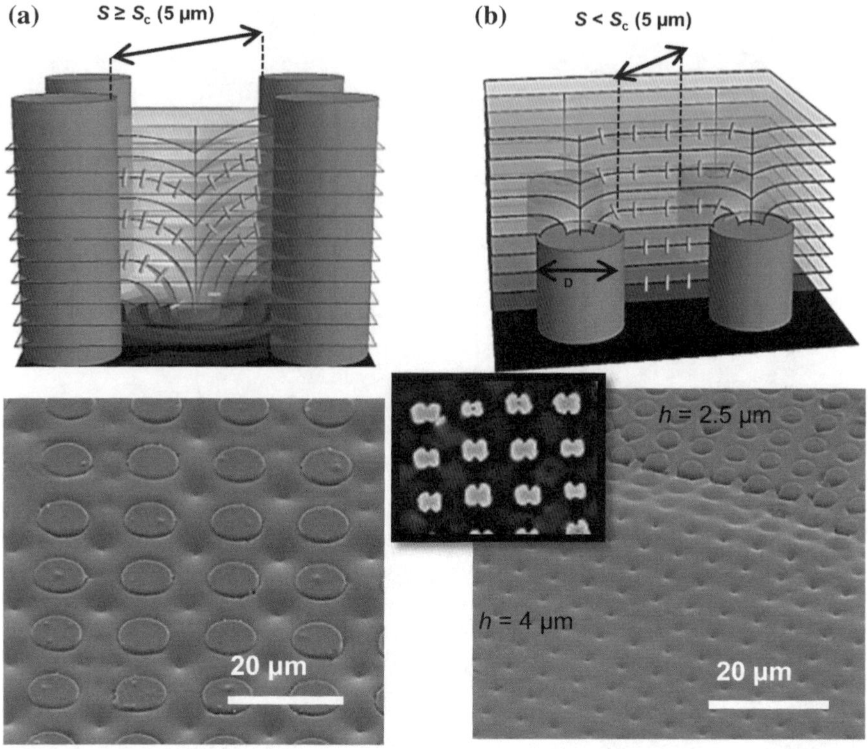

Fig. 2.6 Schematic illustrations of SmA LC confined by a SU-8 square pillar array where FCDs were grown either in between 4 neighboring pillars (**a**) and on top of pillars (**b**) and corresponding SEM and POM images [37]

high LC thicknesses up to 40 μm above the pillar top, demonstrating the unique long-range ordering into the bulk from surface epitaxy. In short, this epitaxial approach offers an entirely new paradigm to promote organization of smectic LC systems in 3D using simple topographic substrates.

In a separate vein, Shojaei-Zadeh et al. investigated the effect of 3D confinement of smectic 8CB in a closed microchannel system with mixed surface anchoring [47]. In the study, the microchannels with closed tops (rather than a free surface) were fabricated by soft lithography with polydimethylsiloxane (PDMS) elastomer, which intrinsically induced homeotropic anchoring of 8CBs and degenerate planar anchoring when oxidized in air plasma. The main advantage of this system was the optical transparency of PDMS, enabling observation of a variety of interesting LC textures in all directions. For instance, Fig. 2.8 shows a rare mixture of toroidal and nontoroidal focal conic textures formed in a hybrid cell with one planar anchoring induced interface and three homeotropic anchoring interfaces.

As the fundamental understandings of TFCD formation on confined templates with trivial geometries begin to estabilish, more questions are stimulated, such as

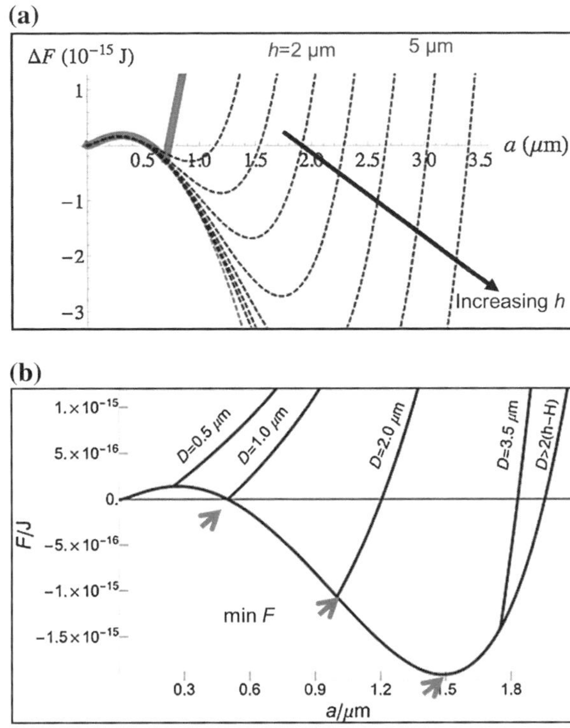

Fig. 2.7 **a** TFCD energy relative to the flat-layer state versus domain radius, a, for a flat substrate, with LC thickness h increasing from uppermost to lowermost curves: $h = 2, 2.5, 3, 3.5, 4, 5$ μm. **b** TFCD energy relative to the flat-layer state versus domain radius for defects on top of the pillar at varying pillar diameters, D using $h - H = 3$ μm. Thermodynamically stable domains form only for $D > D_C \approx 1.0$ μm. For $1.0\ \mu\text{m} < D < 3.0$ μm, the domain radius equals $D/2$. For $D > 3.0$ μm, the domain radius is the same as on the flat substrate, $a = 1.5$ μm. Each curve agrees with the $D > 2(h - H)$ curve for values of a $<D/2$

how SmA LC behaves in a nontrivial geometric template with hierarchical topography, gradient chemistry, various degrees of curvature or any combination of the mentioned. And how 3D defects are formed and organized in a confined space? From technological applications, it requires more profound insights of the topological effect of LCs in complex geometries as well as exploration of novel materials and their property. Kim et al. [48] investigated different shapes of 1D channels (rectangular, V-shaped, and isosceles trapezoidal, see Fig. 2.9A) with degenerate planar anchoring surfaces and reported different FCD textures (alternating fan shape or half FCD, no FCD, complete circular TFCD, respectively) due to variation in surface energy from the side walls and spatial spaces available for LCs to form a complete toroidal structure [48]. Since the layer curvature of a TFCD must be accommodated by a circular defect line on the plane of the bottom channel, growing TFCDs must form in the middle of the trapezoidal channel to circumvent a geometrical mismatch of the inclined walls of trapezoidal channel and the antagonistic boundary conditions. As a result, the TFCD arrays were found much more ordered when confined in a trapezoidal channel than other types of channel. Jeong et al. [49] revealed the in situ transition of the topological defects from a point to a line defect of encapsulated smectic LC when the aspect ratio of the confinement made of homeotropic anchoring induced PDMS prolate spheroids was increased due to mechanical stretching of the elastomer film (see Fig. 2.9B).

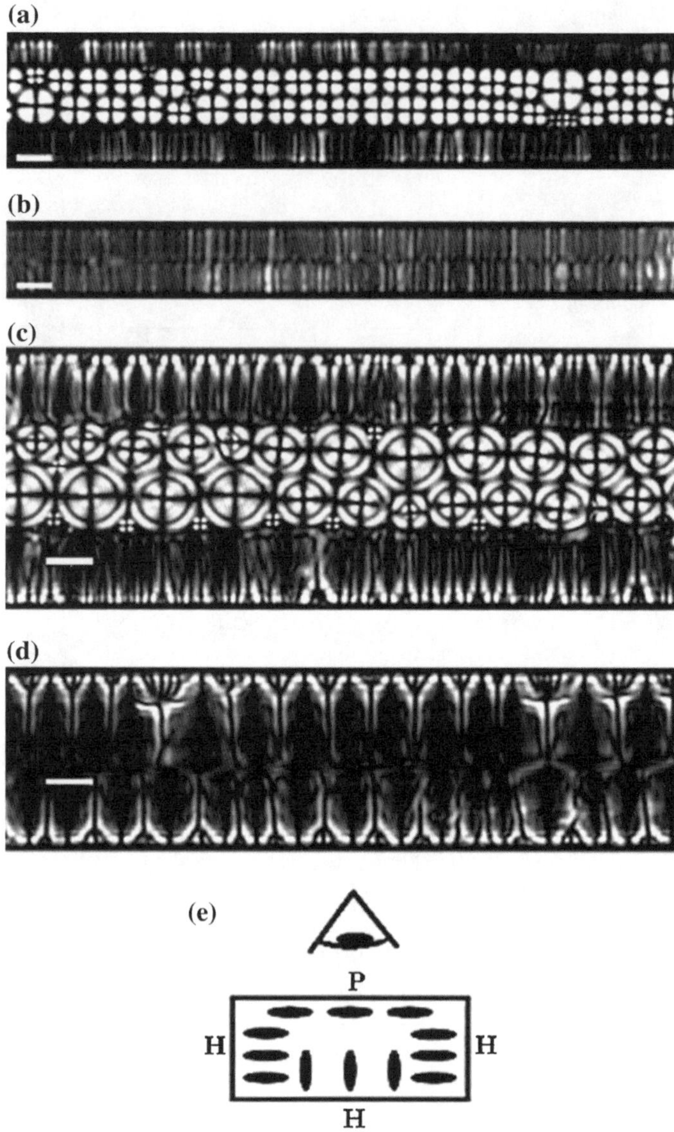

Fig. 2.8 Smectic 8CB confined inside a closed rectangular microchannel with mixed anchoring conditions [47]. Three microchannel walls impose homeotropic anchoring, while the fourth wall imposes planar anchoring as shown in the schematic illustration in (**e**). The microchannels are 20 (**a**), 10 (**b**), 60 (**c**) and 40 (**d**) μm wide and 3.8 (**a**, **b**), 10 μm (**c**, **d**) deep. Scale bars are 5 μm in (**a**) and (**b**) and 10 μm in (**c**) and (**d**). Reprinted with permission [47]. Copyright 2006, American Chemical Society

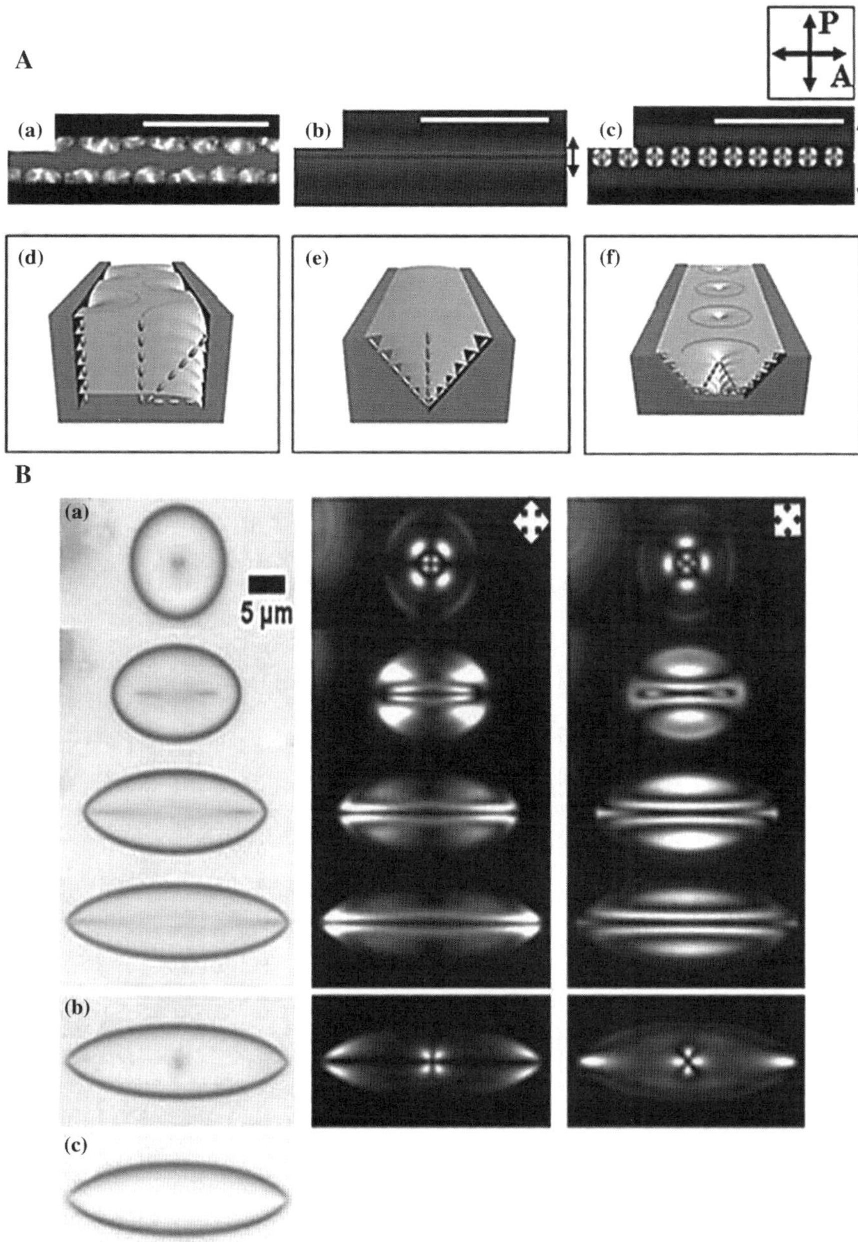

Fig. 2.9 *A* POM images of 8CB in PEI coated (*Aa*) rectangular microchannel, (*Ab*) V-shaped microchannel, and (*Ac*) trapezoidal microchannel (scale bar: 50 mm) [48]. Schematic illustrations of 8CB textures in three types of microchannels in (*Aa–Ac*) [48]. The *yellow rods* represent each 8CB molecules. Reproduced with permission [48]. Copyright 2011, Royal Society of Chemistry. *B* Optical (*first column*) and POM images (*second* and *third columns*) of LC droplet according to the aspect ratio of confined PDMS in smectic-A phase (*Ba*), nematic phase (*Bb*) and isotropic phase (*Bc*) [49]. Reproduced with permission [49]. Copyright 2012, American Physical Society

This work beautifully demonstrated controllability of a variety of confined LC structures using elastic property of PDMS templates, which enabled facile shape transition from spheroid to ellipsoid. The specific optical characteristic and its tunability due to a structural change of LC are highly desirable in development of switchable optical devices such as sensors.

2.2.4 Generation of Focal Conic Domains with Non-zero Eccentricity in Thin Films

Since a FCD is a 3D topological structure whose textures correspond to the molecular configuration of the domain and its placement with respect to the confinement, energetically complied a given boundary condition of molecular anchoring, it is possible to harvest and control defect location and arrangement rather than the typical hexagonal lattice of TFCDs simply by tuning film thickness and anchoring conditions as demonstrated by a number of studies. It is also possible to manipulate the characteristic of the confocal defect lines such as the eccentricity of the ellipse, and thus the structure of defects beyond trivial symmetrical domains with typical toroidal configuration and hexagonal packing through engineering a confinement with predetermined molecular orientation and geometry. The simplest scenario can be seen on a crystalline surface, such as polybdenite and mica, which imposes *unidirectional* planar anchoring. On such surface, SmA LCs organized into 1D arrays of parallel linear cylindrical defects in a thin film as shown in Fig. 2.10a and b [50, 51]. Characterizations of the formed smectic structure based on optical microscopy and x-ray diffraction under grazing incidence confirmed both planar anchoring on these crystalline substrates and a direct connection between the interface structure and anchoring direction of molecules. The results in Fig. 2.10c and d showed that the molecules adapting head-to-tail geometry organized into straight ribbon interfacial textures where the respective positions of each molecule were nearly flat on the MoS_2 substrate.

At a relatively large LC thickness (1.5 μm $<$ h $<$ 3 μm for 8CB), which was larger than that of the 1D line defect array case mentioned above, Zappone et al. reported a fascinating, *biological skin* like texture of smectic defects on a unidirectional planar anchoring surface of a rubbed polyimide substrate and mica (see POM images in Fig. 2.11A, the corresponding schematics in Fig. 2.11B and C, and the AFM data at free surface in Fig. 2.11D) [52]. By combining POM, AFM and fluorescence confocal polarizing microscopy (FCPM) images, the authors described a configuration of 2D hexagonal arrays of non-toroidal FCDs with large eccentricity ($e = 0.707$), tilted away from the normal of the substrate and each with identical anchoring condition [52]. In essence, the resultant texture stemmed from the balance of forces associated with multiple frustrations of the anchoring conditions, layer continuity and layer thickness, where the high energy required connecting the misaligned layers between the neighboring FCDs appeared to be the main driving factor of such assembly and truncation by the interfaces.

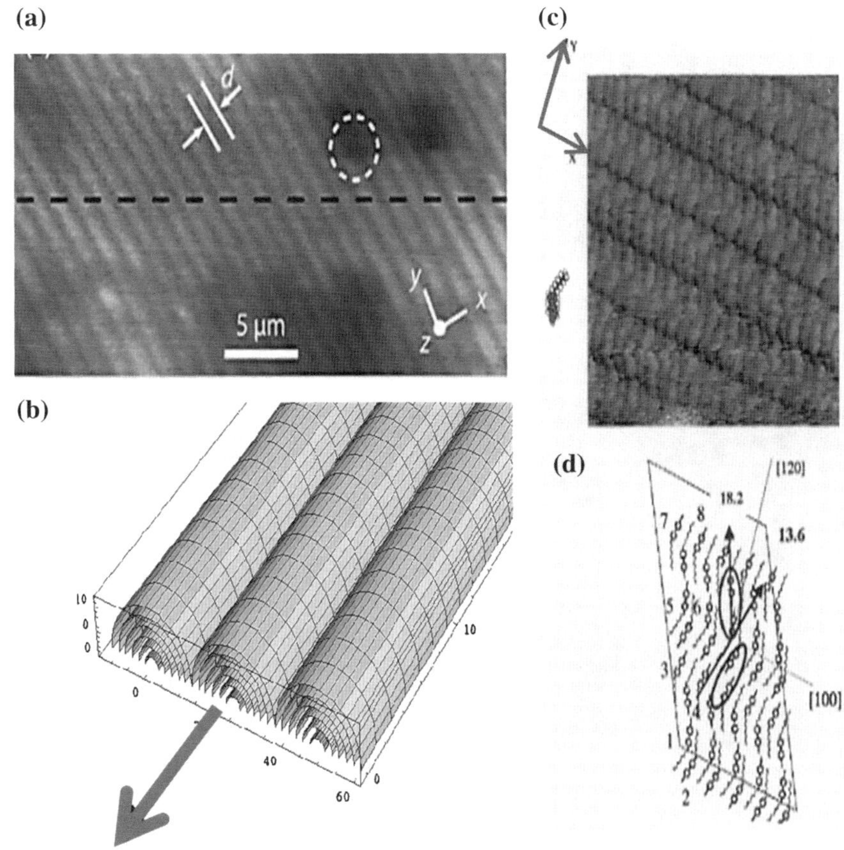

Fig. 2.10 **a** AFM topographic image of arrays of parallel linear domains of SmA 8CB on mica [53]. Reproduced with permission [53]. Copyright 2010, Royal Society of Chemistry. **b** Configuration of arrays of the linear domains extracted from x-ray diffraction experiment [54]. **c** STEM image of absorbed 8CB molecules on MoS_2 and **d** a microscopic "straight ribbons" structure as determined by x-ray diffraction [45]. The crystallographic cell is formed by the eight molecules labeled 1–8 and the two associated dipolar groups of LC are shown at 213.6 and 18.2° away from the direction normal to the ribbons [45]

In a separate study, a topographically curved surface such as microwrinkle grooves of thermally cured polymer was employed to confine smectic LC of 8CB [55]. The wrinkle substrate imposed the appropriate antagonistic boundary condition for FCD formation: planar anchoring at the curved substrate and homeotropic anchoring at the LC/air interface (see Fig. 2.12a). As a result, a uniform linear array of staggered non-zero FCDs was observed along the groove direction as shown in Fig. 2.12b–d provided that the wrinkle wavelength > the critical value of approximately 4 μm. Again, the connection between the domain size and dimension of the confined geometry was apparent in this scenario—the periodicity of FCD increased with the wrinkle wavelength. While the driving force for the

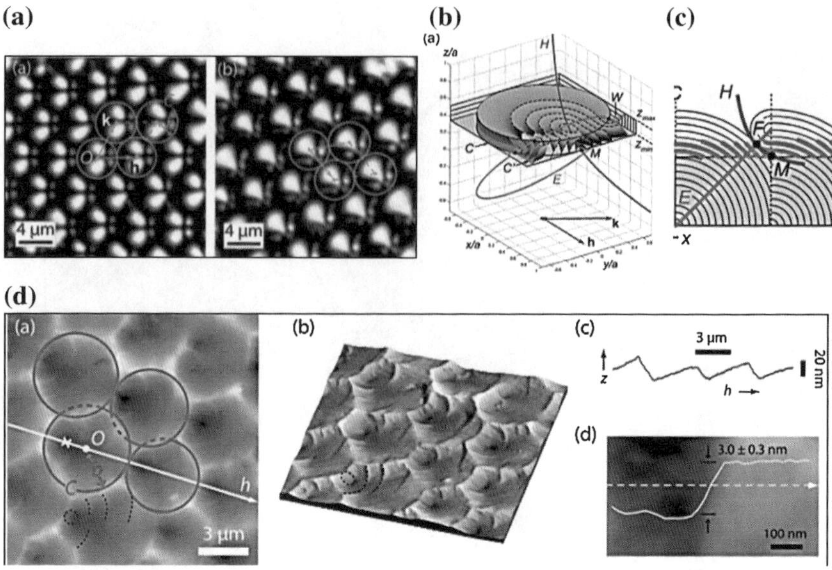

Fig. 2.11 *A* POM images of a lattice of nontoroidal FCDs in a thin 8CB film on a PI substrate rubbed along the h direction. The polarizer was oriented at (*Aa*) 0° and (*Ab*) 45° with respect to *h* [52]. Each FCD is bounded by a circle *C*, centered on *O* with a radius *b*. *B* A schematic diagram of a square FCD constructed according to the hyperbola *H* and the ellipse *E* with respected to the plane of a substrate where the intersection of *H* and *E* occurred at point *M*. *C* Schematic drawing of a vertical cross-section of the FCD lattice in the plane of *H*. *Thick* and *thin lines* represent actual and virtual portions of the layers, respectively. Actual layers are continuous across the cylinder *C* [52]. *Da* AFM image of (*A*). The *white* cross designates the point of minimum depth; *thin dotted lines* show the curved edges of smectic layers [52]. *Db* 3D configuration of (*Da*). *Dc* Line scan revealing the modulation of height *z* along the *h* direction [52]. *Dd* Height steps at the edge of a single layer. The *solid white curve* shows a scan along the *dashed line* [52]. Reproduced with permission [52]. Copyright 2012, Royal Society of Chemistry

observed structure and the effect of curvature from the wrinkle substrate to the non-zero eccentricity of FCDs remained elusive, the observations from the experiments provided sufficient data on LC alignment to qualitatively reconstruct the internal layer configuration of the structure shown in Fig. 2.12e.

Again, using SU8 pillar arrays as templates, we realized a new form of hierarchical growth of FCDs with non-zero eccentricity [56]. By decreasing the pillar height below its critical value as determined in our prior work [37], a transition from confinement of isolated domains to the "pillar edge-pinning" effect of multiple FCDs yielding hierarchical growth of FCDs from F-LCs, tangent to their neighbors, with their hyperbolic focal lines pinned near the pillar edges was observed as a result of minimization of the global free energy of FCDs (see Fig. 2.13a and b). The mechanism of this phenomenon is illustrated in Fig. 2.13b. It was suggested that to construct the specific form of hierarchically organized FCD arrays, the center-to-center spacing of the nearest two diagonal pillars (for a square array of circular pillars) should be approximately 4 times the domain radius

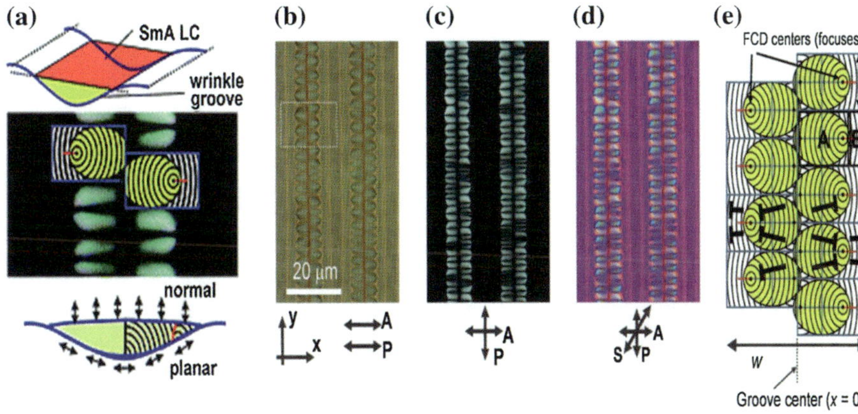

Fig. 2.12 **a** A schematic drawing of 8CB confined in a wrinkle substrate, a portion of POM image of the resulting structure with the overlay image of a possible LC layer configuration, and a drawing of a cross-sectional view of LC filled wrinkle substrate with designation of anchoring at each interface [55]. **b–d** Optical images of 8CB within microwrinkle grooves with $\lambda = 20$ μm in an identical area were obtained at different configurations of the polarizer (*P*) and the analyser (*A*) [55]. A sensitive tint plate (*S*) is used in (*D*), where a blue color implied that the mean LC direction is along the direction of S [55]. **e** A schematic for the linear array of FCDs in a wrinkle groove. A repeating unit is roughly composed of part-A (the ellipses of FCDs in *yellow*) and B (the background in *white*) [55]. A symbol ''*T*'' represents the LC director, with its end pointing outward through the surface of the page. Reproduced with permission [55]. Copyright 2012, Royal Society of Chemistry

such that the nonpatterned region surrounded by 4 neighboring pillars could accommodate multiple domains with tangential ellipses as opposed to just one domain as in the case of the confinement effect by pillar patterns. The size and shape of the pillars (circular, elliptical, triangular, and Y shapes) can be used to control the type of hierarchical FCD arrangement (2 domains per a highly elliptical shaped pillar to 4 domains per a circular shaped pillar); the anisotropy of the pillar shape enables a precise prediction of the locations of FCDs relative to the substrate patterning due to the competing effect between the effective attraction of FCDs to pillar corners and the steric repulsion between domains. Furthermore, these organized FCDs within the edge-pinning regime were found to have non-zero eccentricity, yielding a variety of unique surface profiles of the LC structure (see AFM images in Fig. 2.14a and b). The nontrivial, but apparently smooth, matching of smectic layers between neighboring FCDs on a non-uniform substrate presented an intriguing theoretical problem for which a geometric ansatz was proposed (see the results in Fig. 2.14c and d) and was found in good agreement with the experimental data. To conclude, this work created yet another surface profile of FCD arrays, expanding the range of shapes and symmetries accessible for applications based on self-ordered defects in smectic with non-zero eccentricity.

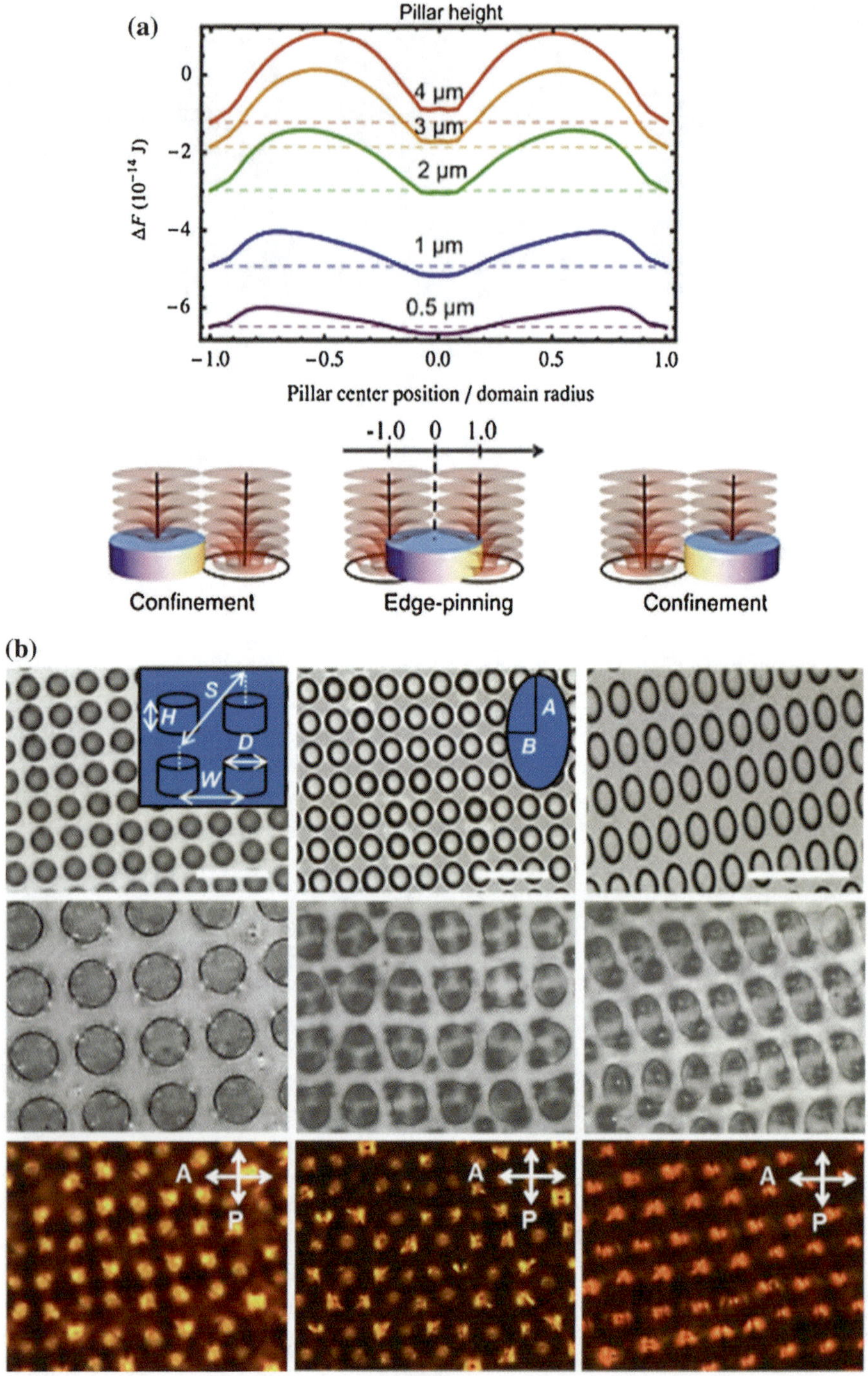
(a)
Pillar height
4 μm
3 μm
2 μm
1 μm
0.5 μm
ΔF (10⁻¹⁴ J)
Pillar center position / domain radius
-1.0 0 1.0
Confinement
Edge-pinning
Confinement
(b)
S
H
D
W
A
B
A
P

◀ **Fig. 2.13** **a** Free energy ΔF relative to the reference state of planar layers as a function of the relative position of circular pillar center along the line connecting the two TFCD centers for different pillar heights ($H = 0.5$–4 μm). The domain radius is set to 5.2 μm at LC thickness $h = 10$ μm on the pillar with radius of 5.72 μm. Schematics illustrate the TFCD arrangements on the pillar with edge-pinning and confinement effects [56]. **b** FCD of F-LC arrangements on 1 μm tall SU8 pillars with variable sizes and shapes (shown in the *first row* of SEM images) were observed under optical microscope (images in the *second row*) and polarized optical microscope (images in the *third row*), respectively. (Scale bars: 20 μm.) (*First column*) Circular pillars with diameter D = 5.5 μm, the center-to-center spacing of the nearest pillars W = 8.5 μm, and the diagonal center-to-center distance of the next-nearest pillars S = 12.0 μm. (*Second column*) Elliptically shaped pillars with major axis length 2A = 6.2 μm, minor axis length 2B = 5.2 μm, W = 7.4 μm (along the shorter lattice vector), and S = 12.2 μm. (*Third column*) Elliptically shaped pillars with 2A = 7.0 μm, 2B = 3.4 μm, W = 6.3 μm (along the shorter lattice vector) and S = 11.8 μm

2.3 Applications of Focal Conic Domain Arrays

Most of today's technologies have relied on top-down processes that cut, mill and shape materials, such as photolithography and etching techniques, to create structures by reducing lateral dimensions of bulk materials into desired shape and order. Top-down approaches are known to be highly effective in manufacturing materials with small footprints of rather simple structures and homogenous composition with high precision. Nevertheless, the top-down processes could be soon overwhelmed by the ever-increasing demand for more complex, multileveled structures with heterogeneous compositions at a smaller lengthscale, as well as the soon-approaching fundamental limits in scalability. In comparison, driven by thermodynamics to form stable structures, self-assembly offers a promising route to mass-produce hierarchical and multi-functional materials in parallel and at a much lower cost. Thus, self-assembly approach as a whole has rapidly gained momentum in research and manufacturing. In thin films, the highly ordered defect domains with adjustable dimensions are now considered as a new class of building blocks to create 3D structures with unique geometry for a variety of applications. Here, we illustrate a number of remarkable and yet explicit applications of TFCD structures, most of which are attributed to Hee-Tae Jung's research group.

The first application of FCDs in thin film was to create superhydrophobic and self-cleaning surfaces like those found in lotus leaves and butterfly wings [57]. Since both dual-scale roughness and low surface energy are prerequisites to obtain superhydrophobicity (water contact angle, WCA $> 150^{\circ}$ and low rolling angle $< 10^{\circ}$), TFCD arrays formed from F-LC molecules comprising of low surface energy fluorinated tail were subject to reactive ion etching (RIE). As seen in Fig. 2.15a and b besides dimple array of 2 μm depth and 10 μm center-to-center distance, secondary features were produced after RIE. The longer the RIE treatment (3–10 min), the rougher the surface (the root mean square roughness: 124.8–199.4 nm) and the higher WCA (120.7–158.5°). The TFCD array surface treated with RIE for 10 min exhibited self-cleaning effect with water droplet rolling angle $<2^{\circ}$.

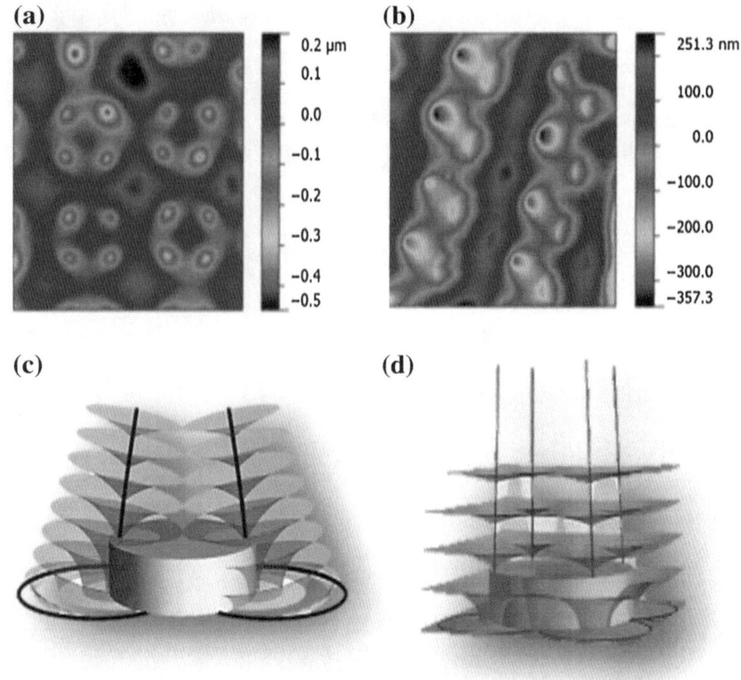

Fig. 2.14 **a** and **b** 3D maps of the surface of the LC films on a circular pillar (**a** *first column* in Fig. 2.13a) and an elliptically shaped pillar (**b** third column in Fig. 2.13b) extracted from AFM measurements based on their height profiles with color representation of relative thickness of the film. **c** Representation of a possible arrangement of FCDs with nonzero eccentricity on circular pillars with the edge-pinning effect. **d** A complete layer construction of four FCDs surrounding a pillar. Adapted with permission [56]. Copyright 2013, National Academy of Sciences, USA

By utilizing of the intrinsic molecular orientations of SmA LCs within the defect structure (conical dimples), Kim et al. [58] demonstrated the application of TFCDs as microlens arrays with graded refractive index. In contrast, conventional microlens arrays have spherical or hemispherical shape with homogeneous refractive index. Alternatively, the TFCD dimple arrays could be utilized as a structural template to prepare polymeric microlens arrays by sequential replica molding. To demonstrate the focusing ability of TFCDs, the film was placed on the sample stage of the optical microscope and illuminated with white light through a projection template (see Fig. 2.15c). The microlens array was scanned in the z-direction and different quality of miniaturized ''F'' images in a hexagonal array were obtained (Fig. 2.16) depending on the scanning distance versus the focal plane of the microlenses.

Indeed, as shown in Fig. 2.17A, the TFCD arrays consist of two imaging elements, microlens and clear windows. The flat regions between the concentric TFCD dimples did not scatter light and therefore functioned as clear windows. Along the line, the Jung group demonstrated the utility of the TFCDs as a selective

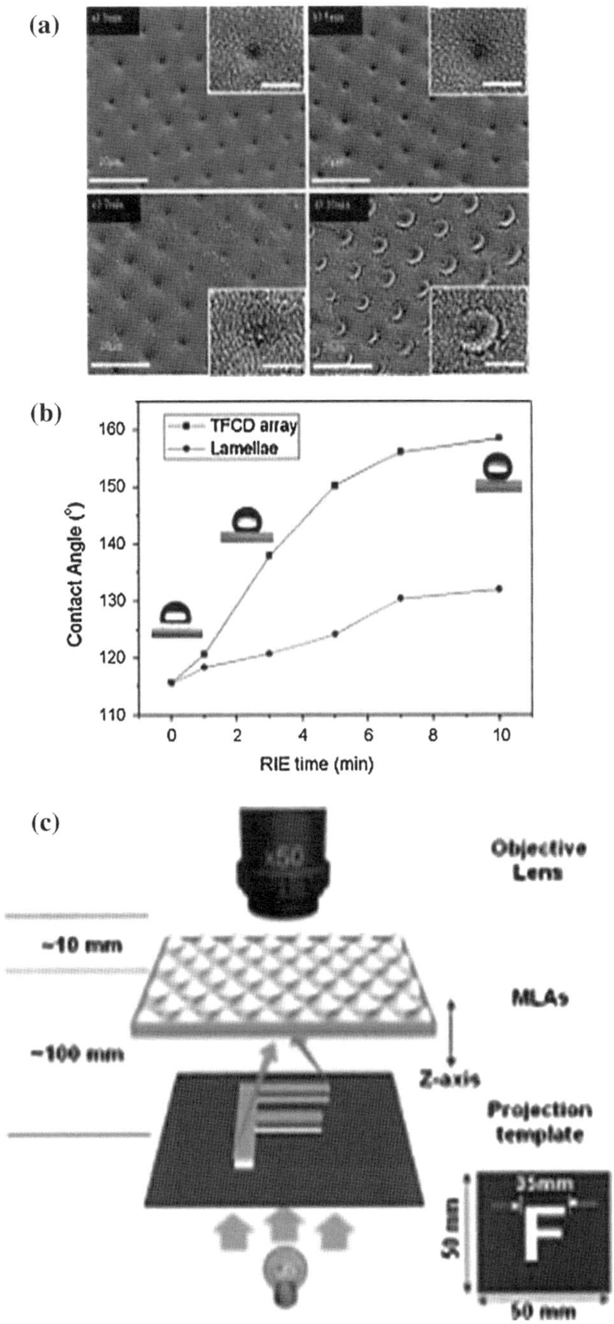

Fig. 2.15 **a** SEM images TFCD surfaces modified with RIE at various treatment times (3, 5, 7 and 10 min) [57]. The inset scale bars are 500 nm. Reproduced with permission [57]. Copyright 2009, WILEY-VCH Verlag GmbH & Co. KGaA, Weinheim. **b** Measured water contact angles of the surface of TFCD arrays and lamellar film as a function of RIE treatment period [57]. **c** The schematic diagram describing the experimental setup of the projection experiment for fabricated MLAs [58]

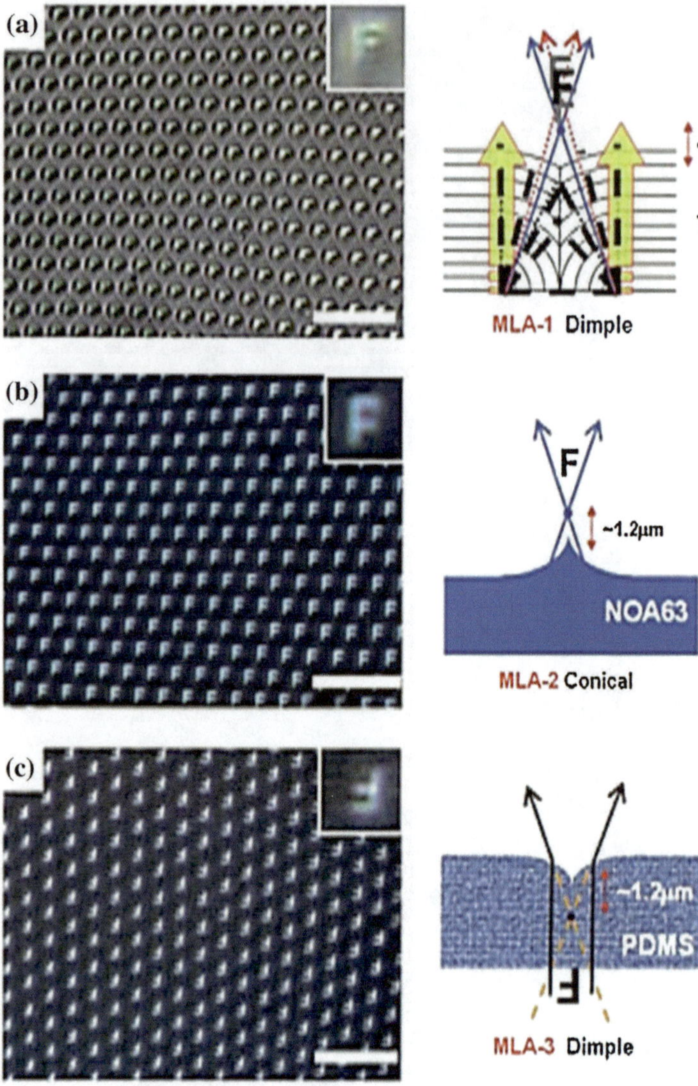

Fig. 2.16 Optical images projected through **a** the TFCD array, **b** the conical NOA63 first mold, and **c** the dimple-shaped PDMS second mold. The scale bars are 50 μm. The right column revealed the corresponding light propagation through each MLA [58]. Reproduced with permission [58]. Copyright 2010, Royal Society of Chemistry

photomask in photolithography [59]. By adjusting (i) LC film thickness, (ii) the illumination dose, (iii) the size of TFCD array mask, and (iv) the tone of the photoresist, a wide range of structures could be produced (see examples in Fig. 2.17B).

Further, by taking advantage of the sharp tips of the dimples, Yoo et al. created metallic nanostructures for study of plasmonic hot spot [60]. Compared to a flat surface and the inverted structure of the FCD arrays (conical shaped structure), the dimple structure revealed a strongest fluorescence signal after drop-casting CdS-ZnSe core–shell quantum dots solution on the tested substrates (see in Fig. 2.18a) [60]. The authors further analyzed the structure-property relationship by using a finite difference time domain (FDTD) simulation to calculate the monochromatic wave irradiations on each structure. As shown in Fig. 2.18b and c, FDTD simulation revealed that the field intensity around the dimpled Au structure increased by 5 times compared to the input, which strongly suggested the effective resonance oscillation of the plasmon wave in the sub-micron hole structure [60]. The tunability of the localized plasmonic coupling and hot spot efficiency at interparticle distances was realized with this FCD template because the geometrical dimension of the dimple structure such as the depth and period was adjustable simply by varying the LC thickness.

The concept of utilizing elasticity in LCs to facilitate organization of functional materials has been established in numerous studies, mostly in a nematic phase [63–65]. Due to a higher degree of complexity, it was not until recently that researchers began to explore the use of smectic LCs and their defect structures as robust templates for self-assembly micron-sized inclusions and nanoparticles. It is known that particles tend to migrate towards the high energy region of linear defects in an anisotropic fluid to minimize the overall energy of the system [62]. Smectic LCs are anticipated to offer a better trapping mode for particles than nematic LCs because of higher distortion energy associated with the defect formation in smectic. Yoon et al. first demonstrated the trapping of fluorescent micron-sized silica particles in the linear defect line of each TFCD in the thin film, resulting in a 2D hexagonal array of fluorescent particles in registry with the location of the defect cores (see Fig. 2.19A).

Because of the potentially unique materials properties at the nanoscale, it will be attractive to synergistically integrate the highly responsive, anisotropic LCs with nanomaterials. Compared to microparticles, however, it is more challenging to trap the nanostructure materials. Dispersion of nanoparticles (NPs) in visco-elastic fluid-like LCs has always been a major problem due to strong attractive forces between NPs. Therefore, most efforts have been devoted to evaluating the sensitivity of the trapping power in nanometer scale. The dispersion of NPs can be controlled by their volume fraction in LCs and surface chemistry or surface anchoring of NPs. Initially, it was expected that NPs might be too small to create a long-range elastic distortion in LCs. However, a number of experiments such as the one by Pratibha et al. proved otherwise [66, 67]. The apparent bumps from elastic deformation of smectic layers and disappearance of FCDs were believed to be the consequences of the inclusion of 7 nm gold nanoparticles in a smectic thin film and attributed to the surface anchoring at NP–LC interfaces [66]. Later studies show that NPs could be organized in various trapping sites of defects including grain boundaries, curvature walls, dislocations and disclinations in the order of reducing trapping efficiency [62, 67]. For instance, Milette et al. [62] reported

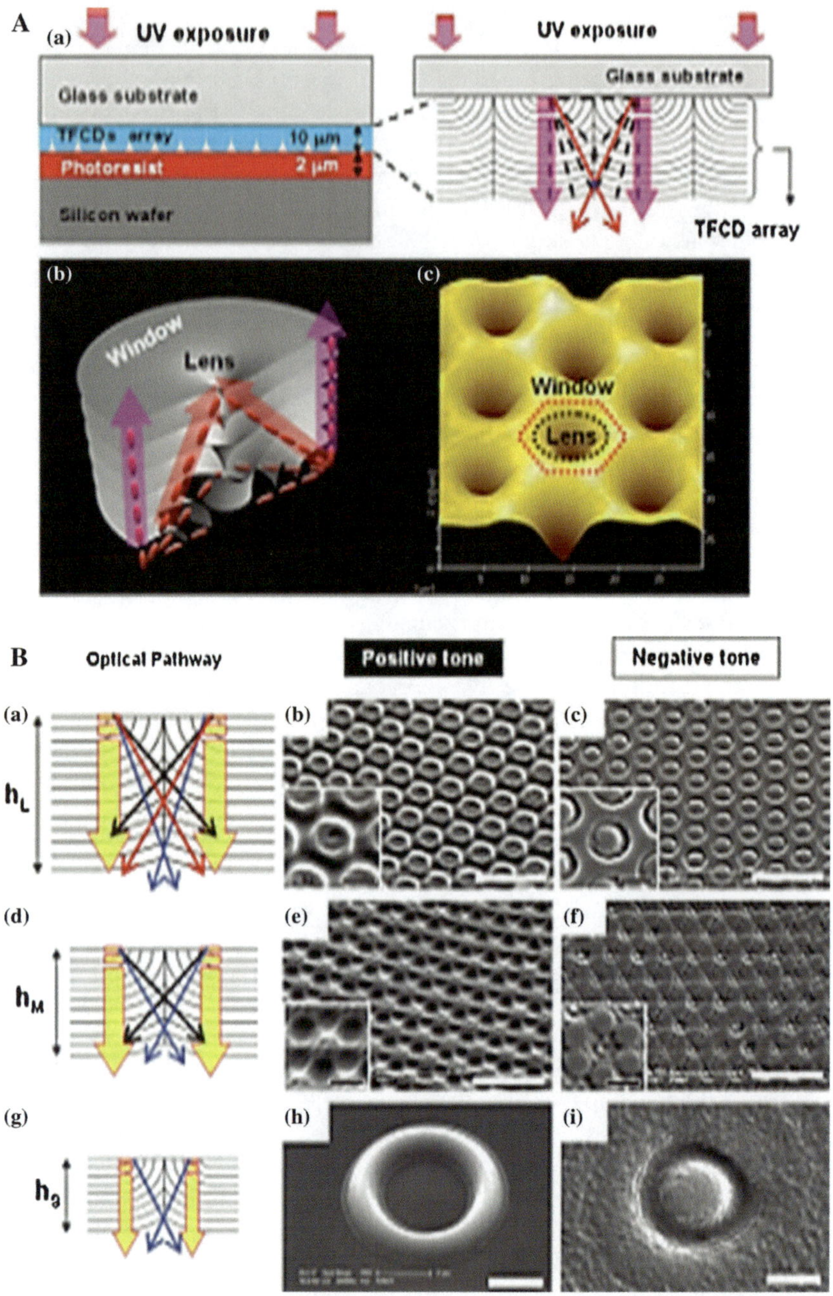
A
(a)
UV exposure
Glass substrate
TFCDs array
10 μm
Photoresist
2 μm
Silicon wafer
UV exposure
Glass substrate
TFCD array
(b)
Window
Lens
(c)
Window
Lens
B
Optical Pathway
Positive tone
Negative tone
(a)
h_L
(b)
(c)
(d)
h_M
(e)
(f)
(g)
h_S
(h)
(i)

Fig. 2.17 *A* Schematic illustrations of a setup of a photolithography process using the TFCD based MLAs photomask and a focusing mechanism of TFCD MLAs where the small red rods represent each smectic molecule [59]. A 3D configuration of the surface profile of TFCD based MLAs from AFM scan. *B* Schematic illustrations of optical pathway through TFCD based MLAs photomasks with various domain sizes (the *top row* 4.5 μm < r < 6.0 μm, the *second row* 3 μm < r < 4.5 μm; the *third row* 2.0 μm < r < 3.0 μm) and the corresponding SEM images of developed patterns from both positive-tone resists (the *second column*) and negative-tone resists (the *third column*) [59]. The scale bars are 20 μm for all images except (*Bh*) and (*Bi*) (2 μm) [59]. Reproduced with permission [59]. Copyright 2010, WILEY–VCH Verlag GmbH & Co. KGaA, Weinheim

organization of linear array of 4.7 nm gold NPs along periodic edge dislocation of 8CBs in a wedged cell (Fig. 2.19B). Both the top and bottom of the cell promoted homeotropic anchoring of 8CB. Because of the variation in the number of layers, edge dislocations in the form of line defects appeared periodically to compromise the high energy penalty from the elastic and interfacial forces. The observation of linear arrays of gold NPs in confined 8CBs confirmed that the high energy core of the edge dislocations provided an effective trapping force that was sensitive to materials down to a few nanometers.

2.4 Conclusions and Perspective

In this chapter, we discuss several key principles to produce various types of FCDs confined in thin smectic films with an exceptionally fine control of the feature size, spatial distribution, symmetry, intrinsic topology and their hierarchical organization. Such confinement is obtained by careful design of LC templates with variable dimensions and symmetries such that they impose both molecular anchoring and geometric restrictions on the 3D smectic structures. The highly ordered FCD arrays have been exploited as a new class of building blocks for a wide range of applications, including functional surfaces [58, 68], directed assembly of soft microsystems [61, 62, 66] and nanopatterns [69], charge enhancement in photovoltaics and transistors [70], and potentially revolutionary technology such as blue phase LC displays. The possibility of patterns generated from this process is not limited to ordered arrays of micro-dimples [69]. As shown earlier, fish skin texture [52], circular ring [58] and alternating hill and valley surfaces could be obtained. While creation of a superhydrophobic surface from an array of dimple like SmA defects in thin film has been demonstrated through simple surface modification [68], the inverse of the texture which features an array of sharp points can function as a stamp to print nanometer patterns in other materials. The dimples of TFCDs now become sharp points after printing, which in turn can be used as lenses and localized surface plasmon resonance hot spots for chemical and biological sensing. Lastly, we speculate some of the most relevant future directions for research.

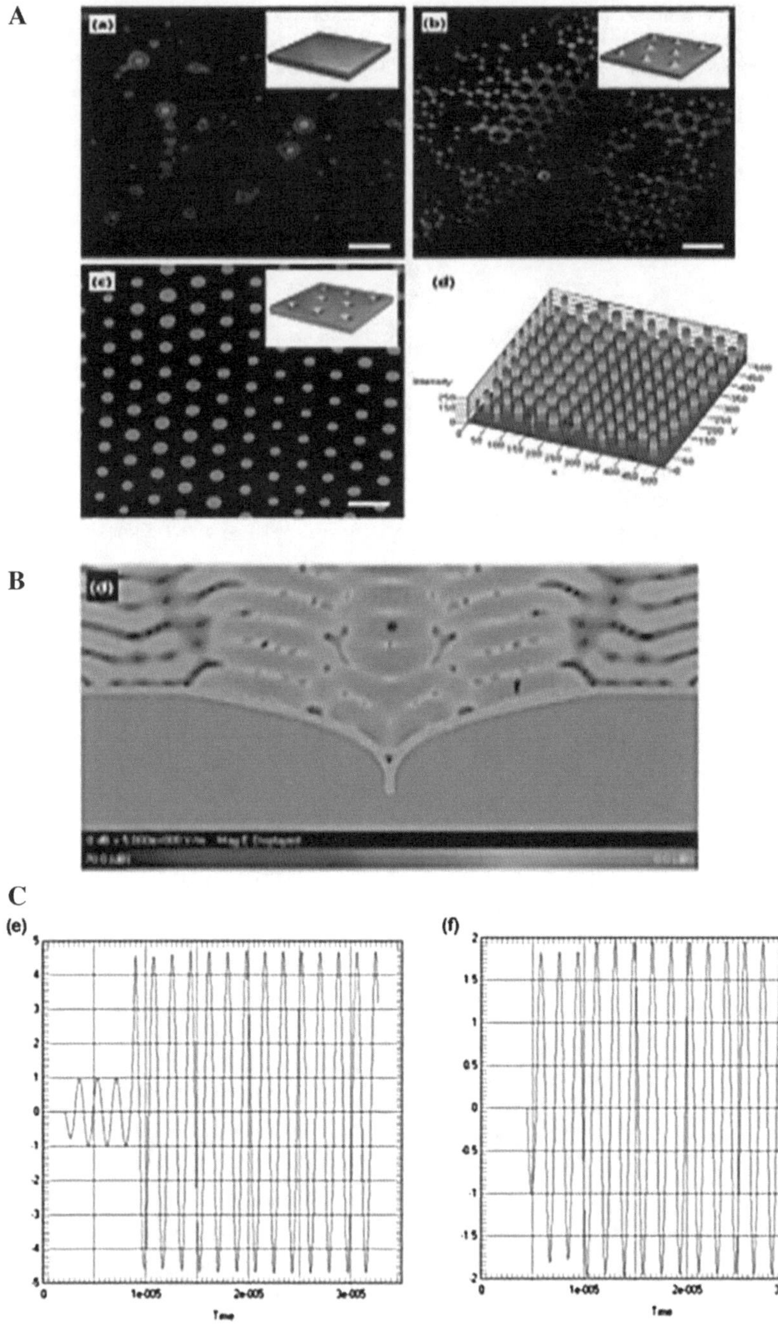

Fig. 2.18 *A* Confocal fluorescence microscope images of the QDs on (*Aa*) a flat gold film, (*Ab*) the conical Au array, and (*Ac*) the dimpled Au array. (*Ad*) 3D visualization of the fluorescence intensity in (*Ac*) [60]. The QD has an emission wavelength of 597 nm. The excitation laser source was a 543 nm HeNe laser. (*B*) Field snapshot image from FDTD simulation results of Au coated dimpled structure. (*C*) Plots for the near-zone field intensity versus time in the dimpled Au structure [60]. Reproduced with permission [60]. Copyright 2013, Royal Society of Chemistry

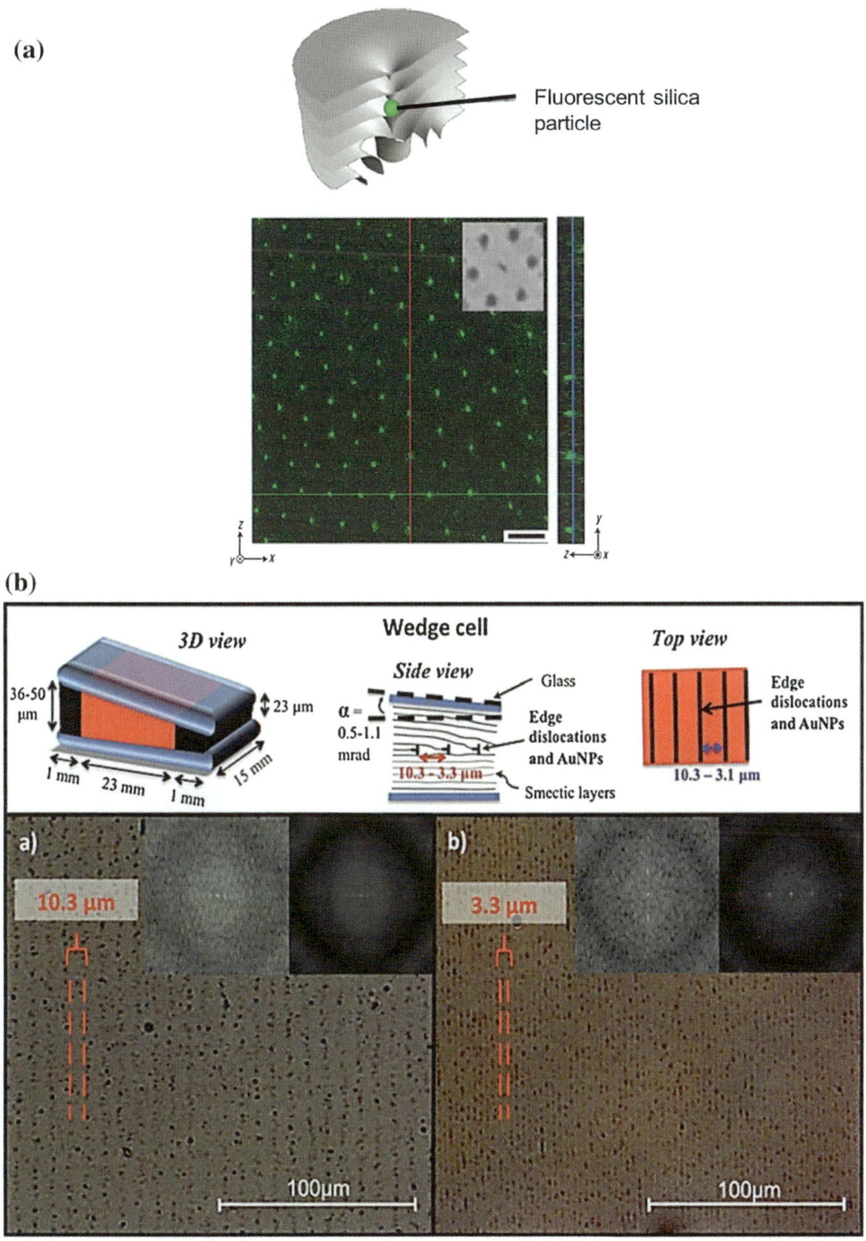

Fig. 2.19 *A* A schematic diagram of a silica particle trapped in the center of a TFCD and a confocal fluorescent microscopy image of fluorescent silica particles trapped in a TFCD array [61]. The scale bar is 10 μm [61]. Reproduced with permission [61]. Copyright 2007, Nature Publishing Group. *B* Schematic illustrations of designed wedge cell for 8CB and the formation of edge dislocations inside the cell [62]. POM images of 0.1 wt% gold nanoparticles dispersed in confined 8CB inside the homeotropic anchoring wedge cell with an angle of (*Ba*) 0.5 mrad and (*Bb*) 1.1 mrad. The *insets* are FFT images of the POM images [62]. Reproduced with permission [62]. Copyright 2012, Royal Society of Chemistry

2.4.1 Complex Templates

Most studies so far focus on confining LCs with uniform boundary conditions at different interfaces in order to produce predictable pattern or topology of liquid crystals with high regularity. It is known that boundary conditions regulate molecular anchoring of LCs at interfaces and in turn influence their macroscopic structures. Therefore, it remains to be studied of LC assembly on a curved surface and a surface with gradient boundary conditions. For the latter, the surface could have variable anchoring strengths, non-uniform pattern height or shape, or a responsive substrate on which the patterns could be dynamically altered, for example, from circular shape to elliptical one using heat or an electric field. This microscopic geometric cue would travel up the layers of LCs and produce micrometer-scale changes on the surface. It would be profoundly interesting to investigate how SmA LC molecules respond to these complex confinements while satisfying heterogeneous boundary conditions by balancing competition between elastic inner force imposed by SmA LC molecules and surface anchoring forces imposed by the template.

2.4.2 Directed Assemblies from Other LC Phases

While we demonstrate the unprecedented control of FCDs from SmA LCs, it will be interesting to broaden the scope of the study and investigate assemblies on variable 1D to 3D templates from other LC phases. For example, confinement of SmC* phase, which is constructed with rotating directors from one layer to the next forming a helix, should yield new 3D orders and chirality due to energetic frustration of SmC* LCs from their equilibrium forms. Cholesteric phase, known as chiral nematic, is another LC phase that deserves much attention. The chiral nematic exhibits a number of unique structural related properties, such as high temperature sensitivity and selective reflection of light, which have already been extensively utilized in today technologies, ranging from LC based thermometers, sensors, advanced displays, clothing, inks to paints. Therefore, the ability to create new chiral nematic structures with a geometric means should immediately benefit its current applications and beyond.

2.4.3 Templating Nanomaterials and Other Applications of SmA LCs

Nanomaterials play important roles to emerging nanotechnologies. However, it remains the major challenge for the preparation of functional materials and devices with engineered order. The ability to control the assembly of SmA LCs in 3D with variable degree of orientational and translational orders could be realized using a variety of long-range periodic and anisotropic structures, including wedge cells, microgroove wrinkle surfaces, 1D microchannels, and 2D pillar arrays.

Because the layers of LCs transmit elastic energy, they can be used to do mechanical work, such as to assemble nanomaterials and induce topological defects in LCs. The feasibility of this scheme has been illustrated in the lab scale. For instance, linear and hexagonal arrays of nanoparticles were elastically trapped at various sites of smectic defects. Single wall carbon nanotubes could be organized into 2D parallel sheets between smectic layers due to specific interactions between π and π interactions between the hexagonal rings of the carbon nanotubes and the aromatic moieties of LC molecules [62, 67, 71].

To conclude, the research on directed assembly of Sm LCs in complex geometries has just begun. The advances on how to control FCDs in thin films could lead to a new paradigm of "*liquid crystal soft lithography*" for micro- and nanofabrications, which in turn will extend applications beyond displays and sensors.

Acknowledgments This work is supported by National Science Foundation (NSF)/MRSEC grant, # DMR11-20901. We are grateful to Randall D. Kamien (RDK) for stimulating our interests in Sm LCs, RDK, Kathleen Stebe, Daniel A. Beller (DAB), Marcello Cavallaro, Jr., Mohamed Amine Gharbi for helpful discussion and collaboration on the work of directed assembly of Sm LCs. We also acknowledge DAB for providing Fig. 2.2a illustration.

References

1. B.W. Lee, N.A. Clark, Alignment of liquid crystals with patterned isotropic surfaces. Science **291**, 2576–2580 (2001)
2. W. Bragg, Liquid crystals. Nature **133**, 445–456 (1934)
3. T.D. Choudhury, N.V.S. Rao, R. Tenent, J. Blackburn, B. Gregg, I.I. Smalyukh, Homeotropic alignment and director structures in thin films of triphenylamine-based discotic liquid crystals controlled by supporting nanostructured substrates and surface confinement. J. Phys. Chem. B **115**, 609–617 (2010)
4. H.K. Bisoyi, S. Kumar, Liquid-crystal nanoscience: an emerging avenue of soft self-assembly. Chem. Soc. Rev. **40**, 306–319 (2011)
5. M.C. Choi, T. Pfohl, Z. Wen, Y. Li, M.W. Kim, J.N. Israelachvili, C.R. Safinya, Ordered patterns of liquid crystal toroidal defects by microchannel confinement. Proc. Natl. Acad. Sci. USA **101**, 17340–17344 (2004)
6. D. Miyajima, F. Araoka, H. Takezoe, J. Kim, K. Kato, M. Takata, T. Aida, Ferroelectric columnar liquid crystal featuring confined polar groups within core-shell architecture. Science **336**, 209–213 (2012)
7. D. Kang, J.E. Maclennan, N.A. Clark, A.A. Zakhidov, R.H. Baughman, Electro-optic behavior of liquid-crystal-filled silica opal photonic crystals: Effect of liquid-crystal alignment. Phys. Rev. Lett. **86**, 4052–4055 (2001)
8. H. Park, E.P.J. Parrott, F. Fan, M. Lim, H. Han, V.G. Chigrinov, E. Pickwell-MacPherson, Evaluating liquid crystal properties for use in terahertz devices. Opt. Express **20**, 11899–11905 (2012)

9. N.A. Clark, S.T. Lagerwall, Submicrosecond bistable electro-optic switching in liquid crystals. Appl. Phys. Lett. **36**, 899–901 (1980)
10. D.J. Gardiner, H.J. Coles, Organosiloxane liquid crystals for fast-switching bistable scattering devices. J. Phys. D Appl. Phys. **39**, 4948–4955 (2006)
11. T. Kato, T. Yasuda, Y. Kamikawa, M. Yoshio, Self-assembly of functional columnar liquid crystals. Chem. Commun. 729–739 (2009)
12. C.M. Paleos, D. Tsiourvas, Supramolecular hydrogen-bonded liquid crystals. Liq. Cryst. **28**, 1127–1161 (2001)
13. D.R. Nelson, *Defects and Geometry in Condensed Matter Physics* (Cambridge University Press, Cambridge, 2002)
14. P. Bhattacharjee, G. Sigl, Origin and propagation of extremely high-energy cosmic rays. Phys. Rep. **327**, 109–247 (2000)
15. M. Trodden, A.-C. Davis, R. Brandenberger, Particle physics models, topological defects and electroweak baryogenesis. Phys. Lett. B **349**, 131–136 (1995)
16. R.H. Brandenberger, Topological defects and structure formation. Int. J. Mod. Phys. A **9**, 2117–2189 (1994)
17. A. Vilenkin, E.P.S. Shellard, *Cosmic Strings and Other Topological Defects* (Cambridge University Press, Cambridge, 2000)
18. M. Kleman, J. Friedel, Disclinations, dislocations, and continuous defects: A reappraisal. Rev. Mod. Phys. **80**, 61–115 (2008)
19. M. Kléman, Defects in liquid crystals. Rep. Prog. Phys. **52**, 555 (1989)
20. T.C. Lubensky, D. Pettey, N. Currier, H. Stark, Topological defects and interactions in nematic emulsions. Phys. Rev. E **57**, 610–625 (1998)
21. C. Domb, *Phase Transitions and Critical Phenomena* (Academic Press, New York, 2000)
22. R.D. Kamien, T. Lubensky, Minimal surfaces, screw dislocations, and twist grain boundaries. Phys. Rev. Lett. **82**, 2892–2895 (1999)
23. T. Lubensky, S. Renn, Twist-grain-boundary phases near the nematic–smectic-A–smectic-C point in liquid crystals. Phys. Rev. A **41**, 4392 (1990)
24. J. Goodby, M. Waugh, S. Stein, E. Chin, R. Pindak, J. Patel, Characterization of a new helical smectic liquid crystal. Nature **337**, 449–452 (1989)
25. S. Meiboom, J.P. Sethna, P. Anderson, W. Brinkman, Theory of the blue phase of cholesteric liquid crystals. Phys. Rev. Lett. **46**, 1216–1219 (1981)
26. H. Stegemeyer, T. Blümel, K. Hiltrop, H. Onusseit, F. Porsch, Thermodynamic, structural and morphological studies on liquid-crystalline blue phases. Liq. Cryst. **1**, 3–28 (1986)
27. G. Srajer, R. Pindak, M. Waugh, J. Goodby, J. Patel, Structural measurements on the liquid-crystal analog of the Abrikosov phase. Phys. Rev. Lett. **64**, 1545–1548 (1990)
28. G. P. Crawford, *Liquid Crystals in Complex Geometries: Formed by Polymer and Porous Networks* (Taylor & Francis, London, 1996)
29. P.S. Drzaic, *Liquid Crystal Dispersions* (World Scientific, Singapore, 1995)
30. P. Poulin, H. Stark, T.C. Lubensky, D.A. Weitz, Novel colloidal interactions in anisotropic fluids. Science **275**, 1770–1773 (1997)
31. I.I. Smalyukh, O.D. Lavrentovich, A.N. Kuzmin, A.V. Kachynski, P.N. Prasad, Elasticity-mediated self-organization and colloidal interactions of solid spheres with tangential anchoring in a nematic liquid crystal. Phys. Rev. Lett. **95**, 157801 (2005)
32. M.G. Friedel, Les états mésomorphes de la matière. Annls. Phys. **18**, 273–474 (1922)
33. M. Kleman, O.D. Lavrentovich, Liquids with conics. Liq. Cryst. **36**, 1085–1099 (2009)
34. J.B. Fournier, G. Durand, Focal conic faceting in smectic-A liquid crystals. J. Phys. II Fr **1**, 845–870 (1991)
35. C. Blanc, M. Kleman, Tiling the plane with noncongruent toric focal conic domains. Phys. Rev. E **62**, 6739–6748 (2000)
36. J.P. Bramble, S.D. Evans, J.R. Henderson, T.J. Atherton, N.J. Smith, Observations of focal conic domains in smectic liquid crystals aligned on patterned self-assembled monolayers. Liq. Cryst. **34**, 1137–1143 (2007)

37. A. Honglawan, D.A. Beller, M. Cavallaro, R.D. Kamien, K.J. Stebe, S. Yang, Pillar-assisted epitaxial assembly of toric focal conic domains of smectic-A liquid crystals. Adv. Mater. **23**, 5519–5523 (2011)
38. Y.H. Kim, D.K. Yoon, M.C. Choi, H.S. Jeong, M.W. Kim, O.D. Lavrentovich, H.-T. Jung, Confined self-assembly of toric focal conic domains (The effects of confined geometry on the feature size of toric focal conic domains). Langmuir **25**, 1685–1691 (2009)
39. W. Guo, S. Herminghaus, C. Bahr, Controlling smectic focal conic domains by substrate patterning. Langmuir **24**, 8174–8180 (2008)
40. G.P. Alexander, B.G.G. Chen, E.A. Matsumoto, R.D. Kamien, Power of the poincare group: elucidating the hidden symmetries in focal conic domains. Phys. Rev. Lett. **104**, 257802 (2010)
41. D. Ki Yoon, R. Deb, D. Chen, E. Körblova, R. Shao, K. Ishikawa, N.V.S. Rao, D.M. Walba, I.I. Smalyukh, N.A. Clark, Organization of the polarization splay modulated smectic liquid crystal phase by topographic confinement. Proc. Nat. Acad. Sci. USA **107**, 21311–21315 (2010)
42. D.K. Yoon, J. Yoon, Y.H. Kim, M.C. Choi, J. Kim, O. Sakata, S. Kimura, M.W. Kim, I.I. Smalyukh, N.A. Clark, M. Ree, H.-T. Jung, Liquid-crystal periodic zigzags from geometrical and surface-anchoring-induced confinement: Origin and internal structure from mesoscopic scale to molecular level. Phys. Rev. E **82**, 041705 (2010)
43. D.J. Gardiner, S.M. Morris, H.J. Coles, High-efficiency multistable switchable glazing using smectic A liquid crystals. Sol. Energy Mat. Sol. C. **93**, 301–306 (2009)
44. W. Guo, C. Bahr, Influence of anchoring strength on focal conic domains in smectic films. Phys. Rev. E **79**, 011707 (2009)
45. E. Lacaze, J.P. Michel, M. Goldmann, M. Gailhanou, M. de Boissieu, M. Alba, Bistable nematic and smectic anchoring in the liquid crystal octylcyanobiphenyl (8CB) adsorbed on a MoS_2 single crystal. Phys. Rev. E **69**, 041705 (2004)
46. J.B. Fournier, I. Dozov, G. Durand, Surface frustration and texture instability in smectic-A liquid crystals. Phys. Rev. A **41**, 2252–2255 (1990)
47. S. Shojaei-Zadeh, S.L. Anna, Role of surface anchoring and geometric confinement on focal conic textures in smectic-A liquid crystals. Langmuir **22**, 9986–9993 (2006)
48. J.H. Kim, Y.H. Kim, H.S. Jeong, E.K. Youn, H.-T. Jung, Highly ordered defect arrays of 8CB (4[prime or minute]-n-octyl-4-cyano-biphenyl) liquid crystal via template-assisted self-assembly. J. Mater. Chem. **21**, 18381–18385 (2011)
49. J. Jeong, M.W. Kim, Confinement-induced transition of topological defects in smectic liquid crystals: From a point to a line and pearls. Phys. Rev. Lett. **108**, 207802 (2012)
50. B. Zappone, E. Lacaze, H. Hayeb, M. Goldmann, N. Boudet, P. Barois, M. Alba, Self-ordered arrays of linear defects and virtual singularities in thin smectic-A films. Soft Matter **7**, 1161–1167 (2011)
51. J.-P. Michel, E. Lacaze, M. Goldmann, M. Gailhanou, M. de Boissieu, M. Alba, Structure of smectic defect cores: x-ray study of 8CB liquid crystal ultrathin films. Phys. Rev. Lett. **96**, 027803 (2006)
52. B. Zappone, C. Meyer, L. Bruno, E. Lacaze, Periodic lattices of frustrated focal conic defect domains in smectic liquid crystal films. Soft Matter **8**, 4318–4326 (2012)
53. B. Zappone, E. Lacaze, H. Hayeb, M. Goldmann, N. Boudet, P. Barois, M. Alba, Self-ordered arrays of linear defects and virtual singularities in thin smectic-A films. Soft Matter **7**, 1161–1167 (2010)
54. J.-P. Michel, E. Lacaze, M. Alba, M. de Boissieu, M. Gailhanou, M. Goldmann, Optical gratings formed in thin smectic films frustrated on a single crystalline substrate. Phys. Rev. E **70**, 011709 (2004)
55. T. Ohzono, Y. Takenaka, J. Fukuda, Focal conics in a smectic-A liquid crystal in microwrinkle grooves. Soft Matter **8**, 2438–2444 (2012)
56. A. Honglawan, D.A. Beller, M. Cavallaro, R.D. Kamien, K.J. Stebe, S. Yang, Topographically induced hierarchical assembly and geometrical transformation of focal conic domain arrays in smectic liquid crystals. Proc. Nat. Acad. Sci. USA **110**, 34–39 (2013)

57. Y.H. Kim, D.K. Yoon, H.S. Jeong, J.H. Kim, E.K. Yoon, H.-T. Jung, Fabrication of a superhydrophobic surface from a smectic liquid-crystal defect array. Adv. Func. Mater. **19**, 3008–3013 (2009)
58. Y.H. Kim, H.S. Jeong, J.H. Kim, E.K. Yoon, D.K. Yoon, H.-T. Jung, Fabrication of two-dimensional dimple and conical microlens arrays from a highly periodic toroidal-shaped liquid crystal defect array. J. Mater. Chem. **20**, 6557–6561 (2010)
59. Y.H. Kim, J.-O. Lee, H.S. Jeong, J.H. Kim, E.K. Yoon, D.K. Yoon, J.-B. Yoon, H.-T. Jung, Optically selective microlens photomasks using self-assembled smectic liquid crystal defect arrays. Adv. Mater. **22**, 2416–2420 (2010)
60. H.-W. Yoo, Y.H. Kim, J.M. Ok, H.S. Jeong, J.H. Kim, B.S. Son, H.-T. Jung, Plasmonic three-dimensional dimpled array from highly ordered self-assembled liquid crystal defects. J. Mater. Chem. C **1**, 1434–1439 (2013)
61. D.K. Yoon, M.C. Choi, Y.H. Kim, M.W. Kim, O.D. Lavrentovich, H.-T. Jung, Internal structure visualization and lithographic use of periodic toroidal holes in liquid crystals. Nat. Mater. **6**, 866–870 (2007)
62. J. Milette, S. Relaix, C. Lavigne, V. Toader, S.J. Cowling, I.M. Saez, R.B. Lennox, J.W. Goodby, L. Reven, Reversible long-range patterning of gold nanoparticles by smectic liquid crystals. Soft Matter **8**, 2593–2598 (2012)
63. I. Muševič, M. Škarabot, U. Tkalec, M. Ravnik, S. Žumer, Two-dimensional nematic colloidal crystals self-assembled by topological defects. Science **313**, 954–958 (2006)
64. C.P. Lapointe, T.G. Mason, I.I. Smalyukh, Shape-controlled colloidal interactions in nematic liquid crystals. Science **326**, 1083–1086 (2009)
65. C. Zamora-Ledezma, C. Blanc, M. Maugey, C.C. Zakri, P. Poulin, E. Anglaret, Anisotropic thin films of single-wall carbon nanotubes from aligned lyotropic nematic suspensions. Nano Lett. **8**, 4103–4107 (2008)
66. R. Pratibha, W. Park, I.I. Smalyukh, Colloidal gold nanosphere dispersions in smectic liquid crystals and thin nanoparticle-decorated smectic films. J. Appl. Phys. **107**, 63511 (2010)
67. D. Coursault, J. Grand, B. Zappone, H. Ayeb, G. Lévi, N. Félidj, E. Lacaze, Linear Self-assembly of nanoparticles within liquid crystal defect arrays. Adv. Mater. **24**, 1461–1465 (2012)
68. Y.H. Kim, D.K. Yoon, H.S. Jeong, J.H. Kim, E.K. Yoon, H.T. Jung, Fabrication of a superhydrophobic surface from a smectic liquid-crystal defect array. Adv. Funct. Mater. **19**, 3008–3013 (2009)
69. Y.H. Kim, D.K. Yoon, H.S. Jeong, H.T. Jung, Self-assembled periodic liquid crystal defects array for soft lithographic template. Soft Matter **6**, 1426–1431 (2010)
70. M. O'Neill, S.M. Kelly, Liquid crystals for charge transport, luminescence, and photonics. Adv. Mater. **15**, 1135–1146 (2003)
71. H.S. Jeong, Y.K. Ko, Y.H. Kim, D.K. Yoon, H.-T. Jung, Self assembled plate-like structures of single-walled carbon nanotubes by non-covalent hybridization with smectic liquid crystals. Carbon **48**, 774–780 (2010)

Chapter 3
Liquid Crystalline 1D and 2D Carbon Materials

Hari Krishna Bisoyi, Sandeep Kumar and Quan Li

Abstract Carbon nanotubes (CNTs) and graphene derivatives have created great expectations in terms of practical applications and advancement of science. Processing through liquid crystalline ordering of CNTs and graphenes is a promising opportunity to develop novel materials and applications with spontaneously aligned one dimensional (1D) (nanotubes) and two dimensional (2D) (graphene) carbon allotropes. This chapter presents the different methodologies that have been developed to achieve nematic liquid crystals of CNTs and graphene derivatives and their subsequent use for the fabrication of field effect transistors (FETs), and high performance carbon fibers etc. by processing via liquid crystalline route. Since carbon-based nanomaterials are being investigated in the right context of contemporary significance, it is optimistically anticipated that the combination of outstanding and unprecedented properties of carbon nanoforms with processability and alignment properties of liquid crystals can lead to some interesting materials and devices with promising properties and performances.

3.1 Introduction

Liquid crystalline phase is a thermodynamically stable state of matter after solid, liquid and gas and is regarded as the fourth state of matter. Liquid crystalline phase possesses less ordering than crystalline state but more ordering than amorphous liquids. Interestingly, liquid crystals (LCs) share the anisotropic (direction dependent) property of crystalline solids and the fluidity of liquids. Owing to its

H. K. Bisoyi · Q. Li (✉)
Liquid Crystal Institute, Kent State University, Kent, OH 44242, USA
e-mail: qli1@kent.edu

S. Kumar
Raman Research Institute, Bangalore, India

Q. Li (ed.), *Nanoscience with Liquid Crystals*, NanoScience and Technology,
DOI: 10.1007/978-3-319-04867-3_3,

appearance between the crystalline solid and isotropic liquid, this phase of matter is often referred to as mesophase (intermediate phase). The constituents of the mesophases are known as mesogens and the mesogens can be organic, inorganic or organo-metallic in nature. LCs have become quintessential materials in our daily life due to their technological applications as the active switching components in flat panel LC display (LCD) devices like mobile phones, computer and television screens, and projectors etc. Recently, LC research and development are undergoing a renaissance due to the successful demonstration of their potential beyond display applications in other areas such as in biomedical applications, microbe detection, sensors, organic optoelectronics, and stimuli responsive devices [1–27]. Scientifically, LC field is a fertile ground to study supramolecular self-assembly of matter and soft matter research. Though LCs were discovered in 1888 and it has been 125 years of their scientific enquiry, they still continue to through us surprises and new challenges to understand this mysterious and intriguing state of matter. New avenues keep opening up with diverse scientific and technological opportunities and challenges.

There are various ways to classify LCs by taking into account the molecular features and supramolecular assemblies. However the most common classification, thermotropic and lyotropic, invokes how the liquid crystalline phases have been obtained. Thermotropic LCs can be obtained either by heating (thermal effect) crystalline solids or by cooling isotropic liquids where the liquid crystalline phase is stable over a range of temperature. These thermotropic LCs are used in flat panel displays in which the liquid crystalline phase has been stabilized around room temperature. On the other hand, lyotropic LCs can be realized by dissolving amphiphilic molecules at appropriate concentrations in suitable solvents. Often water is used as the solvent and such anisotropic aqueous solutions find their application in food and cosmetic industry, soap and detergent industry including drug delivery and gene therapy etc. Moreover, this has significance in biology since the lipid bilayer cell membrane exists in liquid crystalline state which is evident from its ordered and dynamic attributes. Lyotropic liquid crystalline phases can be fabricated and stabilized by dissolving anisometric particles such as rods and discs in appropriate concentrations. In addition to flexible surfactant molecules, it has been observed that tobacco mosaic virus (TMV) and deoxyribo nucleic acid (DNA) exhibit lyotropic liquid crystalline phases in water. Recently, it has been reported that metallic and metal oxide nanorods and nanodisks exhibit liquid crystalline phase behavior under suitable conditions [28–39].

The theoretical aspects of the liquid crystalline phases of the colloids of anisotropic particles (rods and plates) were investigated by Onsager in 1949 [40]. He rationalized the formation of the nematic phase by excluded volume effects working among the particles. As the concentration increases, the colloid transits from the isotropic phase to the isotropic/nematic biphase, and finally to the nematic phase. The original Onsager's theory deals with cylindrical objects with diameter D and height L, which interact only through hard-core repulsions.

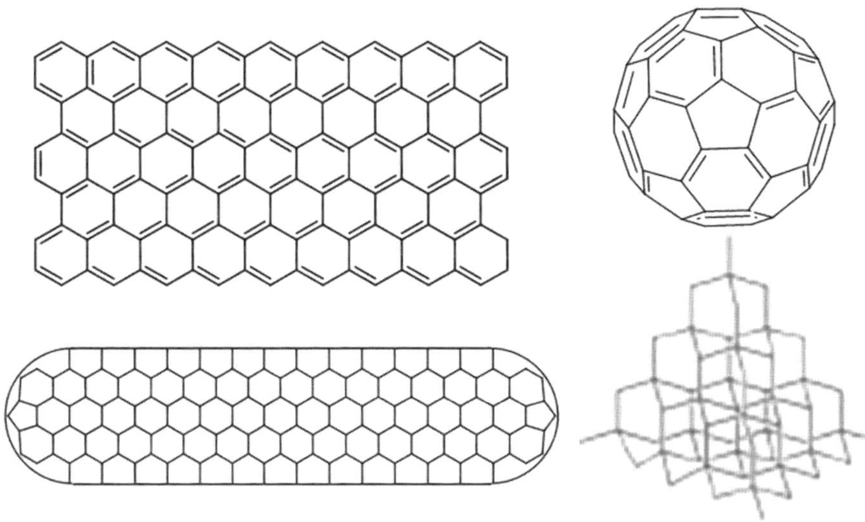

Fig. 3.1 0D, 1D, 2D and 3D carbon allotropes

Objects with L ≫ D and L ≪ D are regarded as rods and discs respectively, and the theory applies for both rods and discs. At a larger aspect ratio and high particle concentration, if the particles partially stop thermodynamical free rotation and lose orientational entropy, the loss of entropy is compensated and outweighed by the gain in excluded volume entropy. Thus an anisotropic colloid transits from isotropic to liquid crystalline state.

Recently, the liquid crystalline ordering of the different allotropes of the wonder element "Carbon" has been in limelight [41–51]. Among the carbon allotropes (Fig. 3.1), fullerenes, carbon nanotubes (CNTs) and graphene derivatives, all of these carbon nanoforms are found to exhibit liquid crystalline phases. The zero dimensional fullerenes have been so tailor-functionalized that they are able to exhibit both thermotropic and lyotropic liquid crystalline phases rendering it an amphotropic building block. One dimensional (1D) CNTs and two dimensional (2D) graphene derivatives upon appropriate functionalization are found to display LC ordering in suitable solvents. The present chapter deals with the liquid crystalline phases of the above mentioned 1D and 2D nanocarbon modifications. Their fabrication, stabilization, study and subsequent applications are discussed below.

3.2 Carbon Nanotube Based LCs

The discovery of CNTs by Iijima in 1991 has sparked tremendous activities in most areas of science and technology owing to their unprecedented physical and chemical properties [42, 43]. CNTs are 1D, well-ordered all-carbon hollow cylinders of graphite with high aspect ratio. Generally CNTs are classified into two categories based on their structure and dimensions: single walled carbon nanotubes (SWNTs), which consist of one layer of cylindrical graphene; and multi-walled carbon nanotubes (MWNTs), which contain several concentric graphene sheets. Recently, the synthesis of double-walled carbon nanotubes (DWNTs) with exactly two concentric carbon sheets has been introduced. SWNTs can be either metallic or semiconducting depending on the sheet direction about which the graphite sheet is conceptually rolled to form a nanotubes cylinder. CNTs can be prepared by several methods, such as chemical vapour deposition (CVD), carbon arc-discharge method, high pressure carbon monoxide method and laser ablation etc. These preparation methods yield random mixtures of tubes of various diameter, chirality, length distribution and different electronic properties. CNTs are chemically extremely inert and practically insoluble. Strong van der Walls interactions between adjacent nanotubes promote clustering into crystalline ropes which in turn aggregate into strongly entangled and unorganized networks or bundles. Synthetic strategies to chemically modify the side wall or tube end by molecular or macromolecular components have been reported to obtain purified and soluble CNTs. Other common techniques used to functionalize CNTs are noncovalent exohedral with polymers and surfactants and endohedral functionalization with fullerenes etc. The combination of superlative mechanical, thermal and electronic properties displayed by CNTs makes them ideal for a wide range of applications such as conductive and high strength composites, catalyst supports in heterogeneous catalysis, energy-storage and energy-conversion devices, field emitters, transistors, sensors, gas storage media, and molecular wires. Depending on their geometry, nanotubes can be metallic or semiconducting, with a strong anisotropy of conductivity being observed; they are primarily conductive along the tubes and hardly conductive across them. Their promising properties open a whole range of applications with devices envisioned in diverse areas including nanoscale electronics, field-emission sources, actuators, sensors for chemical and biological molecular detection, and even for localized drug delivery at the molecular level. Despite the extraordinary promise of CNTs, their realistic application as 1D conductors or semiconductors has been restricted because of difficulties in dispersing and aligning them in the desired direction. Well-aligned CNTs are highly desirable for the preparation of variety of nanodevices particularly where 1D charge migration is important. Recent studies have proved that the alignment of CNTs plays a critical role in the properties of nanotubes based materials [52, 53]. Various methods to align CNTs have been proposed by applying magnetic and electric fields, and mechanical techniques etc. However processing materials with well-controlled CNT alignment still remains a challenge. Two of the main problems faced in large-scale alignment of carbon nanotubes for device applications are their strong tendency towards clustering and aggregation of single tubes into networks and their general insolubility in most of the common organic solvents.

CNTs can be viewed as highly anisometric rigid rod-like particles. Like other anisotropic and 1D molecule, CNTs can form lyotropic liquid crystalline phase

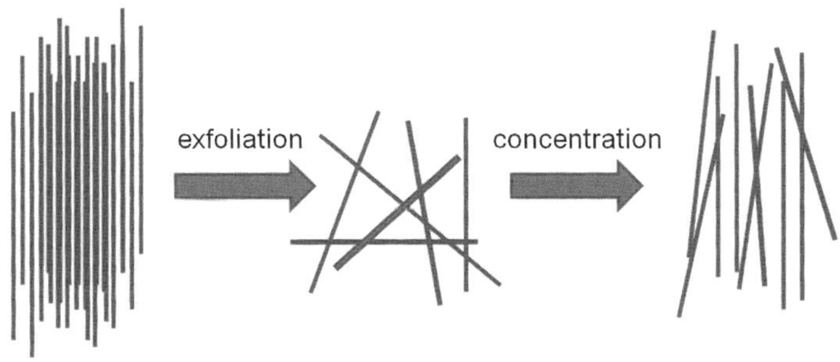

Fig. 3.2 Schematic representation of the fabrication of LC phase from carbon nanotubes

under suitable conditions. The basis for this phenomenon is the classical Onsager argument for liquid crystalline phase formation in concentrated solutions of rigid rods; if both the aspect ratio and concentration of the rods are high enough, the free energy of the system is reduced, resulting in a nematic phase [40]. With an increase of concentration, the system undergoes an isotropic-nematic phase transition via a biphasic region. The above theoretical calculations have shown that to obtain liquid crystalline phases at room temperature, strong van der Walls interactions between CNTs must be drastically reduced. Experimentally, several strategies have been successfully developed to screen out the van der Walls interactions and to increase the repulsive interaction between CNTs to realize liquid crystalline nematic phase [54–63]. This process involves the exfoliation of CNTs down to single tube level in a suitable solvent with the presence of a stabilizer and subsequent concentration of the dispersion (Fig. 3.2).

3.2.1 Acid Functionalized CNTs

Windle group first reported the liquid crystalline phase behavior of carboxylic acid functionalized water soluble MWNTs [44]. They found that when the concentration of the CNT aqueous dispersions reached 4.3 % by volume, there was the phase transition from the isotropic to nematic phase characterized by *Schlieren* texture typical of lyotropic nematic phase. The evaporation of the solvent from the dispersions gave solid samples with very similar microstructure to the original dispersion which has been characterized by scanning electron microscope at a resolution sufficient to view the organization of individual nanotubes. It should be noted that in addition to the Onsager theory, Somoza et al. have predicted the formation of liquid crystalline phases by CNTs as a function of nanotubes length and diameter with continuum-based density-functional theory [63]. Song et al. studied in detail the isotropic-nematic phase transition in dispersions of MWNTs [64–67]. They observed the coexistence of the two phases over a significantly wide range of concentration which was attributed to the polydispersity of nanotubes' dimensions and their straightness. It was found that longer and thicker nanotubes preferentially separate into the anisotropic nematic phase. This observation enabled them to successfully carry out mesogenicity driven fractionation of nanotubes where the long and straight nanotubes segregate to the liquid crystalline phase while shorter nanotubes and impurities with lower mesogenicity preferentially segregate to the isotropic phase. This simple method is very useful for the analysis and purification of nanotubes in aqueous medium. Subsequently they have investigated size-dependence and elasticity of liquid crystalline CNTs. Recently they reported the fabrication of macroscopic fibers of well-aligned MWNTs by wet spinning from the liquid crystalline phase using ethylene glycol as the solvent (Fig. 3.3). The electrical conductivity of these CNT fibers has been found to be highly anisotropic. Nanotubes were found to be highly aligned within these fibers due to the combination of shear forces and the liquid crystalline phase.

3.2.2 Protonated CNTs

Smalley et al. reported the exfoliation of CNTs' bundles with the help of a superacid [68–73]. SWNTs can be dispersed at high concentration in superacid, and the protonation of CNTs' sidewalls eliminates wall–wall van der Waals interactions and promotes the dispersion process. The tubes are stabilized against aggregation by the formation of an electrostatic double layer of protons and negative counter-ions. At very low concentration, CNTs in superacids dissolve as individual tubes. At higher concentration, solvated CNTs form nematic phase with polydomains (Fig. 3.4). Upon introduction of small amounts of water, the liquid crystalline phase separates into needle-shaped strands of highly aligned CNTs. Under anhydrous condition, the liquid crystalline phase can be processed into highly aligned fibers with controlled morphology of pure CNTs without the aid of any surfactants or polymers.

3.2.3 Biopolymer Functionalized CNTs

Badaire et al. reported the formation of liquid crystalline phase of CNTs by noncovalent functionalization with single strand DNA in water [74]. The noncovalent functionalization in water is simpler and does not affect the intrinsic properties of the CNTs. They observed that a nematic-isotropic coexistence is observed for nanotubes concentrations between 2 and 4 wt%. Above 4 wt%, the system forms a single nematic phase of unmodified and freely dispersed nanotubes. They investigated the boundaries of the phase diagram with respect to the aspect ratios of the nanotubes determined by dynamic depolarized light scattering. Poulin et al. uniformly aligned the nematic aqueous suspensions of nanotubes in

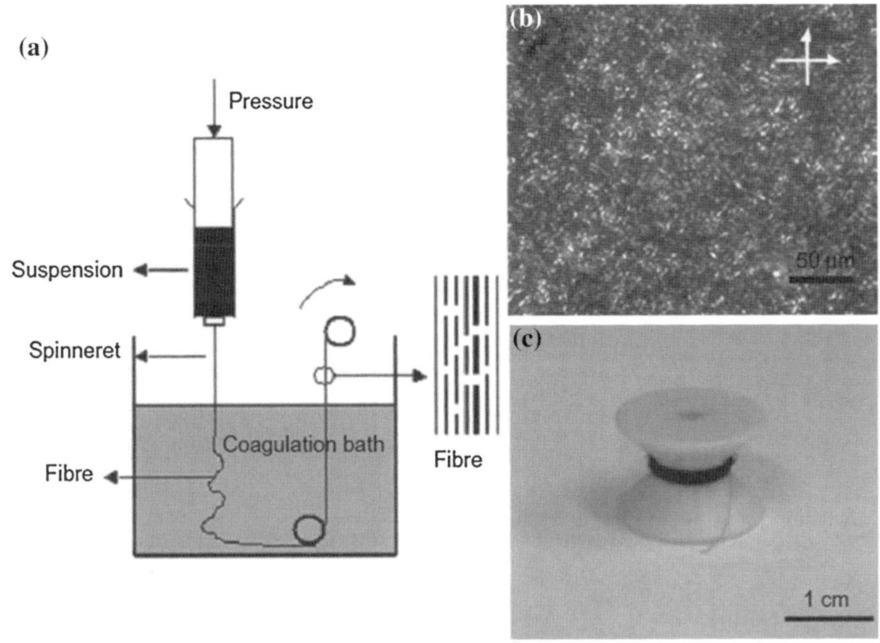

Fig. 3.3 The liquid crystalline spinning process of well-aligned nanotubes fibers in glycol. **a** The schematic of the experimental setup used to make nanotubes fibers. **b** Optical image of 3 wt% MWNT suspension before extruding under crossed polarizers. **c** A 3-m-long MWNT fiber collected on the small winder. Reproduced with permission from [65]. Copy right 2008 Wiley-VCH

thin cells by mechanical shearing [75]. Homogeneous anisotropic thin films were obtained by drying the nematic. The order parameter was measured by polarized Raman spectroscopy and found to be quite weak due to entanglement of the nanotubes. Recently, Puech et al. showed that bile-salt stabilized SWNTs exhibit a first order transition to a nematic liquid crystalline phase [76, 77]. The nematic phase expresses itself in the form of micron-sized spindle shaped nematic droplet also known as tactoids, freely floating in the isotropic host dispersion confirmed by polarization microscopy. Furthermore, the director field was found to be uniform in these nematic droplets. They later reported that shortening the bile salt stabilized nanotubes by high power ultrasonication and centrifugation allows the formation of LCs that can easily be oriented under the form of large macroscopic monodomains. The orientational order parameter of the LC was measured by polarized Raman spectroscopy (Fig. 3.5) which was found to approach the value theoretically expected for LCs made of rigid rods in solution. This study demonstrated that the LC properties of CNTs are extremely sensitive to their morphology and dispersion processing. Liquid crystalline behavior of CNTs dispersed in biological hyaluronic acid solutions has been stabilized and studied [78]. Another biopolymer dispersant gellan gum is used to achieve aqueous dispersions

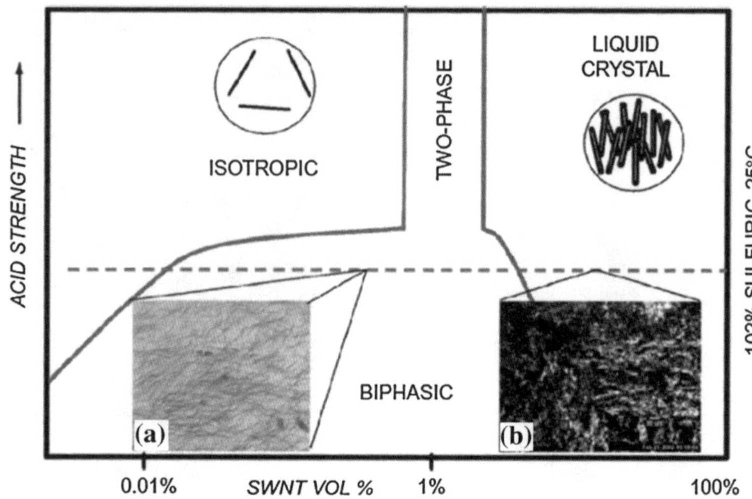

Fig. 3.4 Postulated phase diagram of SWNTs in acids. In 102 % sulfuric acid, the biphasic region extends from a concentration of ~100 ppm to ~8 %. These critical concentrations are expected to vary with the strength of the acid used. Reproduced with permission from [69]. Copy right 2006 American Chemical Society

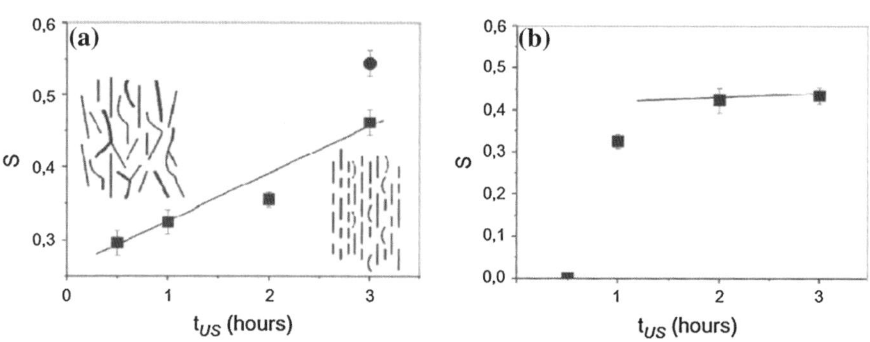

Fig. 3.5 **a** Order parameter of nematic dried films made with CNT dispersions that have been sonicated during different times t_{US}. All of the used dispersions have undergone a single step of 1 h. **b** Order parameter of nematic dried films made with CNT dispersions that has been ultracentrifuged during different times t_{UC}. All of the used dispersions were sonicated during 2 h. Reproduced with permission from [76]. Copy right 2011 American Chemical Society

of highly concentrated SWNTs which exhibit liquid crystalline phase [79]. Large scale aligned nanotube composites membranes have been successfully prepared from such highly concentrated CNT suspensions. Such membranes with improved alignment of CNTs may find potential application in high-performance biocomposite functional materials such as sensors and actuators. Recently, the first lyotropic cholesteric liquid crystalline phase has been realized by dispersing SWNTs

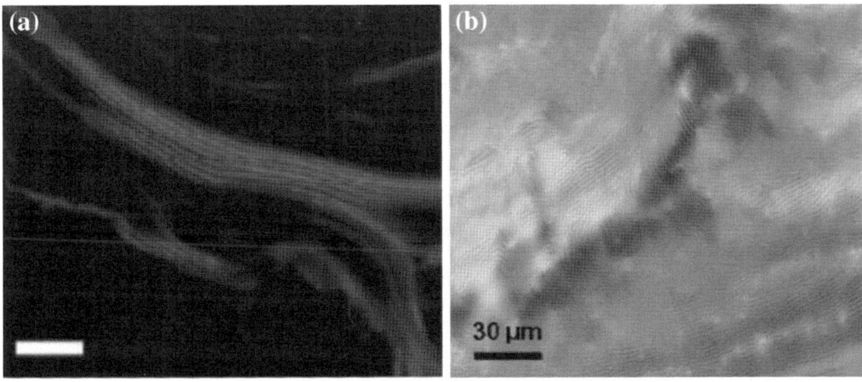

Fig. 3.6 Polarizing optical microscopy images of the chiral nematic phase of carbon nanotubes functionalized with double stranded DNA at low (**a**) and high (**b**) concentrations, respectively. Reproduced with permission from [80]. Copy right 2011 American Chemical Society

in aqueous solutions of double-stranded DNA (dsDNA) [80]. Interestingly, depending on the dispersion methodology, the polydomain nematic phase could also be obtained in these systems. The typical finger print texture of CNT cholesteric LC phase is shown in Fig. 3.6. This controlled phase behavior could open new routes for producing SWNT films with controlled morphology and optical properties of cholesteric films.

3.2.4 Polymer Functionalized CNTs

Noncovalent functionalization of SWNTs is particularly attractive avenue for dispersion because it enables modification of material properties without altering the chemical structure of the nanotubes. Most high concentration dispersions have been obtained in aqueous solutions by mixing SWNTs with surfactants, DNA, or water soluble polymers. Only a few polymers and block copolymers have been reported to disperse SWNTs in organic solvents, usually at low concentrations (0.1 mg/mL). Dispersion of SWNTs at high concentrations in organic solvents, as alternative to water, provides increased flexibility for incorporating various chemical functionalities, tuning morphologies, and as a result controlling the properties of the resulting composite material. Among suitable dispersants, conjugated polymers are promising candidates owing to their ability to undergo π-π stacking with SWNTs. Conjugated polymers are also interesting materials for electronic and photonic applications such as electrodes, light-emitting diodes, thin film transistors, photovoltaic cells and sensors. The judicious addition of small fractions of SWNTs to conjugated polymer active layer has been shown to improve both the efficiency of photovoltaic cells and the mobility of field effect transistors. Mechanical shearing from a nematic lyotropic liquid crystalline dispersion is a promising method to align

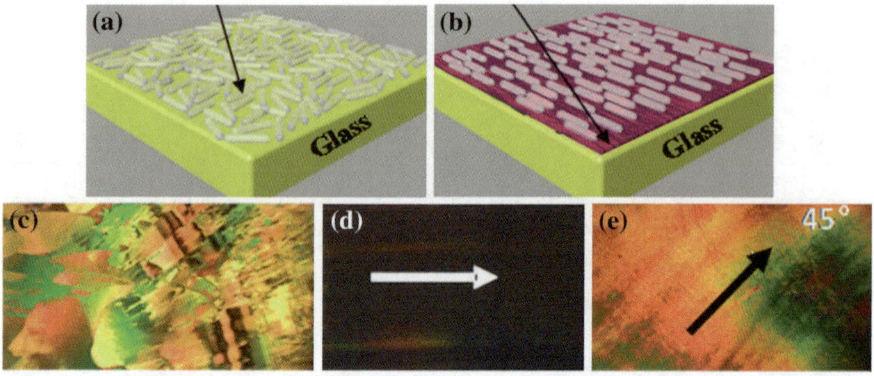

Fig. 3.7 Schematic diagrams for arrays of LC (**a**) on a glass substrate showing random directional local phase, **b** On a shear-aligned composite thin film showing uniformly aligned LC domains. **c** Transmission cross-polarized microscope image of a drop of nematic LC on a bare glass substrate showing many small randomly oriented domains. **d** A drop of LC on a glass substrate with sheared composite thin film with shearing direction parallel to one of the polarizers. **e** Shearing direction 45° to one of the polarizers. Arrows indicate the shearing direction. Reproduced with permission from [81]. Copy right 2009 Wiley-VCH

materials cast from dispersion and usually results in a more enhanced alignment than from non-nematic lyotropic liquid crystalline dispersions. Recently, organic solvent based nematic lyotropic liquid crystalline phase of noncovalently functionalized CNTs has been reported by Bao et al. [81]. They reported a high concentration dispersion of SWNTs in organic solvents by conjugated polymers and for the first time its transition into the nematic lyotropic liquid crystalline phase. The SWNT nematic lyotropic liquid crystalline phase provides a convenient way to align SWNTs over a large area. The polymer-SWNT liquid crystalline composites can be easily aligned through simple shearing by pushing a drop of the polymer/SWNT dispersion with the edge of a glass pipette tip on a glass substrate, resulting in large monodomains of uniform brightness. Anisotropy in electrical conductance was observed in those shear-aligned SWNT films. The conductance of a given film along the shearing direction is always at least twice its conductance perpendicular to the shearing direction. The aligned SWNTs films fabricated from the nematic phase are potentially useful as electrodes and are capable of alignment of liquid crystalline materials and other molecular materials deposited on them. Uniform alignment of the LC was observed as shown in Fig. 3.7. Such alignment could only be achieved by depositing an insulating alignment layer like rubbed polyimide on a conducting electrode. This study shows that CNTs aligned films can act as conducting alignment layer. It may find use in LC displays.

It has been shown that CNTs seed the formation of oriented domains in a liquid crystalline polymer [82]. Using polarized light microscopy it was observed that the molecular alignment in large domains was homogeneous and controlled by the direction of the nanotubes nucleus. CNT films have been generated by deposition onto glass substrates with controlled nanotubes orientation by self-assembly and dip-coating [83]. The obtained nanotube films were used to fabricate optical cells in which planar alignment of nematic LCs was achieved. The LC alignment was amenable to electric field induced switching as observed by polarized light microscopy. Application of an electric field perpendicular to the substrate resulted in homeotropic alignment with director perpendicular to the substrate. Turning off the electric field caused the restoration of macroscopic uniform planar alignment.

Pyrene functionalized poly(methyl methacrylate) (PMMA) were found to be highly efficient to solubilize and disentangle MWNTs in solutions that self-organize as liquid crystalline phases in PMMA and polyethyleneglycol (PEG) 400 matrices (these oligomeric matrices that are still mobile but do not evaporate like a solvent) [84]. High concentration of oriented domains offers the potential to align CNTs at high volume fractions.

3.2.5 Other Methods of Fabricating Liquid Crystalline Phase of CNTs

Stable dispersions of CNTs have been achieved by dissolving purified CNTs in aqueous sodium dodecyl sulphate solution by sonication and centrifugation [85]. Liquid crystalline processing of highly oriented CNTs' arrays for thin-film transistors from the above dispersions has been demonstrated by Tsukruk et al. The liquid crystalline behavior of a CNT solution near the receding contact line during tilted-drop casting was exploited which produced long range nematic like ordering of CNT stripes by confined micropatterned geometry. As the solvent evaporates during titled-drop casting, the CNTs diffuse from the bulk solution to the liquid–solid-air receding contact line by convective flow, causing the concentrated solution to form a nematic liquid crystalline phase. The confined geometry of the micropatterned surface induces a uniform long range orientation of dense CNT films during the surface deposition. It is further demonstrated that the performance of thin-film transistors based on these densely packed and uniformly oriented CNT arrays is largely improved compared to random CNTs as shown in Fig. 3.8. This approach has great potential in processing of high-performance electronic devices based on high density oriented CNT films with good device characteristics.

The noncovalent wrapping of SWNTs with oligo(phenylenevinylene)s with cyanobiphenyl terminates enhanced the dispersibility in chloroform by disentanglement of CNT bundle structure. It was observed that further addition of oligomers caused precipitation from a homogeneous dispersed solution of CNTs [86]. The self-organized precipitates display thermotropic LC properties as revealed by calorimetric, optical and X-ray studies. Figure 3.9 shows the self-organization process of supramolecular oligomer and CNTs complexes.

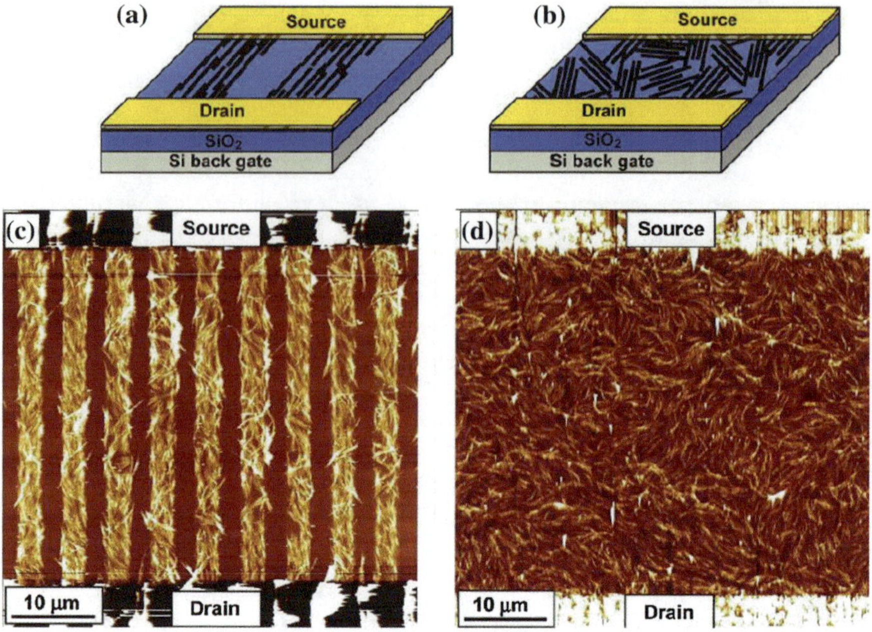

Fig. 3.8 Device configurations of back-gated oriented (**a**) and random CNT-TFTs (**b**) with Au source/drain electrodes. **c** AFM topographical image of a CNT-TFT with uniaxially ordered micropatterned CNT array. **d** AFM topographical image of a CNT-TFT with randomly oriented CNT surface film. Reproduced with permission from [85]. Copy right 2006 American Chemical Society

Recently Jiang et al. reported the formation of liquid crystalline phase by SWNT polyelectrolytes in dimethyl sulphoxide (DMSO) [87]. It was found that small-diameter nanotubes were preferentially functionalized when the polyelectrolytes were dispersed in DMSO.

While both nematic and chiral nematic phases of pure and functionalized CNTs have been observed both in aqueous and organic mediums, however, smectic mesophase has not been observed yet. Since CNTs are actually polydisperse samples, this polydispersity seems to be the preventing factor in smectic phase formation. Moreover, unlike the thermotropic phase behavior of functionalized fullerenes, thermotropic phase behavior of CNTs remains elusive. CNTs have been covalently functionalized with mesogens, however no thermotropic mesomorphism was observed in the pure samples, though interesting results were obtained by blending them with LC hosts [88, 89]. Therefore this challenging area remains to be further explored. Nevertheless, the observation of lyotropic liquid crystallinity of CNTs in both aqueous and organic solvents undoubtedly validates and further strengths our understanding of liquid crystalline phase formation in 1D rod-like objects.

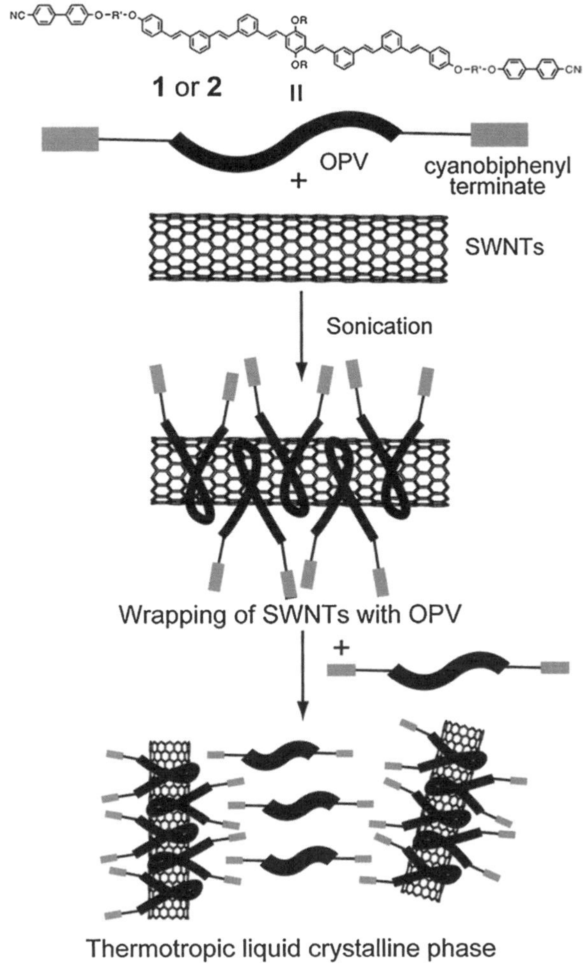

Fig. 3.9 Schematic illustration of self-organizing process for the composite of liquid crystalline oligomers and SWNTs. Reproduced with permission from [86]. Copy right The Royal Society of Chemistry 2009

3.3 Graphene Based LCs

Graphene, the 2D nanoform of carbon, is a single layer of carbon atoms arranged in a honeycomb lattice (Fig. 3.10). Graphite, made up of stacked graphene layers, exists in nature however graphene is not known to exist naturally. This all-atoms-on-the-surface allotrope of carbon is recently found to display outstanding optical, electrical, thermal and mechanical properties. Consequently it has been demonstrated in various materials and device applications to modify physico-chemical properties including gas and energy storage, optoelectronics, energy conversion, catalysis, and biological labeling etc. [90–114]. Two distinct strategies have been undertaken for graphene production, i.e. the bottom–up and the top–down. The bottom-up approach relies on the production of graphene from molecular building

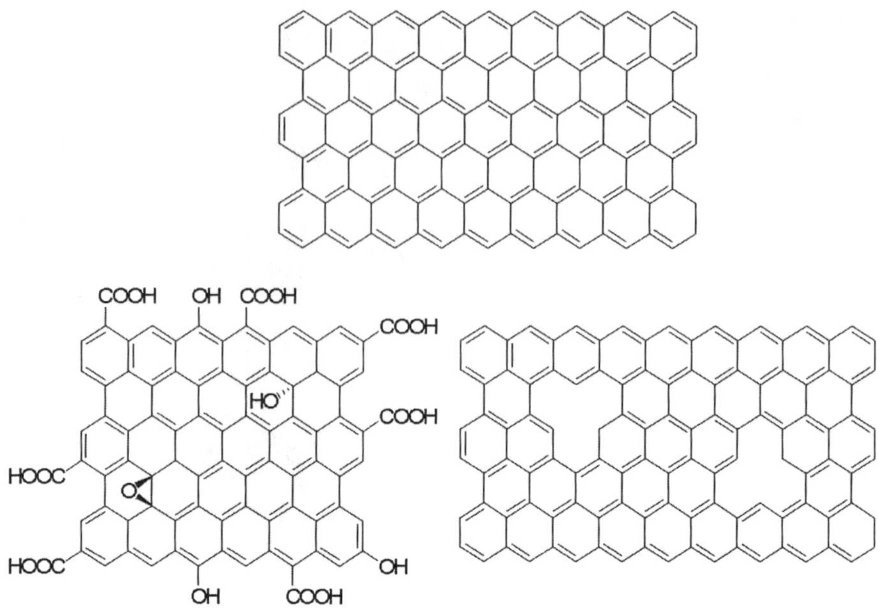

Fig. 3.10 Schematic representation of graphene, graphene oxide and reduced graphene oxide

blocks undergoing chemical reaction to form covalently linked 2D framework. The latter approach occurs via exfoliation of graphite into graphene layers. Top-down liquid phase exfoliation methods are more promising because they are extremely versatile, potentially up-scalable and can be used to deposit graphene in a variety of environments and on different substrates not available using mechanical exfoliation or growth-on-substrate methods. Liquid phase exfoliation can be employed to produce graphene based composites or films, which are key components for many applications such as thin-film transistors, conductive transparent electrodes for indium tin oxide replacement in light-emitting diodes or photovoltaics. Surfactant-assisted and surfactant-free liquid phase exfoliation of graphene in aqueous and organic solutions has been developed.

Graphene oxide (GO) is synthesized by various methods involving oxidation of graphite to different levels. GO is easily processable in liquid media such as water. All the different functional groups including hydroxyl, epoxy, carbonyl and carboxyl groups can be used to covalently and/or noncovalently attach functional units to this scaffold which can tune its chemical and physical properties. While the oxygen containing groups can be removed by reduction, resulting in reduced GO, the severe problem is that reduction cannot remove all structural defects introduced by the process of oxidation. These defects disrupt the band structure and downgrade the electronic properties that make graphene unique. Significantly, most of the knowledge generated on graphene can be extended to other layered crystals such as BN, MoS_2 etc., thereby, opening potential avenues towards 2D materials exhibiting a set of marvelous properties.

Like the liquid crystalline ordering of collection of rod-shaped particles in concentrated dispersions, collections of disc-shaped particles should also exhibit orientational order with sufficient concentration and narrow size and shape distributions [113, 114]. Moreover, disc-like objects are experimentally observed to exhibit nematic, smectic and columnar type mesophases in consistent with theoretical predictions. Since graphene can be considered as disc-like nano-object with very high aspect ratio, it has been shown that upon dispersion in sufficiently high concentrations it exhibits liquid crystalline phase behavior. The following are the different methods to fabricate and stabilize graphene based liquid crystalline phases.

3.3.1 Protonated Graphenes

Pasquali et al. first reported the formation of liquid crystalline phase of graphene in chlorosulphonic acid [45]. It was observed that graphite spontaneously exfoliated into single-layer graphene in chlorosulphonic acid and dissolved at isotropic concentrations as high as 2 mg/mL. Under such conditions graphite exfoliate into the solution owing to the protonation of their surface planes and remains dispersed because of the repulsive interaction among the graphene sheets. However, when the precipitated soluble graphene powder was redispersed in about 20 mg/ml concentration and centrifuged, spontaneous formation of liquid crystalline phase is observed. Moreover, these soluble graphenes yielded isotropic to liquid crystalline phase separation as expected in a solution of polydisperse rigid platelets. The observed liquid crystalline *Schlieren* texture is very similar to typical discotic nematic samples. Many potential applications of graphene including electronics and nanocomposites require that graphene-based materials must be dispersed and processed in a fluid phase. In this context, such high concentration LC solutions could be particularly useful for making flexible electronics and multifunctional fibers since this route does not compromise with the intrinsic properties of graphene. Moreover, it is envisaged that graphene LCs can potentially act as a replacement for traditional carbonaceous mesophase used as precursors for neat carbon fibers and films which could drive down the cost of the ultrastrong carbon composites used in the aerospace, automotive and construction industries etc. However the main drawback of this method is that upon addition of water the graphene layers coagulate and phase separate. Moreover, the processing of such a liquid crystalline phase in the superacid needs very harsh conditions.

3.3.2 Graphene Oxide Based LCs

GO is the oxygenated form of a graphene monolayer platelet with chemical functionalization capability, and extremely large surface area. GO is mass-producible

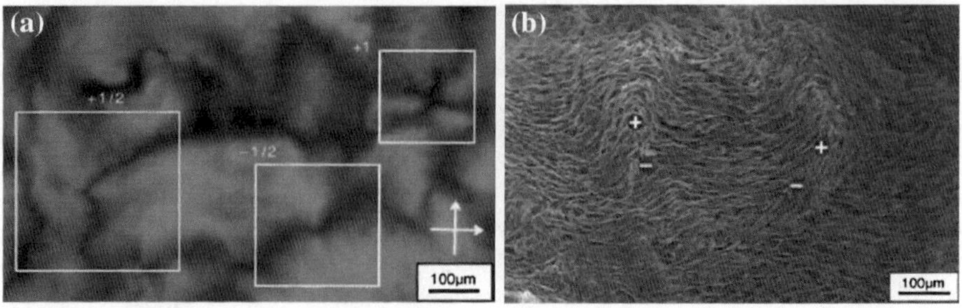

Fig. 3.11 Polarized light Microscopy images of graphene oxide liquid crystals (**a**) and freeze fracture SEM image of nematic phase (**b**). Reproduced with permission from [115]. Copy right 2011 Wiley-VCH

from natural graphite by chemical oxidation and subsequent exfoliation. The hydrophilic surface functional groups, such as epoxide, hydroxyl and carboxyl groups that decorate the basal plane and the edge of graphene oxide enable monolayer exfoliation in common polar solvents including water. Similar to graphene layers, GO layers also exhibit nematic liquid crystalline phase behavior upon dispersion in high concentrations in suitable solvent medium. It has been observed that the electrostatic repulsion from the dissociated surface functional groups such as carboxylate plays a crucial role in the stability of GO LCs since the reduction of repulsive interaction increases the coagulation of graphene oxides.

Kim et al. first introduced GO LCs as a versatile new class of carbon-based LCs [115]. They characterized the nematic phase of aqueous GOs by polarized light microscopy and freeze-fracture scanning electron microscopy (SEM) (Fig. 3.11). Interestingly, the liquid crystallinity could be maintained upon the decoration of the GO platelets with nanoparticles or by including an additional polymer component in the solvent medium. This may significantly broaden the functionality of GO LCs. Moreover, the orientation of GO LCs could be manipulated by a magnetic field or mechanical deformation. The liquid crystallinity of GO offers a versatile route to control the molecular organization and the corresponding properties of the carbon-based materials.

Xu et al. reported the formation and salt-dependent phase behavior of GO LCs [116]. Rheological measurements showed that GO aqueous dispersions displayed typical shear flow and confirm the isotropic-nematic transition. The ordering of GO sheets in aqueous dispersions and in the solid state has been demonstrated by the characterizations of polarized optical microscopy, small angle X-ray scattering, scanning electron microscopy and transmission electron microscopy. The direct, real-time fluorescent inspections by confocal laser microscopy further reveal that the individually dispersed fluorescent GO sheets align with orientational directions along their long axes (Fig. 3.12).

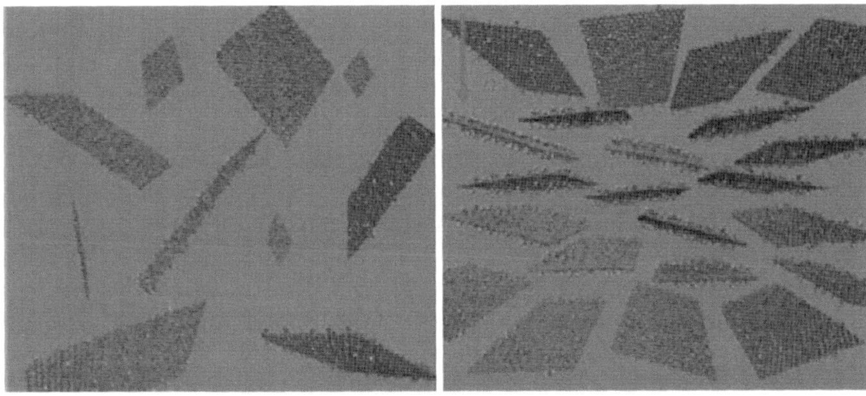

Fig. 3.12 Schematic models for isotropic (*left*) and nematic (*right*) phases of GO aqueous dispersions. GO sheets randomly distribute in isotropic phase and align with an orientational vector (*n*) perpendicular to the sheet planes in the nematic phase. Reproduced with permission from [116]. Copyright 2011 American Chemical Society

Subsequently Gao et al. disclosed that GO sheets at very high concentrations above the nematic phase can exhibit chiral LCs in a twist-grain-boundary phase-like model with simultaneous lamellar ordering and long-range helical frustrations [117, 118]. Such GO LCs were continuously spun into metres of macroscopic GO fibers which upon subsequent chemical reduction furnished neat graphene fibers with high conductivity and good mechanical performance. The flexible, strong graphene fibers were knitted into designed patterns and into directionally conductive textiles (Fig. 3.13). Here the large size of the sheets and their good alignment resulted in better performance of the graphene fibers. The graphene fibers cross-linked by divalent ions displayed record tensile strength with excellent electrical conductivity. Such multifunctional graphene fibers have promise in versatile applications such as next-generation functional textiles, flexible and wearable sensors and supercapacitor devices.

Very recently, the same group has reported lyotropic LCs of polyacrylonitrile-grafted GOs [119]. Polyacrylonitrile chains were covalently and uniformly grafted onto GO surfaces via a simple free radical polymerization process. These functionalized sheets were well-dispersed in polar organic solvents such as dimethylformamide and dimethyl sulphoxide forming nematic and lamellar LCs upon increasing concentrations. Continuous nacre-mimetic fibers have been assembled from these liquid crystalline phases.

Dan et al. reported the fabrication of discotic nematic liquid crystalline phase by giant GO flakes with very high aspect ratio (10^4) in water [120]. It was observed that higher aspect ratio led to a dramatic change in mesomorphic behavior and the phase transition took place at concentrations with one order of magnitude lower than usually observed. The GO flakes were prepared by an improved synthetic method which furnished more regular structures and better intact basal plane

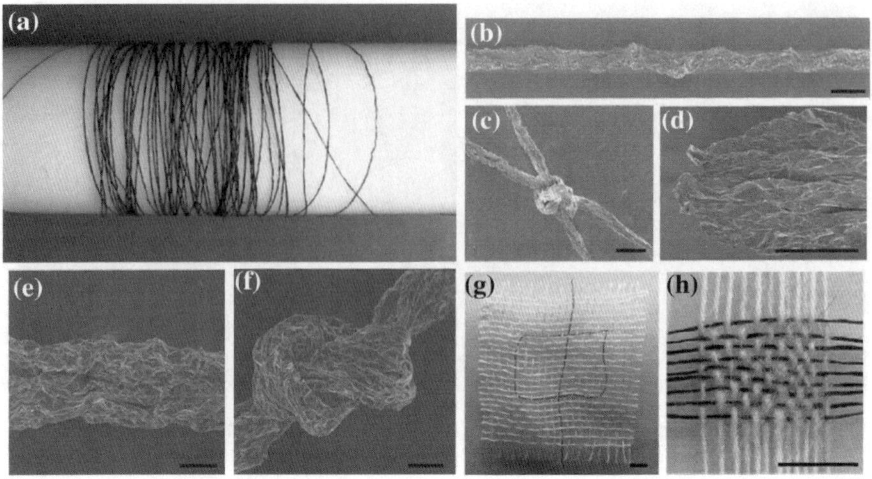

Fig. 3.13 Macroscopic neat GO fibers and chemically reduced graphene fibers. **a** four-metre-long GO fiber wound on a Teflon drum (diameter, 2 cm). SEM image of the fiber (**b**), and its typical knots (**c**). **d** The fracture morphology of GO fiber after tensile tests. The surface winkled morphology (**e**) and the tighten knot (**f**) of graphene fiber. **g** A Chinese character pattern knitted in the cotton network using two graphene fibers. **h** A mat of graphene fibers woven together with cotton threads. Reproduced with permission from [117]. Copyright 2011 Macmillan Publishers Limited

compared to GO produced using other conventional oxidation techniques. The gel-like LCs exhibited an unusual defect-free uniform director alignment over hundreds of nanometers which enabled them to characterize the nematic order parameter, optical birefringence and elastic properties of the mesophase. In order to directly visualize the orientation of GO flakes in the nematic phase, they used two different types of anisotropic colloidal inclusions such as gold triangles and glass microrods. Gold triangular platelets spontaneously oriented with their large-area faces parallel to GO flakes whereas glass microrods spontaneously aligned with their long axes parallel to GO flakes as observed by polarized optical microscopy (Fig. 3.14). The ability of GO LC to disperse and align anisotropic colloidal particles is important from their applications point of view. They have further studied the dipolar and quadrupolar distortions of the director field by suspending colloidal microspheres in the nematic phase. Recently, Senyuk et al. have used nonlinear photoluminescence for fully three dimensional (3D) label-free imaging of structures of isotropic and orientational order of liquid crystalline phases formed by aqueous GO flakes [121]. The huge lateral size, i.e. anisotropy (lateral to thickness) of the mesogenic unit, is one of the most important features. Because of this large anisotropy, nanosheet colloids transit to liquid crystalline phases even at very low concentrations and therefore they are very fluid, allowing macroscopic alignment by weak external forces.

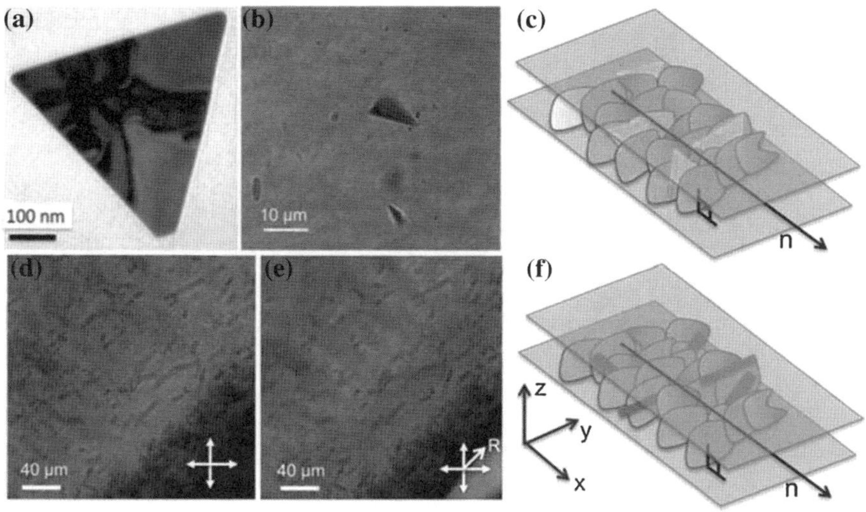

Fig. 3.14 Anisotropic colloidal particles and their spontaneous alignment in GO nematic LC. **a** TEM image of an anisotropic gold triangular platelet. **b** Bright-field transmission-mode images showing triangular platelets with up to 10 μm side length and 5 nm thickness as they spontaneously align parallel to GO flakes of the LC. **c** Schematic diagram of the GO flake and gold platelet alignment. **d–e** Polarizing microscopy images showing spontaneous alignment of glass micro-rods in the GO LC observed under crossed polarizers with (**d**) and without (**e**) an additional retardation plate. **f** Schematic showing the alignment of micro-rods in the LC. Reproduced with permission from [120]. Copyright The Royal Society of Chemistry 2011

Hydration-responsive folding and unfolding in GO liquid crystalline phases has been observed by Guo et al. [122]. They investigated surface anchoring, complex fluid flow and microconfinement of GO LCs. It was observed that ordered GO phases underwent drying and rehydration cycles in which the GO layers folded and unfolded to give unique anisotropic swelling behavior (Fig. 3.15). The detailed investigations of hydration-responsive folding and unfolding opens the potential for LC derived GO phases as smart stimuli-responsive materials and may find applications in dynamic space filling or sealing methods. Moreover, liquid crystalline phases offer a route to long-range order and can be manipulated through flows, fields, surfaces, and microconfinement to produce well-defined equilibrium assemblies that can be thermally converted into carbon materials with precise control of graphene layer structure.

Organic solvent-based GO LCs have been reported by Jalili et al. which overcomes the practical limitations imposed on processing in water [123]. A wide range of common organic solvents including ethanol, acetone, tetrahydrofuran, N, N-dimethylformamide etc. have been found to enable the fabrication of GO liquid crystalline phase. Moreover, the organic solvents aid to disperse and organize large amount of aggregate-free SWNTs, up to 10 wt % without sacrificing the liquid crystalline properties. Interestingly as-prepared liquid crystalline GO-CNT dispersions have been used to achieve self-assembled layer-by-layer multifunctional

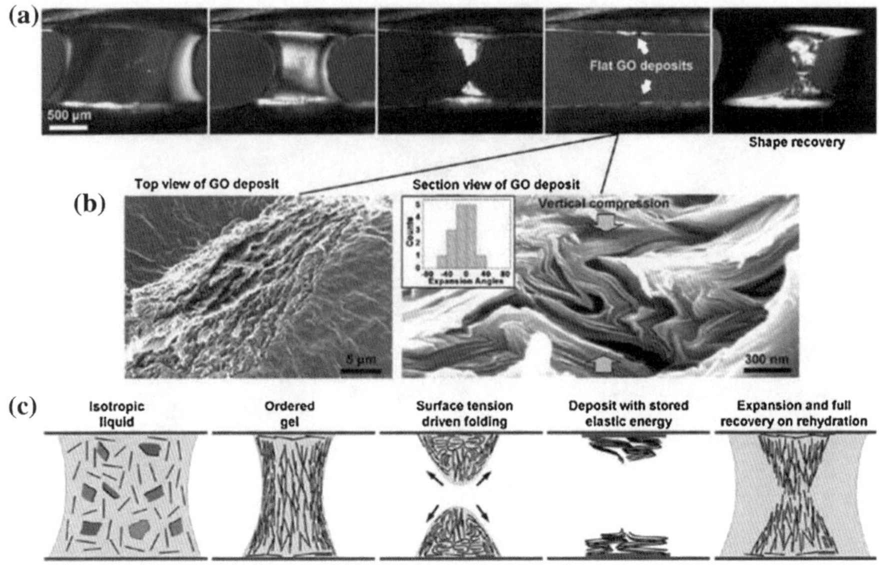

Fig. 3.15 Hydration-responsive folding and unfolding transitions. **a** Optical micrographs of hourglass-shaped droplet of GO suspension positioned in the millimeter gap between two metal needles. *Left* to *right* drying causes surface recession, concentration of the GO layers, and eventually physical gelation due to interlayer interaction. Late stages of drying causes rupture of the hourglass droplet. Rehydration triggers anisotropic growth and recovery of original shape. **b** SEM images of dried deposit. **c** Proposed molecular assembly of GO plates during drying and rehydration. Reproduced with permission from [122]. Copyright 2011 American Chemical Society

3D hybrid architectures (Fig. 3.16) with superior mechanical properties. This study contributes to the fundamental understanding of the solvophobic effect and the parameters affecting the self-assembly process and could provide practical solutions to the processibility of wide range materials that require organic solvents because of solubility issues and water sensitivity etc. Recently, the above research group has successfully demonstrated wet-spinning of graphene fibers and yarns from liquid crystalline dispersions of aqueous GO [124]. Based on rheological and polarized optical microscopy investigations, they have been able to propose the relation between sheet size and polydispersity, concentration, liquid crystallinity and spinnability. The thermal conductivity of continuous spun, fully oriented and robust graphene fibers is found to be much higher than polycrystalline graphite. The simplicity and scalability of the continuous spinning process is expected to enable safe and cost-effective large-scale production of GO fibers for fiber industry.

Kim et al. have developed a solution-phase synthesis method of ultralarge GO sheets involving pre-exfoliation of graphite flakes. Such large GO sheets spontaneously form lyotropic nematic phase at a very low concentration in water [125]. They produced self-assembled brick-like GO nanostructures by simple casting of GO dispersions and drying in ambient conditions. These free-standing GO papers

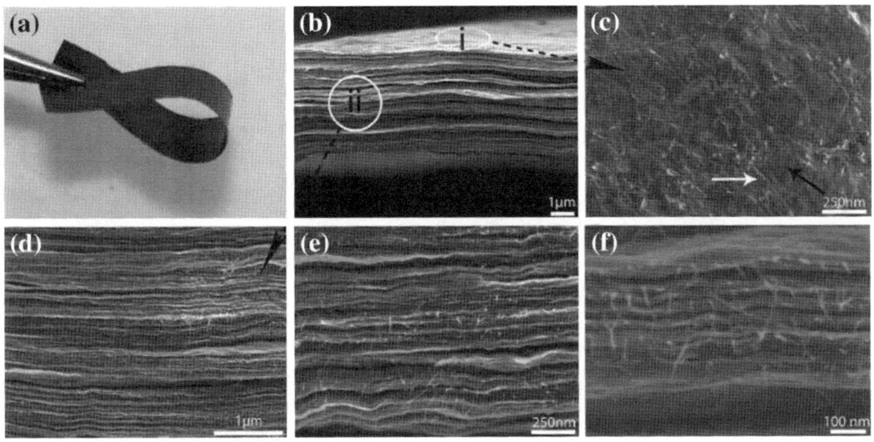

Fig. 3.16 **a** Representative photograph of a flexible free-standing paper of LC GO-SWNT made by cast drying method. **b** SEM image of the cross section of as-cast dried paper. **c** SEM image of the surface of the layer-by-layer composite. Some of the SWNTs are laid on the surface of the paper, while others are placed between layers of GO sheets. **d–f** Cross section of composite paper at different magnifications confirmed the self-oriented nature of the composite as well as maintaining SWNTs debundled after the fabrication of composite. Reproduced with permission from [123]. Copyright 2013 American chemical Society

can be potentially applied in many areas such as electrodes in energy-storage devices and nanocomposites with high degree of orientation.

Recently, Tour group fabricated liquid crystalline phases of GO nanoribbons and chemically reduced graphene nanoribbons in chlorosulphonic acid [126, 127]. The starting material GO nanoribbons were obtained from the oxidative unzipping of MWNTs. The liquid crystalline solutions were spun directly into hundreds of meters of continuous macroscopic fibers with high performances. The better performance of the fibers was attributed to air gap spinning and annealing that produced higher molecular alignment within the fibers as probed by X-ray diffraction and scanning electron microscopy.

3.3.3 Reduced Graphene Oxide Based LCs

Oxidation of graphene is known to downgrade its mechanical and electrical properties severely. Reduced GO (rGO) can be obtained either by thermal or chemical reduction of GO. Chemical reduction of GO to produce rGO allows partial restoration of the electronic conjugation and graphene properties but this recovery is also associated with a gain in hydrophobicity with a concomitant loss of water solubility, which hinders the formation of liquid crystalline phase. Due to the loss of the oxygen containing functional groups from the GO surface, rGO becomes hydrophobic and tends to aggregate in water. Reduction of GO flakes after

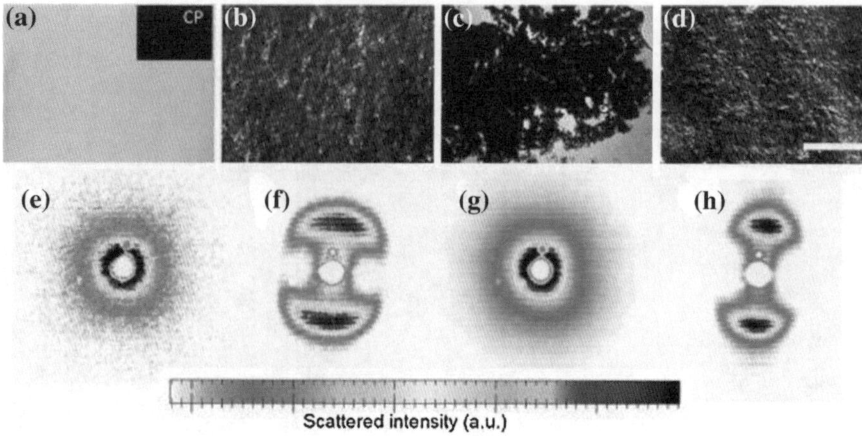

Fig. 3.17 **a** Optical micrographs of an isotropic suspension of diluted GO. **b**, **d** Optical micrographs between crossed polarizers for concentrated GO (**b**) and reduced GO-BS (**d**) suspensions after a single ultracentrifugation. **c** Reduced GO in the absence of BS surfactant: the particles are unstable and form aggregates. **e–h** Corresponding X-ray 2D patterns for all graphene aqueous suspensions shown in the first row. The materials in panels f and h exhibit 'butterflies' patterns arising from nematic ordering. This observation reflects a macroscopic alignment of the director with a small tilt with respect to the capillary tube. Reproduced with permission from [128]. Copyright 2012 American Chemical Society

processing in water cannot always be easily achieved. In particular it can be difficult to perform chemical modifications if fragile electronic devices have been built or if the particles are stacked and not easily accessible for chemical or electrochemical treatment. The direct processing of rGO would in most cases facilitate the use of this material in future applications. Poulin et al. reported an approach that allows the formation of aqueous lyotropic LCs using rGO despite their lack of water solubility [128]. The objective was to combine the advantages of rGO materials and the easy processing of GO flakes. In this approach, the rGO particles were stabilized by bile salts (BSs), which acted as surfactants. Surfactant-stabilized rGO flakes, in contrast with neat rGO, were shown to form LCs in water. From a thermodynamic point of view, the behavior of surfactant-stabilized rGO has been found to be similar to the behavior of surfactant-stabilized colloids. The liquid crystalline phases are observed up to large dilution levels of about 1.0–1.5 wt% of rGO. This diluted boundary of the isotropic to LC transition is due to the large aspect ratio of the rGO flakes. The dilution behavior also shows that van der Waals attractive interactions are efficiently counter-balanced by electrostatic repulsions provided by the surfactants and remaining functional groups at the surface of the rGO flakes. The liquid crystalline structures are characterized by small-angle X-ray scattering experiments that confirm the exfoliation of single-layer graphene flakes (Fig. 3.17). LCs formed with rGO offer therefore opportunities to develop new materials that combine the advantages of liquid crystallinity for processing and ordering with the richer transport properties of rGO compared with GO.

3.3.4 Thermotropic LCs of Synthetic Nanographenes

Müllen and coworkers have synthesized unprecedented large polycyclic aromatic cores which can be considered as nanographenes [129–131]. It is interesting to note that such molecules are defect free, monodisperse structures, possess very high charge carrier mobility and are well soluble in common organic solvents. Upon substitution with suitable number of aliphatic chains with appropriate length such nanographene molecules exhibit thermotropic discotic columnar LC phases. Because of their unique self-organization and promising electronic properties their self assembly in bulk and surfaces has been studied. Furthermore, their application in field effect transistors and solar cells by using different processing techniques both from melt and solutions has been demonstrated [130]. The chemical structures of some of such nanographene molecules with long alkyl chains are shown in Fig. 3.18.

In addition to its liquid crystalline phase behavior, graphene has been demonstrated as a transparent electrode in LCD devices and a potential substitute for metal oxide electrodes [132]. It has been found that the electrooptical characteristics of such devices are superior in nature. In a similar fashion GO has been utilized in its reduced form in LC cells to study the field-induced reorientation of a nematic LC [133, 134]. These preliminary results are very encouraging for future LC devices. The advantages of graphene compared to conventionally used metal oxide electrodes in terms of low resistivity, high transparency and chemical stability hold great promise for large scale exploitation as transparent conductive electrodes.

3.4 Conclusions and Outlook

CNTs and graphene based lyotropic LCs have been fabricated with the help of appropriate physical and chemical functionalization techniques. Both single- and multi-walled CNTs as well as graphenes, GOs and rGOs have been found to function as suitable building blocks of liquid crystalline phases. Their phase behavior has been found to be reminiscent with the theoretical predictions and such studies have provided greater insights and useful guidelines into the understanding of the phase behaviors of 1D and 2D nanoparticles dispersed in suitable solvents. High performance carbon fibers have been obtained from both the 1D and 2D carbon allotropes by liquid crystalline processing. Conductive LC alignment layers have been demonstrated by using liquid crystalline CNTs solutions. Thin film transistors with aligned CNTs have been shown to possess better performance compared to the devices with unaligned CNTs. Though fullerene, CNT and graphene derivatives have been suitably functionalized to exhibit liquid crystalline phases (Fig. 3.19), it will be interesting to explore ways to obtain liquid crystalline diamond. Appropriate functionalization of nanodiamond may provide liquid

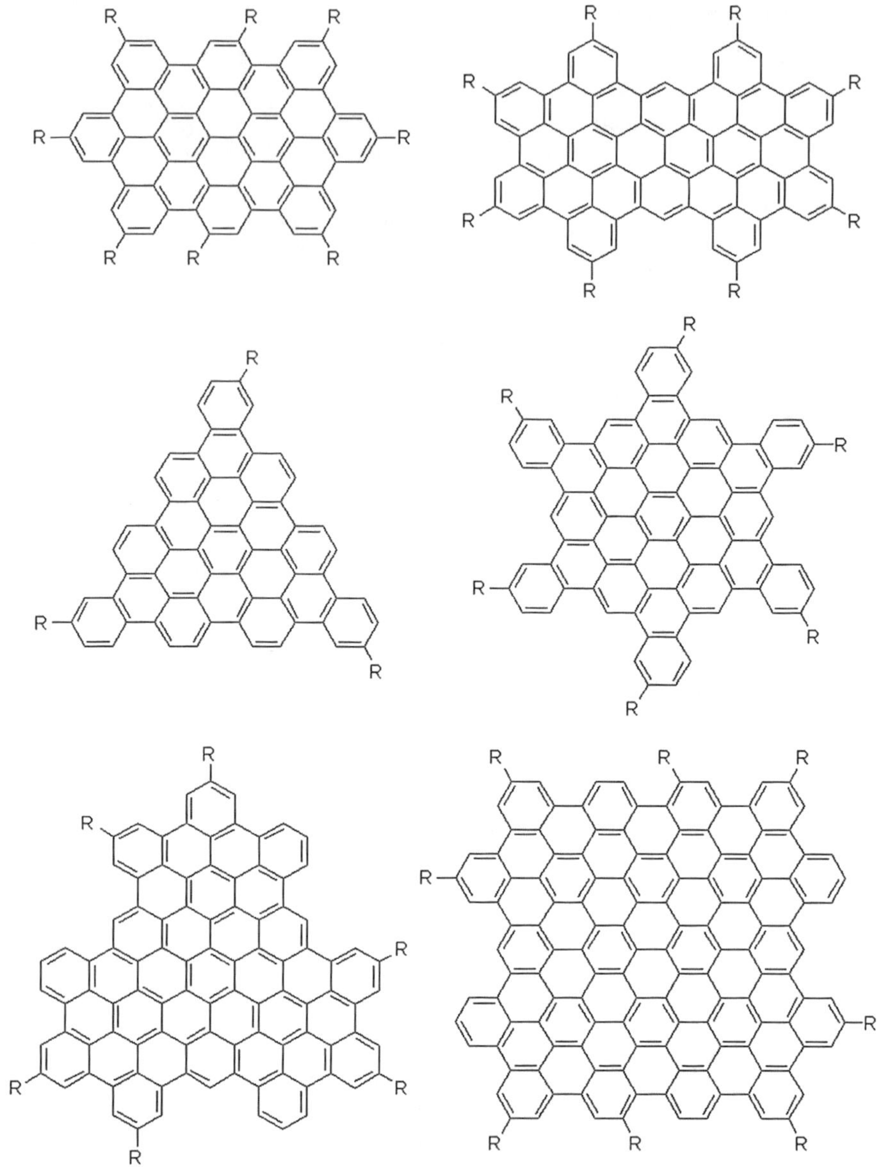

Fig. 3.18 Some synthetic molecular nanographenes exhibiting thermotropic liquid crystalline phase behavior

crystalline "soft materials" from the "hardest material". Since CNTs furnish uniaxial nematic phase and graphene derivatives exhibit uniaxial discotic nematic phase in aqueous dispersions, their homogeneous mixtures in principle has the potential to exhibit the much sought after "biaxial nematic phase at room

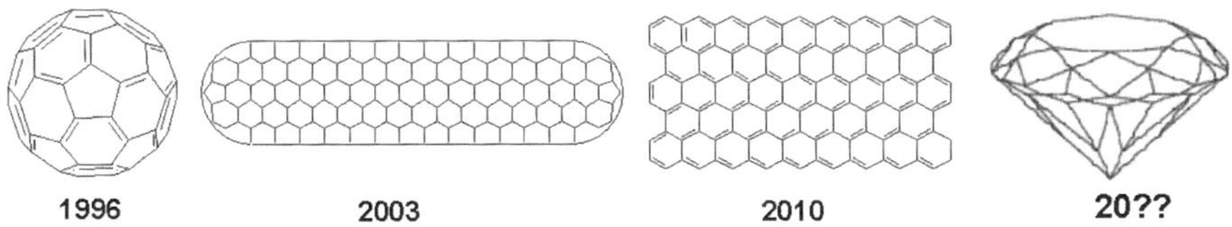

Fig. 3.19 Liquid crystalline carbon allotropes

temperature". A systematic investigation in this direction involving the 1D and 2D carbon allotropes seems worthwhile which may open up new future opportunities to capitalize on. Significantly, most of the knowledge gained on graphene based LCs may be extended to other 2D layered crystals such as BN, MoS_2, WS_2 etc., exhibiting a set of marvelous properties.

Acknowledgments The preparation of this chapter benefited from the support to Quan Li by the Department of Defense Multidisciplinary University Research Initiative (AFOSR MURI FA9550-12-1-00370 and FA9550-06-1-0337), the Air Force Office of Scientific Research (AFOSR FA9550-09-1-0193 and FA9550-09-1-0254), the National Science Foundation (NSF IIP 0750379), the National Aeronautics and Space Adminstration (NASA), the Department of Energy (DOE DE-SC0001412), Ohio Third Frontier, and the Ohio Board of Regents under its Research Challenge program.

References

1. Q. Li (ed.), *Liquid Crystal Beyond Displays: Chemistry, Physics, and Applications* (Wiley, New Jersey, 2012)
2. Q. Li (ed.), *Intelligent Stimuli-Responsive Materials: From Well-Defined Nanostructures to Applications* (Wiley, New Jersey, 2013)
3. T. Kato, N. Mizoshita, K. Kishimoto, Functional liquid-crystalline assemblies: self-organized soft materials. Angew. Chem. Int. Ed. **45**, 38–68 (2006)
4. C. Tschierske, Liquid crystal engineering- new complex mesophase structures and their relations to polymer morphologies, nanoscale patterning and crystal engineering. Chem. Soc. Rev. **36**, 1930–1970 (2007)
5. J.W. Goodby, I.M. Saez, S.J. Cowling, V. Gortz, M. Draper, A.W. Hall, S. Sia, G. Cosquer, S.E. Lee, E.P. Raynes, Transmission and amplification of information and properties in nanostructured liquid crystals. Angew. Chem. Int. Ed. **47**, 2754–2787 (2008)
6. Y. Wang, Q. Li, Light-driven chiral molecular switches or motors in liquid crystals. Adv. Mater. **24**, 1926–1945 (2012)
7. E.-K. Fleischmann, R. Zentel, Liquid-crystalline ordering as a concept in materials science: from semiconductors to stimuli-responsive devices. Angew. Chem. Int. Ed. **52**, 8810–8827 (2013)
8. T. Geelhaar, K. Griesar, B. Reckmann, 125 Years of liquid crystals-a scientific revolution in the home. Angew. Chem. Int. Ed. **52**, 8798–8809 (2013)
9. M. Bremer, P. Kirsch, M. Klasen-Memmer, K. Tarumi, The TV in your pocket: development of liquid-crystal materials for the new millennium. Angew. Chem. Int. Ed. **52**, 8880–8896 (2013)
10. S. Kumar, Self-organization of disc-like molecules: chemical aspects. Chem. Soc. Rev. **35**, 83–109 (2006)

11. M. Mathews, Q. Li, Self-organizing discotic liquid crystals as novel organic semiconductors. Chapter 4 in *Self-Organized Organic Semiconductors: From Materials to Device Applications*, ed. by Q. Li (Wiley, New Jersy, 2011)
12. Y. Li, A. Urbas, Q. Li, Reversible light-directed red, green and blue reflections with thermal stability enabled by a self-organized helical superstructure. J. Am. Chem. Soc. **134**, 9573–9576 (2012)
13. H.K. Bisoyi, S. Kumar, Liquid-crystal nanoscience: an emerging avenue of soft self-assembly. Chem. Soc. Rev. **40**, 306–319 (2011)
14. J.P.F. Lagerwall, G. Scalia, A new era for liquid crystal research: applications of liquid crystals in soft matter nano-, bio- and microtechnology. Curr. Appl. Phys. **12**, 1387–1412 (2012)
15. G.L. Nealon, R. Greget, C. Dominguez, Z.T. Nagy, D. Guillon, J.-L. Gallani, B. Donnio, Liquid-crystalline nanoparticles: hybrid design and mesophase structures. Beilstein J. Org. Chem. **8**, 349–370 (2012)
16. H.K. Bisoyi, S. Kumar, Discotic nematic liquid crystals: science and technology. Chem. Soc. Rev. **39**, 264–285 (2010)
17. J. Ma, Q. Li, Smectic liquid crystal semiconductors. Chapter 5 in *Self-Organized Organic Semiconductors: From Materials to Applications*, ed. by Q. Li (Wiley, New Jersey, 2011)
18. L. Li, S. Kang, J. Harden, Q. Sun, X. Zhou, L. Dai, A. Jakli, S. Kumar, Q. Li, Nature inspired light-harvesting liquid crystalline porphyrins for organic photovoltaics. Liq. Cryst. **35**, 233–239 (2008)
19. M. O'Neill, S.M. Kelly, Ordered materials for organic electronics and photonics. Adv. Mater. **23**, 566–584 (2011)
20. M. Funahasi, Development of liquid-crystalline semiconductors with high carrier mobilities and their application to thin-film transistors. Polymer J. **41**, 459–469 (2009)
21. T.-H. Lin, Y. Li, C.-T. Wang, H.-C. Jau, C.-W. Chen, C.-C. Li, H.K. Bisoyi, T.J. Bunning, Q. Li, Red, green and blue reflections enabled in optically tunable self-organized 3D cubic nanostructured thin film. Adv. Mater. **25**, 5050–5054 (2013)
22. Y. Li, M. Wang, T.J. White, T.J. Bunning, Q. Li, Azoarenes bearing opposite chiral configurations: light-driven dynamic reversible handedness inversion in self-organized helical superstructure. Angew. Chem. Int. Ed. **52**, 8925–8929 (2013)
23. C. Ohm, M. Brehmer, R. Zentel, Liquid crystalline elastomers as actuators and sensors. Adv. Mater. **22**, 3366–3387 (2010)
24. I.W. Hamley, Liquid crystal phase formation by biopolymers. Soft Matter **6**, 1863–1871 (2010)
25. S.J. Woltman, G.D. Jay, G.P. Crawford, Liquid-crystal materials find a new order in biomedical applications. Nat. Mat. **6**, 929–938 (2007)
26. A. Angelova, B. Angelov, R. Mutafchieva, S. Leieur, P. Couvreur, Self-assembled multicompartment liquid crystalline lipid carrier for protein, peptide, and nucleic acid drug delivery. Acc. Chem. Res. **44**, 147–156 (2011)
27. A.M. Lowe, N.L. Abbott, Liquid crystalline materials for biological applications. Chem. Mater. **24**, 746–758 (2012)
28. H.N.W. Lekkerkerker, G.J. Vroege, Liquid crystal phase transitions in suspensions of mineral colloids: new life from old roots. Phil. Trans. R. Soc. A **371**, 20120263 (2013)
29. L. Li, J. Walda, L. Manna, A.P. Alivisatos, Semiconductor nanorod liquid crystals. Nano Lett. **2**, 557–560 (2002)
30. N.R. Jana, L.A. Gearheart, S.O. Obare, C.J. Johnson, K.J. Edler, S. Mann, C.J. Murphy, Liquid crystalline assemblies of ordered gold nanorods. J. Mater. Chem. **12**, 2909–2912 (2002)
31. M. Zorn, S. Meure, M.N. Tahir, Y. Khalavka, C. Sonnichsen, W. Tremel, R. Zentel, Liquid crystalline phases from polymer functionalized semiconducting nanorods. J. Mater. Chem. **18**, 3050–3058 (2008)
32. F.M. van der Kooij, K. Kassapidou, H.N.W. Lekkerkerker, Liquid crystal phase transitions in suspensions of polydisperse plate-like particles. Nature **406**, 868–871 (2000)

33. A.E. Saunders, A. Ghezelbash, D. Smilgies, M.B. Sigman, B.A. Korgel, Columnar self-assembly of colloidal nanodisks. Nano Lett. **6**, 2959–2963 (2006)
34. J.-C.P. Gabriel, F. Camerel, B.J. Lemaire, H. Desvaux, P. Davidson, P. Betail, Swollen liquid-crystalline lamellar phase based on extended solid-like sheets. Nature **413**, 504–508 (2001)
35. F.M. van der Kooij, H.N.W. Lekkerkerker, Formation of nematic liquid crystals in suspensions of hard colloidal platelets. J. Phys. Chem. B **102**, 7829–7832 (1998)
36. J.A.C. Veerman, D. Frenkel, Phase behavior of disklike hard-core mesogens. Phys. Rev. A **45**, 5632–5648 (1992)
37. A. Patti, S. Belli, R. van Roij, M. Dijkstra, Relaxation dynamics in the columnar liquid crystal phase of hard platelets. Soft Matter **7**, 3533–3545 (2011)
38. M.A. Bates, D. Frenkel, Nematic-isotropic transition in polydisperse systems of infinitely thin hard platelets. J. Chem. Phys. **110**, 6553–6559 (1999)
39. N. Miyamoto, T. Nakato, Liquid crystalline inorganic nanosheet colloids derived from layered materials. Isr. J. Chem. **52**, 881–894 (2012)
40. L. Onsager, The effect of shape on the interaction of colloidal particles. Ann. N. Y. Acad. Sci. **51**, 627–659 (1949)
41. H.W. Kroto, J.R. Heath, S.C. O'brien, R.F. Curl, R.E. Smalley, C_{60}: buckminsterfullerene. Nature **318**, 162–163 (1985)
42. S. Iijima, Helical microtubules of graphitic carbon. Nature **354**, 56–58 (1991)
43. S. Iijima, T. Ichihashi, Single-shell carbon nanotubes of 1-nm diameter. Nature **363**, 603–605 (1993)
44. W. Song, I.A. Kinloch, A.H. Windle, Nematic liquid crystallinity of multiwall carbon nanotubes. Science **302**, 1363-1363 (2003)
45. N. Behabtu, J.R. Lomeda, M.J. Green, A.L. Higginbotham, A. Sinitskii, D.V. Kosynkin, D. Tsentalovich, A.N.G. Parra-vasquez, J. Schmidt, E. Kesselman, Y. Cohen, Y. Talmon, J.M. Tour, M. Pasquali, Spontaneous high-concentration dispersions and liquid crystals of graphene. Nat. Nanotech. **5**, 406–411 (2010)
46. R.H. Hurt, Z.-Y. Chen, Liquid crystals and carbon materials. Phys. Today **53**, 39–44 (2000)
47. H.K. Bisoyi, S. Kumar, Carbon-based liquid crystals: art and science. Liq. Cryst. **38**, 1427–1449 (2011)
48. T. Chuard, R. Deschenaux, First fullerene[60]-containing thermotropic liquid crystals. Helv. Chim. Acta **79**, 736–741 (1996)
49. A. Hirsch, The era of carbon allotropes. Nat. Mater. **9**, 868–871 (2010)
50. K.S. Novoselov, A.K. Geim, S.V. Morozov, D. Jiang, Y. Zhang, S.V. Dubonos, I.V. Grigorieva, A.A. Firsov, Electric field effect in atomically thin carbon films. Science **306**, 666–669 (2004)
51. M.F.L. De Volder, S.H. Tawfick, R.H. Baughman, A.J. Hart, Carbon nanotubes: present and future commercial applications. Science **339**, 535–539 (2013)
52. I. Dierking, G. Scalia, P. Morales, D. LeClere, Aligning, reorienting carbon nanotubes with nematic liquid crystals. Adv. Mater. **16**, 865–869 (2004)
53. I. Dierking, G. Scalia, P. Morales, Liquid crystal-carbon nanotubes dispersions. J. Appl. Phys. **97**, 44309-1-5 (2005)
54. C. Zakri, Carbon nanotubes and liquid crystalline phases. Liq. Cryst. Today **16**, 1–11 (2007)
55. S. Schymura, G. Scalia, On the effect of carbon nanotubes on properties of liquid crystals. Phil. Trans. R. Soc. A **371**, 20120261 (2013)
56. M. Rahman, W. Lee, Scientific duo of carbon nanotubes and nematic liquid crystals. J. Phys. D: Appl. Phys. **42**, 63001-1-12 (2009)
57. M. Bravo-Sanchez, T.J. Simmons, M.A. Vidal, Liquid crystal behavior of single wall carbon nanotubes. Carbon **48**, 3531–3542 (2010)
58. H.K. Bisoyi, S. Kumar, Carbon nanotubes in discotic liquid crystals. J. Indian Inst. Sci. **89**, 101–112 (2009)
59. G. Scalia, Alignment of carbon nanotubes in thermotropic and lyotropic liquid crystals. ChemPhysChem **11**, 333–340 (2010)

60. J.P.F. Lagerwall, G. Scalia, Carbon nanotubes in liquid crystals. J. Mater. Chem. **18**, 2890–2898 (2008)
61. C. Zakri, P. Poulin, Phase behavior of nanotubes suspensions: from attraction induced percolation to liquid crystalline phases. J. Mater. Chem. **16**, 4095–4098 (2006)
62. S. Zhang, S. Kumar, Carbon nanotubes as liquid crystals. Small **4**, 1270–1283 (2008)
63. A.M. Somoza, C. Sagui, C. Roland, Liquid-crystal phases of capped carbon nanotubes. Phys. Rev. B **63**, 081403-1-4 (2001)
64. S. Zhang, I.A. Kinloch, A.H. Windle, Mesogenicity drives fractionation in lyotropic aqueous suspensions of multiwall carbon nanotubes. Nano Lett. **6**, 568–572 (2006)
65. S. Zhang, K.K.K. Koziol, I.A. Kinloch, A.H. Windle, Macroscopic fibers of well-aligned carbon nanotubes by wet spinning. Small **4**, 1217–1222 (2008)
66. W. Song, A.H. Windle, Isotropic-nematic phase transition of dispersions of multiwall carbon nanotubes. Macromolecules **38**, 6181–6188 (2005)
67. W. Song, A.H. Windle, Size-dependence and elasticity of liquid-crystalline multiwalled carbon nanotubes. Adv. Mater. **20**, 3149–3154 (2008)
68. V.A. Davis, L.M. Ericson, A.N.G. Parre-Vasquez, H. Fan, Y. Wang, V. Prieto, J.A. Longoria, S. Ramesh, R.K. Saini, C. Kittrell, W.E. Billups, W.W. Adams, R.H. Hauge, R.E. Smalley, M. Pasquali, Phase behavior and rheology of SWNTs in superacids. Macromolecules **37**, 154–160 (2004)
69. P.K. Rai, R.A. Pinnick, A.N.G. Parra-Vasquez, V.A. Davis, H.K. Schmidt, R.H. Hauge, R.E. Smalley, M. Pasquali, Isotropic-nematic phase transition of single-walled carbon nanotubes in strong acids. J. Am. Chem. Soc. **128**, 591–595 (2006)
70. S. Ramesh, L.M. Ericson, V.A. Davis, R.K. Saini, C. Kittrell, M. Pasquali, W.E. Billups, W.W. Adams, R.H. Hauge, R.E. Smalley, Dissolution of pristine single walled carbon nanotubes in superacids by direct protonation. J. Phys. Chem. B **108**, 8794–8798 (2004)
71. V.A. Davis, A.N.G. Parra-Vasquez, M.J. Green, P.K. Rai, N. Behabtu, V. Prieto, R.D. Booker, J. Schmidt, E. Kesselman, W. Zhou, H. Fan, W.W. Adams, R.H. Hauge, J.E. Fischer, Y. Cohen, Y. Talmon, R.E. Smalley, M. Pasquali, True solutions of single-walled carbon nanotubes for assembly into macroscopic materials. Nat.Nanotechnol. **4**, 830–834 (2009)
72. L.M. Ericson, H. Fan, H. Peng, H. Peng, V.A. Davis, W. Zhou, J. Sulpizio, Y. Wang, R. Booker, J. Vavro, C. Guthy, A.N.G. Parra-Vasquez, M.J. Kim, S. Ramesh, R.K. Saini, C. Kittrell, G. Lavin, H. Schmidt, W.W. Adams, W.E. Billups, M. Pasquali, W.-F. Hwang, R.H. Hauge, J.E. Fischer, R.E. Smalley, Macroscopic neat single-walled carbon nanotube fibers. Science **305**, 1447–1450 (2004)
73. M.J. Green, A. Nicholas, G. Parra-Vasquez, N. Behabtu, M. Pasquali, Modelling the phase behavior of polydisperse rigid rods with attractive interactions with application to single-walled carbon nanotubes in superacids. J. Chem. Phys. **131**, 084901-1-10 (2009)
74. S. Badaire, C. Zakri, M. Maugey, A. Derre, J.N. Barisci, G. Wallace, P. Poulin, Liquid crystals of DNA-stabilized carbon nanotubes. Adv. Mater. **17**, 1673–1676 (2005)
75. C. Zamora-Ledezma, C. Blanc, M. Maugey, C. Zakri, P. Poulin, E. Anglaret, Anisotropic thin films of single wall carbon nanotubes from aligned lyotropic nematic suspensions. Nano Lett. **8**, 4103–4107 (2008)
76. N. Puech, C. Blanc, E. Grelet, C. Zamora-ledezma, M. Maugey, C. Zakri, E. Anglaret, P. Poulin, Highly ordered carbon nanotubes nematic liquid crystals. J. Phys. Chem. C **115**, 3272–3278 (2011)
77. N. Puech, E. Grelet, P. Poulin, C. Blanc, P. van der Schoot, Nematic droples in aqueous dispersions of carbon nanotubes. Phys. Rev. E **82**, 20702-1-4 (2010)
78. S.E. Moulton, M. Maugey, P. Poulin, G.G. Wallace, Liquid crystal behavior of single-walled carbon nanotubes dispersed in biological hyaluronic acid solutions. J. Am. Chem. Soc. **129**, 9452–9457 (2007)
79. L. Lu, W. Chen, Large-scale aligned carbon nanotubes from their purified, highly concentrated suspension. ACS Nano **4**, 1042–1048 (2010)

80. G. Ao, D. Nepal, M. Aono, V.A. Davis, Cholesteric and nematic liquid crystalline phase behavior of double-stranded DNA stabilized single-walled carbon nanotubes dispersions. ACS Nano **5**, 1450–1458 (2011)
81. H.W. Lee, W. You, S. Barman, S. Hellstrom, M.C. Leieux, J.H. Oh, S. Liu, T. Fujiwara, W.M. Wang, B. Chen, Y.W. Jin, J.M. Kim, Z. Bao, Lyotropic liquid-crystalline solutions of high-concentration dispersions of single-walled carbon nanotubes with conjugated polymers. Small **5**, 1019–1024 (2009)
82. R.A. Mrozek, B. Kim, V.C. Holmberg, T.A. Taton, Homogeneous, coaxial liquid crystal domain growth from carbon nanotubes seeds. Nano Lett. **3**, 1665–1669 (2003)
83. J.M. Russell, S. Oh, I. LaRue, O. Zhou, E.T. Samulski, Alignment of nematic liquid crystals using carbon nanotube films. Thin Solid Films **509**, 53–57 (2006)
84. S. Meuer, L. Braun R. Zentel, Solubilization of multi-walled carbon nanotubes by alpha-purene functionalized PMMA and their liquid crystalline self-organization. Chem. Commun. 3166-3168 (2008)
85. H. Ko, V.V. Tsukruk, Liquid-crystalline processing of highly oriented carbon nanotubes arrays for thin film transistors. Nano Lett. **6**, 1443–1445 (2006)
86. M. Kimura, N. Miki, N. Adachi, Y. Tatewaki, K. Ohta, H. Shirai, Organization of single-walled carbon nanotubes wrapped with liquid-crystalline Pi-conjugated oligomers. J. Mater. Chem. **19**, 1086–1092 (2009)
87. C. Jiang, A. Saha, C. Xiang, C.C. Young, J.M. Tour, M. Pasquali, A.A. Marti, Increased solubility, liquid-crystalline phase, and selective functionalization of single-walled carbon nanotubes polyelectrolyte dispersions. ACS Nano **7**, 4503–4510 (2013)
88. S. Kumar, H.K. Bisoyi, Aligned carbon nanotubes in the supramolecular order of discotic liquid crystals. Angew. Chem. Int. Ed. **46**, 1501–1503 (2007)
89. H.K. Bisoyi, S. Kumar, Carbon nanotubes in triphenylene and rufigallol-based room temperature monomeric and polymeric discotic liquid crystals. J. Mater. Chem. **18**, 3032–3039 (2008)
90. A.K. Geim, Graphene: status and prospects. Science **324**, 1530–1534 (2009)
91. N. Savage, Super carbon. Nat. Mater. **483**, S30–S31 (2012)
92. F.J.G. de Abajo, Graphene Nanophotonics. Science **339**, 917–918 (2013)
93. P. Avouris, Z. Chen, V. Perebeinos, Carbon-based electronics. Nat. Mater. **2**, 605–615 (2007)
94. O.C. Compton, S.T. Nguyen, Graphene oxide, highly reduced graphene oxide, and graphene: versatile building blocks for carbon-based materials. Small **6**, 711–723 (2010)
95. J. Kim, F. Kim, J. Huang, Seeing graphene-based sheets. Mater. Today **13**, 28–38 (2010)
96. C.N.R. Rao, A.K. Sood, K.S. Subrahmanyam, A. Govindraj, Graphene: the new two-dimensional nanomaterials. Angew. Chem. Int. Ed. **48**, 7752–7777 (2009)
97. M.J. Allen, V.C. Tung, R.B. Kaner, Honeycomb carbon: a review of graphene. Chem. Rev. **110**, 132–145 (2010)
98. J. Kim, L.J. Cote, J. Huang, Two dimensional soft material: new faces of graphene oxide. Acc. Chem. Res. **45**, 1356–1364 (2012)
99. X. Wan, Y. Huang, Y. Chen, Focusing on energy and optoelectronic applications: a journey for graphene and graphene oxide at large scale. Acc. Chem. Res. **45**, 598–607 (2012)
100. D.K. James, J.M. Tour, Graphene: Powder, flakes, ribbons and sheets. Acc. Chem. Res. **46**, 2307–2318 (2013)
101. M. Cai, D. Thorpe, D.H. Adamson, H.C. Schniepp, Methods of graphite exfoliation. J. Mater. Chem. **22**, 24992–25002 (2012)
102. V. Palermo, Not a molecule, not a polymer, not a substrate… the many faces of graphene as a chemical platform. Chem. Commun. **49**, 2848–2857 (2013)
103. Y.H. Hu, H. Wang, B. Hu, Thinnest two-dimensional nanomaterials-graphene for solar energy. ChemSusChem **3**, 782–796 (2010)
104. C. Cha, S.R. Shin, N. Annabi, M.R. Dokmeci, A. Khademhosseini, Carbon-based nanomaterials: multifunctional materials for biomedical engineering. ACS Nano **4**, 2891–2897 (2010)

105. D.S. Su, S. Perathoner, G. Centi, Nanocarbons for the development of advanced catalysts. Chem. Rev. **113**, 5782–5816 (2013)
106. Z. Sun, D.K. James, J.M. Tour, Graphene: synthesis and manipulation. J. Phys. Chem. Lett. **2**, 2425–2432 (2011)
107. D.R. Dreyer, S. Park, C.W. Bielawski, R.S. Ruoff, The chemistry of graphene oxide. Chem. Soc. Rev. **39**, 228–240 (2010)
108. C.K. Chua, M. Pumera, Covalent chemistry on graphene. Chem. Soc. Rev. **42**, 3222–3233 (2013)
109. S. Park, R.S. Ruoff, Chemical methods for the production of graphenes. Nat. Nanotechon. **4**, 217–224 (2009)
110. D.A. Dikin, S. Stankovich, E.J. Zimney, R.D. Piner, G.H.B. Dommett, G. Evmenenko, S.T. Nguyen, R.S. Ruoff, Preparation and characterization of graphene oxide paper. Nature **448**, 457–460 (2007)
111. Y. Zhu, S. Murali, W. Cai, X. Li, J.W. Suk, J.R. Potts, R.S. Ruoff, Graphene and graphene oxide: synthesis, properties and applications. Adv. Mater. **22**, 3906–3924 (2010)
112. L. Rodriguez-Perez, M.A. Herranz, N. Martin, The chemistry of pristine graphene. Chem. Commun. **49**, 3721–3735 (2013)
113. F. Guo, R. Hurt, Supramolecular synthesis of graphenic mesogenic materials. Macromol. Chem. Phys. **213**, 1164–1174 (2012)
114. C. Zakri, C. Blanc, E. Grelet, C. Zamora-Ledezma, N. Puech, E. Anglaret, P. Poulin, Liquid crystals of carbon nanotubes and graphene. Phil. Trans. R. Soc. A **371**, 20120499-1-15 (2013)
115. J.E. Kim, T.H. Han, S.H. Lee, J.Y. Kim, C.W. Ahn, J.M. Yun, S.O. Kim, Graphene oxide liquid crystals. Angew. Chem. Int. Ed. **50**, 3043–3047 (2011)
116. Z. Xu, C. Gao, Aqueous liquid crystals of graphene oxide. ACS Nano **5**, 2908–2915 (2011)
117. Z. Xu, C. Gao, Graphene chiral liquid crystals and macroscopic assembled fibers. Nat. Commun. **2**(571), 1–9 (2011)
118. Z. Xu, H. Sun, X. Zhao, C. Gao, Ultastrong fibers assembled from giant graphene oxide sheets. Adv. Mater. **25**, 188–193 (2013)
119. Z. Liu, Z. Xu, X. Hu, C. Gao, Lyotropic liquid crystal of polyacrylonitrile-grafted graphene oxide and its assembled continuous nacre-mimetic fibers. Macromolecules **46**, 6931–6941 (2013)
120. B. Dan, N. Behabtu, A. Martinez, J.S. Evans, D.V. Kosynkin, J.M. Tour, M. Pasquali, I.I. Smalyukh, Liquid crystals of aqueous, giant graphene oxide flakes. Soft Matter **7**, 11154–11159 (2011)
121. B. Senyuk, N. Behabtu, B.G. Pacheco, T. Lee, G. Ceriotti, J.M. Tour, M. Pasquali, I.I. Smalyukh, Nonlinear photoluminescence imaging of isotropic and liquid crystalline dispersions of graphene oxide. ACS Nano **6**, 8060–8066 (2012)
122. F. Guo, F. Kim, T.H. Han, V.B. Shenoy, J. Huang, R.H. Hurt, Hydration-responsive folding and unfolding in graphene oxide liquid crystal phases. ACS Nano **5**, 8019–8025 (2011)
123. R. Jalili, S.H. Aboutalebi, D. Esrafilzadeh, K. Konstantinov, S.E. Moulton, J.M. Razal, G.G. Wallace, Organic solvent-based graphene oxide liquid crystals: a facile route toward the next generation of self-assembled layer-by-layer multifunctional 3D architectures. ACS Nano **7**, 3981–3990 (2013)
124. R. Jalili, S.H. Aboutalebi, D. Esrafilzadeh, R.L. Shepherd, J. Chen, S. Aminorroaya-Yamini, K. Konstantinov, A.I. Minett, J.M. Razal, G.G. Wallace, Scalable one-step wet-spinning of graphene fibers and yarns from liquid crystalline dispersions of graphene oxide: towards multifunctional textiles. Adv. Funct. Mater. **23**, 5345–5354 (2013)
125. S.H. Aboutalebi, M.M. Gudarzi, Q.B. Zheng, J.-K. Kim, Spontaneous formation of liquid crystals in ultalarge graphene oxide dispersions. Adv. Funct. Mater. **21**, 2978–2988 (2011)
126. C. Xiang, C.C. Young, X. Wang, Z. Yan, C.-C. Hwang, G. Cerioti, J. Lin, J. Kono, M. Pasquali, J.M. Tour, Large flake graphene oxide fibers with unconventional 100 % knot efficiency and highly aligned small flake graphene oxide fibers. Adv. Mater. **25**, 4592–4597 (2013)

127. C. Xiang, N. Behabtu, Y. Liu, H.G. Chae, C.C. Young, B. Genorio, D.E. Tsentalovich, C. Zhang, D.V. Kosynkin, J.R. Lomeda, C.-C. Hwang, S. Kumar, M. Pasquali, J.M. Tour, Graphene nanoribbons as an advanced precursor for making carbon fiber. ACS Nano **7**, 1628–1637 (2013)
128. C. Zamora-Ledezma, N. Puech, C. Zakri, E. Grelet, S.E. Moulton, G.G. Wallace, S. Gambhir, C. Blanc, E. Anglaret, P. Poulin, Liquid crystallinity and dimensions of surfactant-stabilized sheets of reduced graphene oxide. J. Phys. Chem. Lett. **3**, 2425–2430 (2012)
129. W. Pisula, X. Feng, K. Mullen, Charge-carrier transporting graphene-type molecules. Chem. Mater. **23**, 554–567 (2011)
130. J. Wu, W. Pisula, K. Mullen, Graphene as potential material for electronics. Chem. Rev. **107**, 718–747 (2007)
131. C.D. Simpson, J. Wu, M.D. Watson, K. Mullen, From graphite molecules to columnar superstructures-an exercise in nanoscience. J. Mater. Chem. **14**, 494–504 (2004)
132. P. Blake, P.D. Brimicombe, R.R. Nair, T.J. Booth, D. Jiang, F. Schedin, L.A. Ponomarenko, S.V. Morozov, H.F. Gleeson, E.W. Hill, A.K. Geim, K.S. Novoselov, Graphene-based liquid crystal device. Nano Lett. **8**, 1704–1708 (2008)
133. G. Nordendorf, O. Kasdorf, H. Kitzerow, Y. Liang, X. Feng, and K. Mullen, Liquid crystal addressing by graphene electrodes made from graphene oxide. Jpn. J. Appl. Phys. **49**, 100206-1-3 (2010)
134. M. Wahle, O. Kasdorf, H. Kitzerow, Y. Liang, X. Feng, K. Mullen, Electrooptic switching in graphene-based liquid crystal cells. Mol. Cryst. Liq. Cryst. **543**, 187–193 (2011)

Chapter 4
Liquid Crystal-Gold Nanoparticle Hybrid Materials

Chenming Xue and Quan Li

Abstract The interplay between liquid crystals and gold nanoparticles, i.e. liquid crystalline gold nanoparticle materials, is challenging as well as fascinating for creating novel functional materials. The resulting hybrid materials hold great promise in many applications such as displays, optics, optoelectronics, sensors, and metamaterials due to their unique properties. In this chapter, the fundamentals of liquid crystals and gold nanoparticles are introduced, and the intriguing progresses of hybrid materials, although in their early stage, are summarized and discussed. These hybrid materials not only could improve device performances, but also the well-organized gold nanoparticles driven by the intrinsic nature of liquid crystal could contribute to the very interesting research topic of the functional metamaterials, i.e. a class of artificial materials having properties that never exist in nature such as unusual electromagnetic properties (e.g. negative refractive index materials for cloaking devices).

4.1 Introduction

In modern history, particularly in the recent decades, liquid crystals (LCs) have become a very important class of materials. Since the first invention of LC display (LCD), LCs have become the quintessential materials in information displays such as TVs, computer monitors, and digital displays. In the recent development of LC materials, LCs have moved rapidly beyond display applications and are evolving into entirely new scientific frontiers, opening broad avenues for versatile applications such as lasers, photovoltaics, light-emitting diodes, field effect transistors, biosensors, switchable windows, and nanophotonics [1]. All these applications benefit from LC's unique properties, e.g. self-organization and being able to

C. Xue · Q. Li (✉)
Liquid Crystal Institute, Kent State University, Kent, OH, USA
e-mail: qli1@kent.edu

Q. Li (ed.), *Nanoscience with Liquid Crystals*, NanoScience and Technology,
DOI: 10.1007/978-3-319-04867-3_4,

respond to external stimuli such as temperature, electric field, magnetic field, mechanical stress, and surface effect. It is established that LCs can form the unique phase with new properties by doping colloidal particles ranging from nano- to micro-meter sizes with a variety of shapes, involving polymer, silica microbeads, carbon nanotubes, and semiconducting/metal spheres or rods. Combining nano-particles with LCs has been a fast growing research topic in the past two decades, in which a number of interesting phenomena have been demonstrated.

It is known that nanoparticles are typically 1 − 100 nm in at least one dimension. They are made of inorganic or organic materials, which show the characteristics that are quite different from their corresponding bulky ones. Among all nanoparticles, metal nanoparticles, especially gold nanoparticles (GNPs), are of particular interest owing to their diverse and distinct properties. GNPs have plasmonic response which depends on the GNP's size and shape, as well as the dielectric permittivity of the medium surrounding them. Due to their unique plasmonic and optical properties, such as surface plasmon resonance (SPR), surface-enhanced Raman scattering (SERS), enhanced IR absorption, and nonlinear optical, electronic and magnetic properties, GNPs provide many opportunities as well as challenges for insights into fundamental science which opens the door to various applications in optics, sensors, catalysis, and biological imaging devices etc.

Therefore, the combination of LCs and GNPs forming LC/GNP hybrid materials is worthy of exploration. To obtain these hybrid materials, it requires finding a suitable LC medium which is compatible with GNPs or functionalizing GNPs to be compatible with a LC medium. There are two ways to functionalize GNPs: (i) using simple molecules such as *n*-alkyl chains to coat GNPs so that the resultant GNPs can easily dissolve in LC hosts, (ii) utilizing mesogens with flexible structures to coat GNPs either terminally or laterally for better miscibility in a broader range of LC hosts. The properties of LC hosts can be influenced by external stimuli such as applied magnetic field, electric field and surface effects, all of which lead these materials toward their use in stimuli-responsive advanced materials [2]. Functional mesogens can bring novel properties to GNPs, e.g. tuning the SPR and controlling the self-assembly of the GNPs. Based on the LC/GNP materials, new generation electro-optical devices would be achieved. In addition, GNPs as metamaterials can bring unusual electromagnetic properties from its structure and self-assembly rather than from its chemical composition [3, 4]. If the GNPs can disperse in a LC host forming a continuous hybrid structure, not only the plasmonic response can be tuned by changing the component's dielectric constant via reorientation of the LC with an external electric or magnetic field, but also the LC devices would show new optical or magnetic properties easily and rapidly [5–7]. The cutting-edge research topic of developing functional metamaterials using GNPs in LC is thought to be considerably significant but highly challenging. It requires particular GNP shapes, small fluctuations (stability), and highly ordered arrangements (self-organization). The organization of GNPs into ordered systems is of crucial importance for their use in the device applications, realized by a wide variety of "top-down" and "bottom-up" methods which are under continuous development [8–13]. LC materials providing long range orientational order and self-assembling property are ideal for such purpose. Overall, the topic of LC/GNP hybrid materials covers synthesis, analysis, characterization, and modeling, which brings opportunities to create novel physicochemical properties with systematic variation of nanoparticles size, shape, and functionality. With these hybrid materials, the current LC applications such as LCDs could be expanded. On the other hand, since GNPs play an important role in metamaterials, combining GNPs with LCs can offer a potential of creating self-organized GNPs, exhibiting unusual electronic, optical and magnetic properties.

This chapter aims to give an overview of current research of LC/GNP hybrid materials. Fundamentals of LC and GNP materials will be introduced and representative LC/GNP hybrid materials reported in recent years will be summarized and discussed. Their applications will also be presented. This chapter is concluded by outlining future potential and opportunities in this rapidly emerging research field.

4.2 Fundamentals of LCs and GNPs

LC/GNP hybrid materials consist of two crucial parts, LCs and GNPs. There are three types of such hybrid materials: mesogen coated GNPs, simple organic molecules such as *n*-alkyls coated GNPs doped in LC media, mesogen and *n*-alkyl (mixed monolayer) coated GNPs doped in LC media. To make readers understand the chapter better, the fundamentals of LCs and GNPs are first introduced.

The LC state is an intermediate state between the perfectly ordered crystalline state and the totally disordered isotropic (liquid) state. Specifically, LC phases can be observed during the transition in the heating process from the highly ordered crystalline phase with long range orientational, positional and bond orders to the totally disordered isotropic state, or vice versa in the cooling process. In these phases, the long range order of the crystalline state is partially lost and some degree of disorder including translational, rotational and conformational mobility is gained by the molecules [14]. Generally, LCs are classified into two categories: temperature-dependent thermotropic LCs [15, 16] and solvent-dependent lyotropic LCs. For example, typically non-amphiphilic molecules exhibit thermotropic LC phases while amphiphilic molecules show lyotropic LC phases [16, 17]. More specifically, thermotropic LCs usually comprise at least two parts with different structural properties: a rigid anisotropic moiety and flexible segments. Thermotropic LCs are driven by the segregation of the chemically unfavorable units, rigid aromatic cores and flexible alkyl chains. For lyotropic LCs, they are driven by the segregation of hydrophobic and hydrophilic parts of amphiphilic molecules [18, 19]. Additionally, there is a third type called amphotropic molecules, which can form both thermotropic and lyotropic LC phases [15]. Due to the phase separation of the incompatible components, there comes the multi-step melting process which is the characteristic of LC materials [16].

For LC materials including the three types mentioned above, their phases can be classified based on the degree of order and symmetry elements. The least ordered one is nematic phase, which only possesses long range orientational order. As shown in Fig. 4.1a, the molecules in nematic LC phase show an average unidirectional orientation of molecules. They have been well-developed as a significant technology in flat panel LCDs, which is a technological landmark in modern history. The other promising type is the chiral nematic phase (N*), which is characterized by a continuous helical distortion of the director n (helical twist). It also has other appealing applications including temperature sensing by selectively reflecting light of various colors according to its pitch change with temperature. For more ordered LCs with additional positional order, there are smectic (Sm), columnar (Col), and cubic (Cub) phases, based on the shape and chemical characteristics of the LC molecules. Rod-like (calamitic) molecules normally form smectic phase (also called lamellar phase as showing ordered layers). Two most common smectic phases are smectic A phase (SmA) with long molecular axis parallel to the layer normal, and smectic C phase (SmC) with long molecular axis tilted with respect to the layer normal (Fig. 4.1b–c) [20]. By using chiral LC molecules or chiral dopants, their corresponding chiral phases (SmA*, SmC*) have significant applications in electro-optic devices according to their electroclinic effect and surface-stabilized ferroelectric switching effect [21]. In addition to the conventional rod-like LC molecules, there are non-conventional LC phases owing to the particular molecular structures. Recently more research interest has been focused on columnar LC phases based on discotic molecules which are able to aggregate and form columns organizing into arrays (Fig. 4.1d–h). The least ordered form is the discotic nematic phase which does not form columns. With different symmetry types, the columnar LC phases have variations from hexagonal (Col_h), rectangular (Col_r), to oblique (Col_{ob}) [22, 23]. In addition, bent-core (banana-shaped) LCs with unique sterically induced packing have also gained much attention [24].

Metal nanoparticles (MNPs) have a significant position in recent scientific history in terms of preparation, characterization, and applications. Understanding the properties of metal nanoparticles and exploring their potential applications are two major driving forces behind the study of a large variety of metal nanomaterials. Many properties of nanoparticles arise from their large surface-area-to-volume ratio and the spatial confinement of electrons, phonons, and electric fields in and around these nanoparticles. The large surface-area-to-volume ratio in nanoparticles may cause deviations from the usual bulk atomic arrangements. The electron confinement effect in a nanoparticle modifies its spectral properties via shifting the quantum levels and changing transition probabilities. Some properties, such as particle–particle or particle–environment interactions, are affected by the large surface-area-to-volume ratio as well as confinement phenomena. GNPs are a subject of substantial research owing to their attractive properties and applications [25]. Due to the unique properties, GNPs have been extensively investigated and numerous shapes including spheres, rods, cubes, shells, disks, triangles, and stars have been synthesized [26]. Figure 4.2 shows the examples of experimentally

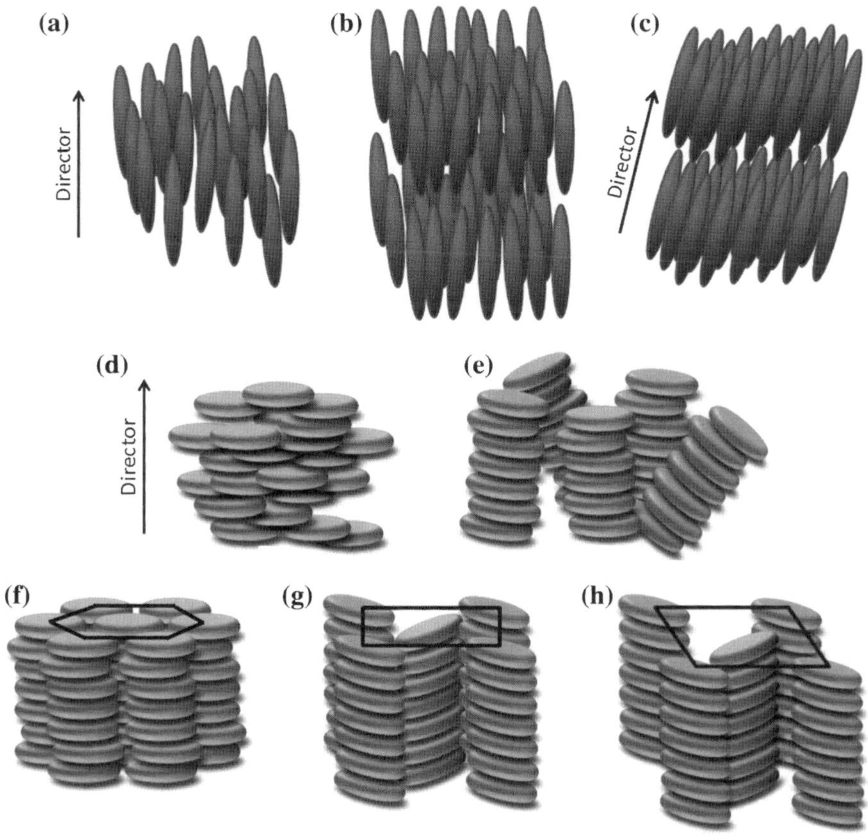

Fig. 4.1 Typical LC phases. **a** nematic (N), **b** smecitc A (SmA), **c** smectic C (SmC), **d** discotic nematic, **e** columnar nematic, **f** columnar hexagnal (Col_h), **g** columnar rectangular (Col_r), **h** columnar oblique (Col_{ob})

synthesized different geometries of GNPs. For GNPs, there are critical shape-dependent optic and electronic properties. For example, electrons decay through radiative (fluorescence), non-radiative (lattice rearrangement), or photothermal (local heating) pathways, where the specific pathway is dependent on the geometry of the nanoparticles and nature of the excitation pulse [27].

In order to disperse GNPs in a LC host, it is necessary to first coat the GNPs with a suitable material that allows them to be more compatible with the LC medium to form a homogenous and stable LC/GNP composite. Simple aliphatic chains as well as LC-alike materials are candidates to attach or adsorb on the surface of the GNPs in order to possess better compatibility with LC hosts. Chemical structure and parameters such as the orientational flexibility and mobility of the molecules on the GNP surface are important. For example, using branched aliphatic molecules can markedly affect packing densities [28, 29] such

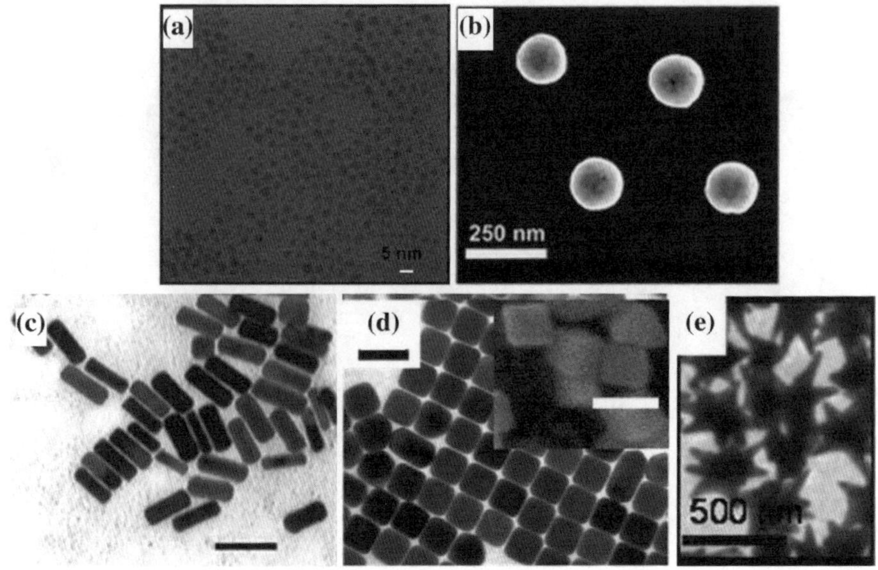

Fig. 4.2 Different shapes of GNPs. TEM images of **a** small spherical GNPs (ca. 2 nm), **b** big spherical GNPs, **c** nanorods (scale bar: 50 nm), **d** nanocubes (scale bar: 100 nm), and **e** nanostars. Adapted with permission from [30] Copyright from Elsvier 2007; [31] Copyright from American Chemical Society (ACS) 2008; [32] Copyright from ACS 2003; and [33] Copyright from ACS 2004

as block-copolymers, supramolecular systems, [34] and dendritic materials [34, 35]. Although complicated organic synthesis is required, the mesogens containing thiol groups including rod, bent-core, and disc like molecules are used to coat GNPs. As a result, these functionalized GNPs could be dispersed into LC media as LC/GNP hybrid materials which are likely to be mesomorphic and of high conductivity, further allowing for the investigation of physical properties.

4.3 LC/GNP Hybrid Materials

4.3.1 Overview

Before introducing LC/GNP hybrid materials, it is noteworthy to mention that the self-organized lyotropic LC templates working as "nano reactors" have played an important role in the synthesis of well-defined metal nanoparticles with different shapes. For the seed mediated method in aqueous solution, the creating process of GNPs is naturally connected with LCs. For example, lyotropic LC phases are useful for synthesizing spherical GNPs [36], gold nanorods (GNRs) [37], gold nanoplates [38], gold nano- and micro-plates [39], and some other metal nanoparticles such as Pb nanoparticles [40] and semiconducting nanorods [41–44].

For the LC/GNP materials, even without surface coating with organic molecules, GNPs can be planted on the substrate surface to influence LC properties. For example, a dynamic, continuous, and reversible voltage-controlled color tuning promoted by gold nanodot arrays formed on a glass substrate of a sandwiched N-LC cell has been invented [5]. In this work the feasibility of electric-field tuning of the plasmonic spectrum of a gold nanodot array in a LC matrix was demonstrated. However, in this case the GNPs cannot be dispersed well in LC medium which has limited interest for future research. As emphasized before, in the study of LC/GNP hybrid system, one prerequisite is that GNPs dissolve well in LC to prevent strong aggregation or precipitation because the GNPs usually tend to aggregate and result in inhomogeneous dispersions which prevent to achieve proper and reproducible results. Therefore, building GNPs protected by functional organic molecules is a rapidly growing fascinating and challenging scientific area of contemporary interest. Coating the surface of GNPs can have several promising effects: (1) affecting the frequency of GNP SPR; (2) allowing GNPs to mix with other media without segregation; and (3) adjusting the GNP organizations that the organic coating can have specific abilities of triggering and tuning self-organization and finally forming self-assembling vehicles for the core GNPs. The external coating layer can also vary from one to more molecule layers, allowing accommodating further functionalization.

The synthesis of the organic molecules is tailored to the particular surface chemistry and morphology characteristics. A suitable anchoring group and structure directing groups are critical in preparing such hybrid GNP structures. There are two main methods of preparing these hybrid GNPs. The first one is direct synthesis of the GNPs in the presence of the organic molecules based on the Brust procedure [45]. The second one is a two-step process in which the GNPs are first synthesized with a protective layer (e.g. a simple alkane thiol), then endure a molecule exchange step offering the product [25]. Normally, the first method gives a GNP surface exclusively coated with the target molecules, whereas the second method gives products with mixed molecules as coating. For the prepared products, the purity of the final sample is one of the most significant factors as the excess free surfactants or molecules could interfere with LC media. Typically, purification of the hybrids is undertaken through methods of repeated centrifugation, chromatography techniques, precipitation, or a combination of these techniques. To characterize the product purity and molecule-grafting rates/ratios, the techniques such as NMR spectroscopy, UV–vis spectroscopy, thermal analysis (DSC/TGA), and elemental analysis are adopted.

The synthesized GNPs have been dispersed into a variety of host LCs and comprehensively characterized by multiple techniques. For example, the transition temperatures of the GNPs coated with mesogen **1** (Scheme 4.1) in 4′-n-octyl-4-cyano-biphenyl (8CB), which exhibits nematic and smectic A phases, were determined as a function of concentration by DSC, as shown in Table 4.1 [46]. For a 5 wt% mixture of **1** coated GNPs in 8CB, by using polarized optical microscope

Scheme 4.1 Chemical structure of mesogen **1**

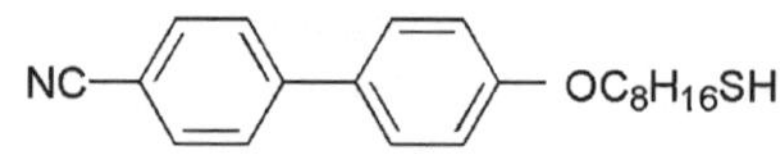

1

Table 4.1 GNPs doping concentration dependent SmA-N and N–I transition temperatures

Concentration (wt%)	SmA-N Transition (°C)	N–I Transition (°C)
0 (pure 8CB)	34.0	41.3
2	33.8	41.2
5	33.5	40.9
10	33.0	40.6
15	32.8	40.7
20	32.8	40.4

(POM), the *Schlieren* and homeotropic defect textures of the nematic phase, and the focal conic and homeotropic defect textures of the smectic A phase could be observed (Fig. 4.3). In addition, the parallel and perpendicular dielectric permittivities of mixtures of the GNPs dispersed in host nematic LCs were determined by using "one-cell method", which was reported by Clark et al. [47–49].

In the following part, examples of recent development in LC/GNP hybrid materials will be briefly introduced and discussed. Different molecules with thiol groups have been used to cap GNPs to mix with LC hosts for resolving the dispersion problem. Either coated by simple *n*-alkyls or mesogens, GNPs could mix with LC media and form LC phases with long-range orientational or positional order. Both molecular structures can affect the type of the phase. The most common and simple choices are *n*-alkyl thiols [46, 50], which bind covalently to the gold surface via Au–S linkages with the head thiol groups, extending *n*-alkyl chains more or less perpendicular to the GNP surface. Mesogenic capping agents are expected to provide better compatibility with similar molecular structures as the LC hosts.

4.3.2 Rod-Like Mesogen Coated GNPs

Generally, rod-like nematic mesogenic molecules can be coated on GNPs via end-on attachment. The LC property of the molecule may be transferred to the hybrid product. This is an effective direct method for bringing GNPs in LC phases. It is noteworthy to point out that all of the LC molecules used so far have a flexible spacer between the rigid rod-like mesogen and the functional head group (i.e. thiol) used to attach on the GNP surface. Flexibility is an important property which allows the molecules to deform easily. This could enable the hybrid GNPs to form anisotropic shape which is crucial for the arrangement in LC phases. The first GNPs covered with rod-like LC molecules were reported by Ikeda et al. in 2001 [51]. The

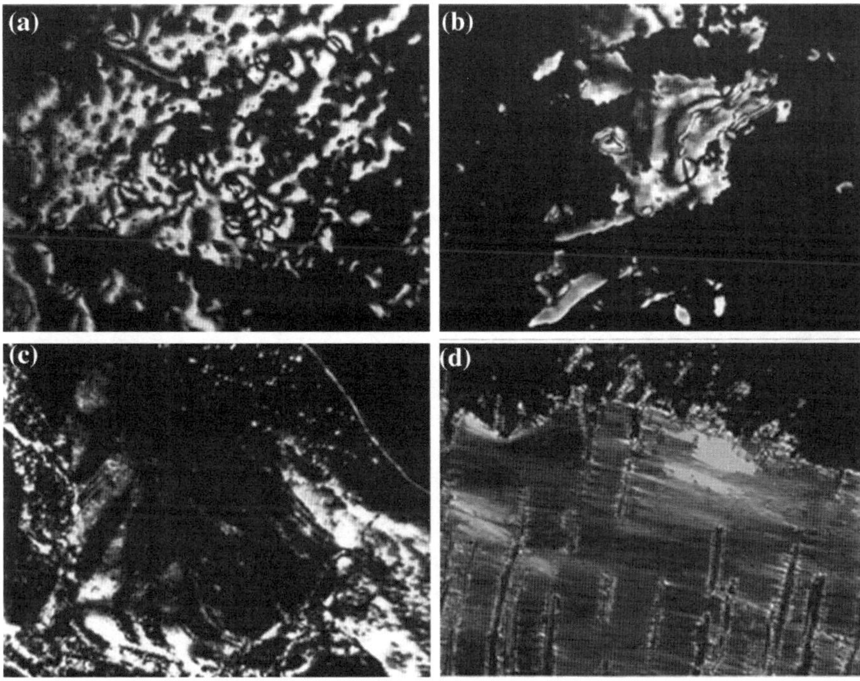

Fig. 4.3 **a**, **b** Homeotropic and *schlieren* textures of the nematic phase of 8CB containing 5 wt% **1** coated GNPs. **c** Sheared nematic texture near to the edge of the cover slip. **d** Focal-conic defect texture of the smectic A phase. Magnification: ×100 [46]. Copyright from Wiley–VCH 2011

authors described the formation of a "mesoscopic" phase for GNPs capped with a monotropic nematic cyclohexylphenoxy thiol derivative. Through direct synthesis of 3 nm GNPs in the presence of a rod-like cyclohexyl phenoxy thiol **2** (Scheme 4.2), a densely packed surface coating could be achieved [51]. The hybrid material not only showed both smectic and nematic phases but also displayed higher transition temperatures and wider LC phase ranges than the free molecules [51]. Many other rod-like mesogenic structures (e.g. **3–9**) were also reported to attach to the GNP surface, respectively (Scheme 4.2) [52–54]. For more examples, GNPs were exclusively capped with cyanobiphenyl thiol molecules **8** and these LC-functionalized GNPs exhibited an unidentified LC phase, which showed a spontaneous one-dimensional (1D) arrangement [53]. Interestingly, after thermal treatment (first isothermal melting at 140 °C for 3 h, second aging at 125 °C for 4 h, and then quenching to room temperature), the sample showed a distinct 1D ordered GNPs, as shown in Fig. 4.4. The length of the array was 13–60 nm and the inter-array distance was about 7 nm. Photochemical activity of lipophilic azobenzene thiols **9** on GNP surface were investigated [54]. There was a chromophore-GNP distance dependent photoisomerization and photodimerization efficiency in these systems. More ordered LC phase were developed by LC/mesogen coated GNPs. As

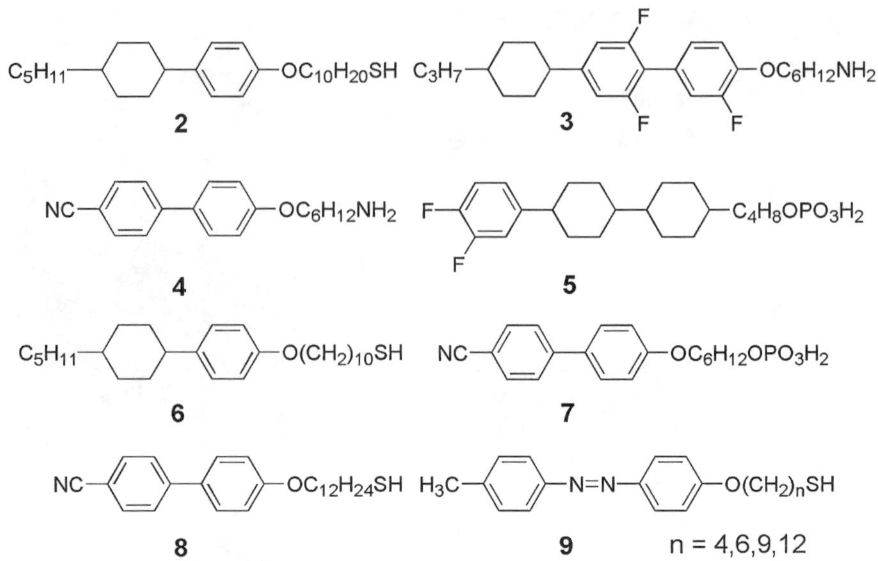

Scheme 4.2 Chemical structures of rod-like mesogen capping agents **2–9**

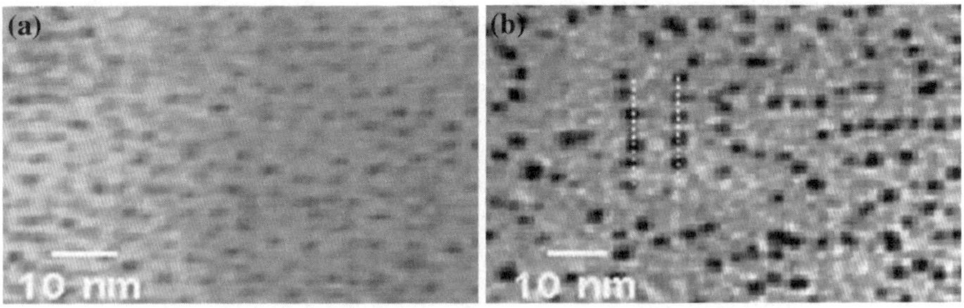

Fig. 4.4 TEM images of GNPs with LC molecules **a** before and **b** after thermal treatment [53]. Copyright from Royal Chemical Society 2005

shown in Fig. 4.5, a tentative structural model was proposed in which the GNPs exhibit a local rectangular (or hexagonal) arrangement. There was a local smectic A_2 order for the head-to-tail regions and local smectic A_d order for the equators.

In addition to end-on attachment, there is also side-on attachment of rod-like molecules onto GNPs. Through laterally substituted thiol, the rod-like nematic LC molecule **10** containing a lateral alkyl thiol was linked to GNPs together with some other *n*-alkane thiols forming a mixed monolayer (Scheme 4.3), which had two components in nearly equal proportions (ca. 1:1) on GNP surface (Fig. 4.6). Both the hexane- and dodecanethiol coated hybrids GNPs exhibited nematic LC behavior which displayed typical marbled or *Schlieren* textures (Fig. 4.7) [55].

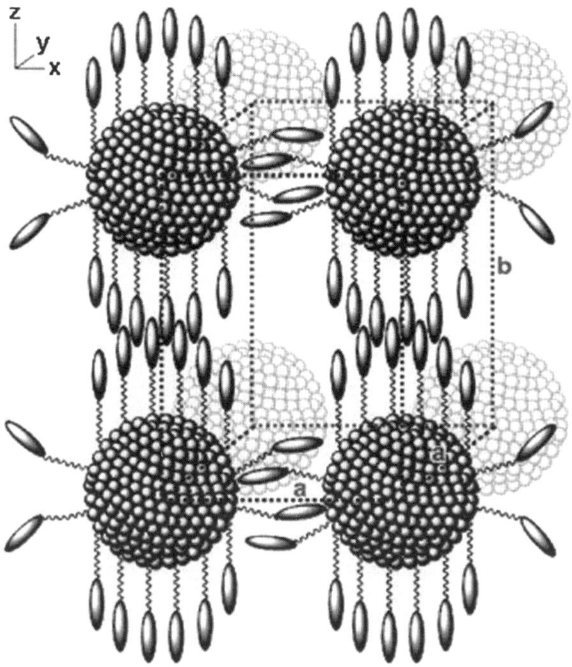

Fig. 4.5 A simplified illustration of the local rectangular arrangement of nanoparticles in a condensed mixed phase. Reprinted with permission from [46]. Copyright from Wiley–VCH 2011

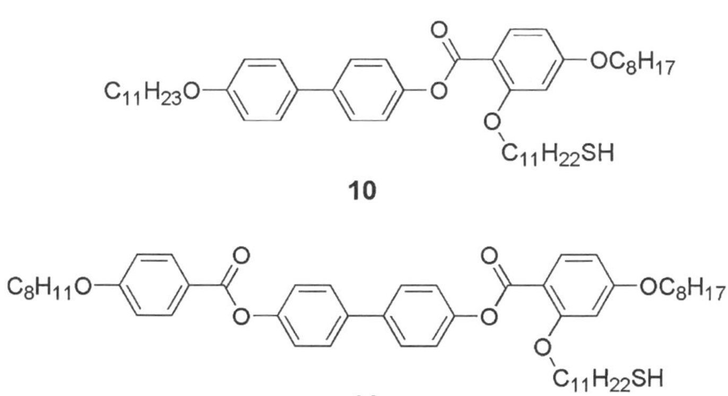

Scheme 4.3 Chemical structures of side-on mesogen capping agents **10–11**

This approach prompted further work with a slightly longer rod-like nematic LC molecule **11** (Scheme 4.3) with side-on thiol [56]. The chain length plays an important role in defect formation, as well as on melting temperature. The shorter chain provides easier defect formation and the longer chain invokes potential

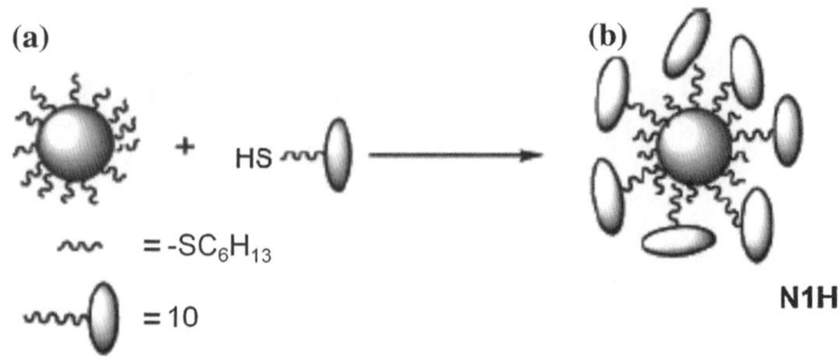

Fig. 4.6 Schematic representations of the GNP place exchange reaction yielding N1H [55]. Copyright from ACS 2006

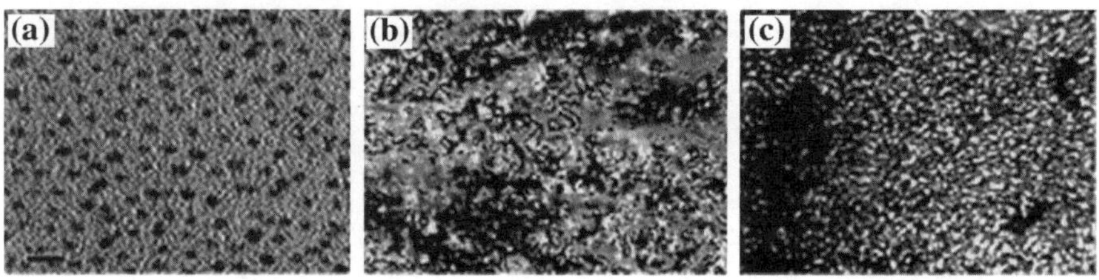

Fig. 4.7 TEM picture of the N1H (*left*) on a graphite surface, scale bar: 5 nm; POM pictures of **10** (*middle*) and N1H (*right*) (magnification for POM pictures ×100) [55]. Copyright from ACS 2006

plastifying effect. The effect has been investigated, and the importance of the connectivity and morphology of the molecules in hybrid GNPs orientation in LC phase has been clearly highlighted [57]. It is expected that by controlling the interparticle distances through molecule design and induced phase changes, one can modulate the electronic and magnetic properties of GNPs assemblies, which is critical for electronic and optoelectronic applications [58].

In a similar mixed monolayer system using **11**, the hybrid LC system exhibited highly ordered bulky arrays of GNPs (Fig. 4.8), investigated by using in-depth X-ray diffraction and transmission electron microscopy (TEM). The particles forming GNP strings could pack in rhombohedral, hexagonal or rectangular columnar lattices, which were surrounded by a nematic continuum formed by the laterally attached LC molecules [59].

To improve the compatibility of GNPs in LC host, both of the simple *n*-alkyl thiol and LC/mesogen molecules are used simultaneously to form mixed monolayer. For example, nematic cyanobiphenyl end-functionalized alkylene thiol covered GNPs and its mixing with *n*-alkyl thiols (n = 6, 12) in two cyanobiphenyl nematic LC hosts (5CB and 8CB) were investigated [60]. By adding small quantities of alkyl thiolate-capped GNPs there was effectively reversed nematic LC orientation which further showed opto-electronic responses [61]. Recently,

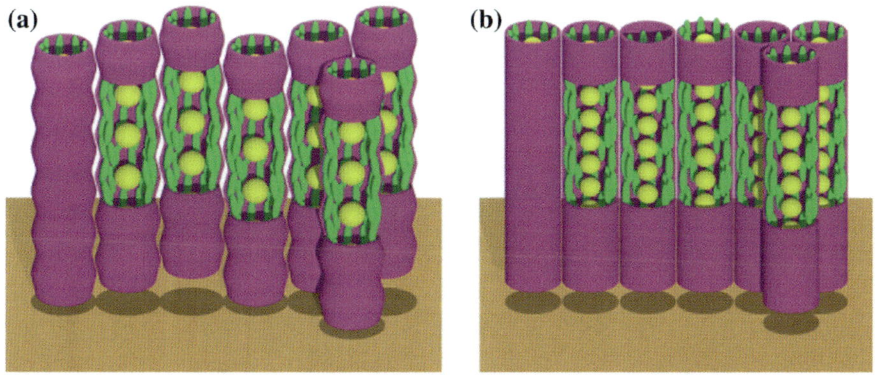

Fig. 4.8 Schematic models of gold string structure: **a** rhombohedral phase **b** hexagonal columnar phase; *yellow*: GNPs, *green*: mesogens [59]. Copyright from Wiley–VCH 2009

Goodby et al. reported that the adjustments in the procedure of synthesizing LC-capped GNPs were critical for achieving composites with well-dispersed GNPs and significantly altered electro-optic response [46]. In addition, chiral dopant-capped GNPs are efficient candidates to transfer chirality to non-chiral LC systems [62, 63], which may lead to the development of novel chiro-optical nanodevices. Chiral LC molecules were used to coat GNPs and there was an interesting work in this field made by Ajayan et al. who demonstrated a single-step, shape-selective, in situ synthesis and self-assembly of GNPs. Chiral glass-forming LC **12** was used as reducing and capping agents (Scheme 4.4). The resulting GNPs displayed LC properties at room temperature which can even form ordered hierarchies at multiple length scales (Fig. 4.9) [66].

4.3.3 Bent-Core Mesogen Coated GNPs

Bent-core LC molecules are one fascinating LC class featured by a banana-shaped or bent-core molecular shape which have attracted a great deal of interest for their exciting electro-optical behavior [65, 66]. When the bent-core molecule **13** was used (Scheme 4.5), a columnar phase of hybrid GNPs were obtained (Fig. 4.10) [50].

4.3.4 Mesogenic Dendron Coated GNPs

One crucial factor for hybrid GNPs is the relative size of the core to the organic shell. Generally, when the thickness ratio of surface molecule to GNP core increases, the shape directing effect of the organic shell becomes more prominent. This could be a useful method to tune the property of LC/GNPs hybrid materials.

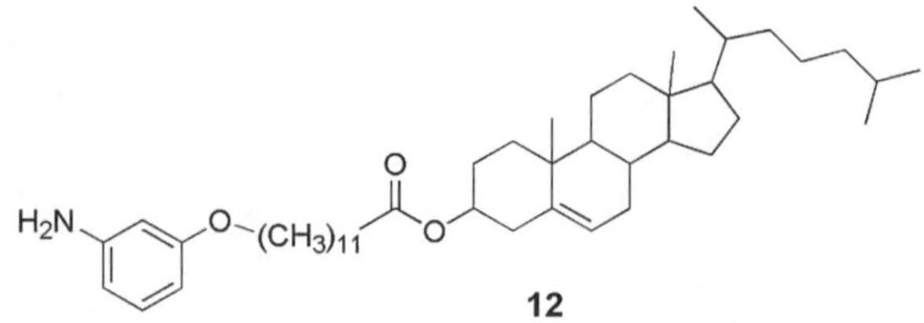

Scheme 4.4 Chemical structure of glassy chiral LC capping agent **12**

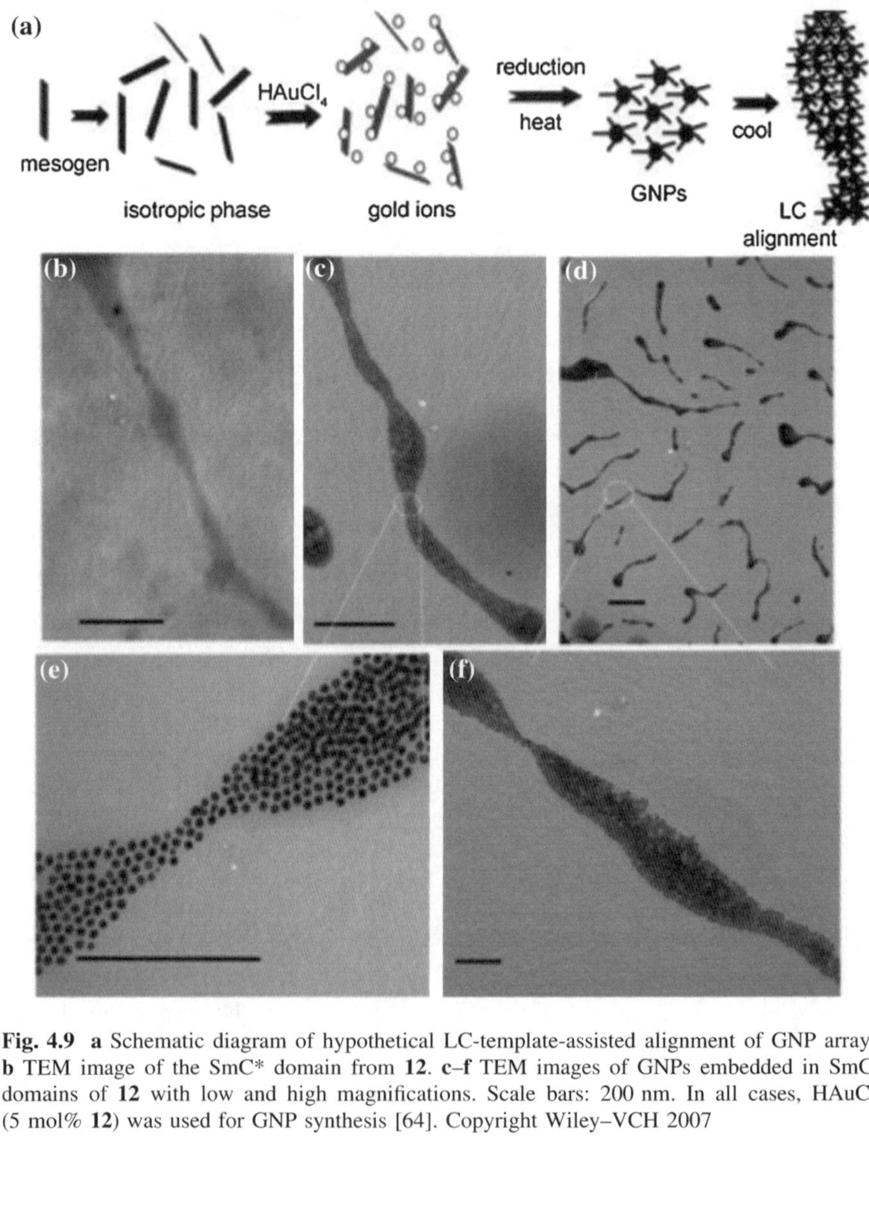

Fig. 4.9 **a** Schematic diagram of hypothetical LC-template-assisted alignment of GNP arrays. **b** TEM image of the SmC* domain from **12**. **c**–**f** TEM images of GNPs embedded in SmC* domains of **12** with low and high magnifications. Scale bars: 200 nm. In all cases, $HAuCl_4$ (5 mol% **12**) was used for GNP synthesis [64]. Copyright Wiley–VCH 2007

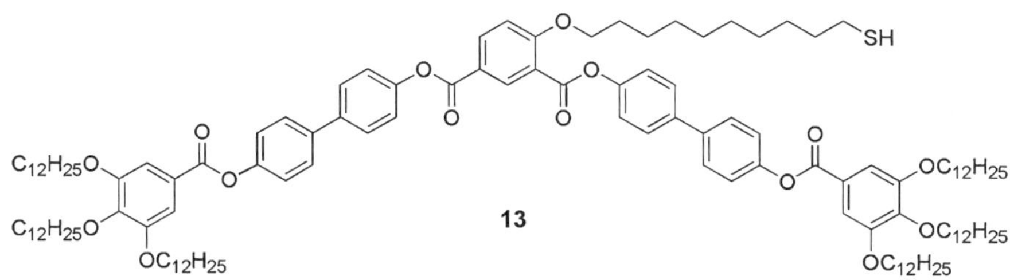

Scheme 4.5 Chemical structure of bent-core mesogen capping agent **13**

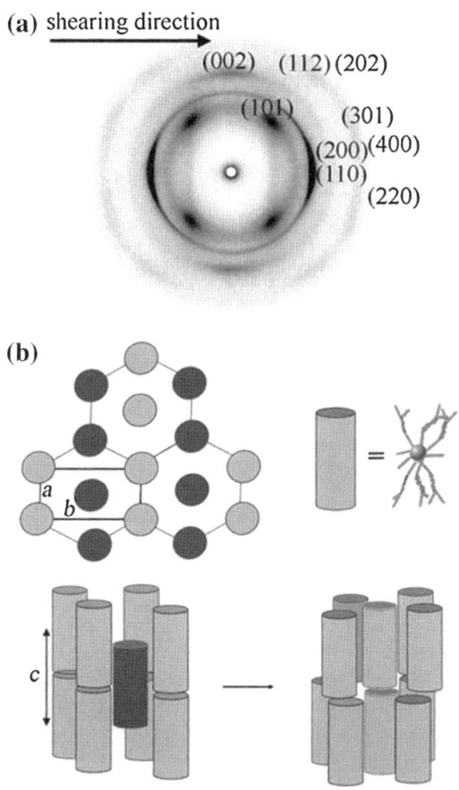

Fig. 4.10 **a** X-ray pattern of the columnar phase of an oriented sample of **13** coated GNPs. A columnar orthorhombic body-centered lattice was chosen for indexing of the signals. **b** Schematic drawing of the arrangement of nanoparticles in the columnar phase, as viewed from above (*top, left*) and from the side (*bottom, left*); **a**, **b**, and **c** are the crystal-lattice parameters. A gold cluster is located at the center of each cylinder, and the cylinder is filled by orientationally and positionally disordered *n*-alkyl and LC thiol molecules redistributed around the gold cluster. The two colors distinguish particles with their mass center (*gold cluster*) positioned at different levels along the **c** direction. *Bottom*: The smectic structure can be obtained from the columnar phase through movement of the particles from the intermediate layer along the column axis and decoupling of the layers [50]. Copyright from Wiley–VCH 2009

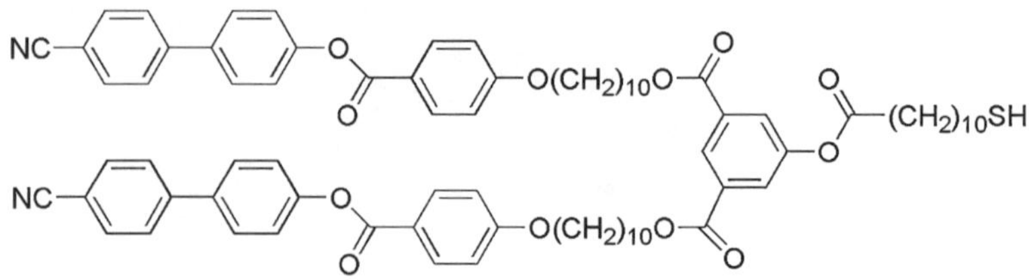

14

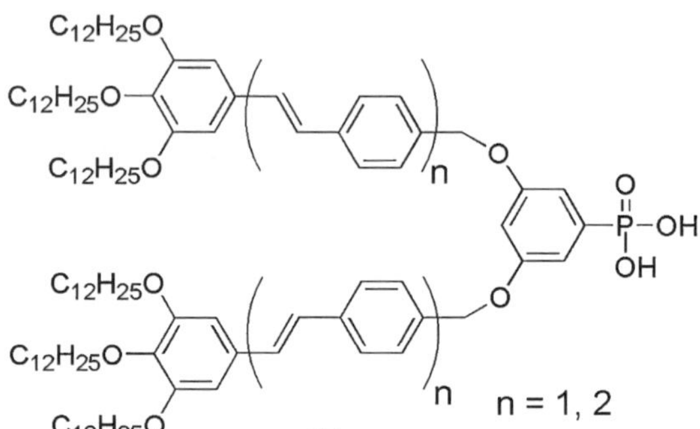

15

Scheme 4.6 Chemical structures of mesogenic dendron capping agents **14–15**

Adjusting the size ratio of the organic shell to the inorganic GNP core can be achieved by changing the sizes of either the organic molecule or the core. For large sized hybrid structures, there have been tested polymer-coated GNP hybrids [67–70]. Besides, due to the ability to control their structure and composition, particularly for the size, through additional generational growth, dendrimers and dendrons are other interesting types of organic materials, which have found applications such as catalyst, molecular electronics, nanomedicine [71], and LC applications [34]. Dendrons including showing LC phase coated GNPs (e.g. **14–15)** (Scheme 4.6) have been investigated [72]. Interestingly, even if the dendrimers **16–18** do not form LC phases themselves (Scheme 4.7), their coated GNP hybrid systems were able to exhibit LC phases. The dendrimer-functionalized GNPs can exhibit a thermotropic cubic phase and 2D hexagonal arrangement on a surface [73]. Remarkably, these hybrid GNPs could self-assemble into a hexagonal lattice on the TEM grid. The mean distances between the first and second neighbors were about 7.6 and 13.7 nm (Fig. 4.11).

4.3.5 Disc-Like Mesogen Coated Gold Nanoparticles

Discotic LC, based on the molecules having disc-like shapes [22, 23, 74], have attracted much attention in recent years owing to their inherent ability to form self-organized columnar structures, which can provide superior charge transport ability

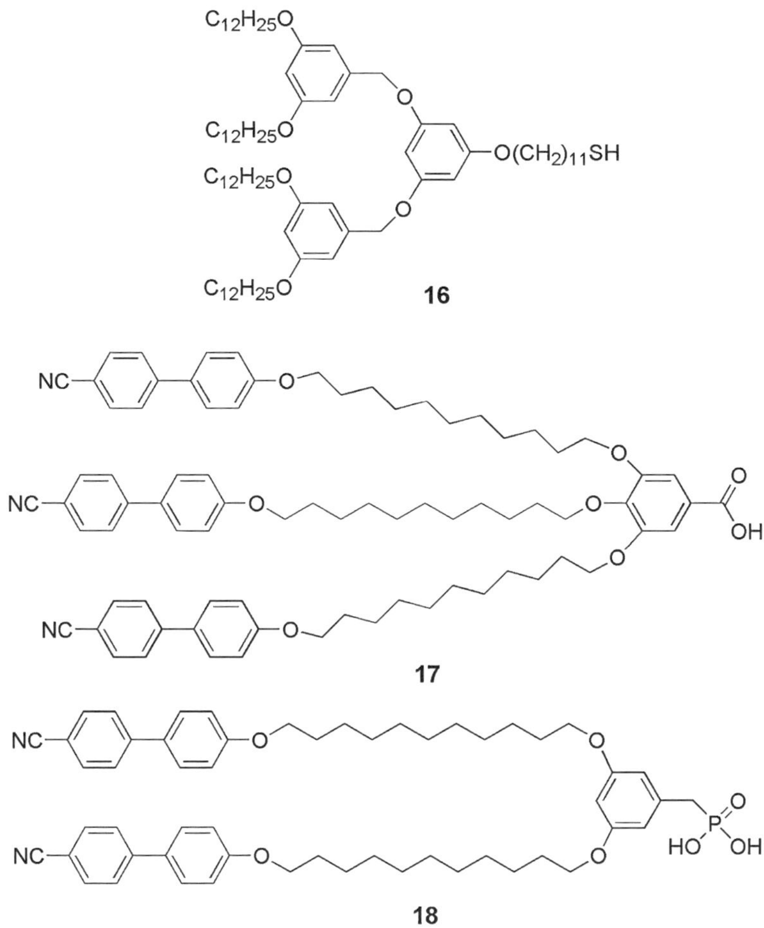

Scheme 4.7 Chemical structures of mesogenic dendron capping agents **16–18**

along the column direction. They have potential uses in opto-electronic devices [22, 74, 75]. There have been some reports but not many in discotic molecules coated GNPs so far. Kumar et al. showed triphenylene-based discotic LC decorated GNPs forming regular hexagonal patterns as indicated by TEM [76]. Although it appeared that discotic LC motifs on the corona of GNPs did not display LC superstructures similar to calamitic moieties, dispersion of small amount of such triphenylene-capped GNPs (**19**, n = 5) (1 % by weight) in the columnar phase of hexaheptyloxy triphenylene (H7TP) was found to increase the DC conductivity of H7TP more than 10^6 times (Scheme 4.8) [76]. There were other similar molecules **19** (n = 4, 8, 12) coated GNPs [80], which were able to self-assemble into well-ordered 2D structures with hexagonal, square-like and

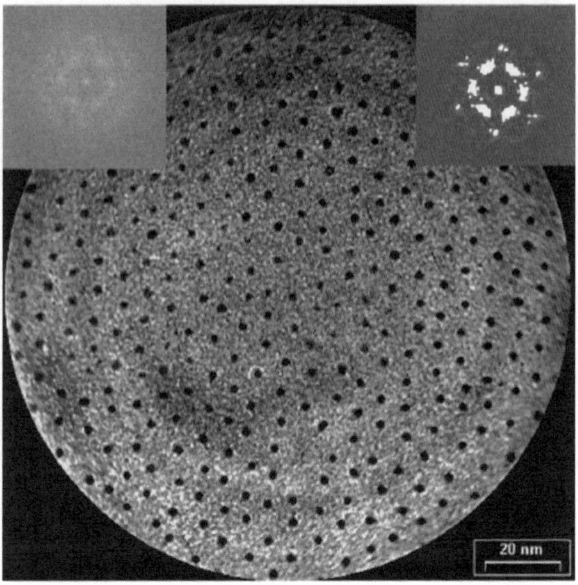

Fig. 4.11 GNPs self-organized in a hexagonal lattice as observed by TEM. Reprinted with permission from [73]. Copyright from Wiley–VCH 2007

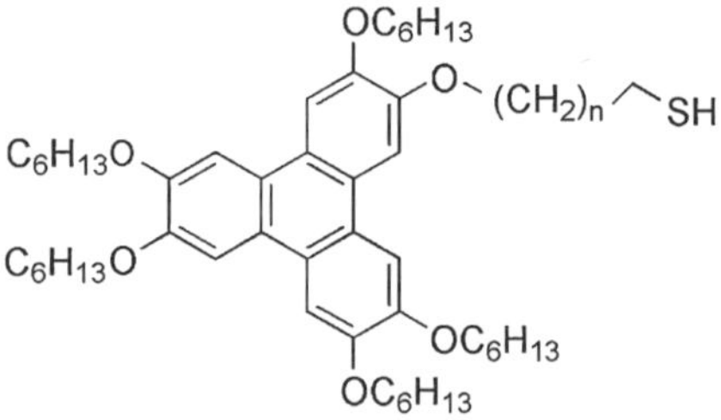

19 n = 4, 8, 12

Scheme 4.8 Chemical structure of disk-like mesogen capping agents **19**

stripe-like arrangements depending on the length of the tethering group and the solvent polarity (Fig. 4.12).

4.3.6 Hybrid Gold Nanorods

Besides spherical GNPs, GNRs are providing many promising applications in optics [78], sensors [79], biological imaging [80] and anticancer agents [81]. Due to their extraordinary shape- and surface chemical environment-dependent optical properties, they are among the most exciting materials to date. They are quite different from the widely investigated spherical GNPs [25], including more distinguished physical properties [37, 82] particularly for their tunable absorption in the visible

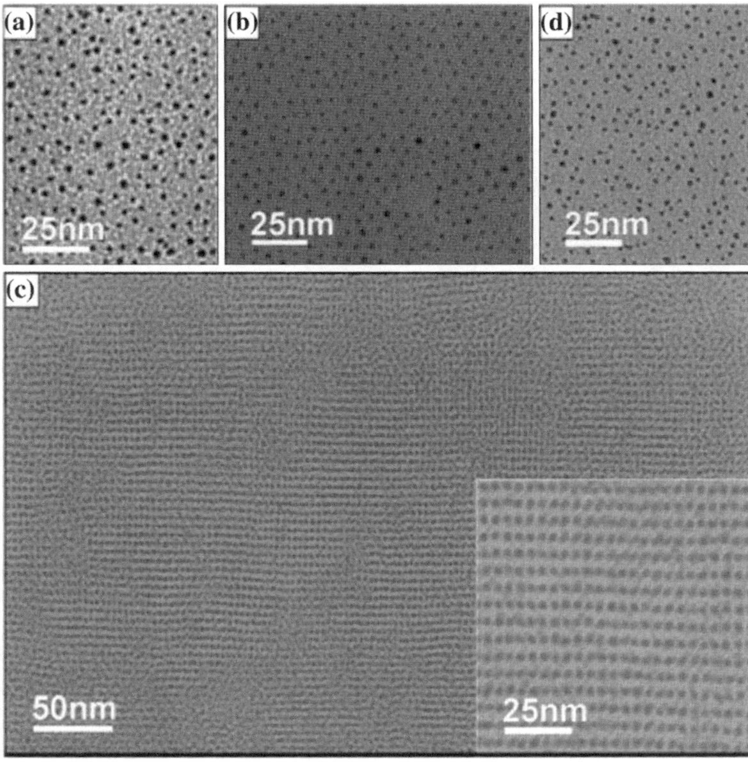

Fig. 4.12 TEM images of **19** (n = 12) coated GNPs prepared from aged solutions stood for 10 days in solutions of **a** 1:1 MeOH/toluene (disordered), **b** 2:1 MeOH/toluene (hcp), **c** 3:1 MeOH/toluene (stripe like; inset was an expansion), **d** 4:1 MeOH/toluene (disordered). Reprinted with permission from [77]. Copyright from ACS 2007

and near IR region. Moreover, since anisotropic metal nanoparticles can give higher sensitivity than spherical ones in surface plasmon shift, GNRs are highly suitable for plasmon sensing with a high-value shape factor (surface curvature) [83]. Nanoparticle shape also plays an important role in surface-enhanced Raman scattering enhancement (SERS). The enhancement factors on the order of 10^4–10^5 have been observed for absorbed molecules on the GNRs, while no such enhancement was observed on GNPs under similar conditions [84]. Moreover, the SPR of GNRs allows them to concentrate and manipulate light depending on their size, shape and proximity. If GNRs are self-assembled, they could couple and enhance local electric fields by a factor as high as 10^5 [85]. The resultant near-field enhancements can be utilized in sensors [86]. Controlling and tuning GNR self-assembly are highly interesting and vital due to their further potential contribution to the preparation of metamaterials with unusual electromagnetic properties [10]. For example, it has been simulated that continually self-assembled GNRs can be an ideal candidate to form aligned wires in a desired medium for cloaking devices [12].

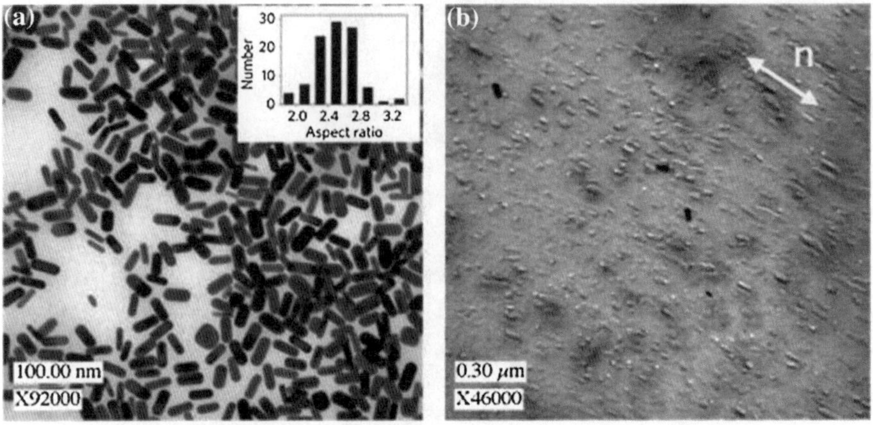

Fig. 4.13 **a** TEM micrograph of GNRs. The inset shows the distribution of measured aspect ratios of GNRs which were of average 53 nm long and 21 nm in diameter. **b** FFTEM image revealing the orientation of GNRs in a lyotropic LC matrix evidenced by unidirectional alignment of GNR replicas in the fracture plane [87]. Copyright from ACS 2010

LC as a host material would be a viable candidate for such purpose, since not only LC can provide a long range orientational order to GNRs but also its orientation capable of responding to an external field such as mechanical stress, electric field, and magnetic field can provide realignment and reconfiguration of GNRs. Lyotropic LC hosts can be quite advantageous in ensuring a high loading of GNRs. For instance, Liu et al. dispersed and aligned GNRs in nematic and hexagonal host phases [87]. Since the director can be macroscopically aligned and later realigned by magnetic fields in the nematic phase or shearing flow in the columnar phase (Fig. 4.13), the plasmonic response could be tuned by changing the nanorod orientation.

It is well established that the surface modification of GNPs can offer versatile means to introduce functionalities, more than just to stabilize the nanoparticles. For example, both fluorescence quenching and enhancement can exist when controlling organic chromophore in appropriate distances to the surface of GNPs [88]. In addition, some functional molecules on the surface of GNPs could work as molecular "glue" to assemble nanostructures [89]. Various thiols have been applied to prepare functional GNRs [90–93]. GNRs assembling in specific patterns, including end-to-end and side-by-side fashions can exhibit unique collective properties which are different from those of both individual GNR and bulk materials [33, 94, 95]. Currently, most molecular 'glues' are in ionic form, which are less stable and solvent dependent, i.e. only compatible with aqueous media [96]. In contrast, non-ionic interactions are able to build stable GNR self-assemblies in organic media, which offered a wider range of potential research avenues. In our study, discotic molecules were used to attach on GNR surface, e.g. perylene **20** [97], porphyrin, and porphyrin metal complex **21** [98, 99] (Scheme 4.9). They

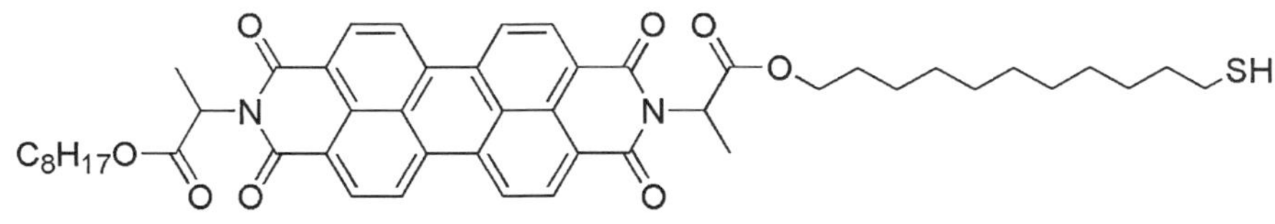

20

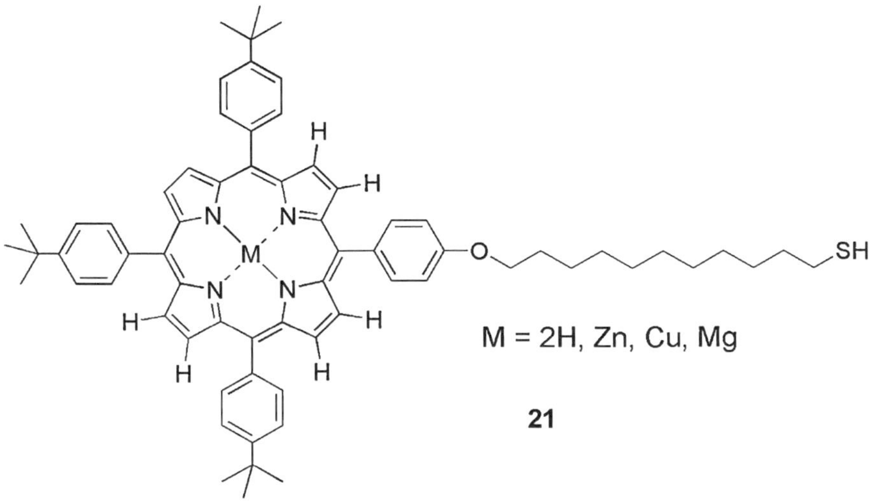

21

Scheme 4.9 Chemical structures of perylene or porphyrin capping agents **20–21**

were effective in fluorescence quenching and altering GNR SPR and guiding to self-assemblies, as shown in Fig. 4.14.

For the prepared **20** and **21** capped GNRs, their chemical structures can be characterized by IR, Raman, EDAX, and typically ^{1}H NMR. When attached on GNR, the ^{1}H NMR spectra of the organic monolayer protected gold clusters generally showed broadened peaks compared to those of free molecules. The broadening effect was the weakest for terminal nuclei and strongest for nuclei closest to the gold core [100]. The reasons for this broadening effect have been raised [101]: (a) the tight packing of protons close to the gold core causes rapid spin–spin relaxation from dipolar interactions; (b) there are different chemical shifts for surface heterogeneities (different nanocrystalline faces: vertexes, edges, terraces) [102] and the chemical shifts vary with core size and defect [103]; and (c) slow rotational diffusion of the clusters (analogous to effects seen for large proteins) depending on the nanoparticle size [100, 104]. For the resulting hybrid structures, they presented unique UV–vis and fluorescence spectra. The chromophores attached on GNR surface altered the SPR of GNRs. Also, the GNRs quenched the fluorescence of the organic chromophores due to the energy transfer between the GNR and chromophores.

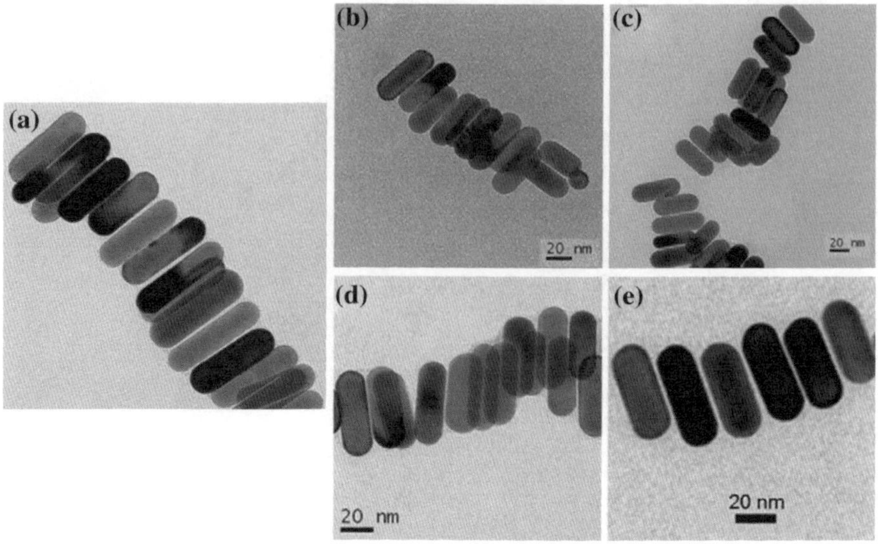

Fig. 4.14 TEM images of side-by-side self-assemblies of GNRs induced by discotic mesogens via π-π intermolecular interactions. **a 21** coated GNRs; **b–e 20** coated GNRs, n = Zn, Cu, Mg, 2H, respectively [97, 98]. [97] Copyright from ACS 2012; [98] Copyright from Elsevier 2013

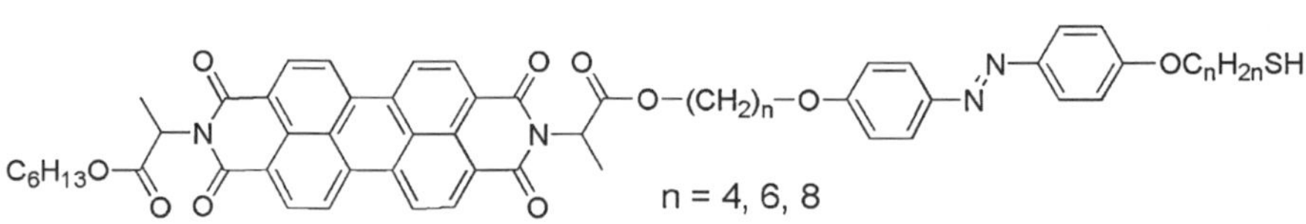

22

Scheme 4.10 Chemical structure of photoresponsive perylene capping agents **22**

Furthermore, inducing GNR alignment in thin films is of particular importance for their practical applications [105, 106]. So far, aligning GNRs in LC host has not been explored much. Although lyotropic LC host has been used for such purpose, GNRs in organic media is more appealing than dispersing them in aqueous media because their low interfacial energies allow for a high degree of control during solution and surface processing. Thus, to enhance GNR's stability and compatibility with organic LC media, hydrophobic thiol monolayer-protected GNRs in thermotropic LCs would be desirable. A series of new thiols **22** (n = 4, 6, 8) comprising a perylene diimide unit (PDI) and a photoresponsive azobenzene moiety linked by flexible alkylene chains of different length were synthesized to stabilize and functionalize GNRs (Scheme 4.10). To the best of our knowledge, this was the first report on photo-modulated self-assembly of GNRs. Here, with π-π intermolecular interactions which can lead to GNR self-assemblies, these GNR

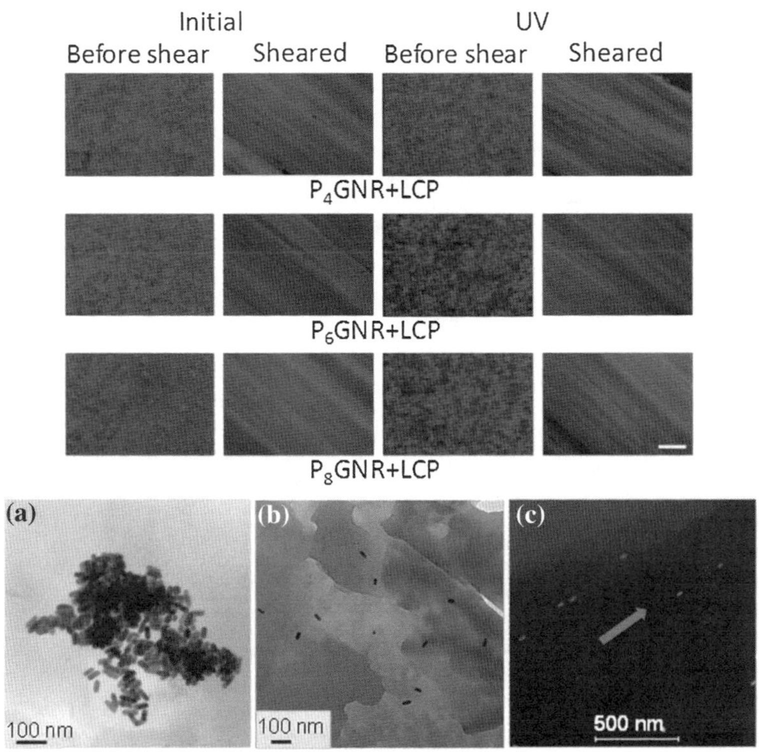

Fig. 4.15 (*Top*) Images of 0.5 wt% P_nGNR (**22**) in LCP thin films under crossed polarized optical microscope. Scale bar: 100 μm. Shearing direction was about 45° to the polarizer [107]. (*Bottom*) TEM images of **22** (n = 8) coated GNRs in LCP. **a** Before UV irradiation. **b** After UV irradiation. **c** After shearing. **c** was from scanning TEM. Copyright from ACS 2013

self-assemblies can be further modulated by photoresponsive azobenzene group, based on the perylene-azo binary structure **22** [107]. Additionally, when mixing in perylene LC, the GNRs with weaker interaction after photo stimuli could disperse and further be aligned (Fig. 4.15). Thus, it brings the chance that the GNR self-assembly can be tuned by light irradiation.

4.4 Applications

To explore novel LC/GNP hybrid materials for a new generation of LC devices, it is important to systematically investigate the structure–property relationships between GNPs and LCs, particularly for alignment and electro-optical responses. LC can organize the dispersed GNPs in various ways depending on the GNP size, shape and how their surface coating interacts with the LC host. For example, an

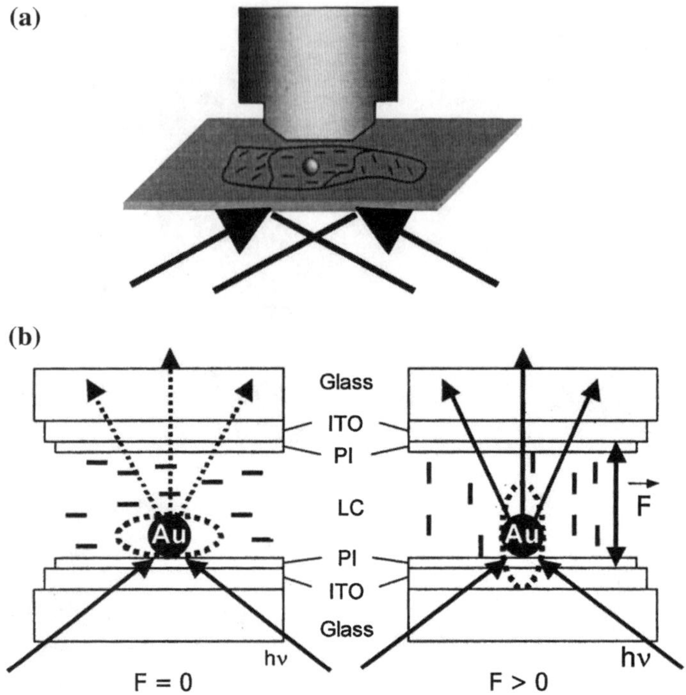

Fig. 4.16 **a** GNPs were illuminated with white light using a darkfield condenser. Only scattered light was detected by the objective lens. The same nanoparticle was imaged before and after coverage with a LC. **b** Sketch of an electrically controlled light scattering device. GNPs were embedded in a LC cell coated with ITO contact and polyimide alignment layers. Application of an electric field F reorients the LC. The optical anisotropy was indicated by the dotted ellipsoid [108]. Copyright from American Institute of Physics 2002

electrically controlled light scattering by GNP in a nematic LC medium was demonstrated (Fig. 4.16) [108]. The idea was to embed GNPs in an electro-optical LC material to induce a spectral shift of the SPR via an electric field. The GNPs become optically spheroidal when covered by an anisotropic LC and the two particle plasmon resonances of the optically spheroidal GNPs can be spectrally shifted by up to 50 meV when 10 kV/cm and higher electric fields were applied.

For the LC/GNP hybrid materials, a most attractive potential application probably is metamaterials. Metamaterials obtain their unusual properties which are unachievable with conventional materials based on their specially designed micro/nano structures, depending on the precise shape, geometry, size, orientation and arrangement. Structural elements of sub-wavelength sizes smaller than the wavelength of light or sound waves are incorporated into metamaterials. Particularly, with LC tailor-designed nanoparticles and their self-organized structures, novel and unique optical properties can be realized, which could be used to produce negative refractive index materials (NIMs) functioning in the optical wavelength regime for cloaking or perfect lenses [12, 13, 109, 110]. Typically, it requires conductive rods or incomplete rings (split ring resonators) nano-structures in regular arrays over large scales. Since the metamaterials typically rely on the subwavelength inclusions ordered on the length scale smaller than the wavelength, fabrication becomes a challenge for optical metamaterials to move to shorter wavelengths. The more interesting visible wavelength systems in this regime require inclusions of tens of nm in size spaced at only several times of their size. Even under the best circumstances at present, fabrication challenge of large area (scale-up) and three-dimensional (3D) patterning have not been figured out. Tremendous potential for LC in metamaterial system has been realized. A wide range of self-assembly approaches are considered and LCs are particularly significant because they have potential for providing self-assembled templates for optical metamaterials in this range, where they could be employed to structure functional inclusions in order to realize designed materials and their effective medium response. Results on the creation of ordered arrays of nanoparticles facilitated by LC surface coating have shown significant progress toward fabricating artificial dielectric materials at optical frequencies (Fig. 4.5) [46]. The research on LC-driven positioning of these particles is very interesting and there are large efforts being devoted to this goal. In order to reach this goal, there are specific factors that need to be considered: the small particle sizes (nanometer range), promising and sufficient alignment of anisotropic metal nanoparticles (e.g. GNRs), well-ordered position/configuration control and the stability of the nanostructures as there were strong fluctuations in LC host. The motivation for this research is to combine the properties of metamaterials with self-organization to create materials with unusual properties. It relies on the synthesis and physical properties of GNPs with various functional adsorbates that will affect the topological shape, and self-organization of the particles, and as a consequence their physical properties. First of all, in LC host, GNPs have been coated with aliphatic chains and/or mesogenic groups, in attempts to create GNPs that were soluble and also mesomorphic in host LC systems [111]. Then, with driving forces either from external or internal stimuli, desirable arrangement of GNPs could be obtained. The bottom-up approach to build well-defined nanostructures is very powerful as it combines the extraordinary variety of morphologies that LC presents with functional entities, relevant to chemical, biological, optoelectronic, and photonic tasks, etc. These types of materials are of interest for the applications mentioned above, and their additional advantage is the possibility to possess anisotropic magnetic, optical and electronic properties [112].

For example, LC-organized split ring resonators is presented in Fig. 4.17 [113], where GNPs were trapped in disclination lines surrounding micron sized silica beads that were organized into an array by the nematic continuum phase. There were other developed interesting types of self-assembled nanostructures and it has been claimed that the next frontier toward commercialization was bringing the fabrication cost down with nanochemistry-based bottom-up self-assembly approaches [114].

When the LC director field is disturbed by nanoparticles, it exerts forces or torques on the nanoparticles to promote an alignment of anisometric particles. One

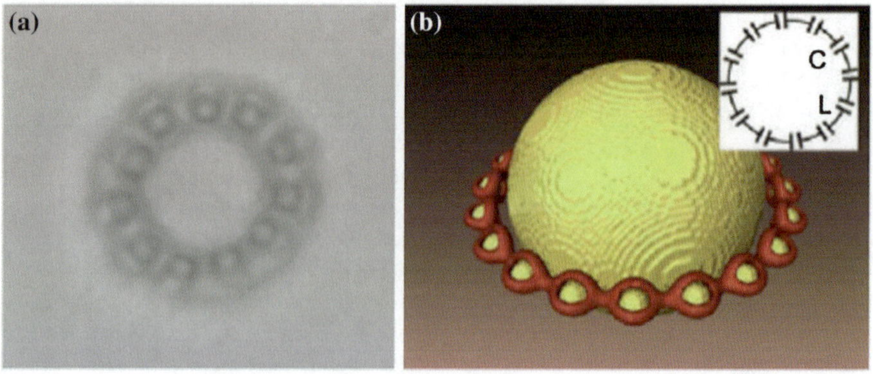

Fig. 4.17 **a** Colloidal GNPs (100 nm diameter) trapped into the Saturn ring disclination line surrounding a large spherical silica particle (1 mm diameter) suspended in a thermotropic N LC host. **b** The calculated stable colloidal ring superstructure for this combination of colloids (inset shows a schematic of the corresponding electrical circuit) [113]. Copyright from American Physical Society 2008

prominent example is the alignment of rod-shaped particles along the nematic director field, e.g. magnetic nanorods, initially considered theoretically by Pierre-Gilles de Gennes and Françoise Brochard [115].

There has been much interest in the investigation of LC phases containing nanoparticles of a broad range of elements, such as silver [116], semiconducting [117–120], silica [121], magnetic [75, 122–124] and ferroelectric nanoparticles [125, 126]. Apart from influencing the optical and electro properties of LCs, nanoparticles are also found to induce and affect the alignment of LCs [127, 128]. This has a significant potential for novel applications in many devices, e.g. the technique can be used to fabricate a flexible LC display requiring a low temperature process [127]. On the other hand, the LC assisted self-organization of metal nanoparticles may contribute to the research in metamaterials.

4.5 Conclusions

While the research of LC/GNP hybrid materials is still in its early stage, the exciting results indicate that these new materials can exhibit cutting edge interest and significance. By coating GNPs with suitably designed molecules, it is possible to adjust the physical properties of the hybrid materials. The properties of these hybrid LC/GNP materials depend both on the nature of the grafted molecules and the GNP cores varying in sizes and shapes. Although simple alkyl chains could provide compatibility of GNPs in LC, it calls for mesogenic molecules for future coating because they can provide compatibility to broader LC hosts and introduce multiple functions. Even without LC host, the mesogen coated GNPs could form

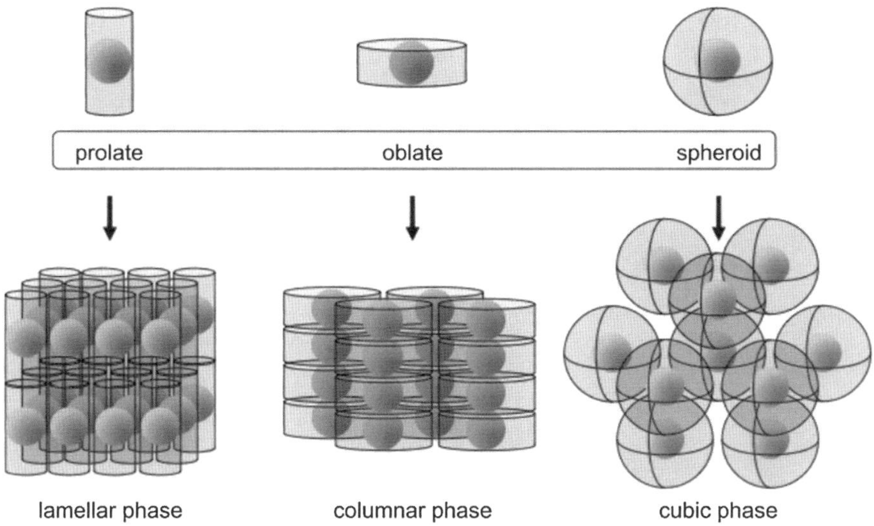

Fig. 4.18 Some of the various hybrid geometries and packing motifs possible upon molecule grafting to the surface of a spherical nanoparticle. The role of the molecules in migrating and deforming at the surface was crucial in providing these structures [129]. Copyright from Beilstein Journal of Organic Chemistry 2012

LC phases themselves, which provides an opportunity to form superlattice structures with stimuli responsive attributes. This is highly interesting for metamaterial fabrication. The surface species must have sufficient flexibility and mobility, and/or orientationally distortion so that the overall hybrid can exhibit the desirable compatibility and anisotropic effect necessary for the formation of LC phases when mixing in LC media. The functional groups of the surface molecules have influences on the phase behavior of the mixed materials. These factors highlight the pivotal role of the coating molecules in influencing the structure of the LC phase, allowing the nanoparticles to be arranged into a variety of phases, such as nematics, smectics, columnar and cubics (Fig. 4.18) [129].

Importantly, it is apparent that LC molecules are not always necessary for inducing LC properties in the GNP hybrids. The LC phases reported for systems involving non-LC molecule precursors indicate these LC hybrid materials are in a more holistic manner, whereby the whole system is more than simply just the sum of its composition parts. By combining the processability, compatibility, the structural and functional versatility of the surface organic coating, and controllable self-assembly of GNPs with ordered structures in LC phases, there appears unique optical and electrical properties including the emergence of new physical properties [130, 131]. The multifunctional LC/GNP hybrids are addressing exciting challenges of this century, including high-density recording media [132, 133], single-electron microelectronics [134] and charge transport devices [135], nanoscale plasmon waveguides [136] and metamaterials [46, 137]. It is anticipated the

LC/GNP hybrids will continually show tremendous opportunities in developing a variety of methods to produce exciting new materials for scientists to advance the field of nanotechnology.

Acknowledgments The preparation of this chapter benefited from the support to Quan Li by the Air Force Office of Scientific Research (AFOSR FA9550-09-1-0254 and FA9550-09-1-0193), the Department of Defense Multidisciplinary University Research Initiative (AFOSR MURI FA9550-06-1-0337 and FA9550-12-1-00370), the Department of Energy (DOE DE-SC0001412), the National Science Foundation (NSF IIP 0750379), the National Aeronautics and Space Adminstration (NASA), Ohio Third Frontier, and the Ohio Board of Regents under its Research Challenge program.

References

1. Q. Li (ed.), *Liquid Crystal Beyond Displays: Chemistry, Physics, and Applications* (John Wiley & Sons, New Jersey, 2012)
2. Q. Li (ed.), *Intelligent Stimuli-Responsive Materials: From Well-Defined Nanostructures to Applications* (John Wiley & Sons, New Jersey, 2013)
3. J.B. Pendry, Negative refraction makes a perfect lens. Phys. Rev. Lett. **85**, 3966–3969 (2000)
4. V.M. Shalaev, W.S. Cai, U.K. Chettiar, H.K. Yuan, A.K. Sarychev, V.P. Drachev, A.V. Kildishev, Negative index of refraction in optical metamaterials. Opt. Lett. **30**, 3356–3358 (2005)
5. P.A. Kossyrev, A.J. Yin, S.G. Cloutier, D.A. Cardimona, D.H. Huang, P.M. Alsing, J.M. Xu, Electric field tuning of plasmonic response of nanodot array in liquid crystal matrix. Nano Lett. **5**, 1978–1981 (2005)
6. V.K.S. Hsiao, Y.B. Zheng, B.K. Juluri, T.J. Huang, Light-driven plasmonic switches based on Au nanodisk arrays and photoresponsive liquid crystals. Adv. Mater. **20**, 3528–3532 (2008)
7. S. Khatua, P. Manna, W.S. Chang, A. Tcherniak, E. Friedlander, E.R. Zubarev, S. Link, Plasmonic nanoparticles-liquid crystal composites. J. Phys. Chem. C **114**, 7251–7257 (2010)
8. Z.Y. Tang, N.A. Kotov, One-dimensional assemblies of nanoparticles: preparation, properties, and promise. Adv. Mater. **17**, 951–962 (2005)
9. M.L. Curri, R. Comparelli, M. Striccoli, A. Agostiano, Emerging methods for fabricating functional structures by patterning and assembling engineered nanocrystals. Phys. Chem. Chem. Phys. **12**, 11197–11207 (2010)
10. K.J. Stebe, E. Lewandowski, M. Ghosh, Oriented assembly of metamaterials. Science **325**, 159–160 (2009)
11. K.J.M. Bishop, C.E. Wilmer, S. Soh, B.A. Grzybowski, Nanoscale forces and their uses in self-assembly. Small **5**, 1600–1630 (2009)
12. W. Cai, U.K. Chettiar, A.V. Kildishev, V.M. Shalaev, Optical cloaking with metamaterials. Nature Photon **1**, 224–227 (2007)
13. V.M. Shalaev, Optical negative-index metamaterials. Nature Photon **1**, 41–48 (2007)
14. P. Collings, M. Hird, *Introduction to Liquid Crystals* (Taylor & Francis, London, 1997)
15. C. Tschierske, Molecular self-organization of amphotropic liquid crystals. Prog. Polym. Sci. **21**, 775–852 (1996)
16. C. Tschierske, Non-conventional liquid crystals: the importance of micro-segregation for self-organization. J. Mater. Chem. **8**, 1485–1508 (1998)

17. R. Lipowsky, E. Sackmann (ed.), *Handbook of Biological Physics*, vol. 1. (Elsevier, Amsterdam, 1995)
18. B. Donnio, S. Buathong, I. Bury, D. Guillon, Liquid crystalline dendrimers. Chem. Soc. Rev. **36**, 1495–1513 (2007)
19. A. Skoulios, D. Guillon, Amphiphilic haracter and liquid crystallinity. Mol. Cryst. Liq. Cryst. **165**, 317–332 (1988)
20. J. Ma, Q. Li, Smectic Liquid Crystal Semiconductors, in *Self-Organized Organic Semiconductors: From Materials to Applications*, ed. by Q. Li (John Wiley & Son, New Jersey, 2011)
21. R.P. Lemieux, Chirality transfer in ferroelectric liquid crystals. Acc. Chem. Res. **34**, 845–853 (2001)
22. M. Mathews, Q. Li, Self-Organizing Discotic Liquid Crystals as Novel Organic Semiconductors, Chapter 4, in *Self-Organized Organic Semiconductors: From Materials to Device Applications*, ed. by Q. Li (John Wiley & Son, New Jersey, 2011)
23. S. Kumar, Self-organization of disc-like molecules: chemical aspects. Chem. Soc. Rev. **35**, 83–109 (2006)
24. R.A. Reddy, C. Tschierske, Bent-core liquid crystals: polar order, superstructural chirality and spontaneous desymmetrisation in soft matter systems. J. Mater. Chem. **16**, 907–961 (2006)
25. M.C. Daniel, D. Astruc, Gold nanoparticles: assembly, supramolecular chemistry, quantum-size-related properties, and applications toward biology, catalysis, and nanotechnology. Chem. Rev. **104**, 293–346 (2004)
26. Y. Xia, Y. Xiong, B. Lim, S.E. Skrabalak, Shape-controlled synthesis of metal nanocrystals: simple chemistry meets complex physics? Angew. Chem. Int. Ed. **48**, 60–103 (2009)
27. P.K. Jain, K.S. Lee, I.H. El-Sayed, M.A. El-Sayed, Calculated absorption and scattering properties of gold nanoparticles of different size, shape, and composition: applications in biological imaging and biomedicine. J. Phys. Chem. B **110**, 7238–7248 (2006)
28. S. Carnie, J.N. Israelachvili, B.A. Pailthorpe, Lipid packing and transbilayer asymmetries of mixed lipid vesicles. Biochim. Biophys. Acta **554**, 340–357 (1979)
29. R. Nagarajan, Molecular packing parameter and surfactant self-assembly: the neglected role of the surfactant tail. Langmuir **18**, 31–38 (2002)
30. X.L. Zhou, J.M. El Khoury, L.T. Qu, L.M. Dai, Q. Li, A facile synthesis of aliphatic thiol surfactant with tunable length as a stabilizer of gold nanoparticles in organic solvents. J. Colloid Interface Sci. **308**, 381–384 (2007)
31. B.E. Brinson, J.B. Lassiter, C.S. Levin, R. Bardhan, N. Mirin, N.J. Halas, Nanoshells made easy: improving Au layer growth on nanoparticle surfaces. Langmuir **24**, 14166–14171 (2008)
32. B. Nikoobakht, M.A. El-Sayed, Preparation and growth mechanism of gold nanorods (NRs) using seed-mediated growth method. Chem. Mater. **15**, 1957–1962 (2003)
33. T.K. Sau, C.J. Murphy, Room temperature, high-yield synthesis of multiple shapes of gold nanoparticles in aqueous solution. J. Am. Chem. Soc. **126**, 8648–8649 (2004)
34. B.M. Rosen, C.J. Wilson, D.A. Wilson, M. Peterca, M.R. Imam, V. Percec, Dendron-mediated self-assembly, disassembly, and self-organization of complex systems. Chem. Rev. **109**, 6275–6540 (2009)
35. I.M. Saez, J.W. Goodby, Chiral nematic octasilsesquioxanes. J. Mater. Chem. **11**, 2845–2851 (2001)
36. P.K. Vemula, V.A. Mallia, K. Bizati, G. John, Cholesterol phenoxy hexanoate mesogens: effect of meta substituents on their liquid crystalline behavior and in situ metal nanoparticle synthesis. Chem. Mater. **19**, 5203–5206 (2007)
37. C.J. Murphy, T.K. San, A.M. Gole, C.J. Orendorff, J.X. Gao, L. Gou, S.E. Hunyadi, T. Li, Anisotropic metal nanoparticles: synthesis, assembly, and optical applications. J. Phys. Chem. B **109**, 13857–13870 (2005)
38. L. Wang, X. Chen, Z. Sun, Y. Chai, Synthesis of gold nanoplates in lecithin lamellar liquid crystals. Can. J. Chem. Eng. **85**, 598–601 (2007)

39. L.Y. Wang, X. Chen, J. Zhan, Y.C. Chai, C.J. Yang, L.M. Xu, W.C. Zhuang, B. Jing, Synthesis of gold nano- and microplates in hexagonal liquid crystals. J. Phys. Chem. B **109**, 3189–3194 (2005)
40. H.M. Yang, M. Yang, Y. Zhang, G.X. Chen, In situ synthesis and lubrication of PbS nanoparticles in lamellar liquid crystal. Colloid J. **66**, 635–641 (2004)
41. S. Saliba, P. Davidson, M. Imperor-Clerc, C. Mingotaud, M.L. Kahn, J.-D. Marty, Facile direct synthesis of ZnO nanoparticles within lyotropic liquid crystals: towards organized hybrid materials. J. Mater. Chem. **21**, 18191–18194 (2011)
42. G. Surendran, F. Ksar, L. Ramos, B. Keita, L. Nadjo, E. Prouzet, P. Beaunier, P. Dieudonne, F. Audonnet, H. Remita, Palladium nanoballs synthesized in hexagonal mesophases. J. Phys. Chem. C **112**, 10740–10744 (2008)
43. R. Guo, B. Zhang, Y. Sun, X. Liu, Lyotropic liquid crystals and applications in synthesis of nanostructured materials. Prog. Chem. **19**, 1695–1702 (2007)
44. T.M. Dellinger, P.V. Braun, Lyotropic liquid crystals as nanoreactors for nanoparticle synthesis. Chem. Mater. **16**, 2201–2207 (2004)
45. M. Brust, M. Walker, D. Bethell, D.J. Schiffrin, R. Whyman, Synthesis of thiol-derivatized gold nanoparticles in a 2-phase liquid–liquid system. J. Chem. Soc. Chem. Commun. **7**, 801–802 (1994)
46. M. Draper, I.M. Saez, S.J. Cowling, P. Gai, B. Heinrich, B. Donnio, D. Guillon, J.W. Goodby, Self-assembly and shape morphology of liquid-crystalline gold metamaterials. Adv. Funct. Mater. **21**, 1260–1278 (2011)
47. M.G. Clark, E.P. Raynes, R.A. Smith, R.J.A. Tough, Measurement of the permittivity of nematic liquid-crystals in magnetic and electric-fields using extrapolation procedures. J. Phys. D Appl. Phys. **13**, 2151–2164 (1980)
48. S.T. Wu, D. Coates, E. Bartmann, Physical-properties of chlorinated liquid-crystals. Liq. Cryst. **10**, 635–646 (1991)
49. S. Murakami, H. Naito, Electrode and interface polarizations in nematic liquid crystal cells. Jpn. J. Appl. Phys. **36**, 2222–2225 (1997)
50. M. Wojcik, W. Lewandowski, J. Matraszek, J. Mieczkowski, J. Borysiuk, D. Pociecha, E. Gorecka, Liquid-crystalline phases made of gold nanoparticles. Angew. Chem. Int. Ed. **48**, 5167–5169 (2009)
51. N. Kanayama, O. Tsutsumi, A. Kanazawa, T. Ikeda, Distinct thermodynamic behaviour of a mesomorphic gold nanoparticle covered with a liquid-crystalline compound. Chem. Commun. 2640–2641 (2001)
52. M. Wojcik, M. Kolpaczynska, D. Pociecha, J. Mieczkowski, E. Gorecka, Multidimensional structures made by gold nanoparticles with shape-adaptive grafting layers. Soft Matter **6**, 5397–5400 (2010)
53. I. In, Y.W. Jun, Y.J. Kim, S.Y. Kim, Spontaneous one dimensional arrangement of spherical Au nanoparticles with liquid crystal ligands. Chem. Commun. 800–801 (2005)
54. J. Zhang, J.K. Whitesell, M.A. Fox, Photoreactivity of self-assembled monolayers of azobenzene or stilbene derivatives capped on colloidal gold clusters. Chem. Mater. **13**, 2323–2331 (2001)
55. L. Cseh, G.H. Mehl, The design and investigation of room temperature thermotropic nematic gold nanoparticles. J. Am. Chem. Soc. **128**, 13376–13377 (2006)
56. L. Cseh, G.H. Mehl, Structure-property relationships in nematic gold nanoparticles. J. Mater. Chem. **17**, 311–315 (2007)
57. J.W. Goodby, I.M. Saez, S.J. Cowling, J.S. Gasowska, R.A. MacDonald, S. Sia, P. Watson, K.J. Toyne, M. Hird, R.A. Lewis, S.-E. Lee, V. Vaschenko, Molecular complexity and the control of self-organising processes. Liq. Cryst. **36**, 567–605 (2009)
58. D.V. Talapin, J.-S. Lee, M.V. Kovalenko, E.V. Shevchenko, Prospects of colloidal nanocrystals for electronic and optoelectronic applications. Chem. Rev. **110**, 389–458 (2010)
59. X. Zeng, F. Liu, A.G. Fowler, G. Ungar, L. Cseh, G.H. Mehl, J.E. MacDonald, 3D ordered gold strings by coating nanoparticles with mesogens. Adv. Mater. **21**, 1746–1750 (2009)

60. H. Qi, B. Kinkead, V.M. Marx, H.R. Zhang, T. Hegmann, Miscibility and alignment effects of mixed monolayer cyanobiphenyl liquid-crystal-capped gold nanoparticles in nematic cyanobiphenyl liquid crystal hosts. Chem. Phys. Chem. **10**, 1211–1218 (2009)
61. H. Qi, B. Kinkead, T. Hegmann, Unprecedented dual alignment mode and Freedericksz transition in planar nematic liquid crystal cells doped with gold nanoclusters. Adv. Funct. Mater. **18**, 212–221 (2008)
62. T.G. Schaaff, G. Knight, M.N. Shafigullin, R.F. Borkman, R.L. Whetten, Isolation and selected properties of a 10.4 kDa Gold: Glutathione cluster compound. J. Phys. Chem. B **102**, 10643–10646 (1998)
63. H. Qi, J. O'Neil, T. Hegmann, Chirality transfer in nematic liquid crystals doped with (S)-naproxen-functionalized gold nanoclusters: an induced circular dichroism study. J. Mater. Chem. **18**, 374–380 (2008)
64. V.A. Mallia, P.K. Vemula, G. John, A. Kumar, P.M. Ajayan, In situ synthesis and assembly of gold nanoparticles embedded in glass-forming liquid crystals. Angew. Chem. Int. Ed. **46**, 3269–3274 (2007)
65. Y. Wang, H. Yoon, H.K. Bisoyi, S. Kumar, Q. Li, Hybrid rod-like and bent-core liquid crystal dimers exhibiting biaxial smectic-A and nematic phases. J. Mater. Chem. **22**, 20363–20367 (2012)
66. J. Etxebarria, M.B. Ros, Bent-core liquid crystals in the route to functional materials. J. Mater. Chem. **18**, 2919–2926 (2008)
67. J.C.P. Gabriel, P. Davidson, Mineral liquid crystals from self-assembly of anisotropic nanosystems. Colloid Chem. **1**(226), 119–172 (2003)
68. E.B. Barmatov, D.A. Pebalk, M.V. Barmatova, Influence of silver nanoparticles on the phase behavior of side-chain liquid crystalline polymers. Langmuir **20**, 10868–10871 (2004)
69. G.A. Shandryuk, E.V. Matukhina, R.B. Vasil'ev, A. Rebrov, G.N. Bondarenko, A.S. Merekalov, A.M. Gas'kov, R.V. Talroze, Effect of H-bonded liquid crystal polymers on CdSe quantum dot alignment within nanocomposite. Macromolecules **41**, 2178–2185 (2008)
70. D. Astruc, E. Boisselier, C. Ornelas, Dendrimers designed for functions: from physical, photophysical, and supramolecular properties to applications in sensing, catalysis, molecular electronics, photonics, and nanomedicine. Chem. Rev. **110**, 1857–1959 (2010)
71. G.R. Newkome, C.N. Moorefield, F. Vögtle, *Dendrimers and Dendrons: Concepts, Synthesis and Applications* (VCH, Weinheim, 2001)
72. S. Frein, J. Boudon, M. Vonlanthen, T. Scharf, J. Barbera, G. Suess-Fink, T. Buergi, R. Deschenaux, Liquid-crystalline thiol- and disulfide-based dendrimers for the functionalization of gold manoparticles. Helv. Chim. Acta **91**, 2321–2337 (2008)
73. B. Donnio, P. Garcia-Vazquez, J.-L. Gallani, D. Guillon, E. Terazzi, Dendronized ferromagnetic gold nanoparticles self-organized in a thermotropic cubic phase. Adv. Mater. **19**, 3534–3539 (2007)
74. L. Li, S. Kang, J. Harden, Q. Sun, X. Zhou, L. Dai, A. Jakli, S. Kumar, Q. Li, Nature inspired light-harvesting liquid crystalline porphyrins for organic photovoltaics. Liq. Cryst. **35**, 233–239 (2008)
75. Q. Li, L. Li, Photoconducting Discotic Liquid Crystals, Chapter 11, in *Thermotropic Liquid Crystals*, ed. by A. Ramamoorthy (Springer, New York, 2007)
76. S. Kumar, S.K. Pal, P.S. Kumar, V. Lakshminarayanan, Novel conducting nanocomposites: synthesis of triphenylene-covered gold nanoparticles and their insertion into a columnar matrix. Soft Matter **3**, 896–900 (2007)
77. Z. Shen, M. Yamada, M. Miyake, Control of stripelike and hexagonal self-assembly of gold nanoparticles by the tuning of interactions between triphenylene ligands. J. Am. Chem. Soc. **129**, 14271–14280 (2007)
78. C. Burda, X.B. Chen, R. Narayanan, M.A. El-Sayed, Chemistry and properties of nanocrystals of different shapes. Chem. Rev. **105**, 1025–1102 (2005)
79. C.Z. Li, K.B. Male, S. Hrapovic, J.H.T. Luong, Fluorescence properties of gold nanorods and their application for DNA biosensing. Chem. Commun. **31**, 3924–3926 (2005)

80. X.H. Huang, I.H. El-Sayed, W. Qian, M.A. El-Sayed, Cancer cell imaging and photothermal therapy in the near-infrared region by using gold nanorods. J. Am. Chem. Soc. **128**, 2115–2120 (2006)
81. C.C. Chen, Y.P. Lin, C.W. Wang, H.C. Tzeng, C.H. Wu, Y.C. Chen, C.P. Chen, L.C. Chen, Y.C. Wu, DNA-gold nanorod conjugates for remote control of localized gene expression by near infrared irradiation. J. Am. Chem. Soc. **128**, 3709–3715 (2006)
82. S. Link, M.A. El-Sayed, Spectral properties and relaxation dynamics of surface plasmon electronic oscillations in gold and silver nanodots and nanorods. J. Phys. Chem. B **103**, 8410–8426 (1999)
83. C. Yu, J. Irudayaraj, Quantitative evaluation of sensitivity and selectivity of multiplex nanoSPR biosensor assays. Biophys. J. **93**, 3684–3692 (2007)
84. B. Nikoobakht, J.P. Wang, M.A. El-Sayed, Surface-enhanced Raman scattering of molecules adsorbed on gold nanorods: off-surface plasmon resonance condition. Chem. Phys. Lett. **366**, 17–23 (2002)
85. P.K. Jain, M.A. El-Sayed, Plasmonic coupling in noble metal nanostructures. Chem. Phys. Lett. **487**, 153–164 (2010)
86. A.M. Funston, C. Novo, T.J. Davis, P. Mulvaney, Plasmon coupling of gold nanorods at short distances and in different geometries. Nano Lett. **9**, 1651–1658 (2009)
87. Q. Liu, Y. Cui, D. Gardner, X. Li, S. He, I.I. Smalyukh, Self-alignment of plasmonic gold nanorods in reconfigurable anisotropic fluids for tunable bulk metamaterial applications. Nano Lett. **10**, 1347–1353 (2010)
88. P. Anger, P. Bharadwaj, L. Novotny, Enhancement and quenching of single-molecule fluorescence. Phys. Rev. Lett. **96**, 113002 (2006)
89. A.K. Boal, F. Ilhan, J.E. DeRouchey, T. Thurn-Albrecht, T.P. Russell, V.M. Rotello, Self-assembly of nanoparticles into structured spherical and network aggregates. Nature **404**, 746–748 (2000)
90. J. Khoury, X. Zhou, L. Qu, L. Dai, A. Urbas, Q. Li, Organo-soluble photoresponsive azo hybrid gold nanorods. Chem. Comm. 2109–2111 (2009)
91. C. Xue, O. Birel, Y. Xue, L. Dai, A. Urbas, Q. Li, pH and temperature modulated aggregation of hydrophilic gold nanorods with perylene dyes and carbon nanotubes. J. Phys. Chem. C **117**, 6752–6758 (2013)
92. Y. Li, D. Yu, L. Dai, A. Urbas, Q. Li, Hydrophobic chiral hybrid gold nanorods. Langmuir **27**, 98–103 (2011)
93. B.S. Avinash, V. Lakshminarayanan, S. Kumar, J.K. Viji, Gold nanorods embedded discotic nanoribbons. Chem. Comm. **49**, 978–980 (2013)
94. Z.Y. Tang, Z.L. Zhang, Y. Wang, S.C. Glotzer, N.A. Kotov, Self-assembly of CdTe nanocrystals into free-floating sheets. Science **314**, 274–278 (2006)
95. S.A. Maier, P.G. Kik, H.A. Atwater, S. Meltzer, E. Harel, B.E. Koel, A.A.G. Requicha, Local detection of electromagnetic energy transport below the diffraction limit in metal nanoparticle plasmon waveguides. Nat. Mater. **2**, 229–232 (2003)
96. F. Caruso, Hollow inorganic capsules via colloid-templated layer-by-layer electrostatic assembly. Top. Curr. Chem. **227**, 145–168 (2003)
97. C.M. Xue, O. Birel, M. Gao, S. Zhang, L.M. Dai, A. Urbas, Q. Li, Perylene monolayer protected gold nanorods: Unique optical, electronic properties and self-assemblies. J. Phys. Chem. C **116**, 10396–10404 (2012)
98. C.M. Xue, O. Birel, Y.N.A. Li, X. Ma, M. Gao, A. Urbas, Q. Li, Porphyrin metal complex monolayer-protected gold nanorods: a parallel facile synthesis and self-assembly. J. Colloid Interface Sci. **398**, 1–6 (2013)
99. C.M. Xue, Y.Q. Xu, Y. Pang, D.S. Yu, L.M. Dai, M. Gao, A. Urbas, Q. Li, Organo-soluble porphyrin mixed monolayer-protected gold nanorods with intercalated fullerenes. Langmuir **28**, 5956–5963 (2012)
100. M.J. Hostetler, J.E. Wingate, C.J. Zhong, J.E. Harris, R.W. Vachet, M.R. Clark, J.D. Londono, S.J. Green, J.J. Stokes, G.D. Wignall, G.L. Glish, M.D. Porter, N.D. Evans, R.W.

Murray, Alkanethiolate gold cluster molecules with core diameters from 1.5 to 5.2 nm: core and monolayer properties as a function of core size. Langmuir **14**, 17–30 (1998)
101. L. Donkers, D. Lee, R.W. Murray, Synthesis and isolation of the molecule-like cluster Au-38(PhCH2CH2S)(24). Langmuir **20**, 1945–1952 (2004)
102. T.G. Schaaff, M.N. Shafigullin, J.T. Khoury, I. Vezmar, R.L. Whetten, Properties of a ubiquitous 29 kDa Au: SR cluster compound. J. Phys. Chem. B **105**, 8785–8796 (2001)
103. O. Kohlmann, W.E. Steinmetz, X.A. Mao, W.P. Wuelfing, A.C. Templeton, R.W. Murray, C.S. Johnson, NMR diffusion, relaxation, and spectroscopic studies of water soluble, monolayer-protected gold nanoclusters. J. Phys. Chem. B **105**, 8801–8809 (2001)
104. R.H. Terrill, T.A. Postlethwaite, C.H. Chen, C.D. Poon, A. Terzis, A.D. Chen, J.E. Hutchison, M.R. Clark, G. Wignall, J.D. Londono, R. Superfine, M. Falvo, C.S. Johnson, E.T. Samulski, R.W. Murray, Monolayers in three dimensions: NMR, SAXS, thermal, and electron hopping studies of alkanethiol stabilized gold clusters. J. Am. Chem. Soc. **117**, 12537–12548 (1995)
105. M.J.A. Hore, R.J. Composto, Nanorod self-assembly for tuning optical absorption. ACS Nano **4**, 6941–6949 (2010)
106. J. Perez-Juste, B. Rodriguez-Gonzalez, P. Mulvaney, L.M. Liz-Marzan, Optical control and patterning of gold-nanorod-poly(vinyl alcohol) nanocomposite films. Adv. Funct. Mater. **15**, 1065–1071 (2005)
107. C. Xue, K. Gutierrez-Cuevas, M. Gao, A. Urbas, Q. Li, Photo-modulated self-assembly of hydrophobic thiol monolayer-protected gold nanorods and their alignment in thermotropic liquid crystal. J. Phys. Chem. C **117**, 21603–21608 (2013)
108. J. Muller, C. Sonnichsen, H. von Poschinger, G. von Plessen, T.A. Klar, J. Feldmann, Electrically controlled light scattering with single metal nanoparticles. Appl. Phys. Lett. **81**, 171–173 (2002)
109. N. Liu, H. Guo, L. Fu, S. Kaiser, H. Schweizer, H. Giessen, Three-dimensional photonic metamaterials at optical frequencies. Nature Mater. **7**, 31–37 (2008)
110. J.B. Pendry, Negative refraction. Contemporary Physics **45**, 191–202 (2004)
111. G. Schmid (ed.), *Nanoparticles. From Theory to Application* (Wiley-VCH, Weinheim, 2004)
112. S. Polarz, Shape matters: anisotropy of the morphology of inorganic colloidal particles — synthesis and function. Adv. Funct. Mater. **21**, 3214–3230 (2011)
113. M. Skarabot, M. Ravnik, S. Zumer, U. Tkalec, I. Poberaj, D. Babic, I. Musevic. Hierarchical self-assembly of nematic colloidal superstructures. Phys. Rev. E **77** (2008)
114. C.M. Soukoulis, M. Wegener, Optical metamaterials-more bulky and less lossy. Science **330**, 1633–1634 (2010)
115. F. Brochard, P.G.D. Gennes, Theory of magnetic suspensions in liquid crystals. J. De Physique **31**, 691–708 (1970)
116. H. Qi, B. Kinkead, T. Hegmann, Effects of functionalized metal and semiconductor nanoparticles in nematic liquid crystal phases, in *Emerging Liquid Crystal Technologies III*, vol 6911, ed. by L.C. Chien (ed.) (2008), pp. 91106–91106
117. L.S. Li, A.P. Alivisatos, Semiconductor nanorod liquid crystals and their assembly on a substrate. Adv. Mater. **15**, 408–411 (2003)
118. L.C.T. Shoute, D.F. Kelley, Spatial organization of GaSe quantum dots: Organic/ semiconductor liquid crystals. J. Phys. Chem. C **111**, 10233–10239 (2007)
119. M. Zorn, S. Meuer, M.N. Tahir, Y. Khalavka, C. Soennichsen, W. Tremel, R. Zentel, Liquid crystalline phases from polymer functionalised semiconducting nanorods. J. Mater. Chem. **18**, 3050–3058 (2008)
120. C. Nobile, L. Carbone, A. Fiore, R. Cingolani, L. Manna, R. Krahne, Self-assembly of highly fluorescent semiconductor nanorods into large scale smectic liquid crystal structures by coffee stain evaporation dynamics. J. Phys.: Condens. Matter **21**, 264013 (2009)
121. J.C. Payne, E.L. Thomas, Towards an understanding of nanoparticle-chiral nematic liquid crystal co-assembly. Adv. Funct. Mater. **17**, 2717–2721 (2007)

122. J.J. Vallooran, S. Bolisetty, R. Mezzenga, Macroscopic alignment of lyotropic liquid crystals using magnetic nanoparticles. Adv. Mater. **23**, 3932–3937 (2011)
123. T.J. Daou, J.M. Greneche, G. Pourroy, S. Buathong, A. Derory, C. Ulhaq-Bouillet, B. Donnio, D. Guillon, S. Begin-Colin, Coupling agent effect on magnetic properties of functionalized magnetite-based nanoparticles. Chem. Mater. **20**, 5869–5875 (2008)
124. A. Demortiere, S. Buathong, B.P. Pichon, P. Panissod, D. Guillon, S. Begin-Colin, B. Donnio, Nematic-like organization of magnetic mesogen-hybridized panoparticles. Small **6**, 1341–1346 (2010)
125. L.M. Lopatina, J.V. Selinger, Theory of ferroelectric nanoparticles in nematic liquid crystals. Phys. Rev. Lett. **102**, 197802 (2009)
126. Y. Reznikov, O. Buchnev, O. Tereshchenko, V. Reshetnyak, A. Glushchenko, J. West, Ferroelectric nematic suspension. Appl. Phys. Lett. **82**, 1917–1919 (2003)
127. S.-C. Jeng, C.-W. Kuo, H.-L. Wang, C–.C. Liao, Nanoparticles-induced vertical alignment in liquid crystal cell. Appl. Phys. Lett. **91**, 061112 (2007)
128. C.-W. Kuo, S.-C. Jeng, H.-L. Wang, C–.C. Liao, Application of nanoparticle-induced vertical alignment in hybrid-aligned nematic liquid crystal cell. Appl. Phys. Lett. **91**, 141103 (2007)
129. G.L. Nealon, R. Greget, C. Dominguez, Z.T. Nagy, D. Guillon, J.L. Gallani, B. Donnio, Liquid-crystalline nanoparticles: Hybrid design and mesophase structures. Beilstein J. Org. Chem. **8**, 349–370 (2012)
130. S. Bedanta, W. Kleemann, Supermagnetism. J. Phys. D Appl. Phys. **42**, 13001 (2009)
131. M.P. Pileni, Nanocrystal self-assemblies: Fabrication and collective properties. J. Phys. Chem. B **105**, 3358–3371 (2001)
132. J. Shi, S. Gider, K. Babcock, D.D. Awschalom, Magnetic clusters in molecular beams, metals, and semiconductors. Science **271**, 937–941 (1996)
133. S.H. Sun, C.B. Murray, D. Weller, L. Folks, A. Moser, Monodisperse FePt nanoparticles and ferromagnetic FePt nanocrystal superlattices. Science **287**, 1989–1992 (2000)
134. D. Greshnykh, A. Froemsdorf, H. Weller, C. Klinke, On the electric conductivity of highly ordered monolayers of monodisperse metal nanoparticles. Nano Lett. **9**, 473–478 (2009)
135. A. Zabet-Khosousi, A.-A. Dhirani, Charge transport in nanoparticle assemblies. Chem. Rev. **108**, 4072–4124 (2008)
136. S.A. Maier, M.L. Brongersma, P.G. Kik, S. Meltzer, A.A.G. Requicha, H.A. Atwater, Plasmonics—a route to nanoscale optical devices. Adv. Mater. **13**, 1501–1505 (2001)
137. E.V. Shevchenko, D.V. Talapin, S. O'Brien, C.B. Murray, Polymorphism in AB(13) nanoparticle superlattices: an example of semiconductor-metal metamaterials. J. Am. Chem. Soc. **127**, 8741–8747 (2005)

Chapter 5
Photoresponsive Chiral Liquid Crystal Materials: From 1D Helical Superstructures to 3D Periodic Cubic Lattices and Beyond

Yannian Li and Quan Li

Abstract Stimuli-responsive self-organized chiral architectures with functional properties hold great promise in design and fabrication of smart soft materials. Chiral liquid crystals (LCs) represent such a promising class of materials due to their unique chiral superstructures and properties as a result of chirality transfer from molecular level to macroscopic liquid crystalline phases. Introducing photochromic molecules into chiral LCs results in self-organized superstructures whose properties can be tuned by light and therefore open the door for their applications in several new directions. For example, the helical superstructure of chiral nematic phase can reflect light selectively according to Bragg's law. The photo-tuning of the reflection wavelength has substantial significance for applications in reflection displays, tunable lasers, photonics, etc. This chapter focus is on the recent progress in photoresponsive chiral LCs such as chiral nematic phases, chiral smectic phases, and blue phases (BPs). We especially discuss their structures, important properties and feasible applications. Photomodulation of chiral nematic phases such as phase transition, handedness inversion, and reflection color change is introduced based on different types of photochromic molecules. For the chiral smectic phase, photomodulation mainly focuses on their ferroelectric nature such as magnitude and sign of spontaneous polarization. BPs as a novel type of three dimensional (3D) photonic crystals are receiving increasing attention, and the light-induced phase transitions and reflection color tuning are also presented.

Y. Li · Q. Li (✉)
Liquid Crystal Institute, Kent State University, Kent, OH, USA
e-mail: qli1@kent.edu

Q. Li (ed.), *Nanoscience with Liquid Crystals*, NanoScience and Technology,
DOI: 10.1007/978-3-319-04867-3_5,

5.1 Introduction

Liquid crystals (LCs) are known as a class of soft materials with an intermediate state between solid and liquid phases, which exhibit unique thermal, mechanical, optical, and electrical properties due to the self-organization of mesogenic molecules with both orientational order of crystalline solid and the fluidity of isotropic liquid [1–4]. LC phases can be triggered by external environment such as temperature variation or the influence of both temperature and solvent, in accordance with LCs being broadly classified as thermotropic or lyotropic LCs. The widely investigated thermotropic LCs composed of rod-like molecules, i.e. calamitic mesogens, exhibit different types of sub-phases according to the degree of ordering. Introducing chirality into LCs leads to novel self-organized superstructures, which are known as chiral liquid crytal phases [3]. For example, chiral nematic phase with helical arrangement of mesogen directors is derived from nematic phase. Similar helical twists are also found in other chiral LC phases, and their helical superstructures provide novel properties such as selective reflection of circularly polarized light (CPL), ferro- and anti-ferro electric polarization.

Being soft matter featuring molecular self-organization with weak intermolecular interactions, LCs can be made sensitive to external stimuli such as temperature, electric field, magnetic field, mechanical stress, light, and/or chemical environment, which opens the door for stimuli-responsive LCs towards many applications. With the maturity of the highly successful application of LC displays (LCDs), which is based on electro-optic response, it turns out that stimuli-responsive LCs have found more broad applications and will continue to attract great interest in organic semiconductors, nonlinear optics, nanophotonics, sensors, metamaterials, etc. [4]. Among the external stimuli, light is particularly attractive due to its advantages of remote, spatial, and temporal control. Thus, photoresponsive LCs, especially where chirality is involved, have served as a novel class of smart materials.

The reports on photoresponsive chiral LCs can be traced back to the 1960s and 1970s, during which a number of examples on phototuning of pitch length in cholesteric phases were demonstrated [5–7]. To date, numerous reports have been found to be related to their physics, chemistry and applications. This chapter mainly deals with the recent progress in this area with a survey of three major classes of chiral LC phases, i.e. cholesteric (N*), chiral smectic C (SmC*) and blue phase (BP) from their structures to properties and applications.

5.2 Chiral Liquid Crystals

Chiral LC phases can be observed either in chiral mesogenic compounds or in achiral LCs doped with chiral molecules. The supramolecular chirality in the former systems originates from the intrinsic chirality of mesogens and forms chiral

LC phase, while the later systems are based on chirality amplification mechanism. There are many types of chiral LC phases such as cholesteric (N*), chiral smectic (Sm*), blue phase (BP), and twist grain boundary (TGB) phase. Among them, N* and SmC* phases are the most studied due to their potential technological applications. BPs are also receiving increasing attention due to their interesting 3D cubic structures and potential applications as 3D photonic crystals.

5.2.1 Chiral Nematic Phase

Chiral nematic phase (N*), also known as cholesteric LC (CLC), is the LC phase with chirality at the molecular level and helical arrangement of layers at the macroscale level. The term "cholesteric" originates from the cholesterol derivatives in which this phase was first observed by Reinitzer in 1888 [8]. Today, various chiral molecules exhibit cholesteric phase and most of them are structurally different from cholesterol. As shown in Fig. 5.1, the spatial orientation of the molecules in CLC phase changes in a helical fashion along a specific direction called helical axis. This helical orientation of the directors is non-superimposable on its mirror image, thus making the phase chiral. The helical superstructure of CLC is characterized by helical pitch and handedness. The pitch (p) is the distance across which the director rotates a full 360° (Fig. 5.1). Handedness describes the direction in which the molecular orientation rotates along the helical axis and it can be expressed as sign (−) and (+), which represent left handedness and right handedness, respectively.

The most important optical property of CLCs is their selective reflection of light. When unpolarized light (UPL) propagates through a CLC medium, only the circularly polarized light (CPL) with the same handedness as the helix is reflected, i.e. left-handed CLCs only reflect left-handed CPL (*l*-CPL) and right-handed CLCs only reflect right-handed CPL (*r*-CPL). The mean reflection wavelength of a cholesteric phase can be determined by $\lambda = np$, where n is the average refraction index of LC material. CLCs can be designed and functionalized to be sensitive to external stimuli. Thus it is possible to tune the important CLC parameters such as the order parameter, helix handedness, and pitch length, which provide opportunities as well as challenges in fundamental science and pave the way to many applications. The nanostructures of CLCs, i.e. CLCs with the pitch length in the range of hundreds of nanometers can reflect light in the visible range and the reflection color tuning has been utilized for tunable color reflectors and filters [9, 10], tunable LC lasers [11, 12], reflection displays [13], and biomedical applications [14, 15].

CLCs can be fabricated by two main methods. The first and simplest way is based on chiral mesogens, which can exhibit the cholesteric phase at a certain temperature range. The cholesteric phase produced by this method consists of pure materials and thus may exhibit advantages such as good uniformity, enhanced stability, and fast response time to stimuli. However, they are typically produced at

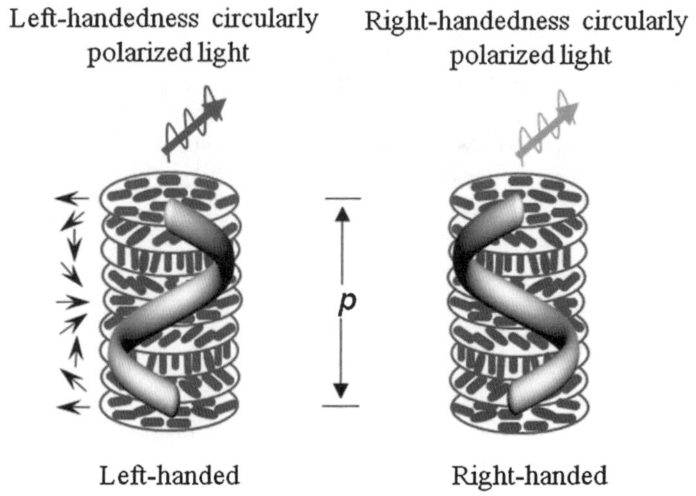

Fig. 5.1 Helical superstructure of cholesteric LCs (CLCs) with different handedness

above-ambient-temperatures and feature physical properties that are difficult to be optimized for device applications. The second strategy is to dope chiral molecules into achiral nematic LCs. The resulting mixture can self-organize into a helical superstructure and form a cholesteric phase [16, 17]. The ability of a chiral dopant to twist an achiral nematic LC phase is defined as helical twisting power (HTP, β) which is quantified as: $\beta = 1/(pc)$, where c is the chiral dopant concentration and p is the pitch length. The main advantage of this method is that the pitch of cholesteric phase can be simply tuned by adjusting the concentration of dopant molecules, especially for reflection color tuning that require the pitch length within the range of hundreds of nanometers. Moreover, requirement of small amount of chiral molecules as dopant reduces synthetic cost. Furthermore, it can be used for various LC hosts to generate CLC mixtures with different properties. It is worthy to point out that this strategy is strongly dependent on the development of chiral dopants having high HTPs and good compatibility with LC hosts. Low HTPs will give rise to the requirement of high dopant concentrations to attain the required properties. It may lead to phase separation, coloration, and altering the desired physical properties such as viscosity and dielectric properties of the host LC phase.

5.2.2 Chiral Smectic C Phase

The most studied chiral smectic phase is ferroelectric SmC* phase [18], which is derived from Smectic C (SmC) phase. As shown in Fig. 5.2, the helical twist in SmC* results from chiral organization of smectic layers as similar to the formation of N* from nematic layers mentioned above. The molecules in each smectic layer

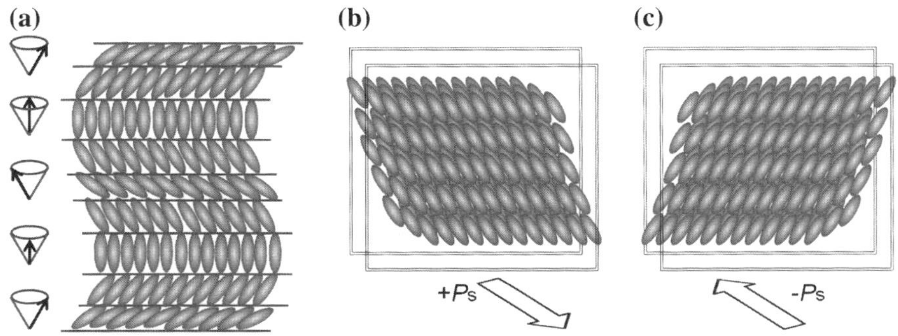

Fig. 5.2 Helical structure of SmC* phase (**a**) and surface stabilized ferroelectric liquid crystals (SSFLC) (**b** and **c**)

have positional ordering with a tilt angle of molecular director with respect to the layer normal. Chirality developed among layers characterized by a shift of the tilt directors from layer to layer with a helical axis along layer normal (Fig. 5.2a). One important property of SmC* phase is that each layer exhibits spontaneous electric polarization (P_S) along the polar axis of layer due to the presence of polar groups. The magnitude of P_S is related to the structure and proportion of chiral molecules in SmC*. Similar to cholesteric phase, P_S can also be left-handed (−) or right handed (+), which is determined by the configuration of the stereogenic center. In the absence of external alignment constraints, the helical superstructure of SmC* does not show ferroelectricity as the vector sum of P_S of each layers is zero over one helical pitch. In 1980, Clark and Lagerwall found that the helical superstructure of SmC* could be unwound in the cell with polyimide-coated alignments conditions and thus formed a surface-stabilized ferroelectric LC (SSFLC) (Fig. 5.2b, c) [19]. The cell thickness is on the order of the helical pitch of SmC*, normally 1–5 μm. The net P_S is perpendicular to the substrate, which can be easily switched to opposite direction by applying an electric filed, producing a bistable ON/OFF light shutter with the help of crossed polarized filter. The critical value of the required electric field is defined as coercive force *Ec*. The switching time is about 10^3 times faster than the twisted nematic device due to the coupling between P_S and electric field. The fast response time and bistability make SmC* LCs promising candidates for display, photonic and nonlinear optic applications [20, 21].

SmC* LCs can be observed in single component chiral mesogen systems and they can also be formulated by doping chiral molecules into achiral SmC host. In the induced SmC* phase, P_S originate from the transverse dipole moments of individual chiral dopant molecule due to the ordering imposed by the SmC host [20]. The polarization power (δ_p) is defined as the ability of the chiral dopant to induce the spontaneous polarization in a non-ferroelectric SmC host. The left-handed and right-handed P_S, i.e. positive and negative P_S can be achieved by chiral molecules with opposite chiral configurations.

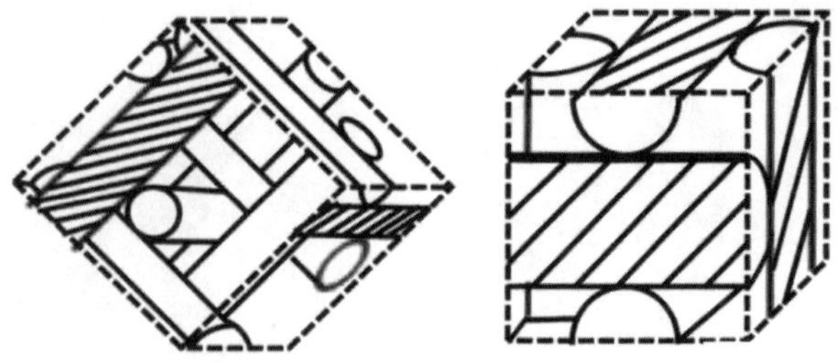

Fig. 5.3 Structures of BP I and BP II

5.2.3 Blue Phase

BPs are an interesting class of chiral LC phases that can self-organize into 3D periodic cubic lattices [11, 22, 23]. BPs generally exist within the temperature range between the isotropic phase and cholesteric phase [3]. The cubic structures are characterized by the double twist cylinders (DTC) which are stabilized by the disclinations. Normally when chirality in LC system is high enough, DTC is more stable than a simple helical structure. However, DTCs cannot be densely packed to fill the entire space and disclination lines appeared at the interstices among the cylinders to form a stabilized 3D cubic structure. There are three types of BPs reported, namely BP III, BP II, and BP I, which are observed during cooling from the isotropic to the cholesteric phase in the order of decreasing temperature. BP I and BP II have body-centered-cubic and simple cubic structures, respectively (Fig. 5.3), while BP III is an amorphous network of disclination lines, which is thermodynamically and kinetically stabilized over crystalline BPs at intermediate chiralities [24].

BP I and BP II are periodic in three dimensions and the periodicity is typically on the order of the wavelength of visible light. These nanostructures enable BPs to reflect light in visible region. In recent years, BPs have been recognized as soft 3D photonic crystals. BPs can be observed in pure chiral mesogens or induced systems of chiral LCs. Compared with simple helical superstructures of chiral nematic phase, the 3D cubic structures of BPs offer several advantages on device performance over CLC materials such as higher contrast ratio, fast electro-optic response and no surface treatment is required.

5.3 Photoresponsive Chiral LCs

As chiral LCs can be formed by chiral mesogens or induced by chiral guest molecules, photoresponsive chiral LCs can be derived from these two systems by photosensitization. One strategy is to photosensitize the existing constituent molecules in chiral LCs, i.e. LC host or chiral dopant. The other strategy is to dope additional photoresponsive agents to the existing chiral LC systems. These two photosensitizing strategies generate many types of photoresponsive chiral LC materials, which can be divided into different subsystems as follows: (a) photoresponsive chiral mesogens; (b) chiral mesogens doped with photoresponsive chiral/achiral molecules; (c) achiral LC host doped with chiral dopant and additional photoresponsive molecules; and (d) achiral LC host doped with photoresponsive chiral dopant. Among these systems, photoresponsive chiral mesogens normally require high processing temperature, while the systems having more than three components may be non-uniform due to their relative complexity. The strategy that utilizes single photoresponsive chiral dopant in achiral LC host is more attractive where the chiral dopant is separately synthesized and the properties of mixture can be readily adjusted by varying the doping concentrations, etc. [20, 25–27].

Photoresponsive molecules are required for the fabrication and photomodulation of photoresponsive CLC materials regardless of being employed as chiral mesogens, achiral LC host, or chiral/achiral dopants. The photoisomerization of molecules leads to change in molecular shape (geometry/conformation) and alter the bulk properties of LC material, which constitutes the basis for the photomodulation in chiral LCs [28, 29]. There are many types of photoresponsive molecules and some examples are shown in Fig. 5.4.

Azobenzenes are a well-known family of photochromic compounds that can isomerize from its *trans* form to *cis* form upon UV irradiation (Fig. 5.4a). The *cis* isomer can be switched back to *trans* form either by visible light or thermal relaxation. The rod shape of the *trans* form can stabilize calamitic LCs, while the *cis* form is bent and normally decreases the order parameters of LC phases. Owing to the dramatic molecular shape and property change between the *trans* and *cis* isomers, azobenzenes are the most investigated photochromic molecules to function as either mesogens or dopants in chiral LCs.

Overcrowded alkenes are a special class of olefins with an inherently helical shape, resulting from extensive steric crowding around the central carbon–carbon double bond (Fig. 5.4b) [30]. Related molecules were first reported by Feringa and coworkers, who used the chiral molecules as chiral dopant in CLC materials [31]. The merit of these molecules as chiral dopants is that the photoisomerization between the two isomers with pseudoenantiomeric relationship often results in the handedness inversion of the cholesteric LCs.

Diarylethenes are another class of fascinating photochromic molecules due to their superior thermal stability and excellent fatigue resistance [32, 33]. They exhibit typical photochromic behavior in organic solvent with transformation from the colorless open-ring form to the colored closed-ring form upon UV irradiation (Fig. 5.4c). The reverse process is thermally irreversible and occurs only by visible light irradiation. Although a few examples of chiral diarylethenes has been reported with mesomorphic property, they are mainly used as chiral dopant for photoresponsive chiral LCs. Compared with azobenzene and overcrowded alkene compounds, they possess the advantage of thermal stability in both isomerized states.

In addition to the above three types of molecules, photoresponsive chiral LCs have also been fabricated with other photochromic molecules such as spirooxazines [34], fulgides [35], butadienes [36], bicyclic ketones [37], and thioindigos [38].

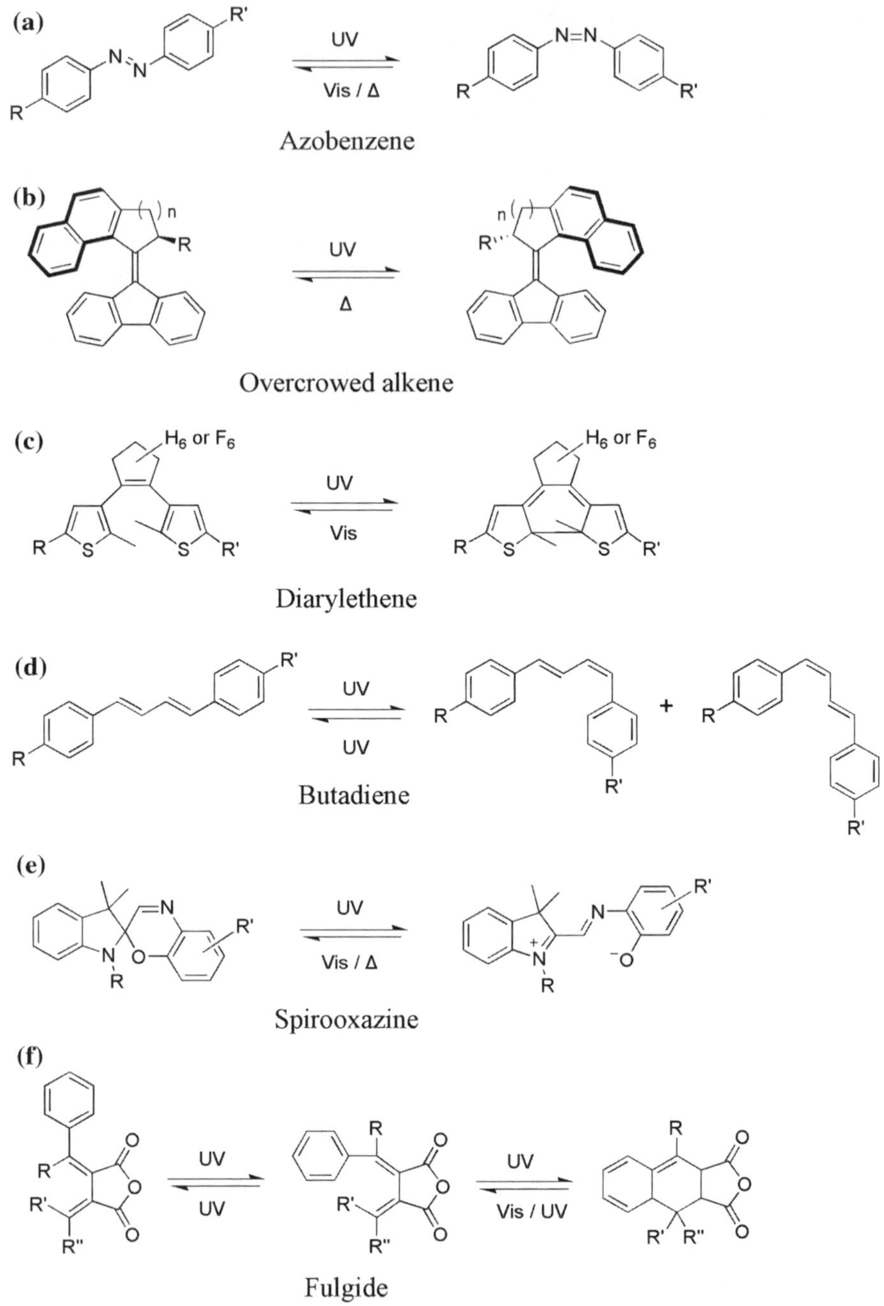

Fig. 5.4 Molecular structures and photoisomerization of some photoresponsive compounds. **a** Azobenzene. **b** Overcrowed alkene. **c** Diarylethene. **d** Butadiene. **e** Spirooxazine. **f** Fulgide

5.4 Photoresponsive Cholesteric LCs

5.4.1 Azobenzene-Based CLCs

Due to the dramatic molecular shape change upon light irradiation, azobenzene molecules are the most popular photochromic molecules that are intensively investigated in CLCs. They can act as achiral/chiral mesogens or dopants to tune the properties of cholesteric LCs, such as order parameter, pitch length, and/or helix inversion. Early examples are mainly focused on achiral azobenzenes that can be mesomorphic itself or used as additive to tune the pitch length or induce the phase transition, while recent progresses mainly rely on the development of chiral azobenzene dopants that exhibit high HTPs for reflection color control or handedness inversion.

5.4.1.1 Phase Transitions Induced by Azobenzenes

It is known that the *trans* isomer of azobenzene has elongated rod shape that is favorable for the stabilization of the LC phases, while the *cis* isomer in bent form is unfavorable and tends to destabilize the LC phases. The *trans–cis* photoisomerization will decrease the order parameters, if significantly enough, which could lead to the destruction of the ordered LC phase and formation of isotropic state. The reverse transition can be achieved with either visible light irradiation or thermal relaxation as a result of *cis–trans* isomerization. This principle was initially used to induce the phase transition from nematic to isotropic in azobenzene LCs [39]. Cholesteric mesophase is intrinsically similar to nematic with additional helical arrangement of nematic layers, thus the photoisomerization of azobenzenes can also bring out the phase transition from the cholesteric state to the photoinduced isotropic (PHI) state.

The phase transitions between cholesteric and PHI were demonstrated initially by doping chiral dopant into azobenzene nematic LCs [40, 41]. For example, Bunning et al. reported the PHI state in multi-component systems containing azobenzene LCs and light-insensitive chiral molecules [41]. The transition behavior can be used to create an imaging platform. The initial state showed cholesteric phase with green reflection color (Fig. 5.5a), which was transformed into PHI state without any reflection (Fig. 5.5b). Information can be written on the cell at the PHI state by scanning with laser beam in a 2D pattern (Fig. 5.5c, d). Very recently, PHI states have also been demonstrated with laser irradiation [42] or by using azobenzene-derived chiral mesogens [43].

The decrease in order parameter in LCs caused by the photoisomerization of azobenzene molecules are also known to induce phase transition from chiral smectic phase to cholesteric phase. This type of transition is mainly observed in chiral mesogenic compounds. In 1999, Joly et al. conducted a systematic research on the phase transition behavior of a series of chiral azobenzene LCs [44]. A photo

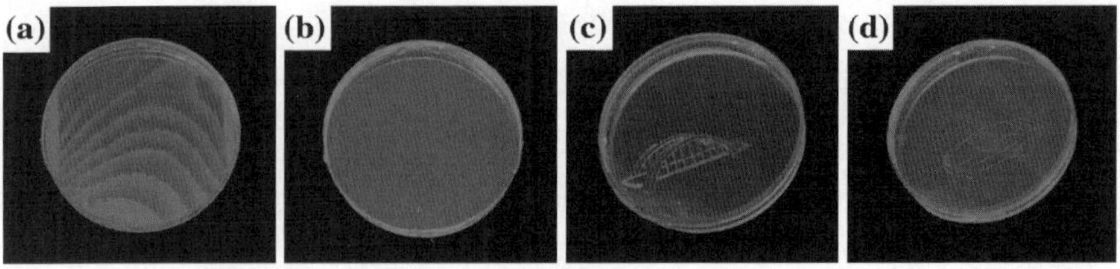

Fig. 5.5 A *green* reflecting azo-CLC mixture at initial state (**a**), PHI state upon UV irradiation (**b**), image written by restoration of reflection with green laser irradiation (**c**–**d**). Reproduced with permission from [41]. Copyright 2007 John Wiley & Sons

illumination-temperature phase diagram was presented to interpret the phase transition sequence including the SmC*-N*-I transition. Tamaoki and coworkers reported a serious of cholesteryl-azobenzene dimesogenic compounds that can transform from Sm phase to N* upon UV irradiation [45–48].

The photoisomerization of chiral azobenzene dopants often leads to the dramatic change in their HTPs. When it is used in combination with another non-photoresponsive chiral dopant with opposite chirality to form a compensated system, the change in HTP can be utilized to reversibly switch the LC phase between compensated nematic and cholesteric [49].

5.4.1.2 Reflection Modulation in Azobenzene-Based CLCs

The tuning of reflection wavelength in CLCs holds great promise for applications in tunable color filters and reflectors, tunable LC lasers and reflection displays. The pitch change occurs in almost all photoresponsive CLCs accompanied by a change in reflection wavelength, and the reflection modulation are normally demonstrated in the range of hundreds of nanometers. In 1971, Sackmann reported a light-induced reversible color change in non-halide based cholestryl mixture doped with azobenzene [6]. The reflection of the CLC mixture blue shifted from 610 to 560 nm upon irradiation with 420 nm light. Although the tuning range is quite narrow, this work is the first demonstration of reversible phototuning of cholesteric reflection colors, providing a new pathway to the photoresponsive CLCs. To date, various material systems have been reported for this purpose, which include: chiral mesogens doped with achiral azobenzene molecules [50–57]; pure photoresponsive chiral mesogens [46, 58]; nematic LC doped with multiple chiral/achiral dopants [41, 53, 54, 59–66] or single photoresponsive chiral dopants [67–78]. Although the achiral azobenzenes in multiple component system were widely adopted at the early stage, the use of single photoresponsive chiral azobenzene dopants is the first choice nowadays for the reflection color modulation due to the advantages over other systems, such as high HTPs, large tuning range and relatively simple formulation.

In CLCs induced by single chiral photoresponsive dopant, the reflection wavelength tuning is mainly determined by the HTP change of the chiral dopant caused by photoisomerization. Therefore this strategy strongly depends on the design and synthesis of chiral dopants which need to exhibit high HTPs as well as large HTP variations upon light irradiation. The design of photoresponsive azobenzene dopants initially focused on the chiral azobenzenes with tetrahedral chiral centers, however, these systems are not very efficient for the reflection wavelength tuning due to the low HTPs of the dopants [67, 68]. With the continuous efforts especially in the past decade, many chiral azobenzene dopants have been developed and used for phototuning the reflection colors of CLCs with large tuning range, very high HTPs, and low doping concentrations.

Tamaoki et al. reported a series of planar chiral dopant which was employed in commercially available nematic LC to achieve phototunable reflection colors [69, 70]. These compounds were designed based on an azobenzenophane compound having conformational restriction on the free rotation of naphthalene moiety to impose an element of planar chirality. Due to the good solubility, moderately high HTPs, and large changes in HTPs during photoisomerization of the dopant, a fast photon mode reversible color control in induced CLCs was achieved.

Binaphthyl derivatives with axial chirality are known as powerful helicity inducers [79]. The unique combinations of binaphthyl with azobenzene units generate a family of axially chiral azobenzene compounds with very high HTPs and significantly broadened the area of reflection color tuning using a single chiral photoresponsive dopant [71, 72]. Li's group has developed a series of binaphthyl-derived axially chiral azobenzene compounds **1–9** (Fig. 5.6) [73–78]. Compound **1–4** exhibited high HTPs as well as good solubility in LC hosts. These light-driven chiral switches were found suitable for dopants in nematic hosts for reflection color tuning with applications in novel optically addressed displays, i.e. photodisplays (Fig. 5.7). This photodisplay is based on the reflection color change induced by **1** with the help of photomask [73, 80]. As shown in Fig. 5.7b, two identical displays were driven by different energies. One is electrically addressed with the standard multiplexing electronics, while the other one is optically addressed. Relatively, the overall size of the display module is reduced in the case of the light-driven one and the cost may be six times less than the electrically driven one. The simplification of the final product can make markets such as security badges, small point of sale advertisements, and other applications that require a very low cost module that are updated frequently. It is worth noting that the photodisplay device can display a high resolution image without the need of attached drive and control electronics substantially reducing the cost of the display unit for use in applications where paper is currently used.

Compounds **5–8** show improved properties with higher HTPs than **1–4**, and they were all found capable of phototuning the reflection color across full visible region, among them the highest HTP of 304 μm^{-1} was achieved with dopant **7** [74, 75]. Very interestingly, it was found that the HTP of dopant **8** was sensitive to the wavelength of exposure during the reverse process as the equilibrium ratio among *trans–trans*, *trans–cis*, *cis–cis* isomers was varied [76]. This compound exhibited a very high HTP of 301 μm^{-1} at the initial state, which significantly decreased to 106 μm^{-1} at the photostationary state (PSS) upon UV irradiation. For

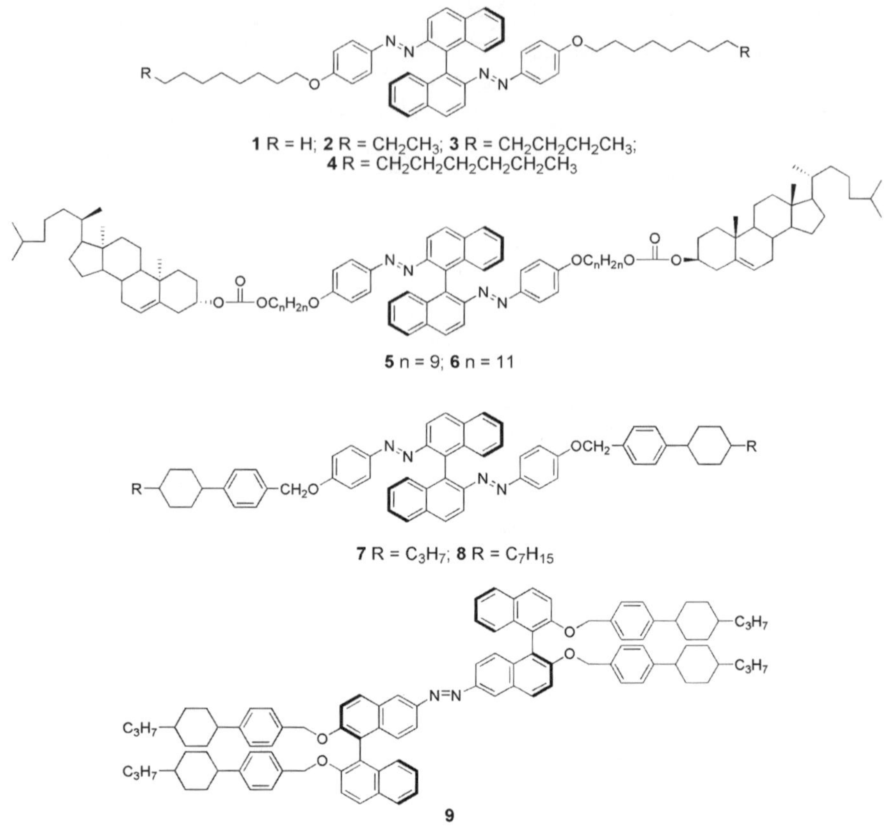

Fig. 5.6 Chemical structures of axially chiral azobenzene dopants **1–9**

the reverse process, using different wavelengths of visible light resulted in PSS_{vis} with different HTP values and reflection colors. Three stable primary colors red, green, and blue were realized at the PSS_{vis} upon irradiation with 440, 450, and 550 nm visible light, respectively. More importantly, the photo-addressed images were written in a homeotropic cell with the capability of electrically switchable imaging. The spatially patterned image can be hidden with an application of an electric field. In Fig. 5.8b-v, the reflectivity of the image in both the outside and photowritten areas disappears with application of a 30 V electric pulse (100 Hz, 1000 ms). The photo-addressed image is stored indefinitely even after the light-driven molecules are thermally relaxed. The hidden image can be restored simply by applying pressure (Fig. 5.8c-vi). The initially formed cholesteric LC texture (Fig. 5.8a-iv) was an imperfect planar texture because the homeotropic alignment layer gives a weak anchoring, yielding a poly-domain structure as schematically shown in Fig. 5.8a-i. The POM image (Fig. 5.8a-vii) confirmed the texture state. When a 30 V pulse (100 Hz, 1000 ms) was applied, the texture converted to a focal conic texture (Fig. 5.8b-viii), which is a scattering, multi-domain structure

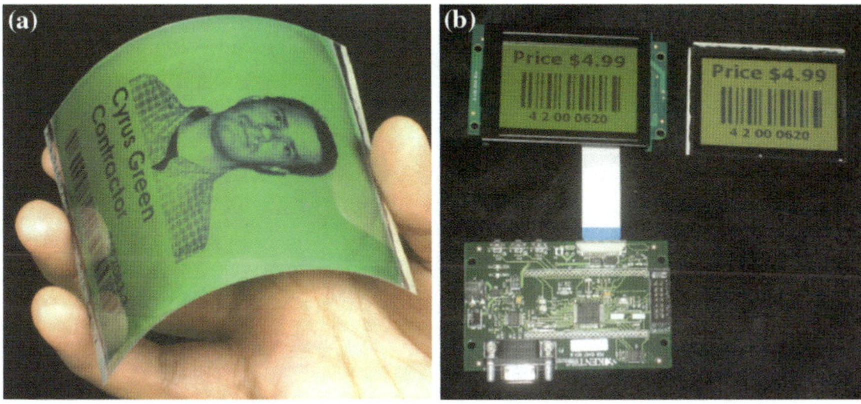

Fig. 5.7 A flexible optically addressed photochiral display (**a**); a conventional display attached bulky and costly electronics (**b**, *left*) compared with an optically addressed display with the same image without the added electronics (**b**, *right*). Reproduced with permission from [77]. Copyright 2008 Society for Information Display

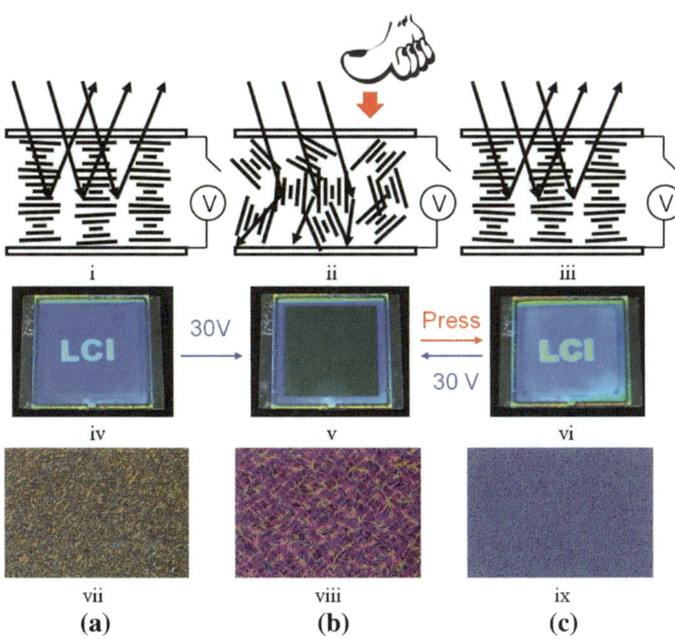

Fig. 5.8 A photoaddressed and multiswitchable cholesteirc LC display of 6.0 wt% chiral dopant 8 in LC host E7 in a 5 μm thick homeotropic aligned cell (1.5 in. × 1.5 in.). *Top* schematic cholesteric textures; *Middle* demonstration of an image; *Bottom* crossed polarized textures. Reproduced with permission from [76]. **a** Planar state. **b** Focal conic state. **c** Planar state. Copyright 2011 John Wiley & Sons

within random alignment of the helices (Fig. 5.8b-ii). The scattered light effectively hides the image that was written through light exposure. When a mechanical force was applied to the cell, a planar texture appeared due to the shear-flow induced alignment effect, restoring the original image.

It is well-known that the *trans–cis* isomerizations of azobenzenes are normally induced by UV light exposure. Recent research pointed out that the photoisomerization can also be achieved by visible light [81, 82]. Dopant **9** in which two binaphthyls were connected with an azo linkage was found to exhibit unprecedented reversible photoisomerization in both organic solvent and LC media only upon visible light irradiation [82]. Furthermore, this compound had very good solubility in the nematic LC hosts, which greatly facilitated the reflection color tuning by using visible light irradiation.

5.4.1.3 Handedness Inversion with Azobenzenes

In induced CLCs containing multiple chiral dopants, the overall HTP (β) of dopants is the sum of individual HTP (β_i) contributions:

$$\text{HTP}\ (\beta) = \sum_i x_i \beta_i$$

where x_i is the molar fraction of i component $(\sum x_i = 1)$ and β_i is the HTP of corresponding component with a sign (+) or (−) for right-handed and left-handed induction, respectively. When a photoresponsive dopant was used together with a light insensitive dopant with opposite handedness, by adjusting the molar ratio of these two dopants, the sign of the overall HTP can be switched by varying the HTP of photoresponsive dopants, i.e. the handedness of this system can be inverted by light.

This strategy has been successfully demonstrated by several groups by using azobenzene dopant combined with another light-insensitive dopant [59, 83, 84]. For example, Kurihara et al. reported a photo-controlled handedness switching of the photoresponsive CLCs consisting of chiral azobenzene (*S*)-**10** and non-photoresponsive (*R*)-**811** (Fig. 5.9) [59]. Chiral azobenzene (*S*)-**10** induced a left-handed helix in achiral nematic E44 whereas (*R*)-**811** induced a right-handed helix. Based on the evaluation of the HTPs of these two dopants, a CLC mixture was formulated by doping 8.4 % of (*S*)-**10** and 11.6 % of (*R*)-**811** into the LC host E44. At the initial state, the dopant (*S*)-**11** is predominant and thus induced a left-handed cholesteric phase. UV irradiation led to the decrease of HTP of (*S*)-**10** and the gradual disappearance of the fingerprint texture indicating the increase of the pitch. Upon further irradiation with UV light, the contribution of (*R*)-**811** became predominant and the mixture exhibited a right-handed cholesteric phase as evidenced by the reappearance of fingerprint texture.

The handedness inversions induced by a single azobenzene chiral dopant have also been reported due to their dramatic shape change and the subtle interactions with host molecules. Some examples of the dopant are shown in Fig. 5.10. Axially

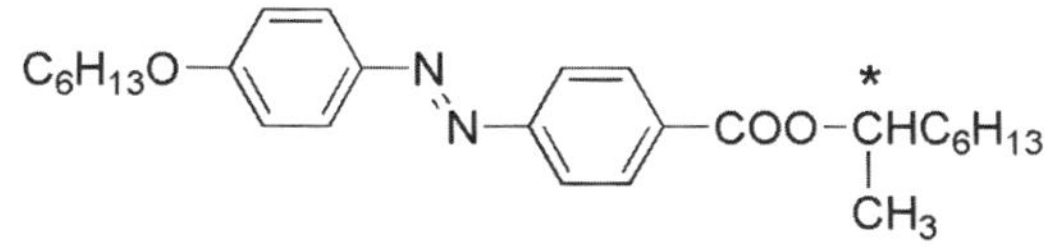

10

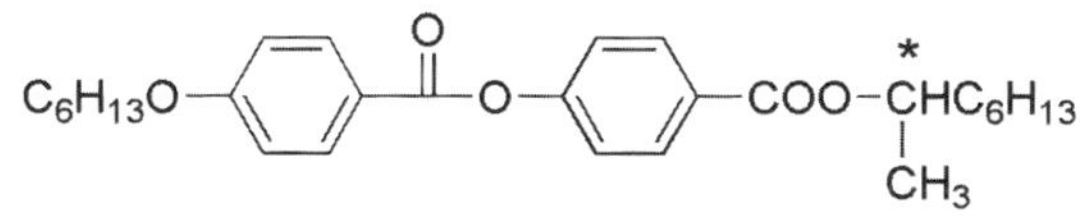

R 811

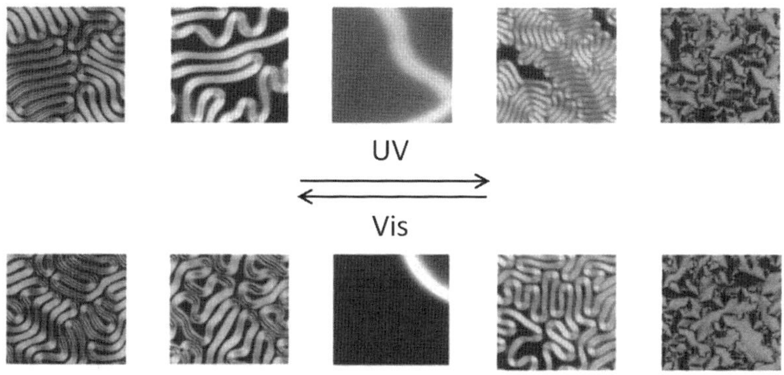

Fig. 5.9 *Top* Molecular structure of chiral azobenzene **10** and non-photoresponsive chiral dopant **811**; *Bottom* Polarized optical micrographs of 11.6 wt% (*S*)-**10** and 8.4 wt% (*R*)-**811** in E44 upon UV and visible light irradiation at 30 °C. The LC mixture was in a 5 μm glass cell without any alignment treatment. Reproduced with permission from [59]. Copyright 2001 American Chemical Society

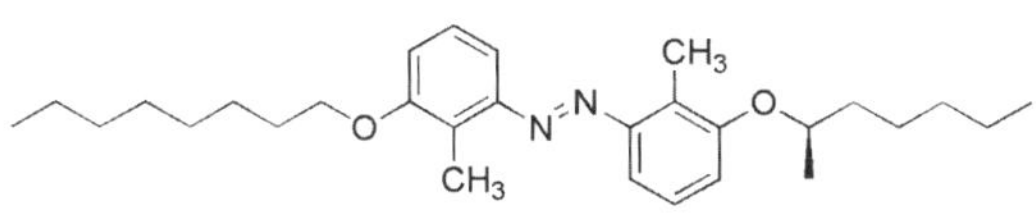

11

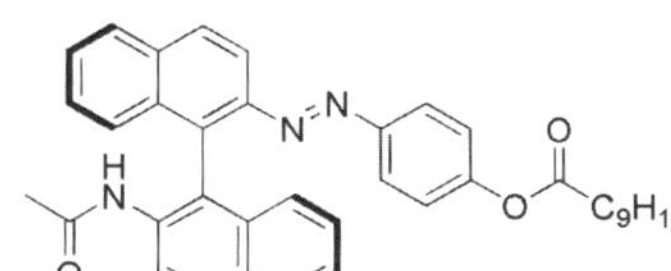

12

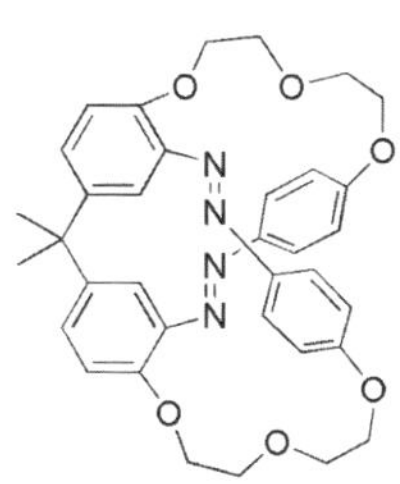

13

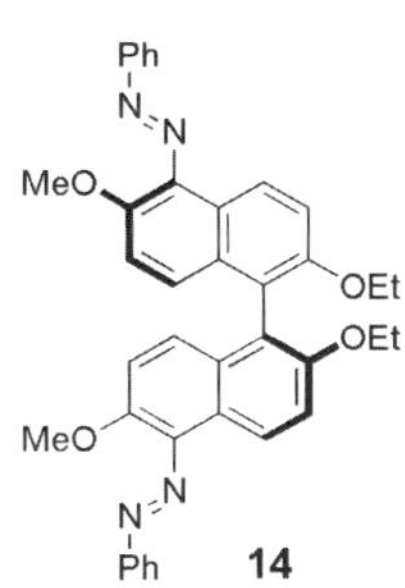

14

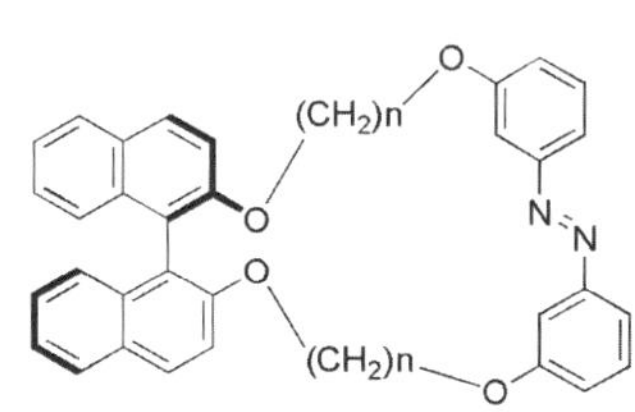

15 n = 3
16 n = 5

Fig. 5.10 Examples of chiral azobenzene dopants capable of promoting helix inversion in CLCs

chiral binaphthyl derivatives are known as powerful helicity inducers [79]. When combined with photochromic units, their conformation can be easily controlled by light. The key structural parameter of binaphthyl conformation is the dihedral angle (θ) of two naphthalenes, which plays a significant role in their cholesteric induction abilities. For the same chiral configuration, the *transoid* conformation ($\theta > 90°$) and the *cisoid* conformation ($\theta < 90°$) induce the helicity with opposite handedness. For example, the *transoid* form of *S* configuration induces a left-handed CLC and the *cisoid* form results in right-handed induction [85, 86]. Thus it is possible to switch the handedness through modulating the conformation of binaphthyl derivatives [71, 87–89].

Li et al. recently reported a fast light-induced reversible handedness inversion of a self-organized helical superstructure, i.e. cholesteric phase, using chiral cyclic azobenzene compound (*R*)-**15** and (*R*)-**16** [89]. These two compounds showed photochemically reversible *trans* to *cis* isomerization in solution and exhibited good solubility, high HTPs, and a large change in HTPs in three commercially available achiral LC hosts. More interestingly, the photoisomerization of the two chiral cyclic dopants were found to be able to induce helix inversion in the cholesteric phase. As shown in Fig. 5.11, a typical oily streak texture observed for the CLC mixture of 10 wt% (*R*)-**15** in nematic LC ZLI-1132 under planar aligned conditions was quickly transformed into a planar N texture upon UV irradiation (Fig. 5.11a, b). As the sample in the N phase was rotated between fixed crossed polarizers, an extinguishing orientation in the cell was found when the orientation of the molecular director was along one of the polarizer directions (Fig. 5.11c). This transient N phase was quickly transformed into an N* phase upon continued UV irradiation for a few more seconds (Fig. 5.11d). The whole switching process was reversible with 440 nm irradiation.

The handedness inversion in these single dopant systems are mainly dependent on chance and found occasionally. It is hard to predict what kind of azobenzene molecules can experience such kind of conformation change of binaphthyls. Very recently, Li et al. reported a novel design strategy based on the chiral conflict and equilibrium shifting between multiple chiral moieties in a single molecule [90]. A series of azobenzene derivatives (*S, R, S*)-**17–19** bearing axially chiral binaphthyl units of opposite chiral configurations were synthesized (Fig. 5.12). They were found to induce the handedness inversion in different LC hosts upon light irradiation with moderate to high HTPs, where the chirality of each chiral moiety does not need to be reversed. Since the *R* and *S* configuration of unbridged binapthyls normally will induced right-handed and left-handed CLCs, the design strategy is to incorporate multiple chiral binaphthyls with opposite chirality into a single molecule functionalized with photoresponsive groups, and hopefully the helix inversion can occur based on the chiral conflict and equilibrium shift between these chiral moieties.

The handedness inversion behaviors upon UV irradiation were observed in both wedge cells and homeotropic cells with the cholesteric (N*)—nematic (N)—cholesteric (N*) phase transition sequence as confirmed by the Cano's lines and fingerprint textures (Fig. 5.12). The mechanism of the helix inversion is proposed

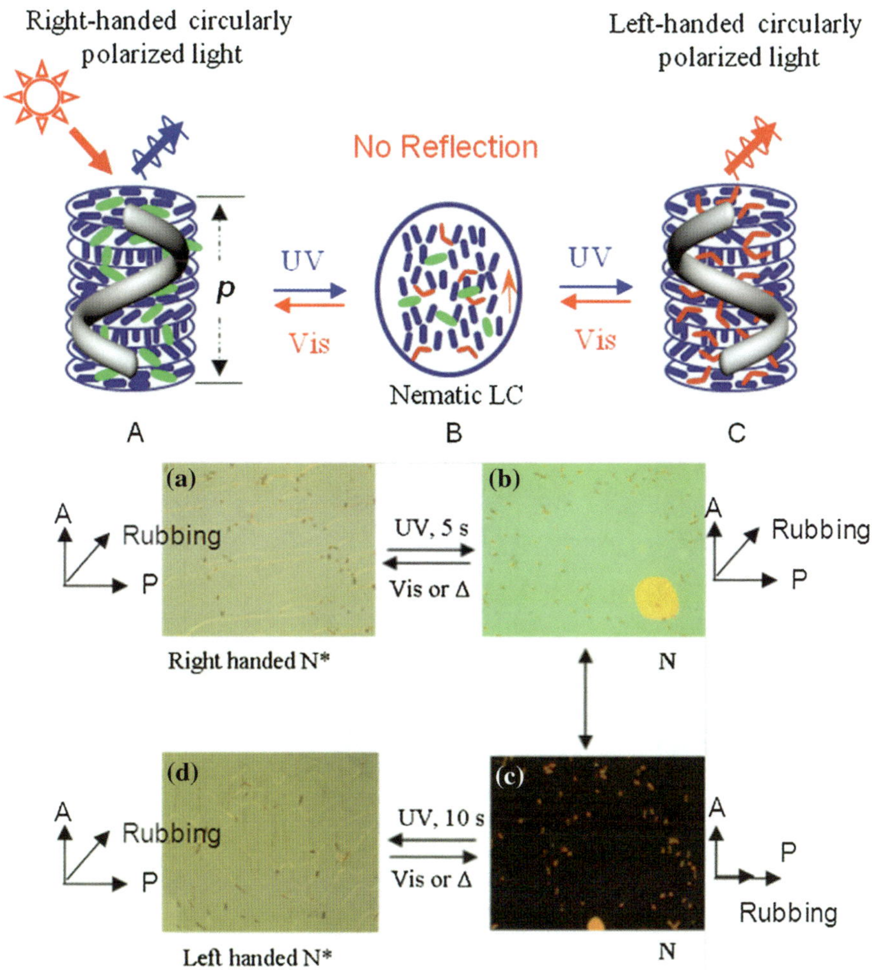

Fig. 5.11 *Top* **a–c** Schematic mechanism of reflection wavelength tuning and handedness inversion of light-driven chiral molecular switch or motor in achiral nematic LC media. Bottom: Polarized optical photomicrographs of a planar aligned N* film containing 10 wt% (*R*)-**15** in ZLI-1132 at room temperature, showing reversible phase transitions occurring by light irradiation of the sample inside a 5 μm cell: **a** oily streak texture of the N* phase before irradiation; **b** N phase obtained by exposure of the sample to UV irradiation; **c** extinguishing orientation of the N cell by rotating between crossed polarizers; **d** regeneration of the oily streak texture of the N* phase upon continued irradiation. Reproduced with permission from [89]. Copyright 2010 American Chemical Society

in Fig. 5.13. The right-handednesses of (*S*, *R*, *S*)-**17–19** at the initial state are attributed to the central axially chiral azobenzene part, whose derivatives normally exhibit very high HTPs with right-handedness [73]. Although the two binaphthyl units induced CLC with left-handedness, the combined effect of these three chiral

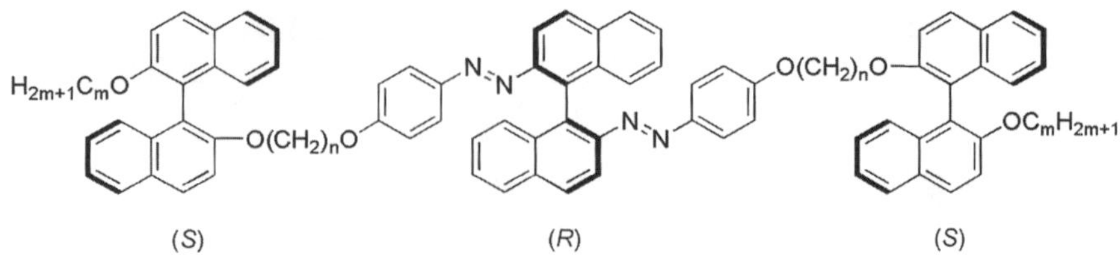

17 m = 18, n = 11; **18** m = 8, n = 6; **19** m = 8, n = 3

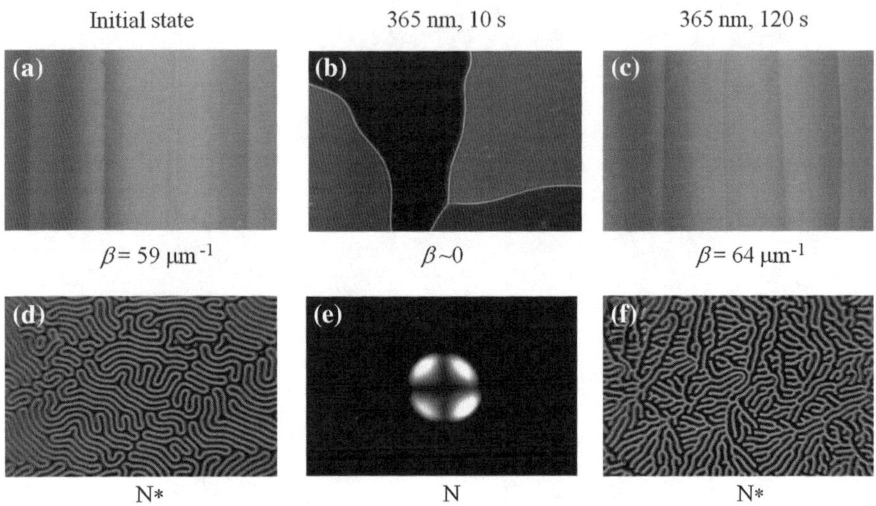

Fig. 5.12 Handedness inversion of 0.27 mol % (*S*, *R*, *S*)-**17** in 5CB observed in wedge cell (**a**–**c**) and homeotropic cell (**d**–**f**) from right-handed CLC (**a**, **d**) to left-handed CLC (**c**, **f**) through transient nematic phase (**b**, **e**). Reproduced with permission from [90]. Copyright 2013 John Wiley & Sons

moieties gave an overall right-handedness, i.e. the azobenzene part played a predominant role in the handedness (Fig. 5.13top). Upon UV irradiation, the right-handedness contribution from the central azobenzene part decreased due to the *trans–cis* isomerization [73], while the two binaphthyl units still maintained comparable left-handed contribution with initial state since they were far away from the azobenzene part. Although minor conformation changes of two binaphthyl units during photoisomerization are possible, when the right-handed contribution decreased significantly enough, the left-handedness became predominant and thus resulted in overall left-handed cholesteric phases (Fig. 5.13bottom).

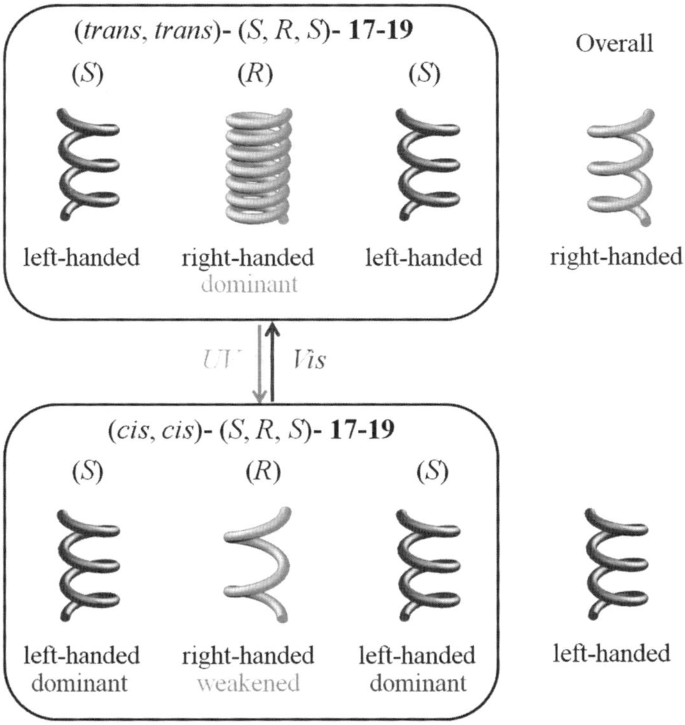

Fig. 5.13 Schematic illustration of the mechanism of handedness inversion in self-organized helical superstructures induced by dopants (*S*, *R*, *S*)-**17–19**. Reproduced with permission from [90]. Copyright 2013 John Wiley & Sons

5.4.2 Overcrowed Alkene-Based CLCs

5.4.2.1 Handedness Inversion with Overcrowed Alkene Dopants

Chiral overcrowded alkenes are known as a family of photoresponsive molecules which are more likely to induce handedness inversion of the cholesteric helix when used as chiral dopants due to the pseudoenantiomeric relationship of the two isomers. A large variety of this type of compounds have been synthesized and characterized by Feringa et al. and other groups [91–102]. Take compound **20** as an example (Fig. 5.14) [92], its initial HTP at (*P*)-*cis* form in nematic M15 is $+5.9\ \mu m^{-1}$, i.e. it induces a right-handed helicity. Irradiation with UV light leads to photoisomerization from *cis* form to *trans* form with a PSS exhibiting an overall HTP of $-9.6\ \mu m^{-1}$. However, the drawback of this compound is the low HTPs at both states. When introducing the fluorene group into this type of molecule, helix inversion with high HTPs at both the states is achieved due to the structural compatibility of the fluorene moiety with the calamitic LC host molecules [100].

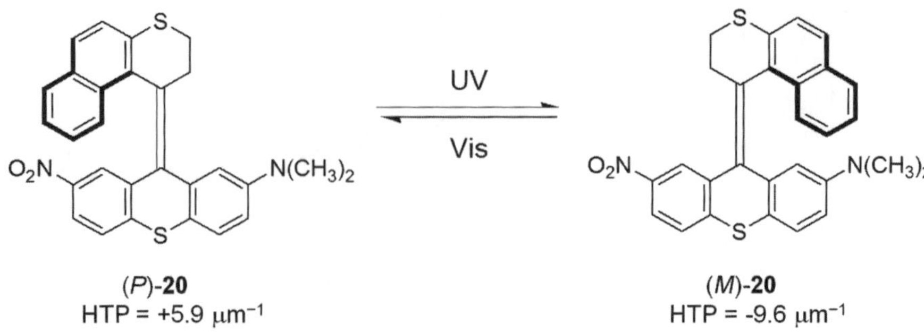

Fig. 5.14 Photoisomerization of overcrowded alkene dopants **20**

5.4.2.2 Mechanical Motion Induced by Overcrowed Alkene Dopants

In 2006, Feringa and coworkers demonstrated the ability to rotate a glass rod at the micrometer scale in photoresponsive CLCs facilitated by a chiral overcrowded alkene dopant **21** (Fig. 5.15) [103]. Upon UV irradiation, the photoisomerization around the central double bond of this dopant led to the helicity inversion of the molecule. A subsequent thermal transformation occurred readily at 20 °C. The photochemical steps combined with following thermal steps generated a full 360° rotary cycle of compound **21** (Fig. 5.15top). When the CLC mixture, composed of 1 wt% of **21** and a LC host, exposed to the air, a unidirectional thin film of this CLC material showed a polygonal fingerprint texture, which is typical texture of cholesteric phase. The UV irradiation resulted in the reorganization of the polygonal texture in a rotational fashion. The rate of rotation gradually decreased until the process halted at PSS. Removing the light caused the rotation to resume with the opposite direction. The rotation of the texture induced by the overcrowded alkene was utilized to move a submillimeter-sized glass rod placed on top of the film. As shown in Fig. 5.15c, the glass rod rotated in the same direction as the cholesteric texture during the photoisomerization or thermal transition steps. Detailed studies revealed that the rotational reorganization resulted from the pitch change accompanied with helix inversion and the direction of this reorganization correlated to the sign of the change in HTPs of the dopants [95, 98, 104]. These works are interesting not only for the demonstration of macro-scale mechanical motion induced by synthetic molecular systems, but also for the correlating rotational movement with the pitch change and helix inversion in CLC.

5.4.2.3 Reflection Color Modulation with Overcrowed Alkene Dopants

Overcrowded alkene dopants are also found capable of tuning the reflection colors in CLCs if the HTP at one state is high [94, 105]. Take compound **21** as an example, its initial HTP at (*P*, *P*)-*trans* form in nematic E7 is +90 μm^{-1}. UV irradiation resulted in helix inversion and the PSS was achieved with a HTP of

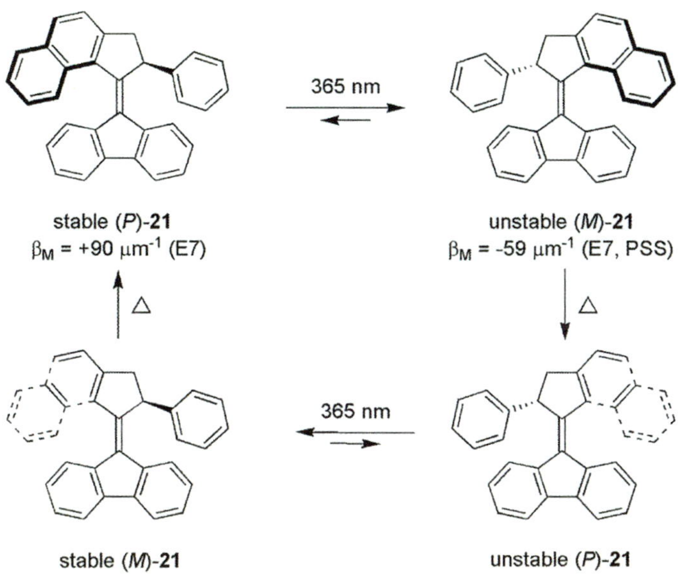

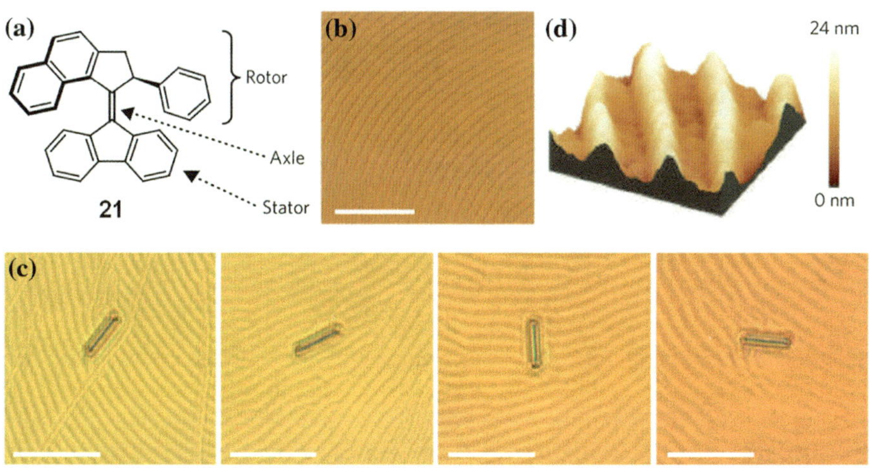

Fig. 5.15 Schematical illustration of photoisomerization of molecular motor. **a** Molecular structure of chiral motor **21**. **b** Polygonal texture of a LC film doped with 1 wt% chiral motor **21**. **c** Glass rod rotating on the LC during irradiation with ultraviolet light. Frames 1–4 (from *left*) were taken at 15 s intervals and show clockwise rotations of 28° (frame 2), 141° (frame 3) and 226° (frame 4) of the rod relative to the position in frame 1. Scale bars, 50 μm. **d** Surface structure of the liquid–crystal film (atomic force microscopy image; 15 μm^2). Reproduced with permission from [103]. Copyright 2006 Nature Publishing Group

$-59\ \mu m^{-1}$. The reverse process only occurred by thermal relaxation and it took more than 30 min to restore the initial state. The high HTP at the initial state with the helix inversion enabled a broad tuning range of reflection wavelength covering the entire visible region with reversibility (Fig. 5.16).

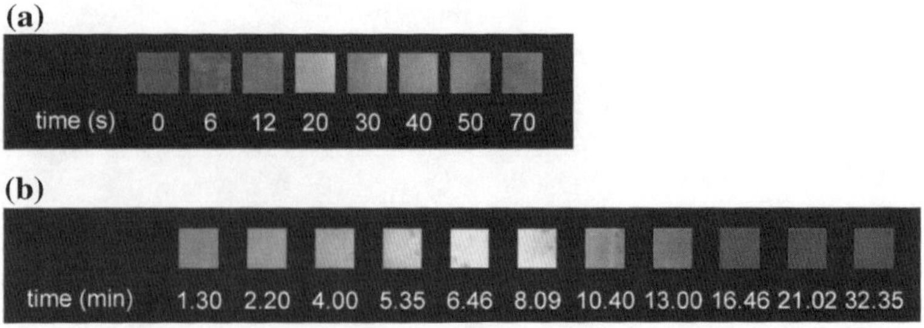

Fig. 5.16 Color change with time of 6.8 wt% of **21** in E7: **a** upon UV irradiation with 365 nm light and **b** during thermal relaxation. Color photographs are taken perpendicular to the surface of the film at room temperature. Reproduced with permission from [94]. Copyright 2006 John Wiley & Sons

Very recently, Sarah et al. investigated the time-programmed helix inversion in phototunable CLCs using overcrowded alkene dopants [106]. As shown in Fig. 5.17, the UV irradiated state showed a blue reflection with central wavelength around 400 nm. When kept in dark, the reflection shifted toward longer wavelength with disappearance of reflection color in 300 s, which was followed by a blue shift in reflection wavelength after helix inversion and the reflection color reappeared at around 1000 s. The total relaxation took about 1800 s and the final reflection color was blue. They also pointed out that the relaxation rate of the cholesteric helix is fully determined by helix inversion of the molecular dopants. Therefore fast relaxation in these systems can be achieved by using dopant with fast relaxation rate.

5.4.3 Diarylethene-Based CLCs

Although a few examples of chiral diarylethene molecules were reported showing cholesteric phase [107, 108], they are normally employed as chiral dopants in photoresponsive CLCs. Unlike the *trans–cis* isomerization of azobenzene and chirality inversion of overcrowed alkenes, the photomodulation of CLCs using diarylethene dopants to achieve satisfied device performance is quite challenging. The variation of HTPs during photocyclization has been utilized to fulfill the phase transition between N and N* [109–113]. At a certain doping concentration, the isomer with higher HTP can efficiently induce a cholesteric phase, while the isomer with lower HTP only generates an apparent nematic phase. The light-induced photocyclization can reversibly switch the LC phase between N and N*. However, the HTPs of diarylethene dopants in these systems are very low [114].

A major breakthrough in designing high HTP diarylethene dopants was achieved by Li et al. with the development of compound **22–25** (Fig. 5.18) [115–118]. The open and closed isomers of compound **22** exhibited high HTPs of 92 and 77 μm^{-1} in 5CB, respectively. The relatively small change in HTP only brought

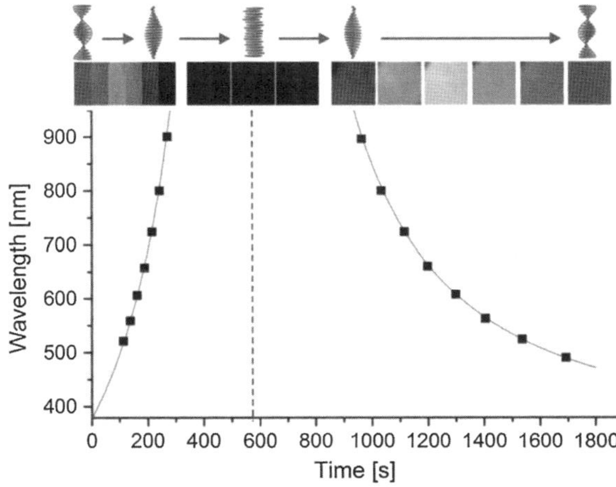

Fig. 5.17 Thermal (helix-inverting) relaxation of CLC doped with overcrowded alkene dopant, as followed by UV/Vis spectroscopy. The data points correspond to the *right edge* of the selective reflection band. The *top panel* shows the reflection colors and pictograms of the cholesteric helix. Reproduced with permission from [106]. Copyright 2013 Royal Society of Chemistry

out a reflection tuning range of ~50 nm. The compound (*S, S*)-**24** was found to possess remarkable changes in HTPs during photoisomerization in addition to the very high HTP of 104 μm^{-1} at the initial state. A doping concentration of 7.7 wt% enabled the phototuning of reflection colors to achieve three primary red, green, and blue colors. Although the phototuning of reflection color has been documented with azobezene and overcrowed alkene dopants, this is the first example of a diarylethene dopant for this purpose with the advantage of thermal stability at both states. As shown in Fig. 5.19, three primary RGB colors was observed simultaneously in a single thin film based on different UV irradiation time facilitated by masking at different areas: red, no irradiation; green, irradiated for 10 s; blue, irradiated for 25 s (Fig. 5.19a–c). After driving the background color to blue by UV irradiation, the red and green reflection colors were recorded through visible light irradiation for different time (Fig. 5.19e–h). Moreover, the optically addressed images can be erased by light irradiation when desired, and the cell is rewritable for many times due to the excellent fatigue resistance.

The structurally similar perfluorocyclopentene based chiral dopant **25** not only exhibited higher HTPs but also showed excellent thermal stability. The half-life time based on the first order kinetics of the thermal decay in absorbance was calculated to be 115 days. A high HTP of 188 μm^{-1} was observed in 5CB when used as a chiral dopant. The increase of HTP during photoisomerization enabled a photo-induced isothermal phase transition from nematic to cholesteric phase at a very low doping concentration of 0.1 mol %. The reflection wavelength of the CLCs with higher doping concentrations was reversibly tuned in the visible region to achieve colors ranging from orange to green and further to blue (Fig. 5.20).

Apart from the HTP tuning for phase transition and color change, a binaphthyl-derived diarylethene dopant was demonstrated inducing the handedness inversion behavior by Akagi et al. [119]. (*R*)-**26** with two un-bridged binapthyls can induce a right-handed CLC at its open form, while could be switched to left-handed upon

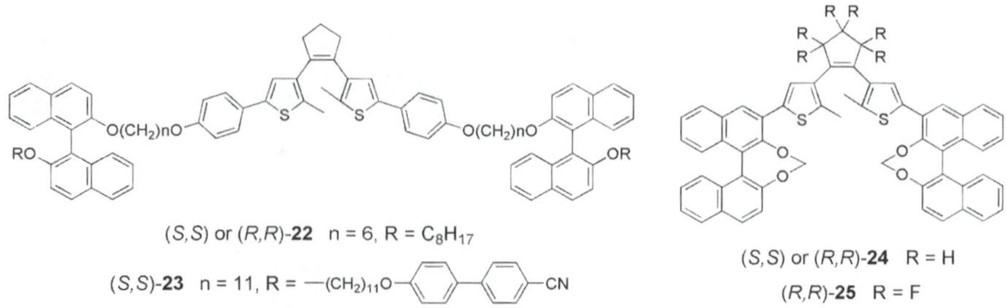

Fig. 5.18 Chemical structures of chiral diarylethene dopants **22-25**

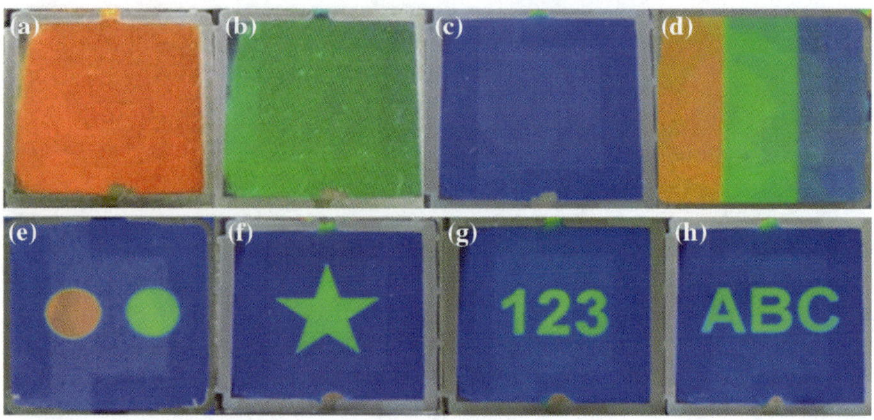

Fig. 5.19 Real cell images of a 8 μm thick planar cell (2.1 cm × 2.5 cm) filled with 7.7 wt% (*S,S*)-**24** in E7. Reproduced with permission from [116]. Copyright 2012 American Chemical Society

Fig. 5.20 Reflection spectra of 1.80 mol % of (*R,R*)-**25** in 5CB upon UV irradiation at 310 nm for 0 s (*orange*), 15 s (*green*), and 60 s (*blue*). Reproduced with permission from [118]. Copyright 2013 Royal Society of Chemistry

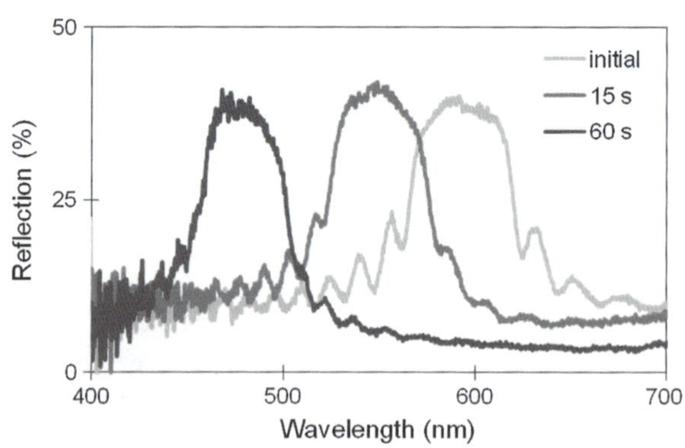

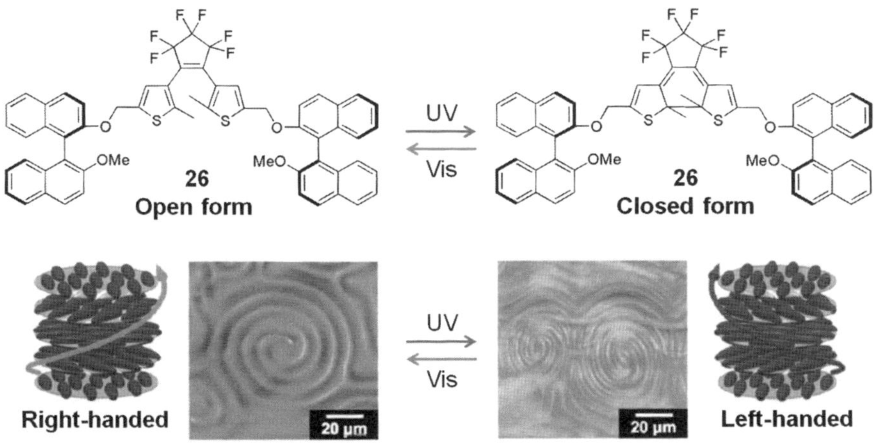

Fig. 5.21 Handedness inversion induced by diarylethene dopant **25**. Reproduced with permission from [119] Copyright 2012 American Chemical Society

UV irradiation (Fig. 5.21). It is believed that the reversible handedness inversion originated from the conformation change of binapthyl moieties as well as the stereospecific intermolecular interactions between the binaphthyl moieties and the surrounding nematic LC molecules. This is the first example of diarylthene dopant capable of handedness inversion in CLCs although it requires specific LC mixture host and only shows low HTPs.

5.4.4 Other CLC Systems Based on Photochromic Molecules

Various photochromic molecules other than above-mentioned azobenzenes, overcrowded alkenes and diarylethenes have also been designed and synthesized for the phototuning of properties of CLCs, such as butadienes [36, 120, 121], spirooxazines [122–124], fulgides [125, 126], and bicyclic ketones [37, 127–129]. Tamaoki et al. reported a series of chiral butadiene mesogens [33, 97, 98], which not only exhibit cholesteric mesophase but also respond to light with reflection color change at the cholesteric state. The photochemically recorded color images can be stored at glassy state with good thermal stability by fast cooling down to ~0 °C. Chiral spirooxazines and fulgides were reported as chiral dopants with the characteristics of fast thermal relaxation and thermal stability, respectively. For bicyclic ketones, their photoresolution is capable of inducing the phase transition from nematic to cholesteric phase [37, 127–129]. The specific characteristics of each type of photochromic molecules are very interesting, making them promising candidates for different application purposes.

5.5 Photoresponsive Chiral Smectic LCs

There are two ways to generate photoresponsive chiral SmC* LCs. The simple one is to add a photochromic molecule to SmC* LCs and the other is based on chirality amplification, i.e. using a photoresponsive chiral dopant in SmC LCs to induce a phototunable SmC* phase. Some examples of mesogens that showing SmC or SmC* are presented in Fig. 5.22.

Similar to the phase transition from nematic to isotropic phase induced by azobenzene molecules, the *trans–cis* isomerization also destabilize the SmC* phase composed of calamitic mesogens and lower the Curie point, which is a transition temperature where the SmC* will transform from ferroelectric to non-ferroelectric. Some examples have been reported with an early demonstration by Ikeda et al. [130–135]. For example, a photoresponsive SmC* was formulated by doping 3 mol % of 4,4′-disubstituted azobenzene **29** into a FLC host **27** and the UV irradiation at 260 nm resulted in the lowering of Curie point and the coercive force *Ec* required to switch the SSFLC due to the destabilization of bent shape *cis* isomers [130] (Fig. 5.23). When the electric field was close to *Ec* before irradiation, the flip of polarization of SmC* was achieved. It is noteworthy that its response time ~500 μs is much faster than normally observed for photochemical N–I phase transitions [131].

A dithienylperfluorocyclopentene derivative was reported by Lemieux et al. for the photochemical switching of polarization in SmC* and this system exhibited the advantage of thermally stability at irradiated state over the azobenzene-doped systems [136, 137]. FLC mixtures **MDW950-PhP** doped with 1.0 and 3.0 mol % of **30** was irradiated with UV light (Fig. 5.23), the photoisomerization of **30** caused a destabilization in SmC* and the P_S versus temperature plots shifted to lower temperatures. The phase transition temperature from ferroelectric SmC* to non-ferroelectric SmA* was decreased by 2 °C upon irradiation of the 3 mol % FLC mixture. The photoswitching was bistable and fatigue resistance (Fig. 5.24).

In addition to the photochemical switching of P_S based on destabilization effect caused by dramatic shape change of dopant molecules, a novel strategy utilizing photoresponsive chiral dopant with similar shape at both states was developed by Lemieux et al. [138–143]. This approach is based on the photoinduced change in transverse dipole moment of thioindigo dopant which maintains a rod-like shape in both forms during photoisomerization (Fig. 5.25). When doped into an achiral non-ferroelectric SmC host, the chiral thioindigo dopant can induce a P_S and the photoisomerization causes a change in P_S without concomitant destabilization effect in SmC* phase.

The thioindigo compounds can undergo *trans–cis* photoisomerization along the central double bond in the visible range, which was revealed by the absorption spectra. The initial design of this type of molecules aimed at incorporating a chiral side-chain to convert the change in transverse dipole moment into a change in P_S [138, 139]. Compound **31** and **32** were able to induce a ferroelectric SmC* phase with a positive sign of P_S while compound **32** had better compatibility than **31** in

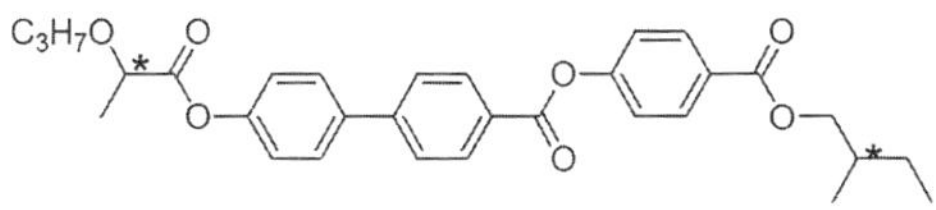

27 Cr 38 SmC* 64 SmA* 110 N* 112 I

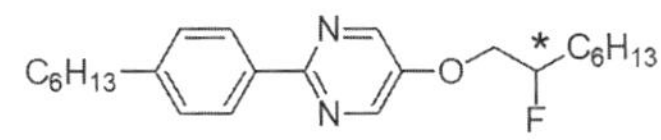

28 SmX* 25 SmC* 45 SmA* 84 I

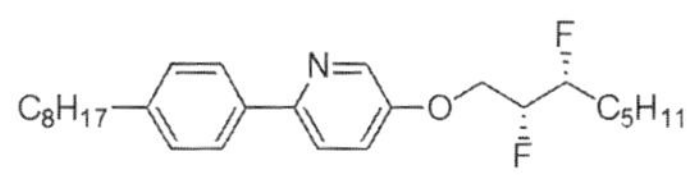

MDW950

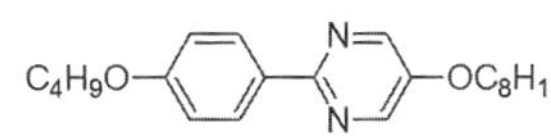

PhP: Cr 58 SmC 85 SmA 95 N 98 I

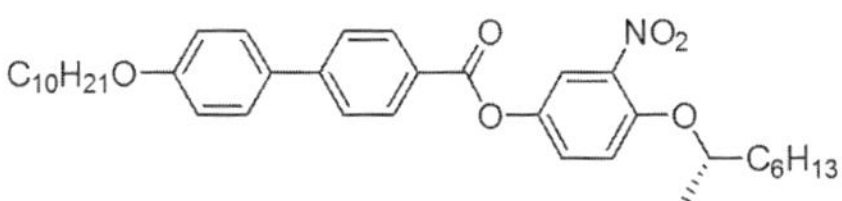

W314

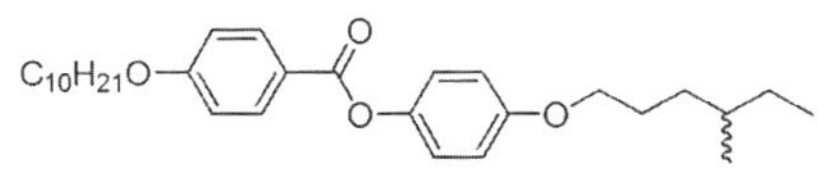

PhB: Cr 35 SmC 70.5 SmA 72 N 75 I

Fig. 5.22 Chemical structures of SmC or SmC* LCs

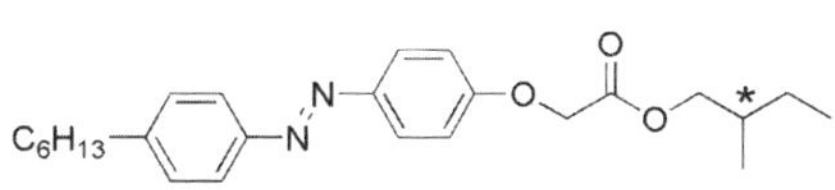

29

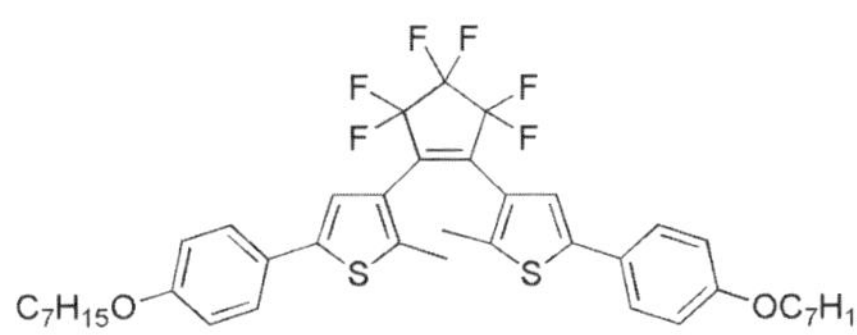

30

Fig. 5.23 Chemical structures of photoresponsive dopants for SmC*

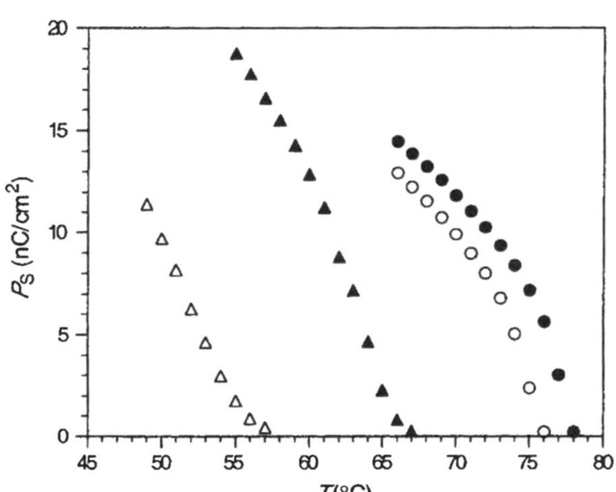

Fig. 5.24 Spontaneous polarization P_S as a function of temperature T for the FLC mixture **MDW950-PhP** doped with **30** at 1 mol % (*circles*) and 3 mol % (*triangles*), before and after irradiation at 313 nm (*open* and *filled symbols*, respectively). Reproduced with permission from [136]. Copyright 2002 American Chemical Society

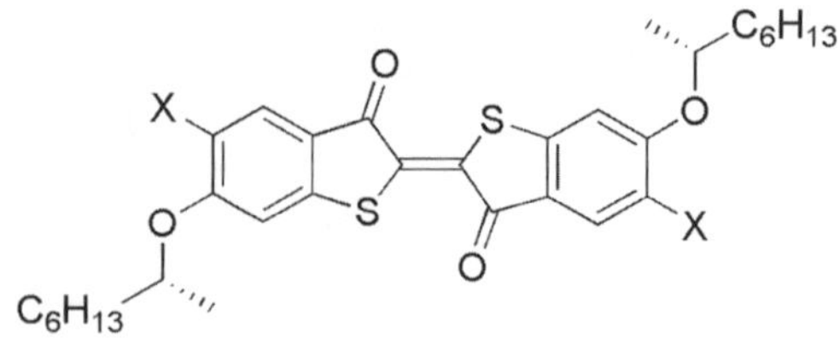

31 X = NO_2 **32** X = Cl **33** X = H

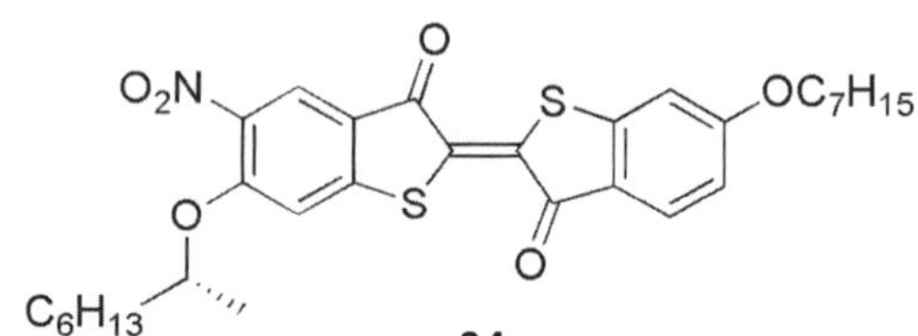

34

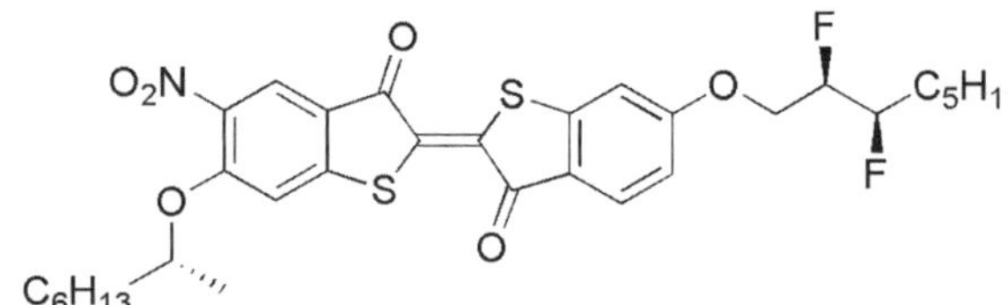

35

Fig. 5.25 Chemical structures of chiral thioindigo dopants

SmC host **PhB**. The plot of P_S as a function of temperature at both states showed that no appreciable shift was found in the Curie point, i.e. no destabilization effect caused by molecular shape change from the photoisomerization observed. The P_S of different concentrations of **32** in **PhB** was found to increase upon irradiation at 532 nm. As shown in Fig. 5.26, irradiation of a 1.9 mol % mixture resulted in an almost fourfold increase, however, the magnitude decreased upon increasing dopant concentration, which might be due to the suppression of *trans–cis* photoisomerization by aggregate formation.

When combined with a photoinert chiral dopant that induced an opposite P_S, dopant **32** was able to switch the sign of P_S by light irradiation (Fig. 5.27) [141]. This is similar with the handedness switching in CLCs using photoresponsive chiral dopant combined with another photoinert chiral dopant with opposite handedness induction. Since **32** induced a positive P_S, chiral diester **36** was selected as photoinert dopant as it induced a negative P_S. The dopant concentration of **32** and **36** were adjusted to 3.0 and 1.3 mol %, respectively. At the initial state, the negative P_S induced by **36** is predominant over positive P_S induced by **32** and the whole system had a net negative P_S in the dark. Upon irradiation at 532 nm, the positive P_S from **32** was increased dramatically, while negative polarization remained unchanged. The increased positive P_S became predominant and the system showed a net positive P_S.

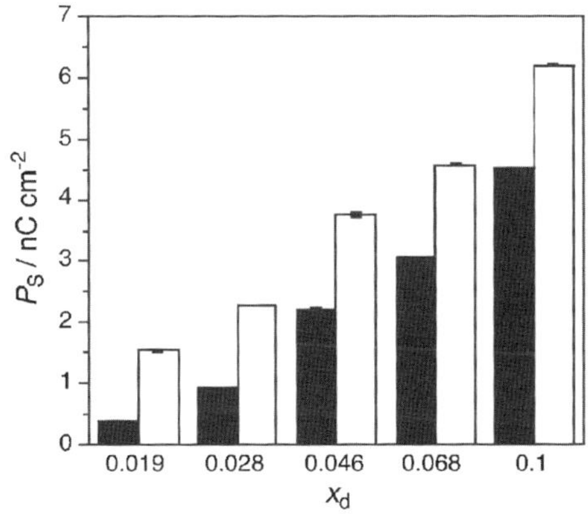

Fig. 5.26 Spontaneous polarization P_S as a function of dopant molar fraction X_d measured at 10 K below the Curie point for mixture of **32** in **PhB** in the dark (*black columns*) and under irradiation at 532 nm (*white columns*). Reproduced with permission from [139]. Copyright 1999 Royal Society of Chemistry

By further design of thioindigo compound that having different chiral units in one single molecule, the photoinduced P_S flip can be realized only by single dopant [142]. In the compound **35**, the (*R*,*R*)-2,3-difluorooctyloxy unit is decoupled from the core and induces a negative P_S, while the (*R*)-2-octyloxy side-change and the thioindigo core are strongly coupled via the nitro group and thus form one stereo-polar unit that result in a positive P_S. The increase in transverse dipole moment of the thioindigo core upon *trans–cis* photoisomerization raises the polarization power of the coupled 2-octyloxy/thioindigo unit above that of the 2,3-difluorooctyloxy unit and inverts the sign of P_S. As shown in Fig. 5.28, the sign of P_S at the initial state in dark was negative confirmed by the relative configuration of the electric field and the switching position of the sample, which revealed that the negative P_S from 2,3-difluorooctyloxy unit plays a predominant role. The photoisomerization inverted the P_S sign from negative to positive due to the increase of positive P_S from 2-octyloxy/thioindigo moiety. The P_S inversion can be reversed by irradiation at 450 nm as a result of *trans*-enriched PSS. The complete reversion to the original polarization state was achieved by thermal *cis–trans* isomerization after the solution was kept in dark for ca. 60 s.

This group also investigated the structural effects on P_S photomodulation in unsymmetrical chiral thioindigo dopants [140]. The newly synthesized unsymmetrical chiral thioindigo dopants showed improved compatibility with SmC LC hosts such as **PhB**. However, this improvement was achieved at the expense of P_S photomodulation. The correlation between the polarization power of the thioindigo dopants in the *trans* form and the calculated transverse dipole moment of the thioindigo cores was consistent with the assumption that the thioindigo core contributed to the induced polarization by virtue of steric coupling the chiral 2-octyloxy side-chain. They also pointed out that the design of chiral thioindigo dopants with high P_S photomodulation had to include a symmetrical thioindigo core.

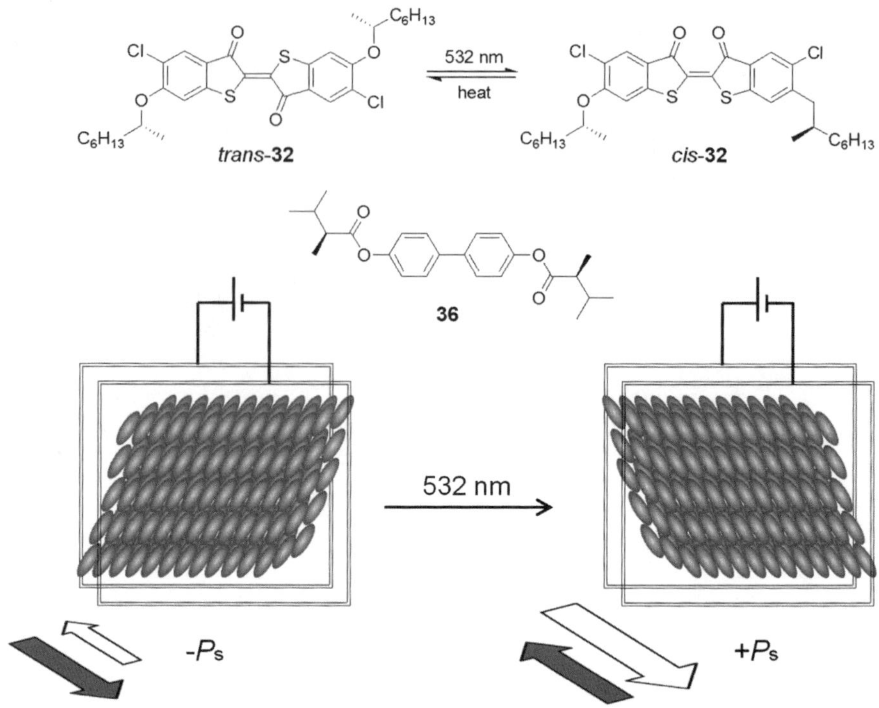

Fig. 5.27 Photoswitching of a FLC mixture with two chiral dopants inducing polarizations with opposite signs: the thioindigo **32** (3 mol %) and the diester **36** (1.3 mol %) in the host **PhB**. The *filled arrow* and *blank arrow* represent the polarizations induced by **36** and **32**, respectively

5.6 Photoresponsive Blue Phase LCs

Compared with phototunable CLCs and SmC* LCs, the BPs that response to light are less investigated and are receiving increasing attention especially in recent years. In 2005, Chanishvili et al. reported the first examples of phototuning the reflection of BPs using *trans–cis* isomerization of azobenzene molecules [144]. The photoresonsive chiral LCs were formulated by doping a chiral dopant into azobenzene-derived nematic LC host. Photoisomerization of azobenzene molecules caused a phase transition from cholesteric to BPs. The reflection wavelength of BPs was also reversibly tuned by light.

Lin et al. also reported an optically switchable band gap of photonic crystal that was based on azobenzene-doped BP (Fig. 5.29) [145]. A CLC mixture was doped with 3 wt% of **M12C** (4-dodecylcyloxy-4′-methoxy-azobenzene), and the BP was present over temperature range of ~ 16 °C, which is larger than that of a typical BP material. The system was kept at BP temperature range and irradiated with 473 nm laser light. Figure 5.29a displayed images of BP platelets with various durations of light irradiation under crossed polarized optical microscope at

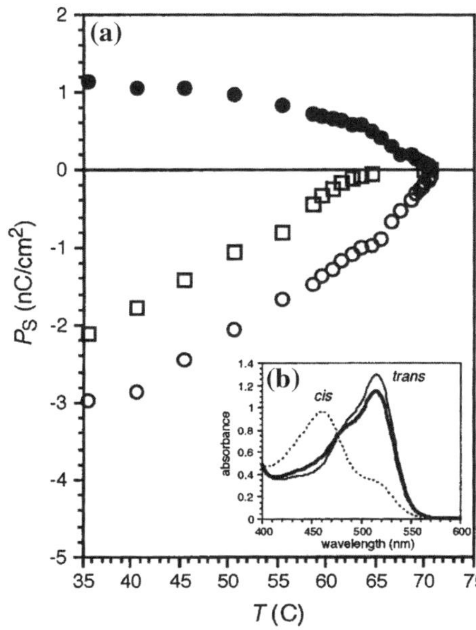

Fig. 5.28 **a** Spontaneous polarization P_S as a function of temperature T for the 1 mol % mixture of **35** in PhB kept in the dark (*open circle*) and under constant irradiation at 510 nm (*solid circle*) and 450 nm (*open square*). **b** UV–vis absorption spectra of 10^{-4} M solution of **35** in the benzene after irradiation at 510 nm (*dashed line*) and 450 nm (*heavy line*) and after the solution was kept in dark for 24 h (*fine line*). Reproduced with permission from [142]. Copyright 2003 American Chemical Society

reflection mode. The reflection color of the BP platelet changed from red-yellow to green within 20 s. The platelets of BP were indexed using a Kossel diagram, which is a widely-employed method to study BPs [146]. The change in the Kossel diagram under irradiation was presented in Fig. 5.29b. The diffraction patterns of the (200) lattice of the BP shrink under light irradiation, indicating that the lattice constant of the BP decreased as the irradiation time increases and the photoisomerization of azobenzene molecules caused the distortion of the cubic unit cell. The reflection wavelength during this process was also collected. The initial reflection was at approximately 610 nm, which experienced a shift to shorter wavelengths under irradiation and reached 550 nm at the PSS (Fig. 5.29c).

The phototuning of BPs can also be fabricated in a pure material system [147]. Das et al. reported a light-induced stable blue phase in photoresponsive diphenylbutadiene based mesogen **37**. This compound was found to exhibit SmA* and N* during heating. When the temperature was kept at 118 °C, the photoisomerization induced an isothermal phase transition from SmA* to N*. Photoirradiation of the SmA* film held at a higher temperature (124 °C) for 100 s resulted in transition to a phase with a characteristic classical BP texture showing in Fig. 5.30. The BP was thermodynamically stable and could be maintained at this state for several hours. The characteristic sharp reflection bands compared to the rather broad reflection bands observed for the chiral nematic phase confirmed the formation of BP. The photoinduced formation of the BP exhibited a reflection centered at 510 nm. Subsequent irradiation led to the blue shift to 480 nm in the reflection band.

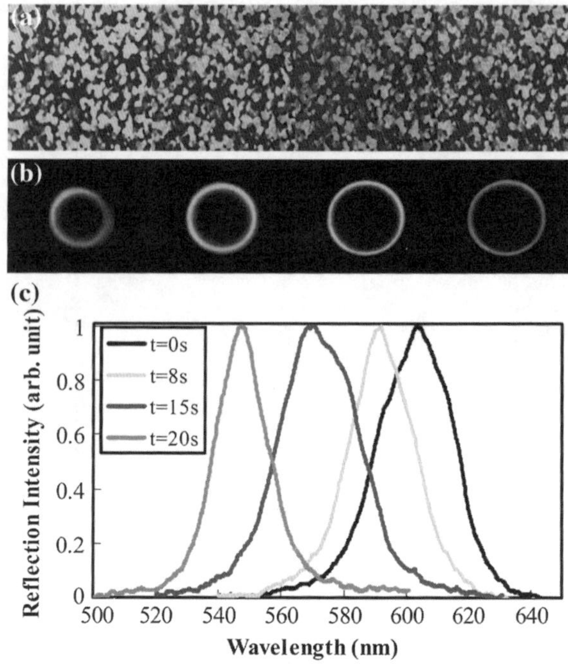

Fig. 5.29 **a** Images under crossed R-POM, **b** Kossel diagram, and **c** Reflection spectrum of BP platelets pumped for various durations. Reproduced with permission from [145]. Copyright 2010 American Institute of Physics

Different with the above BPs based on calamitic mesogens, Takezoe et al. demonstrated an interesting phase transition from BP I to BP III using photo-active bent-core nematic LCs with chiral dopant [148]. The bent-core molecule A131 exhibited a wide nematic range and was doped with 27 wt% of chiral dopant S811. The mixture was kept at 116 °C and the intial state possessed a BP I phase. Upon UV irradiation, the BP I platelets disappeared and were replaced by BP III uniform dark texture, as shown in Fig. 5.31a–c. By ceasing the UV irradiation, the BP III was thermally driven back to a small platelet texture of BP I (Fig. 5.31d–f). The associated pitch change in BP I was also observed and the reflection wavelength could be tuned by light.

The reflection wavelength tuning in aforementioned several systems are quite narrow. Very recently, a breakthrough of full visible range reflection phototuning in BPs was achieved by Li et al. using an axially chiral azobenzene dopant [149]. The initial phase of the doped BP was the BP II with (100) lattice confirmed by the Kossel diagram as shown in Fig. 5.32c (0 s) with a blue reflection color as shown in Fig. 5.32b (0 s). Upon 408 nm light irradiation, the photoisomerization of the azobenzene dopant enlarged the size of the BP lattice and shifted the corresponding reflection toward longer wavelength. Increasing the irradiation time resulted in occurrence of the phase transition from BP II to BP I, which might be attributed to the chirality decrease with dopant isomerization [150]. After the photo-induced phase transition took place, the lattice direction changed from (100) of BP II to (110) of BP I with the corresponding change of reflection wavelength

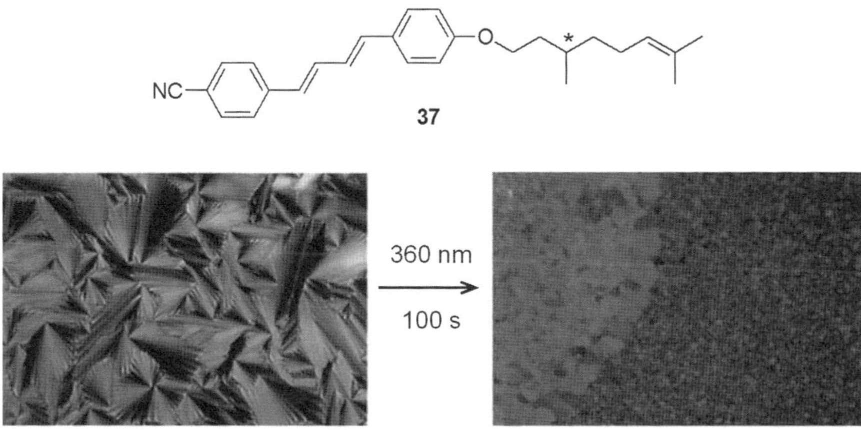

Fig. 5.30 Photoinduced isothermal phase transition from SmA* to BP at 124 °C upon irradiation with 360 nm light. Reproduced with permission from [147]. Copyright 2010 Royal Society of Chemistry

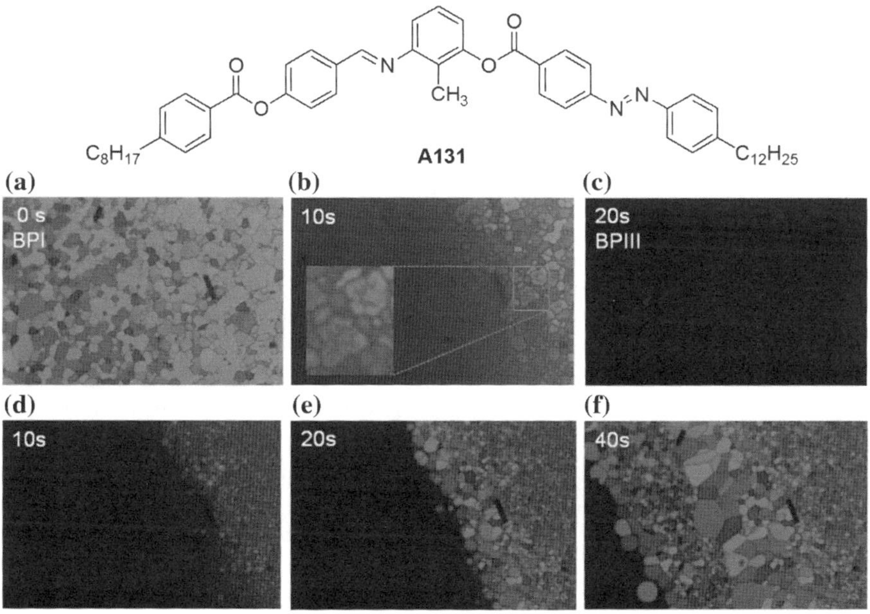

Fig. 5.31 Texture change in a 10 μm thick cell filled with the 27 wt% mixture due to the processes under UV irradiation (**a–c**) and relaxation (**d–f**) after turning off the irradiation at 116 °C. Cracks attributable to the distortion of the cubic lattice are observed particularly on *green* platelets, as shown by an expanded texture in (**b**). Majority of blue domains are observable in the vicinity of the boundary closed to BP III as seen in (**d–f**). Reproduced with permission from [148]. Copyright 2012 Royal Society of Chemistry

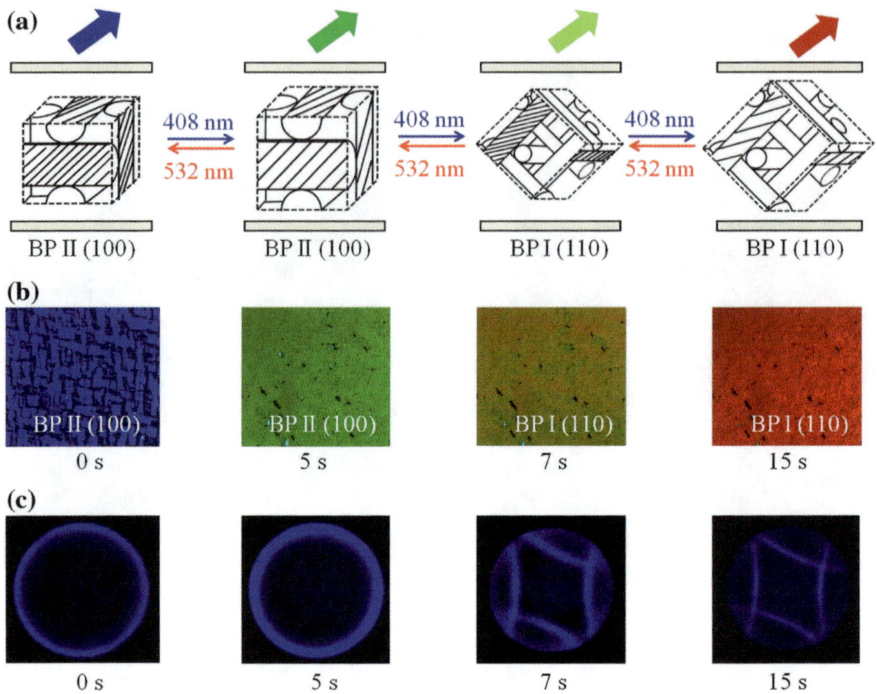

Fig. 5.32 **a** Schematic illustration of the phototuning of the resulting BP. **b** Reflection color images BPs upon 408 nm light irradiation at 40 °C. **c** The corresponding Kossel diagrams with B. Reproduced with permission from [149]. Copyright 2013 John Wiley & Sons

from 520 to 575 nm. Subsequently, a red-shift in reflection upon further irradiation at 408 nm was observed in BP I. If irradiation was stopped at any time, the wavelength of the selective reflection band ceased to change, except by very slow thermal relaxation effect of the dopant towards its initial *trans* form. The rapid reverse process can occur photochemically with 532 nm light irradiation. Figure 5.32b shows BP images with various durations of pumping observed under a crossed polarized optical microscope at reflection mode. Upon 408 nm light irradiation, the reflection color of the BP thin film changed from blue (BP II) to red (BP I) within 15 s. The BP thin films were also indexed using a Kossel diagram as shown in Fig. 5.32c. Both blue and green images in Fig. 5.32b corresponded to (100) lattices of BP II, while both yellow-green and red images represented (110) lattices of BP I. Figure 5.32c also exhibited the variation of the Kossel diagram with different time upon 408 nm irradiation. As the irradiation time increased from 0 to 15 s, the circle patterns of the (100) lattice of the BP II enlarged, switching to diamond-shaped pattern corresponding to (110) lattice of BP I, and gradually shrinking.

Increasing the dopant concentration shifted the initial BP reflection to the UV region, which was tuned across the entire visible region with broad range

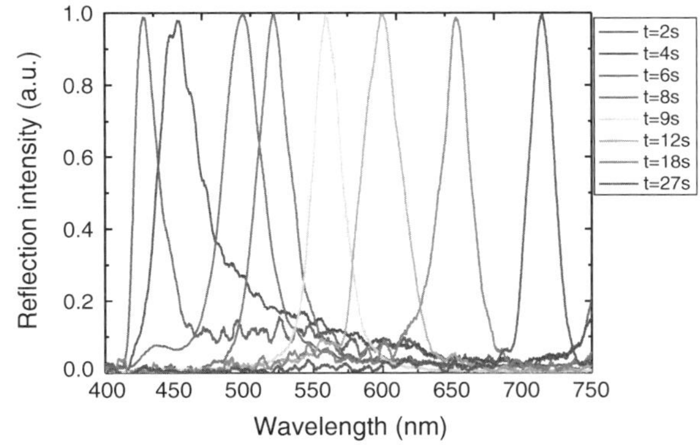

Fig. 5.33 Reflection spectra of BPs doped with axially chiral azobenzene dopant at 37 °C under 408 nm light irradiation with 2, 4, 6, 8, 9, 12, 18, and 27 s (from *left* to *right*). Its reverse process took 33 s upon 532 nm light irradiation. Reproduced with permission from [149]. Copyright 2013 John Wiley & Sons

(Fig. 5.33). Upon 408 nm light irradiation, the reflection bands of BP II red-shifted from the UV to 520 nm. Once the BP II transformed into the BP I, the corresponding reflection bands discontinuously jumped to 560 nm, which was further red-shifted from 560 to 710 nm within BP I state. The reflection spectra shown in Fig. 5.33 clearly indicate that red, green and blue reflections were achieved in this self-organized 3D cubic nanostructured single thin film.

5.7 Conclusions and Outlook

Photoresponsive Chiral LC phases are intriguing due to their unique self-organized superstructures and the abilities of being tuned by light irradiation. This chapter surveyed the recent progress of the photomodulation in three types of chiral LC phases i.e. N*, SmC* and BP. Photoresponsive CLCs are the most investigated with a large amount of literatures describing the light-induced phase transition, reflection color control and helix inversion by using photochromic molecules as chiral/achial LC mesogens or dopants. It is clear from these reports that the development of photoresponsive chiral dopants has received much attention in recent years due to their advantages over other systems. Azobenzene is the most commonly used dopant in cholesteric LCs due to their high HTPs and dramatic molecular shape change in *trans–cis* isomerization, while overcrowded alkene dopants possess the characteristics capable of inducing helix inversion in CLCs. Very recently, the research on chiral diarylethene dopants has achieved progress in the reflection color tuning and helix inversion. The photomodulation in ferroelectric SmC* LCs mainly focused on the phototuning of the magnitude and signs of P_S by using azobenzene, diarylethene and thioindigo compounds as achiral/chiral dopants. Of great interest is chiral thioindigo dopant, whose isomerization leads to polarization power change without stabilization-destabilization effects from the molecular shape change (both states are rod-like). Photoresponsive BPs are less studied chiral LCs compared to cholesteric and SmC* LCs, however, they are receiving increasing attention in recent years due to their unique 3D cubic structures with periodicity of hundreds of nanometers and potential applications as photonic crystals.

The unique structures associated with reversible spectral and polarization behaviors make photoresponsive chiral LCs extremely attractive systems as stimuli-responsive smart materials. Although significant progress has been achieved in recent years, looking forward, it is expected that the improved properties and performance in photomodulation of chiral LCs can be achieved by developing novel material systems, particularly those using photoresponsive chiral dopants with satisfied performance. Furthermore, the multi-functional systems such as those nanostructures capable of reflecting full range of color with both left and right handedness in CLCs and BPs also deserve considerable attentions. The reflection bandwidth and reflectance limits of CLCs and BPs need to be fine tuned for new applications such as broadband polarizers and smart windows working in selective reflection mode controlling the wavelengths of incoming solar light.

Acknowledgments The preparation of this chapter benefited from the support to Quan Li by the Air Force Office of Scientific Research (AFOSR FA9550-09-1-0193 and FA9550-09-1-0254), the Department of Defense Multidisciplinary University Research Initiative (AFOSR MURI FA9550-06-1-0337 and FA9550-12-1-00370), the National Science Foundation (NSF IIP 0750379), the National Aeronautics and Space Administration (NASA), the Department of Energy (DOE DE-SC0001412), Ohio Third Frontier, and the Ohio Board of Regents under its Research Challenge program.

References

1. P.J. Collins, M. Hird, *Introduction to Liquid Crystals-Chemistry and Physics* (Taylor and Francis, London, 1997)
2. D. Demus, J. Goodby, G.W. Gray, H.-W. Spiess, V. Vill (eds.), *Handbook of Liquid Crystals* (Wiley-VCH, Weinheim, 1998)
3. H.-S. Kitzerow, C. Bahr (eds.), *Chirality in Liquid Crystals* (Springer, New York, 2001)
4. Q. Li, *Liquid Crystals Beyond Displays: Chemistry, Physics, and Applications* (Wiley, New Jersey, 2012)
5. W. Haas, J. Adams, J. Wysocki, Interaction between UV radiation and cholesteric liquid crystals. Mol. Cryst. Liq. Cryst. **7**, 371–379 (1969)
6. E. Sackmann, Photochemically induced reversible color changes in cholesteric liquid crystals. J. Am. Chem. Soc. **93**, 7088–7090 (1971)
7. C. Mioskowski, J. Bourguignon, S. Candau, G. Solladie, Photochemically induced cholesteric–nematic transition in liquid-crystals. Chem. Phys. Lett. **38**, 456–459 (1976)
8. F. Reinitzer, Beiträge zur kenntniss des cholesterins. Monatsh. Chem. **9**, 421–441 (1888)
9. N.Y. Ha, Y. Ohtsuka, S.M. Jeong, S. Nishimura, G. Suzaki, Y. Takanishi, K. Ishikawa, H. Takezoe, Fabrication of a simultaneous red–green–blue reflector using single-pitched cholesteric liquid crystals. Nat. Mater. **7**, 43–47 (2008)
10. M. Mitov, N. Dessaud, Going beyond the reflectance limit of cholesteric liquid crystals. Nat. Mater. **5**, 361–364 (2006)
11. W. Cao, A. Muñoz, P. Palffy-Muhoray, B. Taheri, Lasing in a three-dimensional photonic crystal of the liquid crystal blue phase II. Nat. Mater. **1**, 111–113 (2002)

12. S. Furumi, N. Tamaoki, Glass-forming cholesteric liquid crystal oligomers for new tunable solid-state laser. Adv. Mater. **22**, 886–891 (2010)
13. S.-T. Wu, D.-K. Yang, *Reflective Liquid Crystal Displays* (Wiley, West Sussex, 2001)
14. G.T. Stewart, Liquid crystals in biology I. Historical, biological and medical aspects. Liq. Cryst. **30**, 541–557 (2003)
15. S.J. Woltman, G.D. Jay, G.P. Crawford, Liquid-crystal materials find a new order in biomedical applications. Nat. Mater. **6**, 929–938 (2007)
16. R. Eelkema, B.L. Feringa, Amplification of chirality in liquid crystals. Org. Biomol. Chem. **4**, 3729–3745 (2006)
17. S. Pieraccini, S. Masiero, A. Ferrarini, G.P. Spada, Chirality transfer across length-scales in nematic liquid crystals: fundamentals and applications. Chem. Soc. Rev. **40**, 258–271 (2011)
18. G.W. Gray, J.W. Goodby, *Smectic Liquid Crystals* (Leonard Hill, London, 1984)
19. N.A. Clark, S.T. Lagerwall, Submicrosecond bistable electro-optic switching in liquid crystals. Appl. Phys. Lett. **36**, 899–901 (1980)
20. R.P. Lemieux, Chirality transfer in ferroelectric liquid crystals. Acc. Chem. Res. **34**, 845–853 (2001)
21. N.A. Clark, S.T. Lagerwall, *Ferroelctric Liquid Crystals: Principles, Properties and Applications* (Gordon and Breach, Philadelphia, 1991)
22. A. Mazzulla, G. Petriashvili, M.A. Matranga, M.P. De Santo, R. Barberi, Thermal and electrical laser tuning in liquid crystal blue phase I. Soft Matter **8**, 4882–4885 (2012)
23. S. Meiboom, M. Sammon, Structure of the blue phase of a cholesteric liquid crystal. Phys. Rev. Lett. **44**, 882–885 (1980)
24. O. Henrich, K. Stratford, M.E. Cates, D. Marenduzzo, Structure of blue phase III of cholesteric liquid crystals. Phys. Rev. Lett. **106**, 107801(1–4) (2011)
25. Y. Wang, Q. Li, Light-driven chiral molecular switches or motors in liquid crystals. Adv. Mater. **24**, 1926–1945 (2012)
26. R. Eelkema, Photo-responsive doped cholesteric liquid crystals. Liq. Cryst. **38**, 1641–1652 (2011)
27. B.L. Feringa, The art of building small: from molecular switches to molecular motors. J. Org. Chem. **72**, 6635–6652 (2007)
28. N. Tamaoki, T. Kamei, Reversible photo-regulation of the properties of liquid crystals doped with photochromic compounds. J. Photochem. Photobiol. C: Photochem. Rev **11**, 47–61 (2010)
29. T.J. White, M.E. McConney, T.J. Bunning, Dynamic color in stimuli-responsive cholesteric liquid crystals. J. Mater. Chem. **20**, 9832–9847 (2010)
30. B. Feringa, H. Wynberg, Torionally distorted olefins. Resolution of *cis*- and *trans*-4,4'-bi-1,1',2,2',3,3'-hexahydrophenanthrylidene. J. Am. Chem. Soc. **99**, 602–603 (1977)
31. B.L. Feringa, In control of motion: from molecular switches to molecular motors. Acc. Chem. Res. **34**, 504–513 (2001)
32. M. Irie, Diarylethenes for memories and switches. Chem. Rev. **100**, 1685–1716 (2000)
33. H. Tian, S. Wang, Photochromic bisthienylethene as multi-function switches. *Chem. Commun.* 781–792 (2007)
34. G. Berkovic, Spiropyrans and spirooxazines for memories and switches. Chem. Rev. **100**, 1741–1753 (2000)
35. Y. Yokoyama, Fulgides for memories and switches. Chem. Rev. **100**, 1717–1739 (2000)
36. S. Abraham, V.A. Mallia, K.V. Ratheesh, N. Tamaoki, S. Das, Reversible thermal and photochemical switching of liquid crystalline phases and luminescence in diphenylbutadiene-based mesogenic dimers. J. Am. Chem. Soc. **128**, 7692–7698 (2006)
37. M. Zhang, G.B. Schuster, Chirochromism-photochromism by epimerization: search for a liquid crystal phototrigger. J. Am. Chem. Soc. **116**, 4852–4857 (1994)
38. R.P. Lemieux, Photoswitching of ferroelectric liquid crystals using photochromic dopants. Soft Matter **1**, 348–354 (2005)

39. T. Ikeda, O. Tsutsumi, Optical switching and image storage by means of azobenzene liquid-crystal films. Science **268**, 1873–1875 (1995)
40. A. Chanishvili, G. Chilaya, G. Petriashvili, D. Sikharulidze, Light induced effects in cholesteric mixtures with a photosensitive nematic host. Mol. Cryst. Liq. Cryst. **409**, 209–218 (2004)
41. U.A. Hrozhyk, S.V. Serak, N.V. Tabiryan, T.J. Bunning, Photoinduced isotropic state of cholesteric liquid crystals: novel dynamic photonic materials. Adv. Mater. **19**, 3244–3247 (2007)
42. M.-H. Wu, C.-C. Chu, M.-C. Cheng, V.K.S. Hsiao, Reversible phase transition and rapid switching of azobenzene-doped cholesteric liquid crystals with a single laser. Mol. Cryst. Liq. Cryst. **557**, 176–189 (2012)
43. Q. Li, L. Li, J. Kim, H.-S. Park, J. Williams, Reversible photoresponsive chiral liquid crystals containing a cholesteryl moiety and azobenzene linker. Chem. Mater. **17**, 6018–6021 (2005)
44. G. Joly, A. Anakkar, Light induced shifts of ferroelectric mesophase transtions. Liq. Cryst. **26**, 1251–1255 (1999)
45. N. Tamaoki, Y. Aoki, M. Moriyama, M. Kidowaki, Photochemical phase transition and molecular realignment of glass-forming liquid crystals containing cholesterol/azobenzene dimesogenic compounds. Chem. Mater. **15**, 719–726 (2003)
46. V.A. Mallia, N. Tamaoki, Photochemically driven smectic-cholesteric phase transition in an inherently photoactive dimesogen. Chem. Mater. **15**, 3237–3239 (2003)
47. V.A. Mallia, N. Tamaoki, Photoresponsive vitrifiable chiral dimesogens: photo-thermal modulation of microscopic disordering in helical superstructure and glass-forming properties. J. Mater. Chem. **13**, 219–224 (2003)
48. V.A. Mallia, N. Tamaoki, Photoactive dimesogen having different pathways of light driven phase transitions at different temperatures. Chem. Commun. 2538–2539 (2004)
49. M.Z. Alam, T. Yoshioka, T. Ogata, T. Nonaka, S. Kurihara, Influence of helical twisting power on the photoswitching behavior of chiral azobenzene compounds: applications to high-performance switching devices. Chem. Eur. J. **13**, 2641–2647 (2007)
50. U.A. Hrozhyk, S.V. Serak, N.V. Tabiryan, L. Hoke, D.M. Steeves, B. Kimball, G. Kedziora, Systematic study of absorption spectra of donor–acceptor azobenzene mesogenic structures. Mol. Cryst. Liq. Cryst. **489**, 272–583 (2008)
51. U.A. Hrozhyk, S.V. Serak, N.V. Tabiryan, T.J. Bunning, Optical tuning of the reflection of cholesterics doped with azobenzene liquid crystals. Adv. Funct. Mater. **17**, 1735–1742 (2007)
52. S.V. Serak, N.V. Tabiryan, T.J. Bunning, Nonlinear transmission of photosensitive cholesteric liquid crystals due to spectral bandwidth auto-tuning or restoration. J. Nonlinear Opt. Phys. Mater **16**, 471–483 (2007)
53. U.A. Hrozhyk, S.V. Serak, N.V. Tabiryan, T.J. Bunning, Phototunable reflection notches of cholesteric liquid crystals. J. Appl. Phys. **104**, 063102(1–7) (2008)
54. S.V. Serak, N.V. Tabiryan, G. Chilaya, A. Chanishvili, G. Petriashvili, Chiral azobenzene nematics phototunable with a green laser beam. Mol. Cryst. Liq. Cryst. **488**, 42–55 (2008)
55. N. Tamaoki, S. Song, M. Moriyama, H. Matsuda, Rewritable full-color recording in a photon mode. Adv. Mater. **12**, 94–97 (2000)
56. H. Akiyama, V.A. Mallia, N. Tamaoki, Synthesis, liquid-crystalline properties, and photo-optical studies of photoresponsive oligomeric mesogens as dopants in a chiral glassy liquid crystal. Adv. Funct. Mater. **16**, 477–484 (2006)
57. H. Akiyama, A. Tanaka, H. Hiramatsu, J. Nagasawa, N. Tamaoki, Reflection colour changes in cholesteric liquid crystals after the addition and photochemical isomerization of mesogenic azobenzenes tethered to sugar alcohols. J. Mater. Chem. **19**, 5956–5963 (2009)
58. M. Mathews, R.S. Zola, D.-K. Yang, Q. Li, Thermally, photochemically and electrically switchable reflection colors from self-organized chiral bent-core liquid crystals. J. Mater. Chem. **21**, 2098–2103 (2011)

59. S. Kurihara, S. Nomiyama, T. Nonaka, Photochemical control of the macrostructure of cholesteric liquid crystals by means of photoisomerization of chiral azobenzene molecules. Chem. Mater. **13**, 1992–1997 (2001)
60. C. Ruslim, K. Ichimura, Conformational effect on macroscopic chirality modification of cholesteric mesophases by photochromic azobenzene dopants. J. Phys. Chem. B **104**, 6529–6535 (2000)
61. H.-K. Lee, K. Doi, H. Harada, O. Tsutsumi, A. Kanazawa, T. Shiono, T. Ikeda, Photochemical modulation of color and transmittance in chiral nematic liquid crystal containing an azobenzene as a photosensitive chromophore. J. Phys. Chem. B **104**, 7023–7028 (2000)
62. A. Chanishvili, G. Chilaya, G. Petriashvili, R. Barberi, R. Bartolino, M.P. De Santo, Cholesteric liquid crystal mixtures sensitive to different ranges of solar UV irradiation. Mol. Cryst. Liq. Cryst. **434**, 353–366 (2005)
63. T. Yoshioka, T. Ogata, T. Nonaka, M. Moritsugu, S.-N. Kim, S. Kurihara, Reversible-photon-mode full-color display by means of photochemical modulation of a helically cholesteric structure. Adv. Mater. **17**, 1226–1229 (2005)
64. J.H. Liu, P.C. Yang, Optical behaviour of photoimageable cholesteric liquid crystal cells with various novel chiral compounds. Liq. Cryst. **32**, 539–551 (2005)
65. J.H. Liu, P.C. Yang, Y.K. Wang, C.C. Wang, Optical behaviour of cholesteric liquid crystal cells with novel photoisomerizable chiral dopants. Liq. Cryst. **33**, 237–248 (2006)
66. J.H. Liu, P.C. Yang, H.J. Hung, D.J. Liaw, Photochemical tuning capability of cholesteric liquid crystal cells containing chiral dopants end capped with menthyl groups. Liq. Cryst. **34**, 891–902 (2007)
67. S. Kurihara, T. Yoshioka, T. Ogata, A.M. Zahangir, T. Nonaka, Synthesis of chiral azobenzene-based compounds for use in the photochemical tuning of the helical structure of liquid crystals. Liq. Cryst. **30**, 1219–1223 (2003)
68. M.Z. Alam, T. Yoshioka, T. Ogata, T. Nonaka, S. Kurihara, The influence of molecular structure on helical twisting power of chiral azobenzene compounds. Liq. Cryst. **34**, 1215–1219 (2007)
69. M. Mathews, N. Tamaoki, Planar chiral azobenzenophanes as chiroptic switches for photon mode reversible reflection color control in induced chiral nematic liquid crystals. J. Am. Chem. Soc. **130**, 11409–11416 (2008)
70. R. Thomas, Y. Yoshida, T. Akasaka, N. Tamaoki, Influence of a change in helical twisting power of photoresponsive chiral dopants on rotational manipulation of micro-objects on the surface of chiral nematic liquid crystalline films. Chem. Eur. J. **18**, 12337–12348 (2012)
71. S. Pieraccini, G. Gottarelli, R. Labruto, S. Masiero, O. Pandoli, G.P. Spada, The control of the cholesteric pitch by some azo photochemical chiral switches. Chem. Eur. J. **10**, 5632–5639 (2004)
72. S. Pieraccini, S. Masiero, G.P. Spada, G. Gottarelli, A new axially-chiral photochemical switch. Chem. Commun. 598–599 (2003)
73. Q. Li, L. Green, N. Venkataraman, I. Shiyanovskaya, A. Khan, A. Urbas, J.W. Doane, Reversible photoswitchable axially chiral dopants with high helical twisting power. J. Am. Chem. Soc. **129**, 12908–12909 (2007)
74. L. Green, Y. Li, T.J. White, A. Urbas, T. Bunning, Q. Li, Light-driven molecular switches with tetrahedral and axial chirality. Org. Biomol. Chem. **7**, 3930–3933 (2009)
75. J. Ma, Y. Li, T. White, A. Urbas, Q. Li, Light-driven nanoscale chiral molecular switch: reversible dynamic full range color phototuning. Chem. Commun. **46**, 3463–3465 (2010)
76. Q. Li, Y. Li, J. Ma, D.-K. Yang, T.J. White, T.J. Bunning, Directing dynamic control of red, green, and blue reflection enabled by a light-driven self-organized helical superstructure. Adv. Mater. **23**, 5069–5073 (2011)
77. E. Montbach, N. Venkataraman, A. Khan, I. Shiyanovskaya, T. Schneider, J.W. Doane, L. Green, Q. Li, Novel optically addressable photochiral displays. SID Int. Symp. Digest Tech. Papers **39**, 919–922 (2008)

78. N. Venkataraman, G. Magyar, M. Lightfoot, E. Montbach, A. Khan, T. Schneider, J.W. Doane, L. Green, Q. Li, Thin flexible photosensitive cholesteric displays. J. Soc. Inf. Disp. **17**, 869–873 (2009)
79. M. Goh, K. Akagi, Powerful helicity inducers: axially chiral binaphthyl derivatives. Liq. Cryst. **35**, 953–965 (2008)
80. D.-K. Yang, J.L. West, L.-C. Chien, J.W. Doane, Control of reflectivity and bistability in displays using cholesteric liquid-crystals. J. Appl. Phys. **76**, 1331–1333 (1994)
81. A.A. Beharry, O. Sadovski, G.A. Woolley, Azobenzene photoswitching without ultraviolet light. J. Am. Chem. Soc. **133**, 19684–19687 (2011)
82. Y. Wang, A. Urbas, Q. Li, Reversible visible-light tuning of self-organized helical superstructures enabled by unprecedented light-driven axially chiral molecular switches. J. Am. Chem. Soc. **134**, 3342–3345 (2012)
83. I. Gvozdovskyy, O. Yaroshchuk, M. Serbina, Photoinduced nematic–cholesteric structural transitions in liquid crystal cells with homeotropic anchoring. Mol. Cryst. Liq. Cryst. **546**, 1672–1678 (2011)
84. T. Sagisaka, Y. Yokoyama, Reversible control of the pitch of cholesteric liquid crystals by photochromism of chiral fulgide derivatives. Bull. Chem. Soc. Jpn. **73**, 191–196 (2000)
85. L.D. Bari, G. Pescitelli, P. Salvadori, Conformational study of 2,2'-homosubstituted 1,1'-binaphthyls by means of UV and CD spectroscopy. J. Am. Chem. Soc. **121**, 7998–8004 (1999)
86. G. Proni, G.P. Spada, Conformational analysis in solution of C2-symmetric 1,1'-binaphthyl derivatives by circular dichroism spectroscopy and cholesteric induction in nematic mesophases. J. Org. Chem. **65**, 5522–5527 (2000)
87. R.A. van Delden, T. Mecca, C. Rosini, B.L. Feringa, A chiroptical molecular switch with distinct chiral and photochromic entities and its application in optical switching of a cholesteric liquid crystal. Chem. Eur. J. **10**, 61–70 (2004)
88. M. Mathews, N. Tamaoki, Reversibly tunable helicity induction and inversion in liquid crystal self-assembly by a planar chiroptic trigger molecule. Chem. Commun. 3609–3611 (2009)
89. M. Mathews, R.S. Zola, S. Hurley, D.-K. Yang, T.J. White, T.J. Bunning, Q. Li, Light-driven reversible handedness inversion in self-organized helical superstructures. J. Am. Chem. Soc. **132**, 18361–18366 (2010)
90. Y. Li, M. Wang, T.J. White, T.J. Bunning, Q. Li, Azoarenes with opposite chiral configurations: light-driven reversible handedness inversion in self-organized helical superstructures. Angew. Chem. Int. Ed. **52**, 8925–8929 (2013)
91. N.P.M. Huck, W.F. Jager, B. deLange, B.L. Feringa, Dynamic control and amplification of molecular chirality by circular polarized light. Science **273**, 1686–1688 (1996)
92. B.L. Feringa, N.P.M. Huck, H.A. van Doren, Chiroptical switching between liquid crystalline phases. J. Am. Chem. Soc. **117**, 9929–9930 (1995)
93. R.A. van Delden, N. Koumura, N. Harada, B.L. Feringa, Supramolecular chemistry and self-assembly special feature: unidirectional rotary motion in a liquid crystalline environment: color tuning by a molecular motor. Proc. Natl. Acad. Sci. U.S.A. **99**, 4945–4949 (2002)
94. R. Eelkema, B.L. Feringa, Reversible full-range color control of a cholesteric liquid-crystalline film by using a molecular motor. Chem. Asian J. **1**, 367–369 (2006)
95. R. Eelkema, M.M. Pollard, N. Katsonis, J. Vicario, D.J. Broer, B.L. Feringa, Rotational reorganization of doped cholesteric liquid crystalline films. J. Am. Chem. Soc. **128**, 14397–14407 (2006)
96. M.M. Pollard, M. Klok, D. Pijper, B.L. Feringa, Rate acceleration of light-driven rotary molecular motors. Adv. Funct. Mater. **17**, 718–729 (2007)
97. D. Pijper, M.G.M. Jongejan, A. Meetsma, B.L. Feringa, Light-controlled supramolecular helicity of a liquid crystalline phase using a helical polymer functionalized with a single chiroptical molecular switch. J. Am. Chem. Soc. **130**, 4541–4552 (2008)

98. A. Bosco, M.G.M. Jongejan, R. Eelkema, N. Katsonis, E. Lacaze, A. Ferrarini, B.L. Feringa, Photoinduced reorganization of motor-doped chiral liquid crystals: bridging molecular isomerization and texture rotation. J. Am. Chem. Soc. **130**, 14615–14624 (2008)
99. W.-C. Chen, P.-C. Lin, C.-H. Chen, C.-T. Chen, Modulation of photoswitching profiles by 10,11-Dialkoxymethyl substituents in C2-symmetric dibenzosuberane-based helicenes. Chem. Eur. J. **16**, 12822–12830 (2010)
100. J. Vicario, M. Walko, A. Meetsma, B.L. Feringa, Fine tuning of the rotary motion by structural modification in light-driven unidirectional molecular motors. J. Am. Chem. Soc. **128**, 5127–5135 (2006)
101. J. Vicario, A. Meetsma, B.L. Feringa, Controlling the speed of rotation in molecular motors. Dramatic acceleration of the rotary motion by structural modification. Chem. Commun. 5910–5912 (2005)
102. T.J. White, S.A. Cazzell, A.S. Freer, D.-K. Yang, L. Sukhomlinova, L. Su, T. Kosa, B. Taheri, T.J. Bunning, Widely tunable, photoinvertible cholesteric liquid crystals. Adv. Mater. **23**, 1389–1392 (2011)
103. R. Eelkema, M.M. Pollard, J. Vicario, N. Katsonis, B.S. Ramon, C.W.M. Bastiaansen, D.J. Broer, B.L. Feringa, Molecular machines: nanomotor rotates microscale objects. Nature **440**, 163 (2006)
104. A. Kausar, H. Nagano, Y. Kuwahara, T. Ogata, S. Kurihara, Photocontrolled manipulation of a microscale object: a rotational or translational mechanism. Chem. Eur. J. **17**, 508–515 (2011)
105. R.A. van Delden, M.B. van Gelder, N.P.M. Huck, B.L. Feringa, Controlling the color of cholesteric liquid-crystalline films by photoirradiation of a chiroptical molecular switch used as dopant. Adv. Funct. Mater. **13**, 319–324 (2003)
106. S.J. Aßhoff, S. Iamsaard, A. Bosco, J.J.L.M. Cornelissen, B.L. Feringa, N. Katsonis, Time-programmed helix inversion in phototunable liquid crystals. Chem. Commun. **49**, 4256–4258 (2013)
107. K. Rameshbabu, A. Urbas, Q. Li, Synthesis and characterization of thermally irreversible photochromic cholesteric liquid crystals. J. Phys. Chem. B **115**, 3409–3415 (2011)
108. C. Kim, K.L. Marshall, J.U. Wallace, S.H. Chen, Photochromic glassy liquid crystals comprising mesogenic pendants to dithienylethene cores. J. Mater. Chem. **18**, 5592–5598 (2008)
109. C. Denekamp, B.L. Feringa, Optically active diarylethenes for multimode photoswitching between liquid-crystalline phases. Adv. Mater. **10**, 1080–1082 (1998)
110. K. Uchida, Y. Kawai, Y. Shimizu, V. Vill, M. Irie, An optically active diarylethene having cholesterol units: a dopant for photoswitching of liquid crystal phases. *Chem. Lett.* 654–655 (2000)
111. T. Yamaguchi, T. Inagawa, H. Nakazumi, S. Irie, M. Irie, Photoswitching of helical twisting power of a chiral diarylethene dopant: pitch change in a chiral nematic liquid crystal. Chem. Mater. **12**, 869–871 (2000)
112. T. Yamaguchi, T. Inagawa, H. Nakazumi, S. Irie, M. Irie, Photoinduced pitch changes in chiral nematic liquid crystals formed by doping with chiral diarylethene. J. Mater. Chem. **11**, 2453–2458 (2001)
113. T. Yamaguchi, T. Inagawa, H. Nakazumi, S. Irie, M. Irie, Phase transition of a liquid crystal induced by chiral photochromic dopants. Mol. Cryst. Liq. Cryst. **345**, 287–292 (2000)
114. T. van Leeuwen, T.C. Pijper, J. Areephong, B.L. Feringa, W.R. Browne, N. Katsonis, Reversible photochemical control of cholesteric liquid crystals with a diamine-based diarylethene chiroptical switch. J. Mater. Chem. **21**, 3142–3146 (2011)
115. Y. Li, A. Urbas, Q. Li, Synthesis and characterization of light-driven dithienylcyclopentene switches with axial chirality. J. Org. Chem. **76**, 7148–7156 (2011)
116. Y. Li, A. Urbas, Q. Li, Reversible light-directed red, green, and blue reflection with thermal stability enabled by a self-organized helical superstructure. J. Am. Chem. Soc. **134**, 9573–9576 (2012)

117. Y. Li, Q. Li, Photochemically reversible and thermally stable axially chiral diarylethene switches. Org. Lett. **14**, 4362–4365 (2012)
118. Y. Li, M. Wang, A. Urbas, Q. Li, A photoswitchable and thermally stable axially chiral dithienylperfluorocyclopentene dopant with high helical twisting power. J. Mater. Chem. C **1**, 3917–3923 (2013)
119. H. Hayasaka, T. Miyashita, M. Nakayama, K. Kuwada, K. Akagi, Dynamic photoswitching of helical inversion in liquid crystals containing photoresponsive axially chiral dopants. J. Am. Chem. Soc. **134**, 3758–3765 (2012)
120. R.K. Vijayaraghavan, S. Abraham, H. Akiyama, S. Furumi, N. Tamaoki, S. Das, Photoresponsive glass-forming butadiene-based chiral liquid crystals with circularly polarized photoluminescence. Adv. Funct. Mater. **18**, 2510–2517 (2008)
121. N.S.S. Kumar, S. Abraham, K.V. Ratheesh, N. Tamaoki, S. Furumi, S. Das, Indane-1,3-Dione and cholesterol containing butadiene derivatives: photoresponsive liquid crystalline glasses for imaging applications. J. Photochem. Photobiol. A: Chem **207**, 73–78 (2009)
122. H. Hattori, T. Uryu, Photochromic chiral liquid crystalline systems containing spiro-oxazine with a chiral substituent II. Photoinduced behaviour. Liq. Cryst **28**, 1099–1104 (2001)
123. H. Hattori, T. Uryu, Photochromic chiral liquid crystalline systems containing spiro-oxazine with a chiral substituent I. Synthesis and characterization of compounds. Liq. Cryst. **28**, 25–34 (2001)
124. L.-M. Jin, Y. Li, J. Ma, Q. Li, Synthesis of novel thermally reversible photochromic axially chiral spirooxazines. Org. Lett. **12**, 3552–3555 (2010)
125. Y. Yokoyama, S. Uchida, T. Sagisaka, Y. Uchida, T. Inada, Chiral photochromic compounds and control of functions. Enantiomer **3**, 123–132 (1998)
126. Y. Yokoyama, T. Sagisaka, Reversible control of pitch of induced cholesteric liquid crystal by optically active photochromic fulgide derivatives. Chem. Lett. 687–688 (1997)
127. M. Suarez, G.B. Schuster, Photoresolution of an axially chiral bicyclo[3.3.0]octan-3-one: phototriggers for a liquid-crystal-based optical switch. J. Am. Chem. Soc. **117**, 6732–6738 (1995)
128. Y. Zhang, G.B. Schuster, Photoresolution of an axially chiral bicyclo[3.2.1]octan-3-one: phototriggers for a liquid crystal-based optical switch. J. Org. Chem. **60**, 7192–7197 (1995)
129. K.S. Burnham, G.B. Schuster, Transfer of chirality from circularly polarized light to a bulk material property: propagation of photoresolution by a liquid crystal transition. J. Am. Chem. Soc. **121**, 10245–10246 (1999)
130. T. Ikeda, T. sasaki, K. Ichimura, Photochemical switching of polarization in ferroelectric liquid-crystal films. Nature **361**, 428–430 (1993)
131. T. Sasaki, T. Ikeda, K. Ichimura, Photochemical control of properties of ferroelectric liquid crystals: photochemical flip of polarization. J. Am. Chem. Soc. **116**, 625–628 (1994)
132. H.G. Walton, H.J. Coles, D. Guillon, G. Poeti, Photomechanically induced ferroelectricity in smectic liquid crystals. Liq. Cryst. **17**, 333–349 (1994)
133. T. Sasaki, T. Ikeda, Photochemical control of properties of ferroelectric liquid crystals. 3. Photochemically induced reversible change in spontaneous polarization and electrooptic property. J. Phys. Chem. **99**, 13013–13018 (1995)
134. L. Komitov, O. Tsutsumi, C. Ruslim, T. Ikeda, K. Ichimura, K. Yoshino, Optical recording using a photochromic ferroelectric liquid crystal. J. Appl. Phys. **89**, 7745–7749 (2001)
135. G. Joly, A. Anakkar, M. Ismaili, P. Cluzeau, N. Isaert, H.T. Nguyen, Chiral azobenzene liquid crystals under illumination: Thickness influence and spontaneous polarisation variations. Ferroelectrics **277**, 67–74 (2002)
136. K.E. Maly, M.D. Wand, R.P. Lemieux, Bistable ferroelectric liquid crystal photoswitch triggered by a dithienylethene dopant. J. Am. Chem. Soc. **124**, 7898–7899 (2002)
137. K.E. Maly, P. Zhang, M.D. Wand, E. Buncel, R.P. Lemieux, Reversible photocyclization of achiral dithienylperfluorocyclopentene dopants in a ferroelectric liquid crystal: bistable SSFLC photoswitching. J. Mater. Chem. **14**, 2806–2812 (2004)

138. L. Dinescu, R.P. Lemieux, Photomodulation of the spontaneous polarization of a ferroelectric liquid crystal: harnessing the transverse dipole modulation of a chiral thioindigo dopant. J. Am. Chem. Soc. **119**, 8111–8112 (1997)
139. L. Dinescu, K.E. Maly, R.P. Lemieux, Design of photonic liquid crystal materials: synthesis and evaluation of new chiral thioindigo dopants designed to photomodulate the spontaneous polarization of ferroelectric liquid crystals. J. Mater. Chem. **9**, 1679–1686 (1999)
140. J.Z. Vlahakis, R.P. Lemieux, Photoswitching of ferroelectric liquid crystals using unsymmetrical chiral thioindigo dopants: structural effects on polarization photomodulation. J. Mater. Chem. **14**, 1486–1494 (2004)
141. L. Dinescu, R.P. Lemieux, Optical Switching of a ferroelectric liquid crystal spatial light modulator by photoinduced polarization inversion. Adv. Mater. **11**, 42–45 (1999)
142. J.Z. Vlahakis, M.D. Wand, R.P. Lemieux, Photoinduced polarization inversion in a ferroelectric liquid crystal using an ambidextrous chiral thioindigo dopant. J. Am. Chem. Soc. **125**, 6862–6863 (2003)
143. J.Z. Vlahakis, M.D. Wand, R.P. Lemieux, Photoswitching of ferroelectric liquid crystals using unsymmetrical chiral thioindigo dopants: photoinduced inversion of the sign of spontaneous polarization. Adv. Funct. Mater. **14**, 637–642 (2004)
144. A. Chanishvili, G. Chilaya, G. Petriashvili, P. Collings, Trans-cis isomerization and the blue phases. Phys. Rev. E **71**, 051705 (1–5) (2005)
145. H.-Y. Liu, C.-T. Wang, C.-Y. Hsu, T.-H. Lin. J.-H. Liu, Optically tuneable blue phase photonic band gaps. Appl. Phys. Lett. **96**, 121103 (1–3) (2010)
146. P. Cladis, T. Garel, P. Pieranski, Kossel diagrams show electric-field-induced cubic-tetragonal structural transition in frustrated liquid-crystal blue phases. Phys. Rev. Lett. **57**, 2841–2844 (1986)
147. R.K. Vijayaraghavan, S. Abraham, D.S.S. Rao, S.K. Prasad, S. Das, Light induced generation of stable blue phase in photoresponsive diphenylbutadiene based mesogen. Chem. Commun. **46**, 2796–2798 (2010)
148. H.-C. Jeong, K.V. Le, M.-J. Gim, S.-T. Hur, S.-W. Choi, F. Araoka, K. Ishikawa, H. Takezoe, Transition between widened BPs by light irradiation using photo-active bent-core liquid crystal with chiral dopant. J. Mater. Chem. **22**, 4627–4630 (2012)
149. T.-H. Lin, Y. Li, C.-T. Wang, H.-C. Jau, C.-W. Chen, C.-C. Li, H.K. Bisoyi, T.J. Bunning, Q. Li, Red, green and blue reflections enabled in an optically tunable self-organized 3D cubic nanostructured thin film. Adv. Mater. **25**, 5050–5054 (2013)
150. D.K. Yang, P.P. Crooker, Chiral-racemic phase diagrams of blue-phase liquid crystals. Phys. Rev. A **35**, 4419–4423 (1987)

Chapter 6
Glassy Liquid Crystals as Self-Organized Films for Robust Optoelectronic Devices

H.-M. Philp Chen, Jane J. Ou and Shaw H. Chen

Abstract This chapter will review recent advances in glassy liquid crystals (GLCs) as self-organized nano- to microstructured organic solids and their potential for optoelectronic device applications. Conformational multiplicity serves as the kinetic and thermodynamic foundation for glass formation and stability against crystallization from the glassy state as well as solubility in common solvents to facilitate material synthesis, purification, and spin-casting into films. From the molecular perspective, conformational multiplicity has been bestowed by chemically bonding mesogenic groups to volume-excluding cores via flexible linkages, and alternatively by functionalizing conjugated oligomers with aliphatic side groups. Two distinct material classes have emerged for the realization of defect-free GLC films across a large area while assuring long-term morphological stability. Solvent-vapor annealing of spin-cast films on photo-alignment layers at room temperature managed to achieve the same extent of orientational order as thermal annealing on rubbed polyimide alignment layers. Varied device concepts have been demonstrated, including circular polarizers, optical notch filters and reflectors; photonic switching and optical memory; polarized fluorescence and electroluminescence; and robust organic solid-state lasers including spatial resolution for multiple lasing wavelengths on demand within a single film.

H.-M. Philp Chen
Department of Photonics and Institute of Display, National Chiao-Tung University, Hsinchu 30010, Taiwan

J. J. Ou
Department of Chemical Engineering, University of Rochester, Rochester, NY 14623-1212, USA

S. H. Chen (✉)
Department of Chemical Engineering and Laboratory for Laser Energetics, University of Rochester, Rochester, NY 14623-1212, USA
e-mail: shch@LLE.rochester.edu

Q. Li (ed.), *Nanoscience with Liquid Crystals*, NanoScience and Technology,
DOI: 10.1007/978-3-319-04867-3_6,

6.1 Concept and Relevance of Glassy Liquid Crystals

Liquid crystals (LCs) are a class of organic compounds that are fluid with a uniaxial, lamellar, helical or columnar arrangement at the supramolecular level. In the fluid state, molecular orientation can respond to an applied field reversibly in milliseconds, as in LC displays and spatial light modulators. Without incurring molecular orientation, LCs may respond to photonic or electronic stimuli in submicron through femtoseconds, as in the nonlinear optics setting. Additionally, LCs can also function in passive devices, such as polarizers and optical notch filters. In applications where molecular orientation with an applied field is not required, solid state is preferred for temporal stability and environmental durability. Whereas glass formation appears to be a privilege of polymeric materials, their generally high melt viscosity and broad distribution of relaxation times are problematic for processing into uniform self-organized, nano- to microstructured films across a large area. A low-molar-mass LC fluid film can in principle be thermally annealed to maximize order through defect annihilation before cooling to below its glass transition temperature (T_g) to obtain a glassy liquid crystal (GLC) film. In theory, all liquids should enter a glassy state at a sufficiently rapid cooling rate. In practice, however, crystallization is likely to occur, which modifies the long-range molecular order intended for device applications. To overcome these hurdles, molecular systems have been actively pursued for the creation of defect-free GLC films.

The differential scanning calorimetric (DSC) thermograms compiled in Fig. 6.1 differentiate three distinct single-component LCs. To delete thermal history prior to each scan, a sample is heated to beyond its isotropization point and then cooled to room temperature before gathering the second heating scan. Thereafter, the second cooling scan is collected. Heating a conventional LC causes a first-order transition from a crystalline solid to a liquid crystal at its melting point, T_m, followed by a transition to an isotropic liquid at T_c. In contrast, a kinetically stable GLC undergoes a second-order transition from a mesomorphic solid to a liquid crystal fluid at T_g without altering its molecular order, followed by a transition to an isotropic liquid at T_c. Intermediate between a conventional LC and a stable GLC is an unstable GLC, which tends to crystallize from the mesomorphic melt above T_g upon heating with subsequent melting to a liquid crystal at T_m and then clearing at T_c. Typical cooling thermograms are also included in Fig. 6.1, where both the conventional LC and the kinetically stable GLC exhibit hysteretic thermal reversibility. The same cooling scan as that of a stable GLC delineates the best-case scenario for an unstable GLC that allows a liquid crystalline mesophase to be preserved below T_g albeit with inferior stability. The DSC Set (c) in Fig. 6.1 provides the first hint of kinetic stability with a GLC film's long-term morphological behavior further tested for practical application. From a fundamental perspective, the phase transition from a liquid crystal to a GLC on cooling adds a new dimension to the hitherto challenging theoretical treatments of glass transition from an isotropic liquid to an isotropic solid [1, 2].

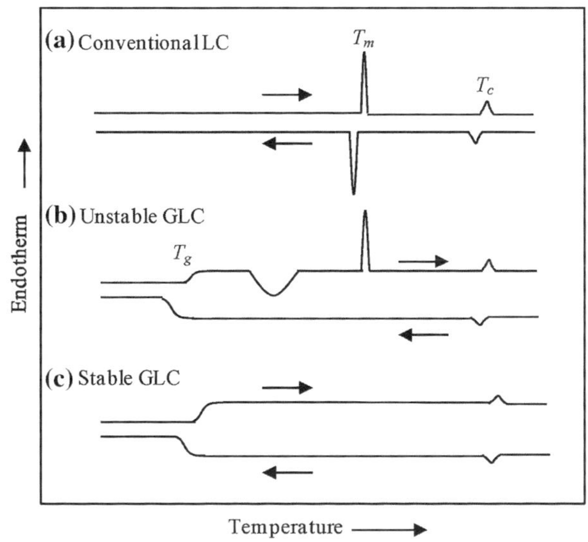

Fig. 6.1 Representative DSC thermograms of single-component samples of **a** a conventional LC, **b** an unstable GLC, and **c** a kinetically stable GLC. Samples were preheated to isotropic liquid to delete their thermal histories followed by cooling to room temperature for gathering the compiled second heating and cooling scans at finite rates

6.2 Prior Empirical Approaches to GLCs

It is little surprise that a low T_g accompanied by poor morphological stability resulted from the first attempts at GLCs [3, 4]. In parallel to low-molar-mass GLCs, liquid crystalline polymers have been actively explored [5–7]. Compared to liquid crystalline polymers, GLCs are advantageous in their superior chemical purity and favorable rheological properties [8, 9]. The molecular structures of GLCs reported early on can be categorized into: (i) laterally or terminally branched, one-string compounds with a T_g mostly around room temperature [10]; (ii) twin molecules with an above-ambient T_g, but generally lacking morphological stability [11–14]; (iii) cyclosiloxanes functionalized with mesogenic and chiral pendants [15–17]; (iv) carbosilane dendrimers exhibiting a low T_g [18–20], and (v) macrocarbocycles with mesogenic segments as part of the ring structure [21]. Illustrated as (**1**) through (**5**) in Fig. 6.2, representative molecular structures do not seem to convey a systematic design concept.

6.3 Modular Approaches to GLCs

Two modular approaches are highlighted in what follows. The core-pendant strategy was introduced previously [23, 24]. Emphasis here is placed on new molecular systems, effective synthesis methodologies, and emerging device

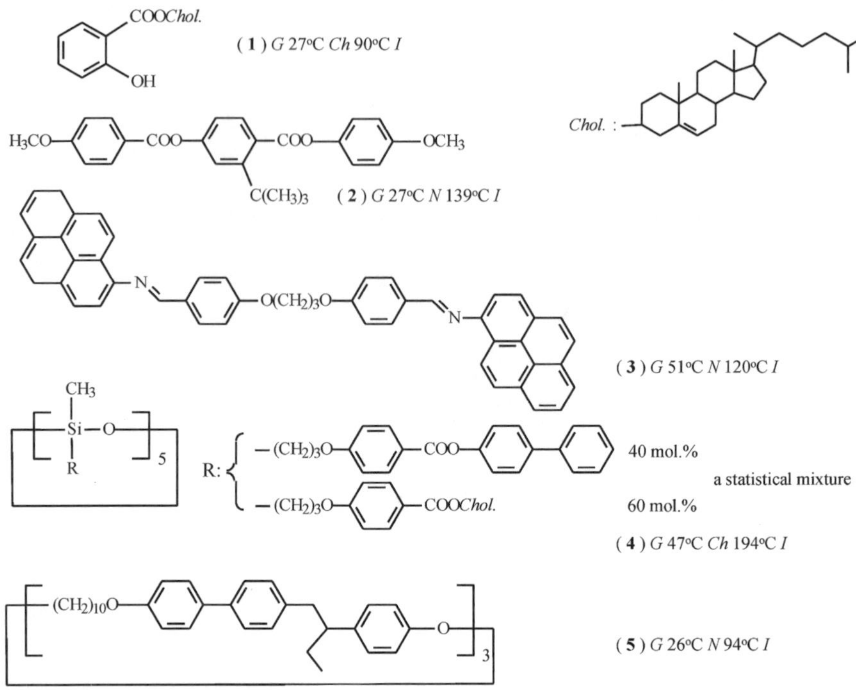

Fig. 6.2 Representative GLCs from early efforts accompanied by DSC data from second heating scans. *G* glassy, *Ch* cholesteric, *N* nematic, *I* isotropic. Used with permission [22]

applications. From the materials perspective, our focus is on single-component, enantiotropic liquid crystals exhibiting LC mesomorphism on both heating and cooling to strive for morphological stability, as opposed to monotropic liquid crystals that exhibit mesomorphism on cooling but not heating. To prevent spontaneous crystallization, a modular approach has been established where mesogenic and chiral pendants are bonded via flexible linkages to a volume-excluding core. While the core and pendants are crystalline as separate chemical entities, their hybrids readily form GLCs on cooling thanks to the conformational multiplicity furnished by the flexible linkages. The increased entropy as a result of conformational multiplicity frozen in the glassy state also contributes to a high Gibbs energy barrier to crystallization. Nonetheless, there exists an optimum range of flexible length, e.g. $-(CH_2)_n-$ with n = 2–5, outside of which spontaneous crystallization is inevitable. The first-generation GLCs are highlighted as (**6**) through (**11**) in Fig. 6.3 to demonstrate the versatility of the core-pendant strategy calling upon cyclohexane, bicyclooctene, adamantane, and cubane as the volume-excluding cores to which nematic and chiral pendants are attached [24].

To substantially expand the material database, single volume-excluding cores employed in Fig. 6.3 plus benzene are enlarged with both linear and angular nematogenic pendants [25–27]. With rigid and high-melting cyanoterphenyl pendants

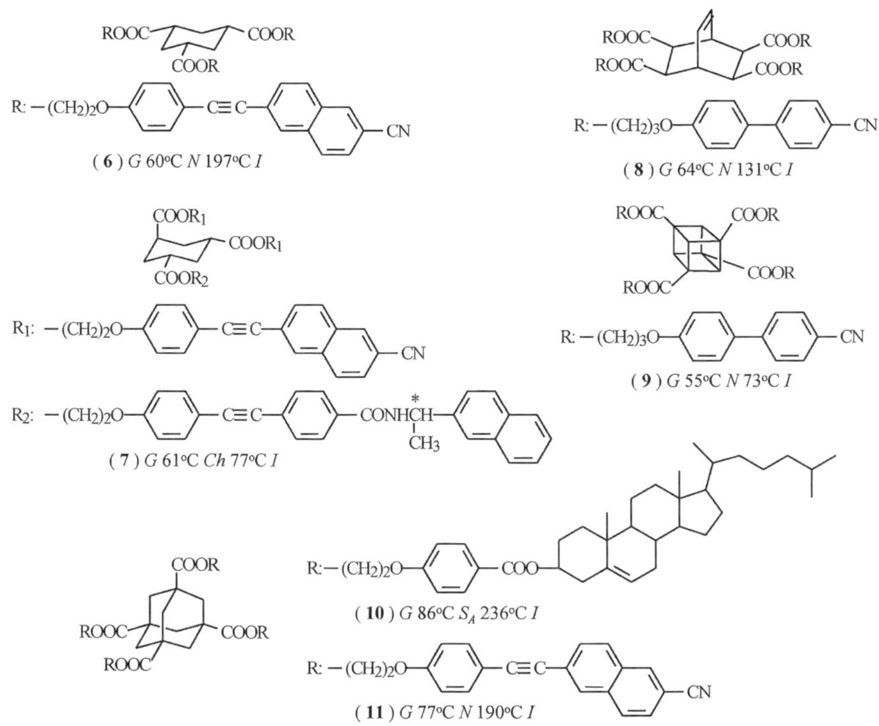

Fig. 6.3 First-generation GLCs based on the core-pendant strategy accompanied by DSC data from second heating scans. In addition to the symbols defined in Fig. 6.2, S_A denotes smectic A mesophase. Used with permission [24]

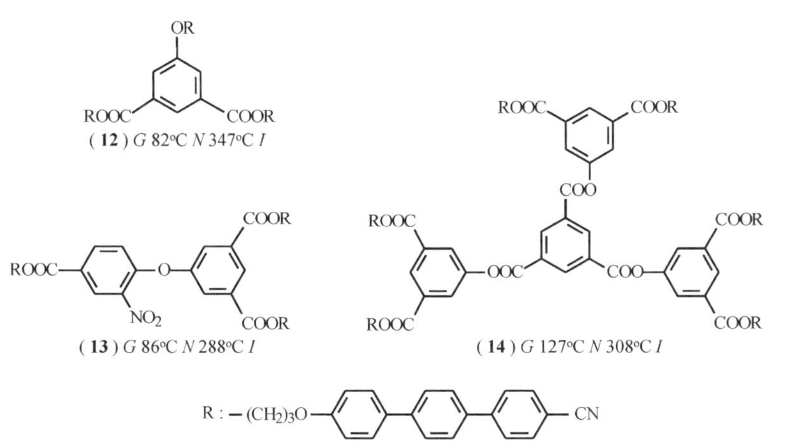

Fig. 6.4 High-temperature nematic GLCs with benzene and extended volume-excluding cores accompanied by DSC data from second heating scans

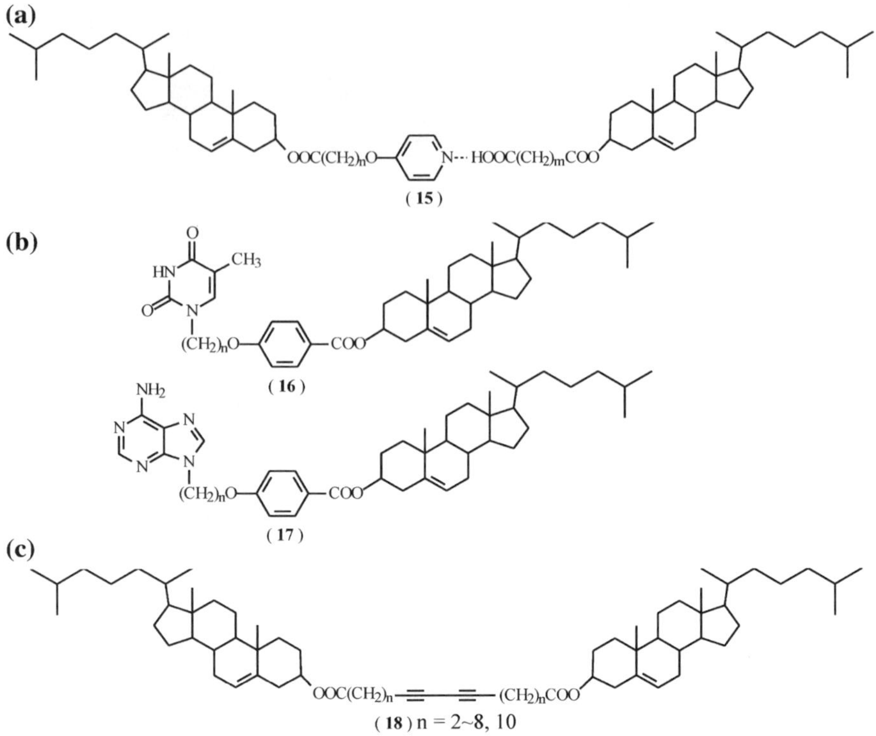

Fig. 6.5 Cholesteric GLCs consisting of **a** and **b** hydrogen-bonded, and **c** chemically bonded dimeric cholesterols

attached to benzene and more extended cores, morphologically stable nematic GLCs with elevated T_g and T_c have been achieved, e.g. (**12**) through (**14**) in Fig. 6.4.

Intermolecular hydrogen bonding forms the basis for another modular approach illustrated with cholesteric LCs. Following Kato and Frechet's work [28] on the formation of nematic and smectic liquid crystals through hydrogen bonding to extend the molecular aspect ratio, Itahara et al. [29] and Tamaoki et al. [30] reported cholesteric GLCs using functionalized cholesterols, e.g. (**15**) and (**16**) + (**17**) in Fig. 6.5. The resultant hydrogen-bonded binary systems, however, have been reported to be monotropic, albeit exhibiting higher transition temperatures with broader mesomorphic fluid temperature ranges compared to pure components. Akin to the 1:1 hydrogen-bonded complex is a chemically linked monotropic cholesteryl dimer, e.g. (**18**) in Fig. 6.5 with temperature-dependent reflective coloration in the fluid state [14]. A similar cholesteryl dimer linked by dipheylbutadiene was demonstrated for photo- and thermomodulation of reflective coloration and a modest degree of circularly polarized fluorescence [31].

6.4 Optical Properties of Cholesteric LC Films

When sandwiched between a pair of glass substrates coated with rubbed polyimide alignment layers followed by thermal annealing, the cholesteric LC film consists of a helical stack of quasinematic layers. Under polarizing optical microscopy, a cholesteric mesophase is identifiable with two distinct textures, focal conic and Grandjean, depending on the orientation of the helical axis with respect to the substrate. In the Grandjean mode, the helical axis is perpendicular to the substrate. From one quasinematic layer to the next, the director makes an incremental rotation about the helical axis, resulting in a right- or left-handed supramolecular structure selectively reflecting the right- and left-handed circularly polarized incident, respectively. The helical stack is further characterized by pitch length, p, defined as the distance along the axis over which the director completes a 360° rotation. The selective reflection wavelength, $\lambda_R = p\ n_{avg} \cos\theta$, where θ is the off-normal incidence angle and $n_{avg} = (n_e + n_o)/2$ with n_e and n_o denoting the underlying quasinematic layer's extraordinary and ordinary refractive index, respectively. Moreover, the spectral width of selective reflection at normal incidence, $\Delta\lambda = p\ \Delta n$, where $\Delta n = n_e - n_o$ [32]. A single-handed cholesteric LC film is capable of selective reflection and circular polarization simultaneously, and a stack of right- and left-handed films acts as an optical notch filter and reflector of incident unpolarized light. To prevent the susceptibility of helical pitch length and optical quality to extrinsic temperature variability and physical perturbations encountered with fluid films, cholesteric GLCs are intrinsically advantageous. Defined ad hoc as defect-free films across a 2.54 cm-diameter fused silica substrate, ~10-μm-thick monodomain GLC films can be readily prepared between alignment-coated substrates with thermal annealing above T_g prior to cooling to ambient temperature to preserve high quality helical stacking in solid state. Nano- to microstructures can be visualized with atomic-force or scanning-electron microscopy of freeze-fractured cross-sections.

6.5 Synthesis of Core-Pendant Cholesteric GLCs

Most cholestric GLCs consist of both nematic and chiral groups attached to volume-excluding cores. The commonly practiced statistical approach involves reacting nematic and chiral precursors at a chosen stoichiometric ratio with a volume-excluding core carrying multifunctional groups. Despite the ease of implementation, the separation and purification of the targeted cholesteric GLC from a multicomponent mixture at the end of the reaction demands intensive effort. To alleviate the work-up procedures, four distinct deterministic strategies have been formulated. A common denominator of the first three deterministic approaches entails selective functionalization of a volume-excluding core followed by sequential reactions with the nematic and chiral precursors. The fourth deterministic approach is the most desirable, as it combines nematic and chiral moieties into a single precursor, thus obviating the needs for synthesis of separate chiral and nematic precursors and selective functionalization of the core. Further details are provided in what follows.

6.5.1 Statistical Synthesis

Esterification of a tricarboxylic acid with nematic and chiral precursors, Nm_1-OH:Ch_1-OH at a 2:1 molar ratio as annotated in Reaction Scheme 6.1, was relatively straightforward. Subsequent isolation and purification produced a nematic and a cholesteric GLC, (**19**) and (**20**), at a 20 and 31 % yield, respectively. The elaborate work-up procedures involved dissolution-precipitation cycles, repeated liquid column chromatograpy complemented by recrystallization as needed [33].

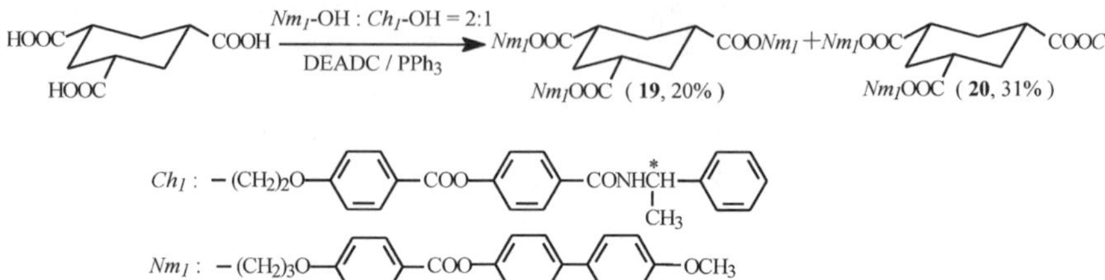

Scheme 6.1 Statistical synthesis of cholesteric and nematic GLCs. *DEADC* diethyl azodicarboxylate, PPh_3 triphenylphosphine

6.5.2 Deterministic Synthesis

The key intermediate in Reaction Scheme 6.2, viz. 2,4-dioxo-3-oxa-bicyclo[3.3.1]nonane-7-carboxylic acid chloride, was synthesized for esterification with nematic and chiral precursor alcohols in two consecutive steps [34] to produce a nematic GLC, (**21**), and a chiral molecular glass, (**22**). Whereas the work-up procedures were substantially simplified compared to the statistical approach, the synthesis of the key intermediate and its stability remained challenging. The 14 % overall yield of an enantiomeric chiral molecular glass turned out to be inferior to that from statistical synthesis.

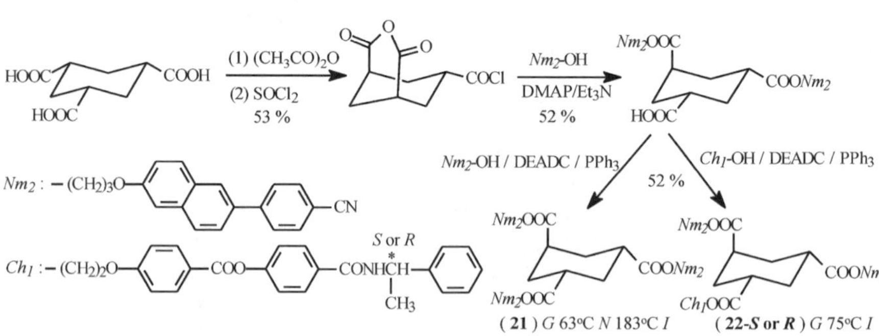

Scheme 6.2 Deterministic synthesis of cholesteric and nematic GLCs via a bicyclic intermediate with anhydride and acid chloride functionalities. *DMAP* 4-dimethylamino pyridine

Monohydrolysis of trimethyl esters is the key to two alternative deterministic approaches as described in Reaction Scheme 6.3 [35]. About 50 % overall yield of cholesteric GLCs, (**23**) and (**24**), using pig liver enzyme as the first step in Approach (a), and a 23 % overall yield of (**25**) involved base catalysis in Approach (b), the former being superior to statistical synthesis with much less laborious work-up procedures.

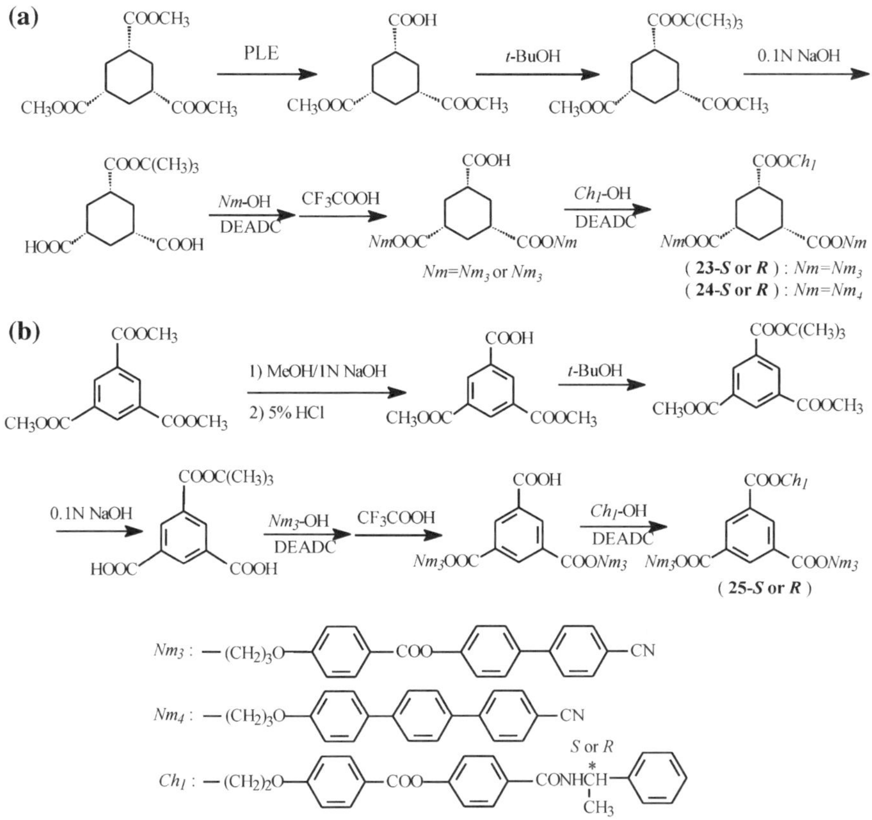

Scheme 6.3 Deterministic synthesis of cholesteric GLCs starting with monohydrolysis of **a** 1,3,5-cyclohexanetricarboxylic acid trimethyl ester with pig liver enzyme, *PLE*; and **b** 1,3,5-benzenetricarboxylic acid trimethyl ester

In comparison to the synthesis of separate chiral and nematic precursors, the idea of a hybrid chiral-nematic precursor, exemplified as Ch_2 in Reaction Scheme 6.4, is quite appealing. Moreover, the synthesis of a cholesteric GLC can be completed in a single step at a 66 % yield [36], which compares favorably with the preceding statistical and deterministic approaches. Of particular significance is that (**26**) with (*S*)-3-bromo-2-methylpropanol incorporated as the chiral linkage produced a left-handed cholesteric GLC film with λ_R located at 413 nm.

Scheme 6.4 Deterministic synthesis of a cholesteric GLCs consisting of hybrid chiral-nematic pendant groups

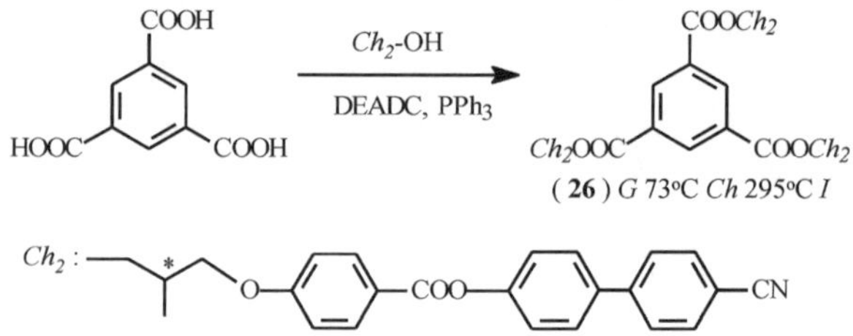

6.6 Hairy Rods for Preparation of Conjugated GLC Films

Rigid conjugated polymers and monodisperese oligomers are normally tough to dissolve in common solvents or melt into liquids, thus obscuring both lyotropic and thermotropic liquid crystalline mesomorphism. These challenges have been overcome by functionalization with aliphatic groups to generate conformational multiplicity, producing what are referred to as hairy rods. As a result, they can be readily processed into solid films through spin-casting from solution or melt-processing. Traditional condensation polymerization yields polydisperse conjugated polymers with relatively high molecular weights, which are typically purified by dissolution-precipitation cycles. Monodisperse conjugated oligomers with low to medium molecular weights can be synthesized deterministically following a divergent-convergent approach [37–39]. Moreover, oligomers are amenable to purification by recrystallization, column chromatography, and vacuum sublimation to arrive at superior chemical purity. These inherent material traits are imperative to gaining fundamental insight into how chemical structure affects electronic, photonic, and morphological properties. From a practical standpoint, chemical purity and structural uniformity are critical to device performance. In the absence of chain entanglements or defects (e.g., bends and kinks), relatively short and uniform chains are conducive to the formation of monodomain GLC films. Oligomers are, however, more prone to crystallization than relatively high molecular weight polymers. Strategies have been developed for engineering oligomers for glass transition while resisting crystallization, forming morphologically stable GLC films.

6.7 Optoelectronic Devices Utilizing GLCs

To exploit both the ease of processing across a large area without defects and the stability against crystallization, GLCs have been appraised extensively for a diversity of device concepts in our recent studies.

6.7.1 Circular Polarizers, Optical Notch Filters and Reflectors

Enantiomeric chiral glasses (**22-*R*** and **-*S***) and nematic GLC (**21**) were mixed at a 42:58 mass ratio to produce cholesteric GLCs with a T_g and T_c at 67 and 131 °C, respectively. With (**22**-*R* and **-*S***) as the chiral components, the two binary mixtures with (**21**) were prepared into an 8-μm-thick, right- and left-handed cholestric GLC film, respectively, as characterized in Fig. 6.6a and b. Noted in Fig. 6.6a are a 4 % leakage of RCP (right-handed circularly polarized) incident through a right-handed film and a 104 % transmission of LCP (left-handed circularly polarized) incident, which are artifacts arising from: (i) handedness preservation upon reflection from a cholesteric GLC film, and (ii) handedness reversal upon reflection from a glass-air interface. Similar observations in the opposite sense are made of Fig. 6.6b. In any event, the excellent optical quality of cholesteric GLC films per se is evidenced by the 100 % difference between the transmissions of the RCP and LCP incidents with square tops. As an optical notch filter, the stack of the right- and left-handed films with their directors——defined as the rubbing direction on the polyimide alignment layers—oriented perpendicular to each other produced an attenuation of 3.7 optical units (see Fig. 6.6c) [34], the highest contrast ratio ever achieved with organic materials that can be further improved by optimizing materials properties, alignment layers, and glass substrates bounding the GLC films. Enantiomerically pure (**26-*R*** and **-*S***) synthesized following Reaction Scheme 6.4 yielded a right- and left-handed circular polarizer, respectively, both centered at 413 nm [36]. A mixture of the two enantiomers at unequal proportions will result in cholesteric GLC films with desired handedness at any λ_R value longer than 413 nm, depending on which enantiomer is chosen as the major component. Compared to rugate and holographic filters, cholesteric GLCs can be readily processed into robust large-area films for laser protection, reflective head-up displays, and for enhanced power efficiency with improved contrast ratio in liquid crystal displays.

6.7.2 Modulating Circular Polarization and Reflective Coloration

As prescribed in Sect. 6.4, a constant-pitch cholesteric LC film typically gives rise to a selective reflection bandwidth around 100 nm in the visible region as determined by the Δn value. The bandwidth can be broadened with a gradient pitch through film thickness, according to the Hajdo-Erigen theory [40]. Photoracemizable (*R*)-dinaphtho[2,1D:1′,2′-*f*] [1, 3] dioxepin was mixed with a nematic and cholesteric GLC at a molar ratio of (**27**):(**28**):(**29**) = 1.00:0.23:0.13 as depicted in Fig. 6.7a. With a T_g and T_c at 65 and 156 °C, respectively, the ternary mixture was prepared into a

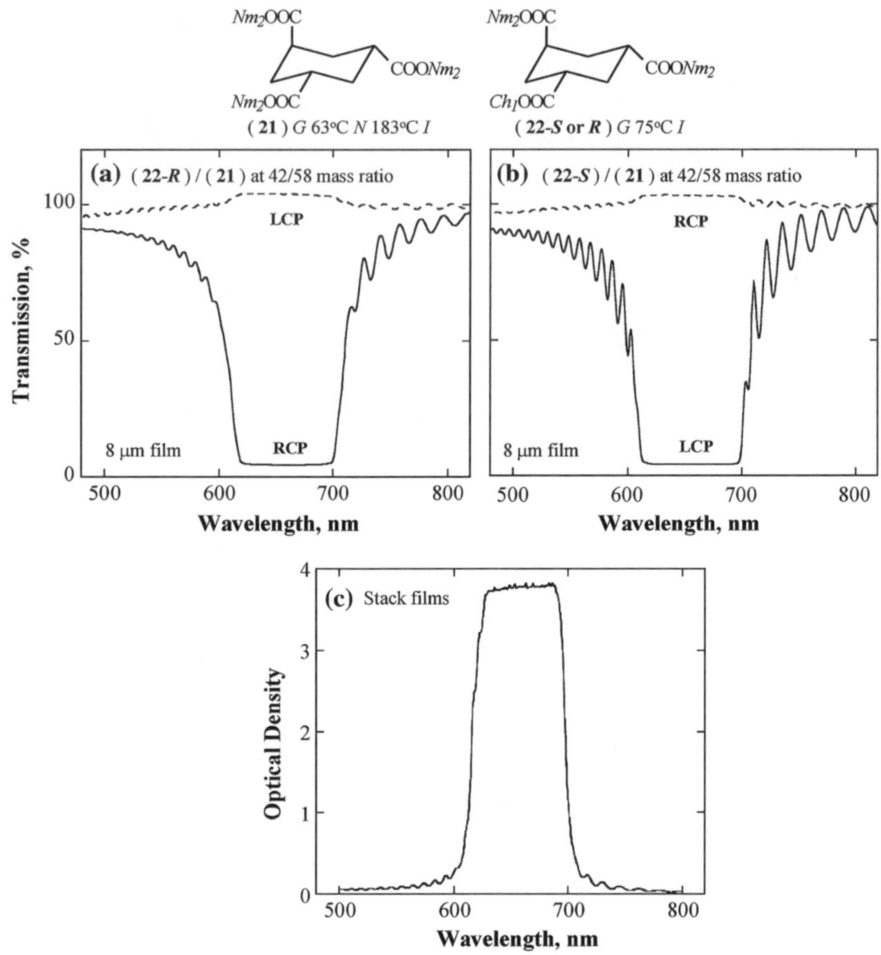

Fig. 6.6 Mixtures (**22**):(**21**) at a 42:58 mass ratio for **a** *right*-handed and **b** *left*-handed circular polarizers, **c** the stack of which acts as an optical notch filter. Used with permission [34]

22-μm-thick constant-pitch, left-handed cholesteric GLC film exhibiting a circular polarization bandwidth of 125 nm based on the LCP and RCP spectra prior to irradiation (*i.e.* t = 0 h) in Fig. 6.7b, where circularly polarized incidents were employed. In the fluid state at 100 °C, the film was irradiated with a 334 nm UV source at 140 μW/cm^2 for 2 h to introduce a pitch gradient followed by cooling to room temperature to preserve it in the glassy state. As a result, the circular polarization bandwidth was broadened to 425 nm, a factor of 3.4 over the initial value, as also shown in Fig. 6.7b. That pitch loosing occurred as a result of partial racemization indicates that (**28**) with (*S*)-1-phenylethylamine and (**29**) with an (*R*)-configuration

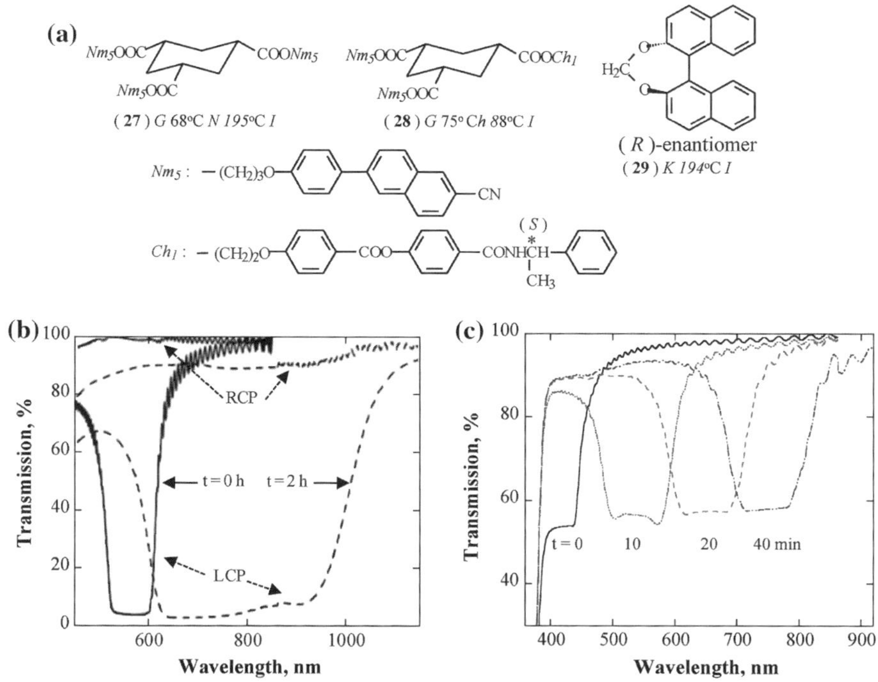

Fig. 6.7 **a** Molecular structures of the three components for **b** a 22-μm-thick gradient-pitch glassy liquid crystal film with (**27**):(**28**):(**29**) at 1.00:0.23:0.13 molar ratio through irradiation at 334 nm (140 μW/cm^2) and 100 °C for 2 h using circularly polarized incidents for transmission measurement, and **c** 8-μm-thick constant-pitch films with (**27**):(**28**):(**29**) at 1.00:1.65:0.40 molar ratio through irradiation at 334 nm (70 μW/cm^2) and 120 °C for durations as indicated using unpolarized incidents for transmission measurement. Used with permission [41]

contribute constructively to a left-handed cholesteric film. The spatial modulation of phororacmization can be understood as follows. The relative molar extinction coefficients at the irradiation wavelength of 334 nm were estimated at (**27**):(**28**):(**29**) = 7.0:3.5:1.0, indicating that the two hosts absorb predominantly over the dopant and that the irradiation intensity at 334 nm diminished by a factor of 10^5 after the first micron into the exposed film surface. Therefore, photo-induced chiral conversion from the (*R*)- to (*S*)-isomer of (**29**) of the dopant took place to a decreasing extent into the first micron of the film. Since the rest of the film remained essentially in the dark, where no chiral conversion took place, counter diffusion between the (*R*)- and (*S*)-enantiomers ensued to set up a decreasing extent of racemization from the irradiated film surface. At the end of photochemical processing, the resultant pitch gradient was frozen in the solid film through glass formation, as validated by atomic force microscopic imaging of its freeze-fractured cross-section [41, 42].

To demonstrate tunable reflective coloration via photoracemization to varying degrees, a ternary film was prepared at a molar ratio of (**27**):(**28**):(**29**) = 1.00:1.65:0.40 with a T_g and T_c at 68 and 134 °C, respectively. The uniform extent of photoracemization through the film thickness was assured by reducing the film thickness from 22 to 8 μm and the irradiation intensity at 334 nm from 140 at 70 $\mu W/cm^2$ while raising the irradiation temperature from 100 to 120 °C. The resultant cholesteric pitch was lengthened by photochemical processing over an increasing duration. The modulated selective reflection bands were preserved in the glassy state by cooling to room temperature with reflective coloration displayed in Fig. 6.7c as characterized by unpolarized incident. In essence, varied reflective colors can be recorded in solid films using a set of the same films irradiated to varying degrees of photoracemization [22].

6.7.3 Circularly Polarized Fluorescence

At a decreasing wt% of (**20**) in the (**19**):(**20**) mixtures from 100 through 95, 51 to 38 without encountering phase separation, the λ_R values of 8-μm-thick films increased from 410 through 434, 890 to 1210 nm labeled as A thorough D in Fig. 6.8a with a monotonic elevation in T_c from 147 to 235 °C and a modest depression in T_g from 77 to 67 °C [33]. With (*S*)-1-phenylethylamine as the chiral building block, left-handed cholesteric GLC films emerged. Without affecting T_g, T_c, and λ_R values, 14-μm-thick cholesteric GLC films doped with blue-emitting Exalite 428 at 0.2 wt% were examined to access their dissimilar factor, g_e. Here $g_e \equiv 2(F_L - F_R)/(F_L + F_R)$, in which F_L and F_R denote, respectively, the left- and right-handed circularly polarized fluorescence intensity with a note that the absolute g_e value at 2 signifies pure circularly polarized fluorescence. It is demonstrated in Fig. 6.8a that unpolarized excitation at 370 nm of the dopant for emission within the stop-band resulted in strong circularly polarized fluorescence that undergoes handedness crossover [43]. Furthermore, nearly pure circularly polarized fluorescence is achieved, as shown in Fig. 6.8b, using a 35-μm-thick Film A. Strongly circularly polarized light sources may find applications in color-image projection and stereoscopic displays. Under the same photoexcitation of Exalite 428 in a 11.5-μm-thick cholesteric fluid film with λ_R = 30.6 μm, both the circularly polarized fluorescence theory and experiment placed the maximum absolute g_e value at 0.65 [44].

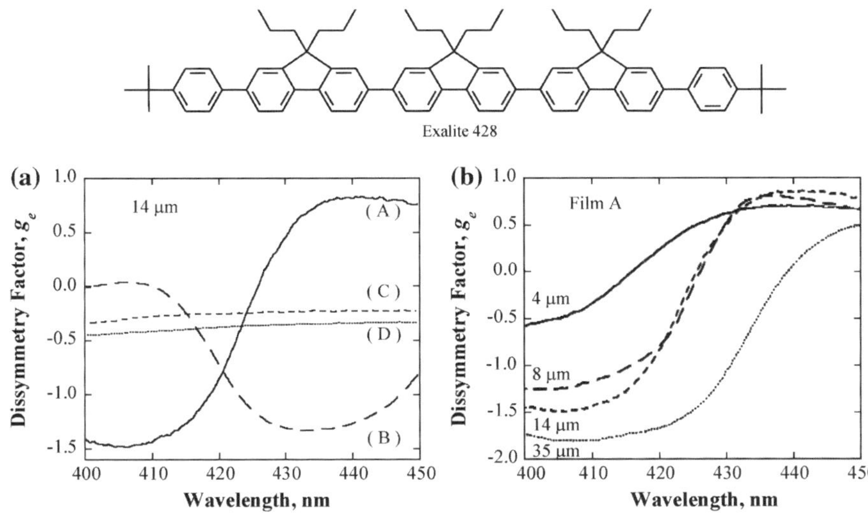

Fig. 6.8 Glassy liquid crystal Films *A–D* defined by wt% of (**20**) in the (**19**):(**20**) mixtures at 100, 95, 51, and 38, respectively, all doped with Exalite 428 at 0.2 wt% for photoexcitation at 370 nm: dissymmetry factors for **a** 14-μm-thick films, and **b** Film A as a function of film thickness. Used with permission [43]

6.7.4 Photoswitchable Nematic GLC Film

In place of the inert cores characteristic of most GLCs, a nematic GLC shown as (**30**) in Fig. 6.9a consists of a photoresponsive core to which two sets of nematogen are chemically bonded with a double-headed arrow locating the nematic director. Of all the prospective diarylethenes, 1,2-bis(2,4-dimethyl-5-phenylthiophen-3-yl)perfluoro-cyclopentene was chosen for its favorable thermal stability, quantum yields, response time, and fatigue resistance. This dithienylethene core undergoes ring closure and opening reversibly with irradiation at 365 and 577 nm, respectively, in a uniaxially oriented, 4-μm-thick nematic GLC film between CaF_2 substrates, as demonstrated in Fig. 6.9b. To test the idea of non-destructive readout using Abbé refractometry with a 780 nm laser beam, a 14-μm-thick nematic GLC film was prepared between FK5 and SF57 substrates for the measurement n_e and n_o. Optical characterization following repeated ring closure-opening cycles led to the observations (i) that $n_e = 1.657 \pm 0.001$ and $n_o = 1.5528 \pm 0.0005$ with $\Delta n \equiv n_e - n_o = 0.105 \pm 0.002$ for the open form, and (ii) that $n_e = 1.7159 \pm 0.0004$ and $n_o = 1.5567 \pm 0.0004$ with $\Delta n = 0.1592 \pm 0.0002$ for the closed form. The photoinduced changes can be summarized as follows: $\Delta n_e = 0.059 \pm 0.001$, $\Delta n_o = 0.0039 \pm 0.0006$, and $\Delta(\Delta n) = 0.054 \pm 0.002$. The measured n_e and n_o values were accompanied by random errors from one cycle to the next, suggesting the

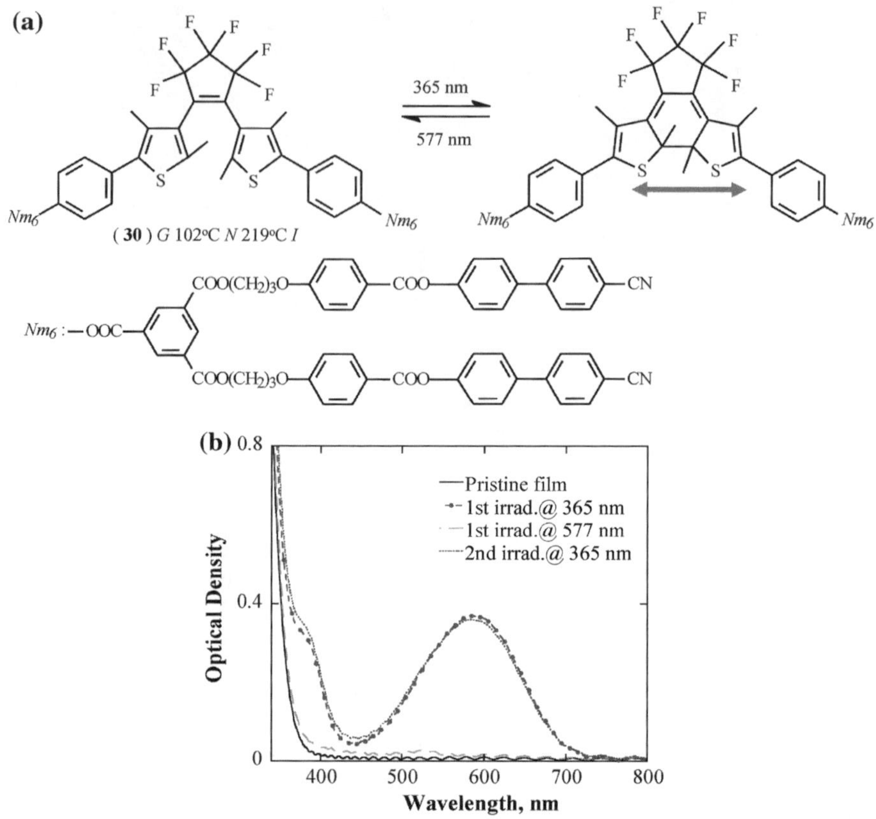

Fig. 6.9 **a** Molecular structure of a nematic GLC (**30**) with a dithienylethene core undergoing ring closure and opening with irradiation at 365 and 577 nm, respectively, and **b** reversible photochromism using a 4-μm-thick film at room temperature. Used with permission [45]

absence of photodegradation [45]. Thus, photochromic nematic GLCs hold promises for nondestructive rewritable optical memory as well as photonic switching, including attenuators, beam deflectors, filters, and polarizers.

6.7.5 Photoswitchable Cholesteric GLC Film

Two sets of chiral-nematic hybrid are chemically bonded to the same dithienylethene core as for the photoswitchable nematic GLC (**30**), resulting in a smectic C*-cholesteric GLC as depicted in Fig. 6.10. A sample of (**31**) was placed between two alignment-coated fused silica substrates for heating to 230 °C followed by cooling to 207 °C, where shearing was applied to induce alignment. The

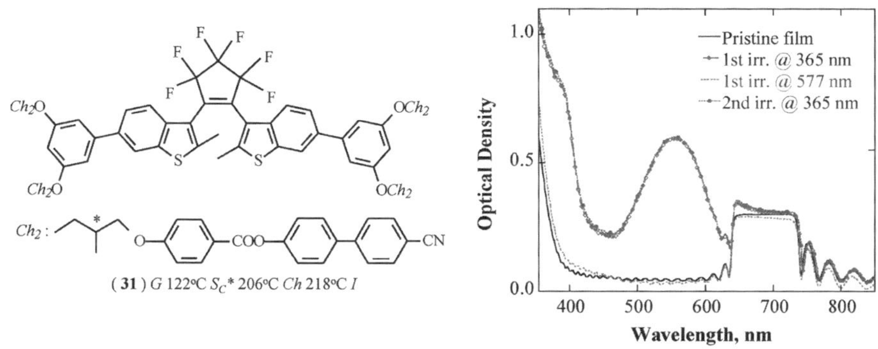

Fig. 6.10 Molecular structure of a smectic C*-cholesteric GLC (**31**) for photoswitchable absorptive coloration via ring closure and opening on top of reflective coloration in a 5-μm-thick film. Used with permission [46]

5-μm-thick film was then annealed for 1 h before quenching in liquid nitrogen to bypass the smectic C* mesophase. With (*S*)-3-bromo-2-methylpropanol as the chiral building block for the flexible linkage, a left-handed cholesteric film was identified with a selective reflection band centered at 686 nm. In addition, ellipsometric analysis resulted in a helical pitch length of 439 nm and an average refractive index of 1.564 beyond 600 nm into the near infrared region. As part of the ellipsometric analysis, an orientational order parameter of 0.65 was also obtained for the quasi-nematic layers comprising the cholesteric film. The feasibility of selective coloration coupled with photoswitchable absorptive coloration via ring closure and opening reactions of the dithienylethene core is demonstrated in Fig. 6.10 using the resultant photochromic cholesteric GLC film [46].

6.7.6 Linearly Polarized Fluorescent Organic Light-Emitting Diodes

Synthesized by the divergent-convergent approach, monodisperse oligofluorenes can be readily processed into monodomain nematic GLC films for blue-light emission [38]. Both T_g and T_c values and stability of glassy films against crystallization depend on the number of fluorene units and the structures of aliphatic side groups. For example, the mixed aliphatic structures in dodecafluorene, (**32**) in Fig. 6.11, contribute to a T_g and T_c at 123 and above 375 °C, respectively, and a morphologically stable nematic GLC film. As part of the fluorescent organic light-emitting diode (FOLED), included as the inset in Fig. 6.11, a uniaxially oriented dodecafluorene film was spin-cast on a rubbed PEDOT:PSS conductive layer followed by thermal annealing at 10 °C above T_g for ½ h before cooling to room

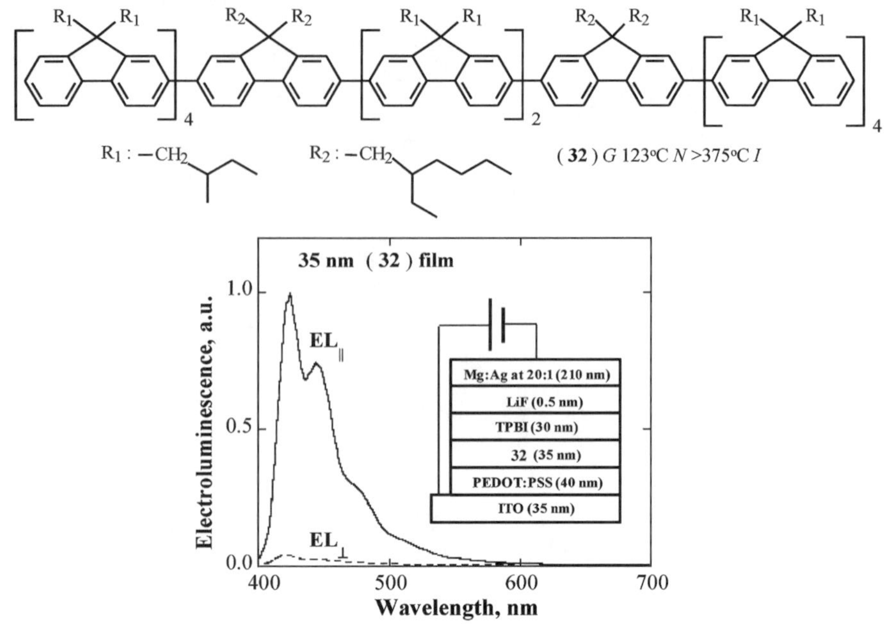

Fig. 6.11 Molecular structure of a dodecafluorene (**32**) and the FOLED device structure for generating linearly polarized deep-blue electroluminescence spectra, where $\mathbf{EL}_{\|}$ and $\mathbf{EL}_{\perp}$ represent the emission intensity parallel and perpendicular, respectively, to the rubbing direction. Used with permission [47]

temperature. The deep blue emission is qualified by blue CIE (Commission Internationale de L'Eclairage) coordinates at (0.159, 0.062). The electroluminescence spectra shown in Fig. 6.11 indicate the maximum dichroic ratio of 31 with a turn-on voltage less than 4 V and a luminescence yield of 1.1 cd/A at a current density of 20 mA/cm^2, the best set of performance data to date for deep-blue polarized FOLED [47].

For emissions at longer wavelengths than deep blue, varied units are inserted at the center of blue-emitting oligofluorenes, as illustrated by the molecular structures shown in Fig. 6.12a [48, 49]. Oligomers (**34**) through (**37**) are used as acceptors in blue-emitting (**33**) as the donor from which Förster energy transfer takes place spontaneously to generate emission wavelengths characteristic of the acceptors' central units. A major advantage of this donor–acceptor approach is the enhanced device efficiency as a result of minimizing concentration quenching between light-emitting acceptor molecules. While (**33**) and (**36**) are morphologically stable nematic GLCs, (**34**), (**35**) and (**37**) are nematic GLCs with all their T_g values above 100 °C but a tendency to crystallize. Nonetheless, morphologically stable nematic GLC films containing up to 4 wt% of the acceptors in (**33**) were readily prepared by spin coating on rubbed PEDOT:PSS

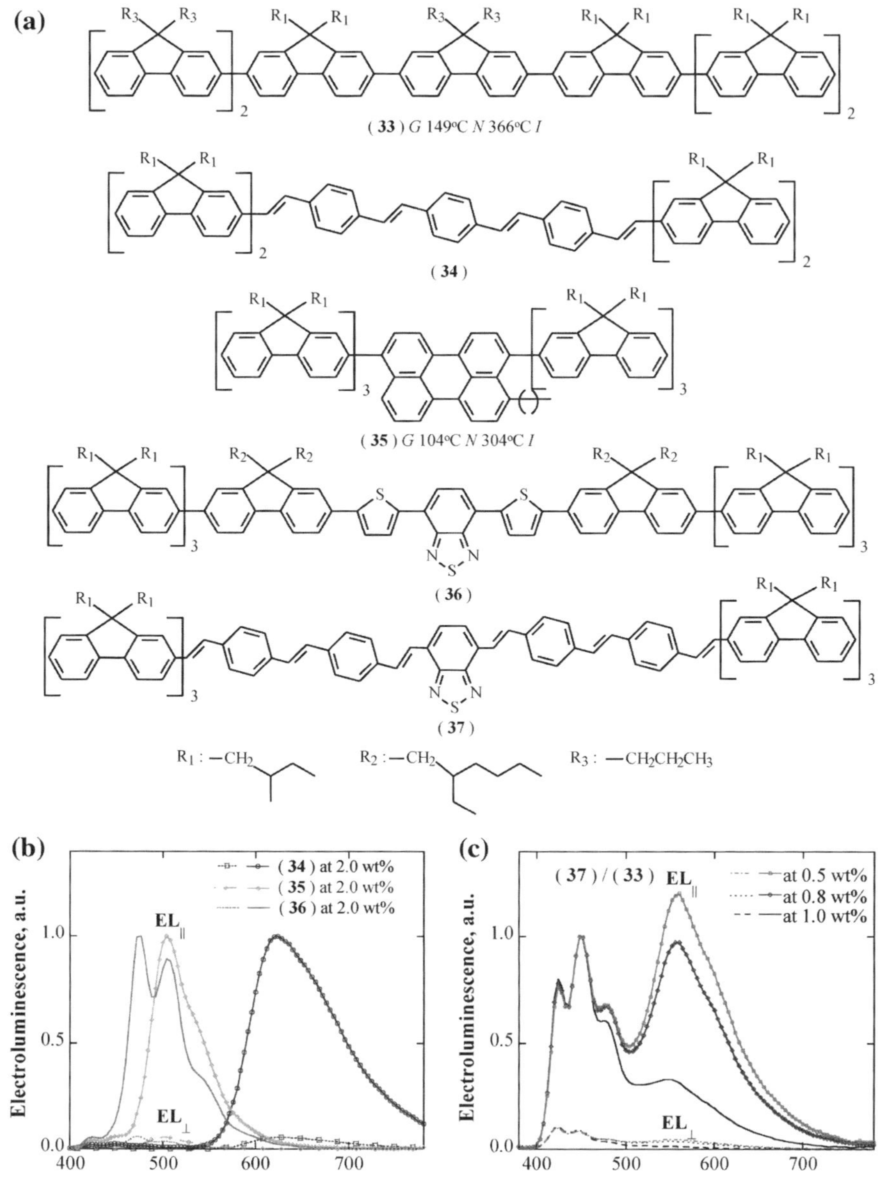

Fig. 6.12 **a** Molecular structures of donor (**33**) and acceptors (**34**) to (**37**); linearly polarized electroluminescence spectra of FOLED devices shown as the inset in Fig. 6.11 comprising **b** (**34**) to (**36**) mixed with (**33**), and **c** (**37**) mixed with (**33**) at the indicated compositions as the emitting layer. Used with permission [49]

conductive layers followed by thermal annealing at 160 °C for ½ h. The T_g and T_c values of the lightly doped mixtures were depressed by no more than 3 °C from those of the pure (**33**) with a T_g and T_c at 149 and 366 °C, respectively. The same FOLED device structure as described in Fig. 6.11 was adopted for the donor–acceptor emitting layers. According to the spectra complied in Fig. 6.12b, emission dichroic ratios from 16 to 26 were achieved with turn-on voltages less than 4 V and luminescence yields from 1.5 to 6.4 cd/A at a current density of 20 mA/cm^2. At the same current density, (**37**) doped in (**33**) at 0.8 wt% yielded nearly pure white-light emission with CIE coordinates at (0.315, 0.333), based on the spectra shown in Fig. 6.12c, at a turn-on voltage less than 4 V with a luminescence yield of 4.5 cd/A and an overall emission dichroic ratio of 16 across the entire spectrum. Furthermore, the CIE coordinates were found to be independent of current density up to 100 mA/cm^2. Compared to the previously reported green- and red-emitting polarized FOLEDs, the donor–acceptor approach resulted in superior polarization ratios and luminescence yields in addition to the first polarized white-light FOLED [49]. Linearly polarized OLEDs are potentially useful as energy-efficient backlights for liquid crystal displays, electroluminescent displays with reduced glare and increased contrast, projection displays, and for stereoscopic imaging systems.

6.7.7 Linearly Polarized Phosphorescent Organic Light-Emitting Diodes

With a glassy-nematic pentafluorene, (**38**), as the donor and a discotic Pt(II) complex, (**39**), as the acceptor, both depicted in Fig. 6.13 with their thermotropic transition temperatures, a polarized phosphorescent OLED (PhOLED), ITO/rubbed PEDOT:PSS (30 nm)/(**38**):(**39**) at 3:1 mass ratio (55 nm)/TPBi (45 nm)/LiF (0.5 nm)/Al (150 nm), was fabricated following similar procedures to those performed for polarized FOLEDs except thermal annealing of the mixed emitting layer at 120 °C for 1 h. At a current density of 10 mA/cm^2, a turn-on voltage of 3–4 V was observed with a luminescence yield of 2.4 cd/A. Emission anisotropy presented in Fig. 6.13 suggests that the columnar axes of self-organized Pt(II) complexes are oriented largely parallel to the substrate surface and normal to the rubbing direction [50]. This work represents the first proof-of-concept polarized PhOLED with a modest overall performance compared to polarized FOLEDs that have been much more intensively explored since the mid-1990s. Nevertheless, phosphorescence is intrinsically superior to fluorescence in terms of internal

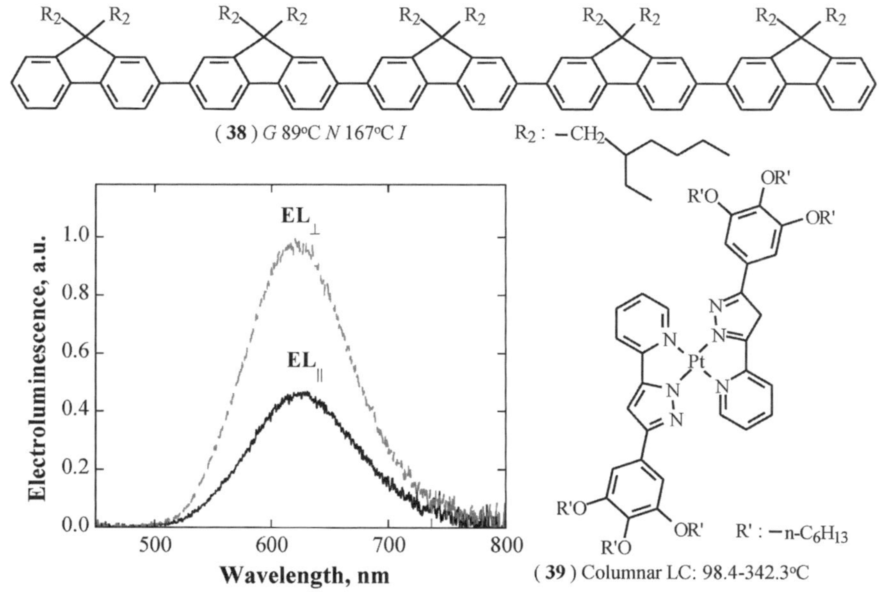

Fig. 6.13 Molecular structures of the donor and acceptor constituting the emitting layer in a PhOLED, ITO/rubbed PEDOT:PSS (30 nm)/(**38**):(**39**) at 3:1 mass ratio (55 nm)/TPBi (45 nm)/ LiF (0.5 nm)/Al (150 nm) with its linearly polarized electroluminescence spectra. Used with permission [50]

quantum yields, 100 versus 25 %. Work is in progress to elucidate the nanostructures and film morphology responsible for the observed polarized electroluminescence and ways to improve device performance.

6.7.8 Circularly Polarized Fluorescent Organic Light-Emitting Diodes

Monodisperse oligofluorenes carrying enantiomeric aliphatic side groups are capable of forming monodomain cholesteric GLC films exhibiting circular dichroism and circularly polarized fluorescence orders-of-magnitude higher than those originating in polymeric helices in view of supramolecular (i.e., cholesteric stacks) over macromolecular chirality (i.e., polymer helices) [37, 51]. A typical nanofluorene identified as (**40**) in Fig. 6.14 was incorporated as a 70-nm-thick glassy cholesteric film in a circularly polarized FOLED having the same device structure as shown in Fig. 6.11 except the emitting layer, resulting in a turn-on voltage less than 5 V and a luminescence yield of 0.94 cd/A at a current density of

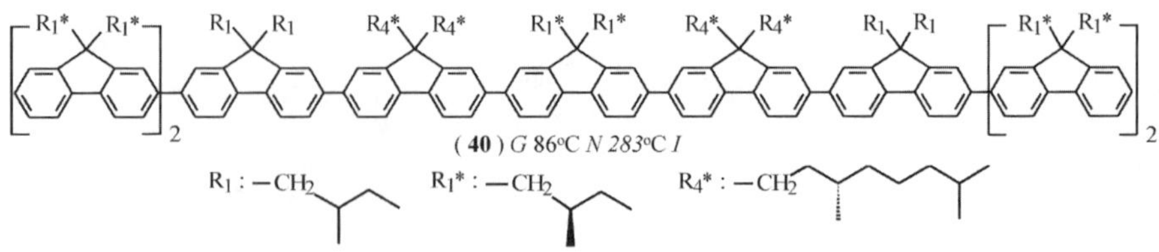

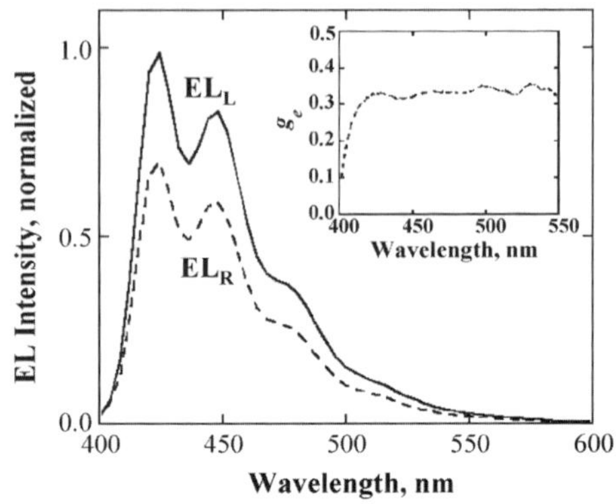

Fig. 6.14 Molecular structure of a nonafluorene (**40**) as the emitting layer in the FOLED device shown as the inset in Fig. 6.11 with its circularly polarized electroluminescence spectra and the calculated dissymmetry factor, where **EL**$_L$ and **EL**$_R$ represent the left- and right-handed circularly polarized electroluminescence, respectively, and the dissymmetry factor is as defined for circularly polarized fluorescence in Sect. 6.7.3. Used with permission [51]

20 mA/cm^2. The CIE coordinates at (0.157, 0.068) correspond to deep blue light emission, and the g_e value of 0.35, as defined in Sect. 6.7.3 and plotted in the inset of Fig. 6.14, represents the best value to date [51].

6.7.9 Cholesteric GLC Film as Robust Solid-State Laser

Cholesteric LC films constitute a one-dimensional photonic band-gap, serving as a resonator for the embedded light emitters to undergo lasing. Solid films are superior to fluid films by preventing perturbations to helical stacks during photoexcitation and lasing. With a mixture comprising (**42**):(**43**) = 76.0:24.0 mass ratio to which (**41**) was doped at 2.0 wt%, a 22-μm-thick monodomain cholesteric GLC film was prepared to tune the fluorescence maximum of **(41)** to the lower energy edge of the host film's stop-band. Note that the resultant ternary mixture has the same T_g as the those of the two major components at 92 °C in the host with a stop-band centered at 590 nm. Using the apparatus described elsewhere [52] with 532 nm irradiation at 35 ps pulse duration, a sharp lasing peak appeared at 635 nm. The temporal stability of the lasing output from a cholesteric GLC film

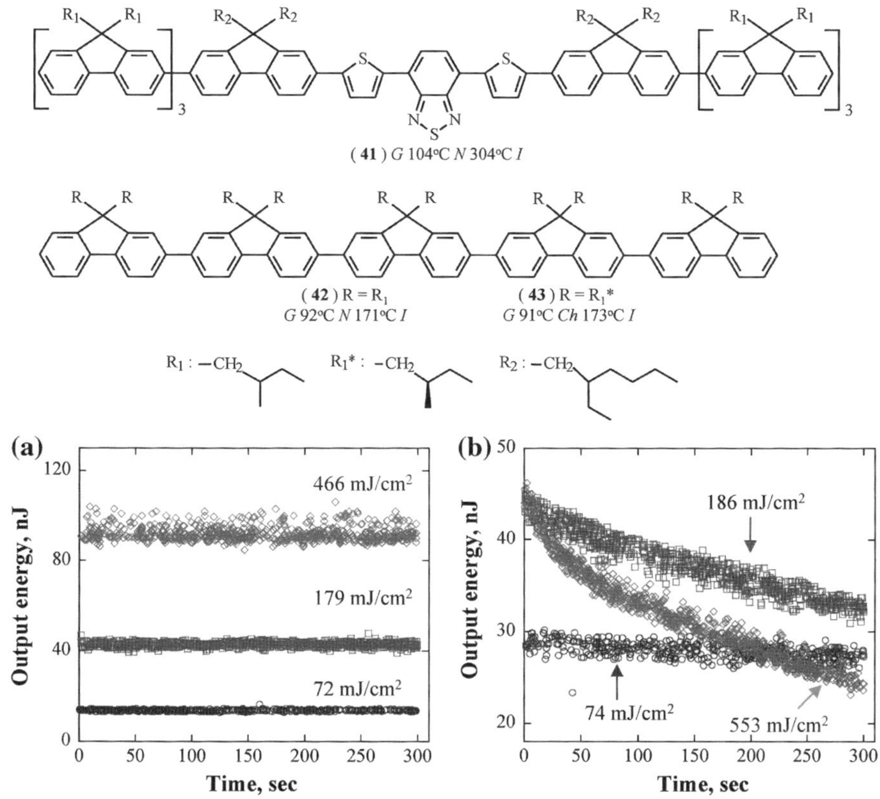

Fig. 6.15 Molecular structures of a *red*-emitting dopant (**41**), nematic (**42**) and cholesteric (**43**) GLCs mixed to serve as the host. Temporal evolution of lasing output with specified pump fluences at 10 Hz from **a** a cholesteric GLC film prepared with (**42**):(**43**) at a 76.0:24.0 mass ratio, and **b** a cholesteric fluid film, CB-15:ZLI-2244-000 at a 35.6:64.4 mass ratio, both films containing 2.0 wt% (**41**). Used with permission [53]

was validated by increasing the pump fluence from 72 to 466 mJ/cm^2, as shown in Fig. 6.15a. A fluid cholesteric LC film containing 2.0 wt% (**41**) in CB-15:ZLI-2244-000 at a mass ratio of 35.6:64.4 was also subjected to the stability test. The results reported in Fig. 6.15b with pump fluences from 74 to 533 mJ/cm^2 reveal the decaying lasing output with time most likely caused by heating via optical pumping, light-induced pitch dilation, and/or laser-induced fluid flow, any or all of which would have disrupted the cholesteric helical stacking and the orientational order of dopant molecules [53]. Robust cholesteric GLC lasers are potentially useful for laser spectroscopy, medical diagnostics and treatments, and laser projection displays.

6.7.10 Spatially Resolved Lasing from a Cholesteric GLC Film

Instead of a single lasing peak, as reported in Sect. 6.7.9, a laterally gradient-pitch host film consisting of a cascade of Grandjean-Cano bands induced by thermal processing can serve to produce multiple lasing wavelengths. Two cholesteric GLC mixtures containing (**42**):(**43**) at 71.0:29.0 and 80.0:20.0 mass ratios, both doped with (**41**) at 2.5 wt%, were melted on a fused silica substrate coated with a robbed polyimide alignment layer. Another substrate with a parallel alignment coating was pressed on top against 14-μm-thick spacers to define film thickness while enforcing the formation of a sharp interface between the two cholesteric GLC melts. Molecular diffusion across this interface was thermally activated at 220 °C for 62 h, followed by cooling through the T_g at 92 °C to room temperature for the induced lateral pitch gradient to be frozen in solid state. As a result of the laterally varying chiral concentration, the emerging Grandjean-Cano lines originating from the strong surface anchoring are identified as white vertical lines in Fig. 6.16a.

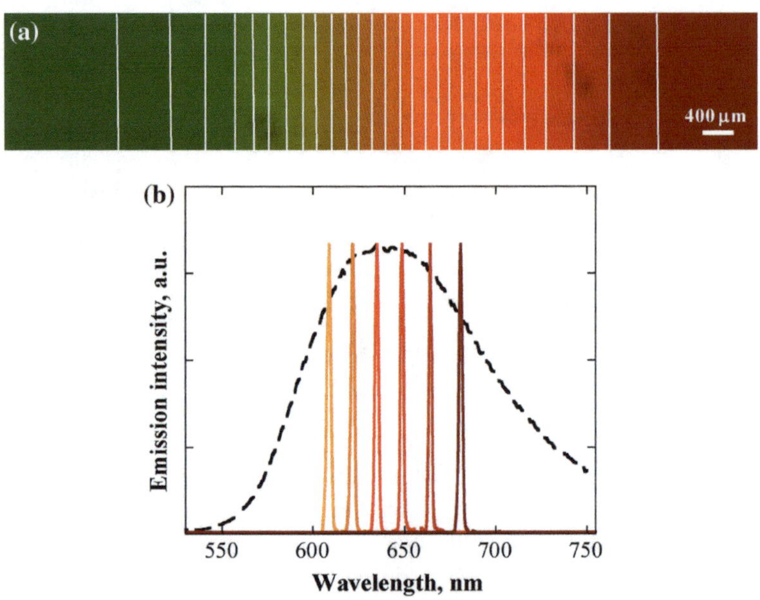

Fig. 6.16 **a** Polarizing optical micrograph of a 14-μm-thick, thermally processed cholesteric GLC film with lateral pitch gradient, where Grandjean-Cano bands are located by *white vertical lines*; the Grandjean-Cano bandwidth increases from 130 μm at the center to 1200 μm at both edges. **b** Lasing from (**41**) at 2.5 wt% present in the gradient-pitch film shown in **a** with optical pumping using a 532 nm laser for 35 ps at a 10 Hz repetition rate; the six lasing peaks from *right* to *left* acquired from the 10, 12, 14, 16, 18, and 20th Grandjean-Cano bands from the *right* with the *dashed curve* representing fluorescence of a 14-μm-thick nematic GLC film of (**42**) containing 2.5 wt% (**41**) with unpolarized photoexcitation at 525 nm. Used with permission [54]

Each Grandjean-Cano band is characterized by a single pitch length giving rise to an individual stop-band, and a series of them present a set of lasing peaks as shown in Fig. 6.16b using the same set-up as in Sect. 6.7.9. The observed lasing thresholds, 6.6–7.6 mJ/cm^2, are the lowest of all spatially resolved cholesteric LC lasers reported to date. The slope efficiency from 0.2 to 1.5 % observed herein compares favorably with the maximum value of 0.5 % reported for gradient-pitch fluid cholesteric LC lasers [54]. In a nutshell, cholesteric GLCs are in a unique position to generate multiple lasing wavelengths on demand within a single film.

6.8 Solvent-Vapor Annealing of Conjugated Oligomers

Monodomain glassy-nematic oligofluorene films are attainable by thermal annealing of spin-cast films above their respective T_g values on both rubbed polyimide alignment and photoalignment layers comprising coumarin-containing polymers [55, 56]. Further attempts have been made to produce monodomain nematic GLC films on rubbed polyimide layers via equilibration with solvent vapor at room temperature. Representative oligomer and polymer structures tested are compiled in Fig. 6.17. To illustrate the first idea of solvent-vapor annealing, a 30-nm-thick film of (**44**) spin-cast from chlorobenzene was exposed to the saturated vapor of chlorobenzene at room temperature for ½ min to yield an orientational order parameter of 0.82, a value identical to that achieved with thermal annealing at 10 °C above the T_g of (**44**) at 102 °C, both on rubbed polyimide alignment layers [57]. Note that the preparation of polyimide layers entails thermal curing at 250 °C for 1 h. Photoalignment is an attractive alternative from the energy-savings perspective. Unlike rubbing polyimide to enable orientation of liquid crystals, photoalignment is a non-contact method that precludes potentially adverse effects such as mechanical damage, generation of dust particles and electrostatic charges. Approximately 10-nm-thick spin-cast films of (**45**) were irradiated with linearly polarized source between 300 and 330 nm under argon at room temperature to 31 % coumarin dimerization. About 90-nm-thick films of conjugated oligomers (**33**) and (**44**) were spin-cast from chloroform on the resultant photoalignment layers. As controlled experiments, 15-nm-thick rubbed polyimide alignment layers were also used for solvent-vapor annealing of (**33**) and (**44**) spin-cast films. To prevent dewetting, (**44**) and (**33**) films were exposed to chloroform vapor at 95 and 90 % saturation, respectively. The orientational order parameter values turned out to be 0.82 and 0.74 for (**44**) and (**33**), respectively, independent of the underlying alignment layers. Nevertheless, it took 5–10 s to achieve asymptotic values of orientational order on rubbed polyimide alignment layers, compared to 6–8 min on photoalignment layers. In contrast, the polymer analogue (**46**) could not be oriented at all on either rubbed polyimide or photoaligment layers with solvent-vapor annealing for up to 24 h [58]. Conjugated oligomers readily processed into well oriented nematic GLC films are potentially

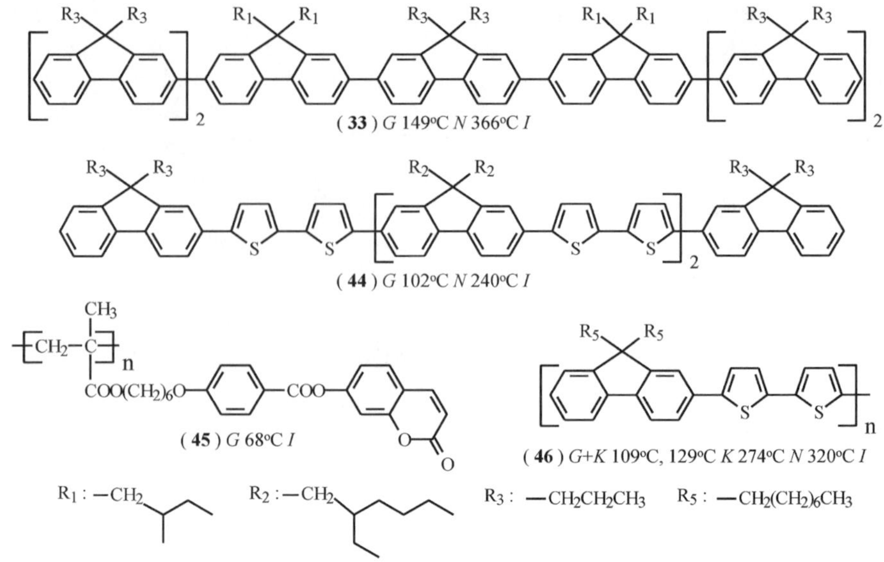

Fig. 6.17 Molecular structures of conjugated oligomers (**33**) and (**44**), a polymer analogue (**46**), and a coumarin-containing polymer (**45**), for an evaluation of solvent-vapor annealing at room temperature

useful as energy-efficient backlights in liquid crystal displays, anisotropic charge transport to suppress crosstalk in logic circuit and pixel switching elements, and polarization-sensitive photodiodes for sensing applications.

6.9 Conclusions

Glassy liquid crystals are designed to preserve nano- to microstructured molecular self-organization underlying liquid crystal fluids through glass formation upon cooling while sustaining long-term morphological stability. Although fundamental understanding is lacking, it stands to reason that conformational multiplicity not only facilitates glass formation during film processing but also increases the Gibbs energy barrier to crystallization because of the increased entropy locked in the resultant glassy state. Moreover, solubility in common solvents can be enhanced by conformational multiplicity to facilitate material synthesis, purification and film preparation via spin-casting. This general theme is implemented in (i) the core-pendant strategy in which geometrically disparate cores and pendants are chemically bonded through flexible linkages, and (ii) the hairy rod strategy in which flexible side groups are attached to conjugated oligomers. We remark in passing that conformational multiplicity has also proven crucial to the development of morphologically stable amorphous hosts for phosphorescent organic light-emitting diodes [59, 60]. With a theme on GLCs herein, key accomplishments are recapitulated as follows:

1. Morphologically stable GLCs have been constructed with cyclohexane, benzene, bicyclooctene, adamantane, cubane, and the extended cores derived therefrom to which mesogenic and chiral groups are linked via ethylene or propylene spacers. A T_g and T_c close to 130 and 350 °C, respectively, have been achieved while remaining glassy liquid crystalline at room temperature for over one year in most cases.
2. In view of the unique roles of cholesteric GLCs in material synthesis and device application, the commonly practiced statistical synthesis and the continuing effort in deterministic synthesis are highlighted. All factors considered, a hybrid chiral-nematic precursor is much preferred over separate chiral and nematic precursors in terms of synthesis and subsequent product separation and purification.
3. Oligofluorenes and oligo(fluorene-*co*-bithiophene)s to which aliphatic groups are attached exhibit a T_g and a T_c close to 150 and above 375 °C, respectively. Unlike typical high molecular weight conjugated polymers, these oligomers are readily soluble in common solvents for spin-casting into nematic and cholesteric GLC films that resist crystallization when left at room temperature for over one year.
4. Cholesteric and nematic GLC films following the core-pendant strategy are promising for circular polarizers, optical notch filters and reflectors with spectral position and bandwidth adjustable by chemical and photochemical means; circularly polarized fluorescence from a doped light-emitter; photonic switching and optical memory with cholesteric and nematic GLC films comprising a dithienylethene core.
5. Nematic and cholesteric GLC films comprising hairy rods have been used to demonstrate linearly polarized full-color and white-light as well as circularly polarized blue-emitting fluorescent organic light-emitting diodes (OLEDs); linearly polarized phosphorescent OLED; and robust organic solid-state lasers with spatially resolved Grandjean-Cano bands for emission at multiple wavelengths on demand within a single film.
6. Orientation of liquid crystals is traditionally accomplished on rubbed polyimide alignment layers that require high-temperature curing. On photoaligment layers prepared at room temperature, monodomain glassy-nematic conjugated oligomer films arise from solvent-vapor annealing conducted at room temperature, resulting in the same orientational order as that achieved on polyimide alignment layer with thermal annealing above T_g.

Acknowledgments This chapter summarizes the contributions by former doctoral students, post-doctoral fellows, and research associates in S. H. Chen's laboratory since year 1990: Drs. H.-M. P. Chen, Y. Geng, D. Katsis, S. W. Culligan, A. C.-A. Chen, C. Kim, A. Trajkovska, J. J. Ou, J. U. Wallace, L. Zeng, S. K.-H. Wei, H. Shi, J. C. Mastrangelo, and B. M. Conger. Professor Anthamatten's critical reading of the manuscript with constructive suggestions is greatly appreciated. We are grateful for the generous support spanning over two decades at the University of

Rochester by the U.S. National Science Foundation, Department of Defense, Department of Energy, New York State Energy Research and Development Authority, the Japanese Ministry of International Trade and Industry, and Eastman Kodak Company. Additional funding was provided by the U.S. Department of Energy Office of Inertial Confinement Fusion under cooperative agreements with Laboratory for Laser Energetics at the University of Rochester. The support of DOE does not constitute an endorsement by DOE of the views expressed herein.

References

1. P.G. Dibenedetti, F.H. Stillinger, Supercooled liquids and the glass transition. Nature **410**, 259–267 (2001)
2. K.F. Freed, The decent into glass formation in polymer fluids. Acct. Chem. Res. **44**, 194–203 (2011)
3. K. Tsuji, M. Sorai, S. Seki, New finding of glassy liquid crystal—a non-equilibrium state of cholesteryl hydrogen phthalate. Bull. Chem. Soc. Jpn. **44**, 1452 (1971)
4. M. Sorai, S. Seki, Glassy liquid crystal of the nematic phase of N-(*o*-hydroxy-*p*-methoxybenzylidene)-*p*-butylaniline. Bull. Chem. Soc. Jpn. **44**, 2887 (1971)
5. C.B. McArdle (ed.), *Side Chain Liquid Crystal Polymers* (Chapman and Hall, New York, 1989)
6. N.A. Platé (ed.), *Liquid-Crystal Polymers* (Plenum Press, New York, 1993)
7. A. Ciferri (ed.), *Liquid Crystallinity in Polymers: Principles and Fundamental Properties* (VCH, New York, 1991)
8. H. Shi, S.H. Chen, M.E. De Rosa, T.J. Bunning, W.W. Adams, Dynamic mechanical properties of cyclohexane-based glass-forming liquid crystals and a linear side-chain polymer analogue. Liq. Cryst. **20**, 277–282 (1996)
9. M.E. De Rosa, W.W. Adams, T.J. Bunning, H. Shi, S.H. Chen, Dynamic mechanical relaxation behavior of low molecular weight side chain cyclic liquid crystalline compounds near the glass transition temperature. Macromolecules **29**, 5650–5657 (1996)
10. W. Wedler, D. Demus, H. Zaschke, K. Mohr, W. Schafer, W. Weissflog, Vitrification in low-molecular-weight mesogenic compounds. J. Mater. Chem. **1**, 347–356 (1991)
11. G.S. Attard, C.T. Imrie, Liquid-crystalline and glass-forming dimers derived from 1-aminopyrene. Liq. Cryst. **11**, 785–789 (1992)
12. H. Dehne, A. Roger, D. Demus, S. Diele, H. Kresse, G. Pelzl, W. Weissflog, Sulphur ligated siamese twin mesogens. Liq. Cryst. **6**, 47–62 (1989)
13. G.S. Attard, C.T. Imrie, F.E. Karasz, Low molar mass liquid-crystalline glasses: preparation and properties of the α-(4-cyanobiphenyl-4′-oxy-)-ω-(1-pyrenimine-benzylidene-4′-oxy)alkanes. Chem. Mater. **4**, 1246–1253 (1992)
14. N. Tamaoki, G. Kruk, H. Matsuda, Optical and thermal properties of cholesteric solid from dicholesteryl esters of diacetylenedicarboxylic acid. J. Mater. Chem. **9**, 2381–2384 (1999)
15. F.H. Kreuzer, D. Andrejewski, W. Haas, N. Haberle, G. Riepl, P. Spes, Cyclic siloxanes with mesogenic side groups. Mol. Cryst. Liq. Cryst. **199**, 345–378 (1991)
16. F.H. Kreuzer, R. Maurer, P. Spes, Liquid-crystalline silsesquioxanes. Makromol. Chem., Macromol. Symp. **50**, 215–228 (1991)
17. K.D. Gresham, C.M. McHugh, T.J. Bunning, R.J. Crane, H.E. Klei, E.T. Samulski, Phase behavior of cyclic siloxane-based liquid crystalline compounds. Polym. Sci.: Part A: Polym. Chem. **32**, 2039–2047 (1994)
18. K. Lorenz, D. Hölter, B. Stühn, R. Mülhaupt, H. Frey, A mesogen-functionalized carbosilane dendrimer: a dendritic liquid crystalline polymer. Adv. Mater. **8**, 414–416 (1996)
19. S.A. Ponomarenko, I. Boiko, V.P. Shibaev, R.M. Richardson, I.J. Whitehouse, E.A. Rebrov, A.M. Muzafarov, Carbosilane liquid crystalline dendrimers: From molecular architecture to supramolecular nanostructures. Macromolecules **33**, 5549–5558 (2000)

20. I.M. Saez, J.W. Goodby, R.M. Richardson, A liquid crystalline silsesquioxane dendrimer exhibiting chiral nematic and columnar mesophases. Chem. Eur. J. **7**, 2758–2764 (2001)
21. V. Percec, M. Kawasumi, P.L. Rinaldi, V.E. Litman, Synthesis and characterization of cyclic liquid crystalline oligomers based on 1-(4-hydroxy-4'-biphenylyl)-2-(4-hydroxyphenyl)butane and 1,10-dibromodecane. Macromolecules **25**, 3851–3861 (1992)
22. S.H. Chen, H.M.P. Chen, D. Katsis, J.C. Mastrangelo, Glassy liquid crystals for tunable reflective coloration. ACS Symp. Ser. **888**, 290–306 (2004)
23. S.H. Chen, H. Shi, J.C. Mastrangelo, J.J. Ou, Thermotropic chiral nematic side-chain polymers and cyclic oligomers. Prog. Polym. Sci. **21**, 1211–1233 (1996)
24. S.H. Chen, H. Shi, B.M. Conger, J.C. Mastrangelo, T. Tsutsui, Novel vitrifiable liquid crystals as optical materials. Adv. Mater. **8**, 998–1001 (1996)
25. S.H. Chen, J.C. Mastrangelo, T.N. Blanton, A. Bashir-Hashemi, Novel glass-forming liquid crystals. IV. Effects of central core and pendant group on vitrification and morphological stability. Liq. Cryst. **21**, 683–694 (1996)
26. F.Y. Fan, J.C. Mastrangelo, D. Katsis, S.H. Chen, T.N. Blanton, Novel glass-forming liquid crystals. V. Nematic and chiral-nematic systems with an elevated glass-transition temperature. Liq. Cryst. **27**, 1239–1248 (2000)
27. F.Y. Fan, S.W. Culligan, J.C. Mastrangelo, D. Katsis, S.H. Chen, T.N. Blanton, Novel glass-forming liquid crystals. 6. High temperature glassy nematics. Chem. Mater. **13**, 4584–4594 (2001)
28. T. Kato, M.J. Fréchet, New approach to mesophase stabilization through hydrogen-bonding molecular interactions in binary mixtures. J. Am. Chem. Soc. **111**, 8533–8534 (1989)
29. T. Itahara, M. Sunose, T. Kameda, T. Ueda, Control of liquid-crystalline properties by base pairing of adenine and thymine. ChemPhysChem **3**(4), 378–379 (2002)
30. A. Takahashi, V.A. Mallia, N. Tamaoki, Novel supramolecular hydrogen-bonded cholesteric mesogens: liquid crystalline, thermoptical and glass-forming properties. J. Mater. Chem. **12**, 1582–1587 (2003)
31. R.K. Vijaaraghvan, A. Abraham, H. Akiyama, S. Furumi, A. Tamaoki, S. Das, Photoresponsive glass-forming butadiene-based chiral liquid crystals with circularly polarized photoluminescence. Adv. Funct. Mater. **18**, 2510–2517 (2008)
32. S. Chandrasekhar, *Liquid Crystals* (Cambridge University Press, New York, 1992)
33. D. Katsis, H.P. Chen, J.C. Mastrangelo, S.H. Chen, T.N. Blanton, Vitrified chiral-nematic liquid crystalline films for selective reflection and circular polarization. Chem. Mater. **11**, 1590 (1999)
34. H.M.P. Chen, D. Katsis, J.C. Mastrangelo, S.H. Chen, S.D. Jacobs, P.J. Hood, Glassy liquid-crystal films with opposite chirality as high performance optical notch filters. Adv. Mater. **12**, 1283–1286 (2000)
35. H.M.P. Chen, D. Katsis, S.H. Chen, Deterministic synthesis and optical properties of glassy chiral-nematic liquid crystals. Chem. Mater. **15**, 2534–2542 (2003)
36. C. Kim, K.L. Marshall, J.U. Wallace, J.J. Ou, S.H. Chen, Novel cholesteric glassy liquid crystals comprising benzene functionalized with hybrid chiral-nematic mesogens. Chem. Mater. **20**, 5859–5868 (2008)
37. Y. Geng, A. Trajkovska, D. Katsis, J.J. Ou, S.W. Culligan, S.H. Chen, Synthesis, characterization, and optical properties of monodisperse chiral oligo(fluorene)s. J. Am. Chem. Soc. **124**, 8337–8347 (2002)
38. Y. Geng, S.W. Culligan, A. Trajkovska, J.U. Wallace, S.H. Chen, Monodisperse oligofluorenes forming glassy-nematic films for linearly polarized blue emission. Chem. Mater. **15**, 542–549 (2003)
39. J.U. Wallace, S.H. Chen, Fluorene-based conjugated oligomers for organic photonics and electronics. Adv. Polym. Sci. **212**, 145–186 (2008)
40. L.E. Hajido, A.C. Erigen, Theory of light reflection by cholesteric liquid crystals possessing a pitch gradient. J. Opt. Soc. Am. **69**, 1017–1023 (1979)
41. S.H. Chen, J.C. Mastrangelo, R.J. Jin, Glassy Liquid-crystal films as broadband polarizers and reflectors via spatially modulated photoracemization. Adv. Mater. **11**, 1183–1186 (1999)

42. S.H. Chen, R.J. Jin, D. Katsis, J.C. Mastrangelo, S. Papernov, A.W. Schmid, Photoracemization broadening of selective reflection and polarization band of glassy chiral-nematic films. Liq. Cryst. **27**, 201–209 (2000)
43. S.H. Chen, D. Katsis, J.C. Mastrangelo, A.W. Schmid, T. Tsutsui, T.N. Blanton, Circularly polarized light generated by photoexcitation of luminophores in glassy liquid-crystal films. Nature **397**, 506–508 (1999)
44. D. Katsis, A.W. Schmid, S.H. Chen, Mechanistic insight into circularly polarized photoluminescence from a chiral-nematic film. Liq. Cryst. **26**, 181–185 (1999)
45. S.H. Chen, H.M.P. Chen, Y. Geng, S.D. Jacobs, K.L. Marshall, T.N. Blanton, Novel glassy-nematic liquid crystals for nondestructive optical memory and photonic switching. Adv. Mater. **15**, 1061–1065 (2003)
46. C. Kim, K.L. Marshall, J.U. Wallace, S.H. Chen, Photochromic glassy liquid crystals comprising mesogenic pendants to dithienylethene cores. J. Mater. Chem. **18**, 5592–5598 (2008)
47. S.W. Culligan, Y. Geng, S.H. Chen, K. Klubek, K. Vaeth, C.W. Tang, Strongly polarized and efficient blue organic light-emitting diodes using monodisperse glassy-nematic oligo(fluorene)s. Adv. Mater. **15**, 1176–1180 (2003)
48. Y. Geng, A.C.A. Chen, J.J. Ou, S.H. Chen, K. Klubek, K.M. Vaeth, C.W. Tang, Monodisperse glassy-nematic conjugated oligomers with chemically tunable polarized light emission. Chem. Mater. **15**, 4352–4361 (2003)
49. A.C.A. Chen, S.W. Culligan, Y. Geng, S.H. Chen, K. Klubek, K. Vaeth, C.W. Tang, Organic polarized light-emitting diodes through förster energy transfer using monodisperse conjugated oligomers. Adv. Mater. **16**, 783–788 (2004)
50. S.H. Liu, M.S. Lin, L.Y. Chen, Y.H. Hong, C.H. Tsai, C.C. Wu, A. Poloek, Y. Chi, A.C.A. Chen, S.H. Chen, H.F. Hsu, Polarized phosphorescent organic light-emitting devices adopting mesogenic host-guest systems. Org. Electron. **12**, 15–21 (2011)
51. Y. Geng, A. Trajkovska, S.W. Culligan, J.J. Ou, H.M.P. Chen, D. Katsis, S.H. Chen, Origin of strong chiroptical activities in films of nonafluorenes with a varying extent of pendant chirality. J. Am. Chem. Soc. **125**, 14032–14038 (2003)
52. K. Dolgaleva, S.K.-H. Wei, S.G. Lukishova, S.H. Chen, K. Schwertz, R.W. Boyd, Enhanced laser performance of cholesteric liquid crystals doped with oligofluorene dye. J. Opt. Soc. Am. B. **25**, 1496–1504 (2008)
53. S.K.-H. Wei, S.H. Chen, K. Dolgaleva, S.G. Lukishova, R.W. Boyd, Robust organic lasers comprising glassy-cholesteric pentafluorene doped with a red-emitting oligofluorene. Appl. Phys. Lett. **94**, 4111 (2009)
54. S.K.-H. Wei, S.H. Chen, Spatially resolved lasers using a glassy cholesteric liquid crystal film with lateral pitch gradient. Appl. Phys. Lett. **98**, 111112 (2011)
55. A. Trajkovska, C. Kim, K.L. Marshall, T.H. Mourey, S.H. Chen, Photoalignment of a nematic liquid crystal fluid and glassy-nematic oligofluorenes on coumarin-containing polymer films. Macromolecules **39**, 6983–6989 (2006)
56. C. Kim, J.U. Wallace, A. Trajkovska, J.J. Ou, S.H. Chen, Quantitative assessment of coumarin-containing polymer film's capability for photoalignment of liquid crystals. Macromolecules **40**, 8924–8929 (2007)
57. L. Zeng, F. Yan, S.K.-H. Wei, S.W. Culligan, S.H. Chen, Synthesis and processing of monodisperse oligo(fluorene-*co*-bithiophene)s into oriented films by thermal and solvent annealing. Adv. Funct. Mater. **19**, 1978–1986 (2009)
58. S.K.-H. Wei, L. Zeng, K.L. Marshall, S.H. Chen, Room-temperature processing of π-conjugated oligomers into uniaxially oriented monodomain films on coumarin-based photoalignment layers. J. Polym. Sci. Part B: Polym. Phys. **49**, 725–731 (2011)
59. T.Y.-H. Lee, Q. Wang, J.U. Wallace, S.H. Chen, Temporal stability of blue phosphorescent organic light-emitting diodes affected by thermal annealing of emitting layers. J. Mater. Chem. **22**, 23175–23180 (2012)
60. Q. Wang, J.U. Wallace, T.Y.-H. Lee, L. Zeng, J.J. Ou, S.H. Chen, Charge carrier mobility through vacuum-sublimed glassy films of *s*-triazine- and carbazole-based bipolar hybrid and unipolar compounds. Org. Electron. **14**, 2925–2931 (2013)

Chapter 7
Directing Self-Organized Columnar Nanostructures of Discotic Liquid Crystals for Device Applications

Hari Krishna Bisoyi and Quan Li

Abstract The columnar nanostructures of discotic liquid crystals are fascinating supramolecular systems with promising electronic and optoelectronic properties. Due to their remarkable performance in organic field effect transistors and organic photovoltaic devices, they have been regarded as a new generation of soft organic semiconductors. However, in order to realize the full potential of these intriguing materials, the discotic molecules need to be suitably oriented in the device structures directed by different stimuli. The devices fabricated with oriented materials have been demonstrated to perform better than using non-oriented materials. Over the years, different techniques have been developed to direct the appropriate alignment of the columnar phase of discotic liquid crystals on and in-between substrates. This chapter discusses the different methods used for the alignment control of discotic columnar phases parallel (planar) and perpendicular (homeotropic) to the substrates up to a macroscopic length scale.

7.1 Introduction

Liquid crystals (LCs) simultaneously exhibit the anisotropic property of crystalline solids and flow property of liquids. In the liquid crystalline phase, the molecules diffuse like in liquids but they maintain some degree of orientational order while doing so. The combination of order and mobility in LCs makes them fascinating and promising for practical applications. LCs exhibit extreme sensitivity to small external perturbations such as electric field, magnetic field, and surface effect. The most common and commercial application of LCs is in flat panel LC displays

H. K. Bisoyi · Q. Li (✉)
Liquid Crystal Institute, Kent State University, Kent, OH 44242, USA
e-mail: qli1@kent.edu

Q. Li (ed.), *Nanoscience with Liquid Crystals*, NanoScience and Technology,
DOI: 10.1007/978-3-319-04867-3_7,

(LCD) such as mobile phones, computer, laptop and television monitors, projectors etc. which influence the comfort of our everyday modern life [1–11]. LCs also play very important role in biological and biomedical science, materials science and nanoscience etc. Consequently a large number of beyond-display applications of LCs are rapidly evolving [12–17]. LCs are used as templates for the fabrication of mesoporous materials and serve as model systems for biomembranes [18–20]. Their biomedical applications such as in drug delivery, phospholipid labeling and in microbe detection have been demonstrated [21–24]. Moreover, LCs can potentially be used as functional materials for electron, ion transporting, sensory, optical materials in addition to their use as organizing media for assembling different types of nanoparticles [25–29].

Recently discotic LCs (DLCs), i.e. LCs formed by disk-shaped molecules, are drawing greater research attention [30–50]. The columnar phases of DLCs are regarded as model systems for the study of energy and charge migration in self-organized supramolecular assemblies. The columnar nanostructures have been used as active components in electronic and optoelectronic devices such as organic field effect transistors (OFET) and organic photovoltaic (OPV) cells. Due to their superior performance in such devices they have been regarded as a new generation of soft organic semiconductors. With the advent of modern organic chemistry, it is possible to engineer the molecular structure of discotics to tune their optical and electronic properties at will. Consequently a variety of DLCs have been designed and synthesized with tailored physical properties and complementary electronic properties. This has been achieved by varying the nature and size of the central aromatic cores (Scheme 7.1), nature of linking groups, and nature, number and length of peripheral flexible chains. However the various molecular designs turnout to be the first step in the realization of the full potential of these promising materials. The molecular design can tune the individual molecules properties, however the electronic characteristics of their ensembles poorly reflect their full potentials. It has been recognized that the molecules be suitably oriented with higher order over macroscopic areas in order to have optimized performance in devices architecture [51–62]. Macroscopic orientation of the π-conjugated components in organic devices is of central importance for determining the direction of charge-carrier mobility. The successful molecular design can be undermined by the inferior performance of the devices due to lack of appropriate orientation of the functional nanostructure.

DLCs generally exhibit two types of phases: nematic phase and columnar phase (Fig. 7.1). Discotic nematic phases are rarely observed while columnar phases are commonly found. In the nematic phase, the molecules possess only orientational ordering; however in the columnar phase, the disc-like molecules spontaneously self-assemble into infinite one dimensional (1D) stacks, which in turn self-organize to various two dimensional (2D) liquid crystalline lattices. The hexagonal columnar phase is the most commonly encountered phase in DLCs. Some discotic compounds are also known to exhibit lamellar and cubic phases.

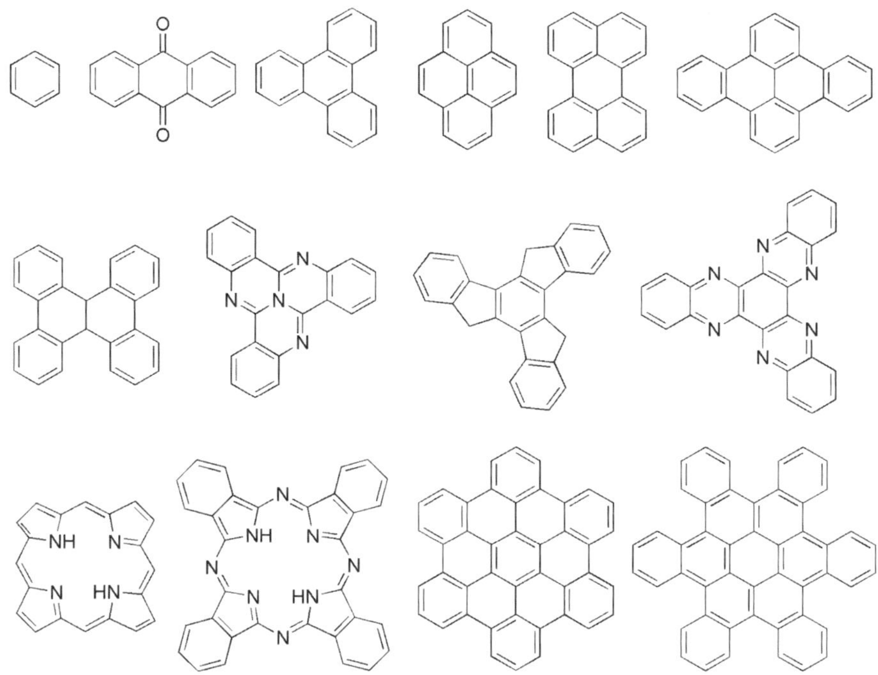

Scheme 7.1 Common central aromatic cores of discotic liquid crystals

Fig. 7.1 Discotic nematic and columnar hexagonal phases exhibited by discotic LCs

The negative birefringence films formed by polymerized nematic DLCs have been commercialized as optical compensation films to enlarge the viewing angle and enhance the contrast ratio of commonly used twisted nematic LCDs [63–65]. High strength and high performance carbon fibers for industrial applications have been obtained by processing from discotic nematic mesophase. Various molecular architectures have been designed and synthesized to exhibit the rarely occurring discotic nematic phase over a wide range of temperature. Subsequently a variety of

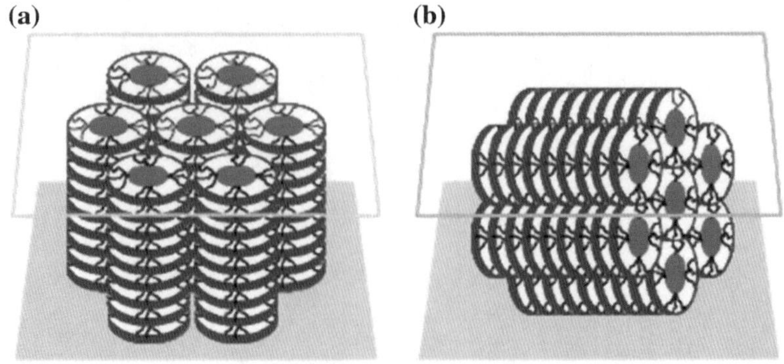

Fig. 7.2 Homeotropic (**a**) and planar alignment (**b**) of discotic columnar phases

different alignment techniques have been developed to align the less viscous discotic nematic phase in suitable and desired orientations [66–78].

Conjugated polymers and organic single crystals are two semiconducting classes of materials with different virtues. Organic single crystals have high charge carrier mobilities but are usually impractical whereas conjugated polymers have good processability but low mobility. DLCs exhibit mobilities approaching those of single crystals and are easy-to process. Thus semiconducting DLCs bring the two advantageous characteristics, i.e. high charge carrier mobility of single crystals and processability of polymers in one class of materials. Discotic columns with 2–5 nm in diameter and few hundred nanometers of length can self-assemble between electrodes with a single crystal-like packing. Charge carrier mobilities are several orders of magnitude higher along the column axes than across them and the columnar phases exhibit self-healing of structural defects due to their dynamic nature. However, the suitable processing methods with controllable molecular orientation and with high order limit their use [51–62]. Suitable alignment of the columnar phases in appropriate orientation is required to characterize and comprehend their properties. Moreover, aligned samples are necessary to evaluate the charge carrier mobility values, energy migration efficiencies, conductivity anisotropy etc. which are critical for their integration into electronic devices as active organic semiconductors. It has been found that dispersed nanomaterials can be aligned and reoriented in the desired direction by translating the alignment of host DLCs [79–82].

The columnar phases of DLCs display two types of alignments: one is homeotropic alignment and the other type is planar alignment (Fig. 7.2). In the homeotropic alignment, the supramolecular columns are perpendicular to the substrate surfaces while in the planar alignment, the columns are parallel to the surfaces. In the homeotropic alignment the molecular cores/planes of the discotics lie flat or "face-on" with respect to the substrate surface whereas in the planar alignment the discotic cores are oriented orthogonal or "edge-on" to the substrate plane. It is important that in the planar alignment configuration, all the columns orient uniformly along one direction yielding uniaxial planar alignment. In devices, the discotic columns have to bridge the gap between the electrodes all within a defect-free and long range ordered film. Homeotropic alignment of columns is required for solar cell and organic light emitting diode (OLED) applications while uniaxial planar alignment is essential for OFET applications of DLCs. The development of alignment control technologies of discotic columnar phases is driven by their remarkable anisotropic semiconducting properties. The alignment techniques developed for calamitic (rod-like) liquid crystalline phases are hardly effective for the alignment control of highly viscous discotic columnar phases. Further conventional deposition methods such as spin-coating or drop-casting cannot be used to achieve long-range molecular order because they often lead to polydomain formation. The self-organizing ability of discotic molecules has been often combined with different homeotropic or planar alignment control techniques for device fabrications. In the following the different alignment control techniques developed for both homeotropic and planar orientation are discussed.

7.2 Homeotropic Columnar Orientation

In the homeotropic alignment of DLCs, the column axes orient vertical with respect to the substrate planes. Therefore homeotropically aligned samples typically do not show birefringence in the polarizing optical microscope (POM) between crossed polarizers since the optic axis in this case coincides with the columnar axes. In order to distinguish between the homeotropically aligned columnar phase which can be confused with isotropic phase, additional experiments are often employed. Observation of dendritic textures under light microscopy without polarizers is characteristic of homeotropic samples. Moreover, conoscopy which displays the Maltese cross for homeotropic samples under POM is a reliable method to optically ascertain the alignment. In homeotropic alignment case where there is no optical anisotropy detectable by polarizing microscopy, differential interference contrast (DIC) is suitable to visualize the growth of the mesophase, especially to enhance the boundaries of the LC domains. By tilting the sample with respect to the light beam by nearly 45° the dendritic area of the homeotropically aligned sample appear birefringent under crossed polarizers. X-ray diffraction is another method which confirms the homeotropic alignment by displaying six symmetrically positioned spots by directing the X-ray probe beam along the column axes.

The homeotropic alignment of the discotic columnar phases can be obtained in between two substrates, i.e. in cells by slowly cooling from the isotropic phase. Obtaining homeotropic alignment on a single substrate is very challenging due to the presence of air-interface however a few studies have demonstrated successful vertical alignment of discotic layers on a substrate surface. This is quite important when subsequent layers are needed to be deposited onto the homeotropic layer for device fabrication. The following are the different techniques used for the homeotropic alignment of DLCs.

7.2.1 Thermal Alignment Method

On a solid surface, the way that disc-like molecules anchor to the surface and the orientation of the discotic columns are determined by many factors such as geometrical shape of molecules, intra and intermolecular interaction, surface topography of substrate, physicochemical properties of the substrate, molecule-solid surface interaction etc. Homogeneous and homeotropic alignments typically have unequal interface free energies that can be exploited through a judicious choice of the annealing temperature and/or cooling rates in order to promote one alignment at the expense of the other. The surface energetic state of the substrate has been shown to be a crucial factor that decides the type of the discotic molecule anchoring and the orientation of the discotic columns. These indicate that the nature of the surface on which a discogen assembles has huge impact on the molecular stacking of discotic molecules. Many DLC molecules orient homeotropically in a cell by slow cooling from the isotropic liquid to the columnar phase without any special surface treatment for alignment. However, owing to the high viscosity of the columnar phases, the alignment often obtained is polydomain in nature. As will be seen in the following, large area monodomain alignment has been achieved by the combined effect of temperature and surface effects.

Vauchier et al. have reported alignment of some discotic triphenylene derivatives in the columnar phase on perfect cleavage surfaces of apophyllite, a lamellar tetragonal silicate and the perfect cleavage of muscovite, a monoclinic mica [66]. Subsequently Perova et al. have investigated the orientation of the columnar phase of triphenylene and truxene based discotic liquid crystals **1** and **2** on amorphous Si and polycrystalline ZnSe substrates by infrared (IR) spectroscopy [83–85] (Scheme 7.2). The IR spectra are highly sensitive to the local structure and the dynamics of the discotic liquid crystalline molecules. The infrared dichroic ratio has been used for determining the effect of the structure of the substrate on the alignment in the columnar phase of DLCs. They found that the core of triphenylene and truxene align face-on to the Si substrate whereas these cores are tilted at an angle to the surface of the ZnSe substrate. These results demonstrate that the structure of substrates plays an important role in determining the type of alignment for discotics. They further studied the influence of a surface on the alignment of DLCs and the stability of the alignment as a function of temperature and time. They reported the observation of an anchoring transition in discotics using polarized Fourier transform infrared (FTIR) spectroscopy. Orientation of the DLC and the stability of its alignment in the discotic columnar phase were studied on different (Si, ZnSe, ZnS and CaF_2) clean as well as treated substrates. In the cells made up of clean Si and CaF_2 substrates, an anchoring transition from edge-on to face-on has been observed while annealing in the columnar phase. Similar anchoring transition effects have been observed in ZnSe and ZnS cells when the substrates are coated with nylon polymer. These observations have been described by taking into account the contributions from the topology, steric, polar and dispersive interactions.

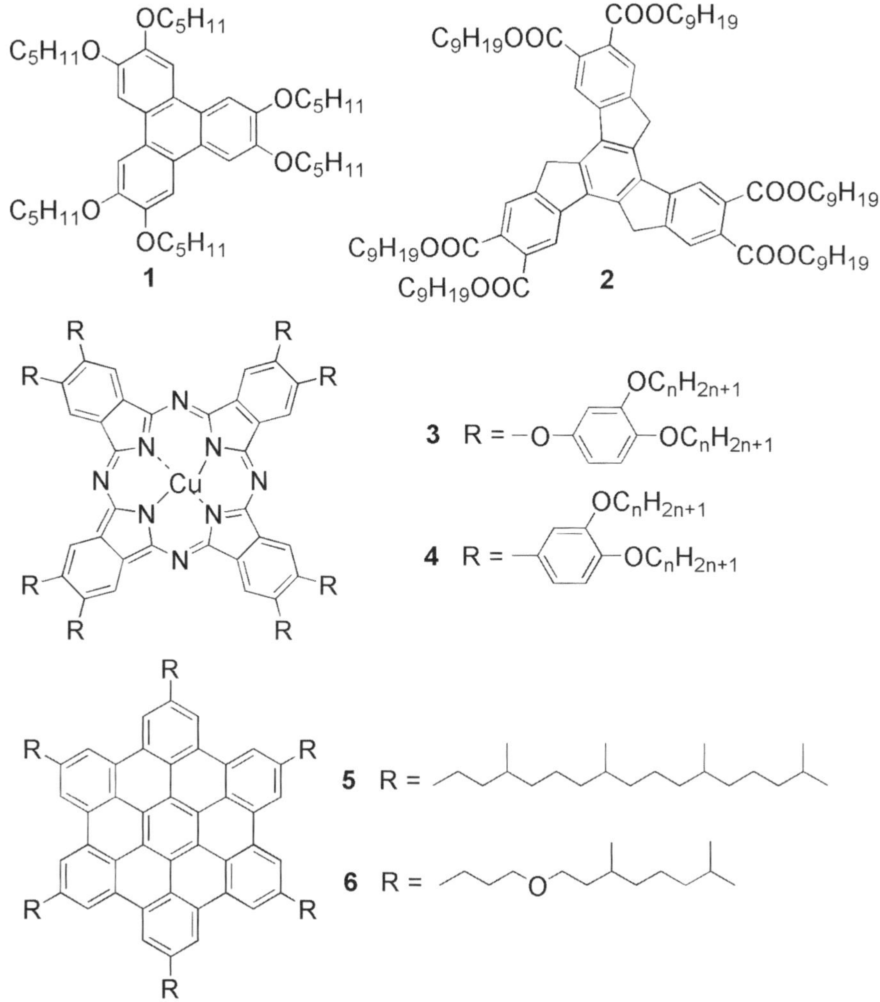

Scheme 7.2 Chemical structures of some DLCs aligned by thermal method

Ohta and coworkers developed the phthalocyanine (Pc) derivatives **3** and **4** which exhibited spontaneous uniform homeotropic alignment in the columnar tetragonal mesophase [86–88]. It has been observed that introduction of oxygen atoms has furnished large area alignment of Pc **3** compared to its non-oxygenated counterpart molecules. Uniform monodomain homeotropic alignment without domain boundaries could be achieved only for the columnar tetragonal mesophase between soda-lime or quartz glass plates. It has been speculated that the lone pairs of the intervening oxygen atoms may coordinate to the dangling bonds of silicon atoms on the surface of glass and/or quartz. The first disk adhered to the surface can act like a nucleation seed and may trigger stacking of the disks on top to form perpendicular columns. It is interesting that the columnar tetragonal mesophase of octaazaphthalocyanine derivatives shows remarkable spontaneous homeotropic alignment in a large area, without disclinations and polydomains, in the temperature range from room temperature to the decomposition temperature. Recently sandwich structures of phthalocyanine derivatives were prepared between ITO and Indium cells to investigate the device properties; the difference in performance before and after heating the cells is attributed to alignment induced by thermal treatment [89]. Pisula et al. have found that the introduction of ether linkages within the side chains enhances the affinity of the discotics molecules **6** towards polar surfaces, resulting in homeotropic self-assembly [90]. Successful homeotroic alignment of the all-hydrocarbon hexa-*peri*-hexabenzocoronene **5** (HBC) on ITO confirms that specific molecular affinity for surfaces is not a prerequisite for the face on organization [91]. However, the incorporation of heteroatoms such as oxygen near the rigid core of discotic molecules has been shown to enhance the homeotropic alignment. The exact mechanism leading to homeotropic alignment is not clear. The nucleation of the discotic mesophase most likely first occurs on the solid substrate, leading to a face-on arrangement of the molecules.

Groups around Bock and Grelet extensively investigated the alignment control and reorientation of different DLCs on single substrates as well as in between two substrates [92–97]. Highly ordered homeotropic growth of open thin films of columnar LC **7** and **8** on substrate has been obtained by controlling the growth kinetics during thermal annealing (Scheme 7.3). It is observed that there is a competition between planar and homeotropic orientations of the LCs, if a thermal annealing process is performed at elevated temperature inside the columnar phase range, the homeotropic orientation switches to a planar alignment after a few hours. This anchoring transition takes place more quickly if the annealing temperature is close to the isotropic temperature due to low viscosity. The conditions of the control between homeotropic and planar orientations have been determined and explained in terms of the involved surface tensions. Subsequently they found that homeotropic alignment of columnar phase can be obtained on indium tin oxide electrodes with and without surface treatment by thermal annealing. They reported an efficient alignment process in order to achieve face-on orientation of columnar LC films of **9** confined between a glass slide and a thin metallic electrode. The face-on orientation of the discotic compound is obtained by anchoring transition of a columnar phase from a degenerate planar orientation to the homeotropic alignment in the mesophase. Uniform homeotropic bilayer thin films of two DLCs **7** and **10** with complementary electronic properties have been obtained by sequential thermal annealing of their co-deposited mixtures through the liquid-to-LC phase transitions on a substrate [98]. For this purpose a pair of room temperature columnar discotic materials has been designed such that they have selective solubility, low miscibility, and different transition temperatures. The homeotropic orientation of the open bilayer has been characterized by both optical microscopy and X-ray scattering. This is the first example of an organic heterojunction based on two oriented columnar LC layers and could potentially be used in organic solar cells.

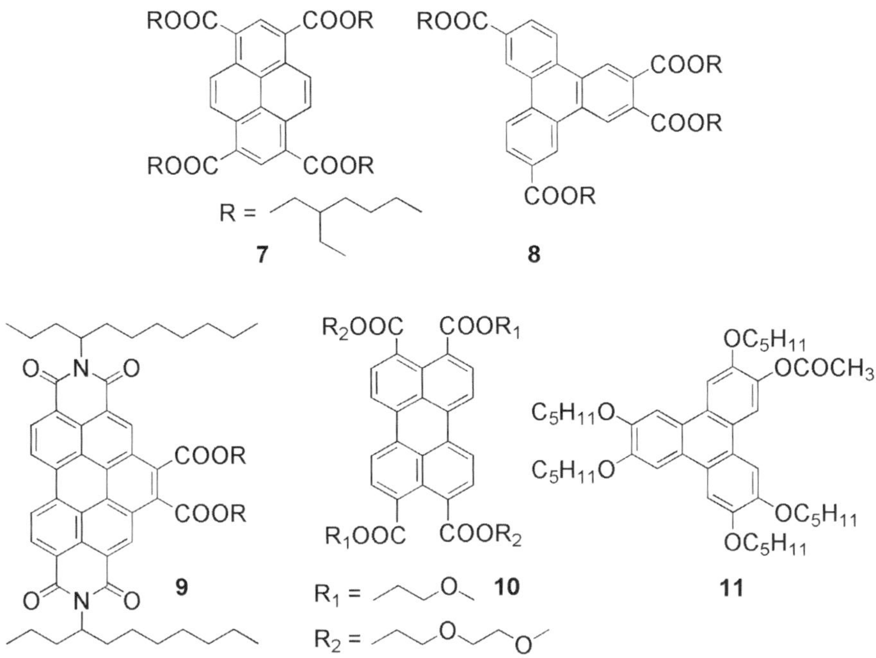

Scheme 7.3 Chemical structures of some discotic compounds exhibiting homeotropic alignment

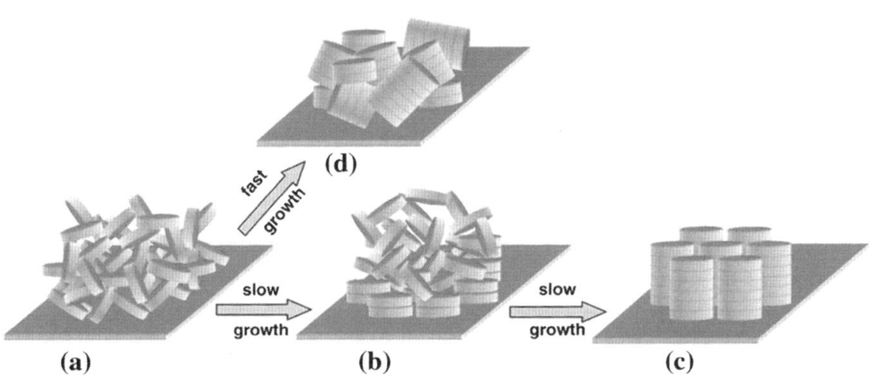

Fig. 7.3 Schematic presentation of the homeotropic alignment of discotic liquid crystals on a single substrate at a slow or a fast growth rate. Reproduced with permission from [99]. Copyright 2009 Elsevier

A mixed ether ester triphenylene derivative **11** has been observed to spontaneously align homeotropically on a single substrate in the presence of an air-interface (Fig. 7.3) [99]. This observation is very significant since DLCs generally tend to align in planar manner on single substrates in the presence of air-interface.

Charge transfer (CT) complex of a triphenylene derivative exhibits spontaneous homeotropic alignment on glass substrate [100]. Interestingly, these alignments are stable at room temperature. Similarly the blend of donor phthalocyanine and acceptor perylenetetracarboxidiimide derivatives has been found to self-align homeotropically between two electrodes [101]. The columnar hexagonal phase of a mixture of two phthalocyanine mesogens exhibit homeotropic alignment in the absence of a specific substrate-LC interaction [102]. Star-shaped discotic pentamers and heptamers have been found to form macroscopic monodomains of columnar stacks which self-align homeotropically between glass substrates on cooling from the isotropic phase [103]. It has been reported that thin films deposited from solution by hollow capillary writing technique exhibit macroscopic uniaxial domains with an in-plane arrangement of discotic stacks [104]. The orientation of the thin open supported films can be changed from in-plane to homeotropic by annealing below the LC/isotropic phase boundary. However rapid cooling leads to predominantly in-plane alignment. These observations have been attributed to heterogeneous nucleation at the film/substrate interface.

7.2.2 Chemical Structure Modification

The surface affinity of discotic molecules can be tuned by modifying their chemical structures. The chemical nature of peripheral side chains of DLCs plays a vital role in this aspect. Such molecular engineering strategies have been used to obtain preferential orientation of DLCs on different surfaces. Monobe et al. found that the introduction of a perfluoroalkyl group into the peripheral chains of triphenylene mesogens **12** (Scheme 7.4) confers strong tendency towards homeotropic orientation of the columnar hexagonal phase on a variety of surfaces [105]. The alignment behavior was investigated on glass substrates modified with polyimide, cetyltrimethylammonium bromide (CTAB) and indium tin oxide (ITO) by polarizing optical microscope. The above results were compared with those of the corresponding hydrocarbon counterparts. It was clearly observed that the introduction of fluoromethylene groups into the peripheral chains of discotic triphenylenes could easily give rise to a spontaneous homeotropic alignment in the columnar hexagonal phase. Subsequently it has been reported that spontaneous homeotropic alignment could be easily attained by the introduction of an appropriate number and length of the fluoromethylene chains in the peripheral parts of the discogens [106].

Novel nanoscale discotic liquid crystalline porphyrins **14** with partial alkyl chain perfluorination has been designed and synthesized in our group which exhibits exceptionally enhanced tendency to spontaneously self-assemble into homeotropically ordered nanostructures [107]. The defect-free homeotropically aligned fluorinated porphyrin thin films were fabricated and characterized. In the films thinner than 10 μm in glass cells, it shows strong tendency towards homeotropic

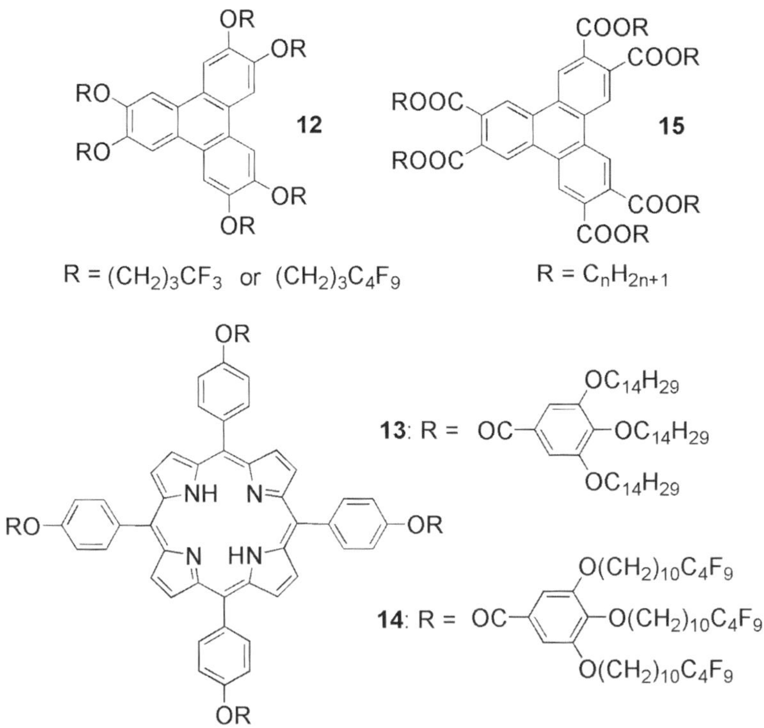

Scheme 7.4 Chemical structures of some discotic compounds aligned by chemical structure modification

alignment compared with its non-fluorinated counterpart **13** on cooling from the isotropic liquid phase (Fig. 7.4). Moreover, the homeotropic alignment of the hexagonal columnar phase remains stable down to room temperature. The experimental results show that a modified substrate surface such as ITO-coated glass and polyimide-coated glass with and without mechanical rubbing seems to have no bearing on obtaining defect-free homeotropic alignment. Promisingly enough, in addition to homeotropic alignment, the porphyrins LC can also easily yield a uniform homogeneous alignment by simple mechanical shearing a homeotropically aligned cell in the mesophase. Hence this material has the potential to be used in both organic photovoltaic and thin film transistor devices [108].

Liquid crystalline triphenylene hexacarboxylic esters **15** have been recently synthesized and the columnar phase has been observed to align homeotropically on different substrate surfaces [82]. The alignment behavior has been studied by polarizing optical microscopy and X-ray diffraction techniques. POM shows a dark view whereas optical microscopy image displayed a dendritic texture typical of columnar hexagonal assembly. The hexagonal columnar phase displays

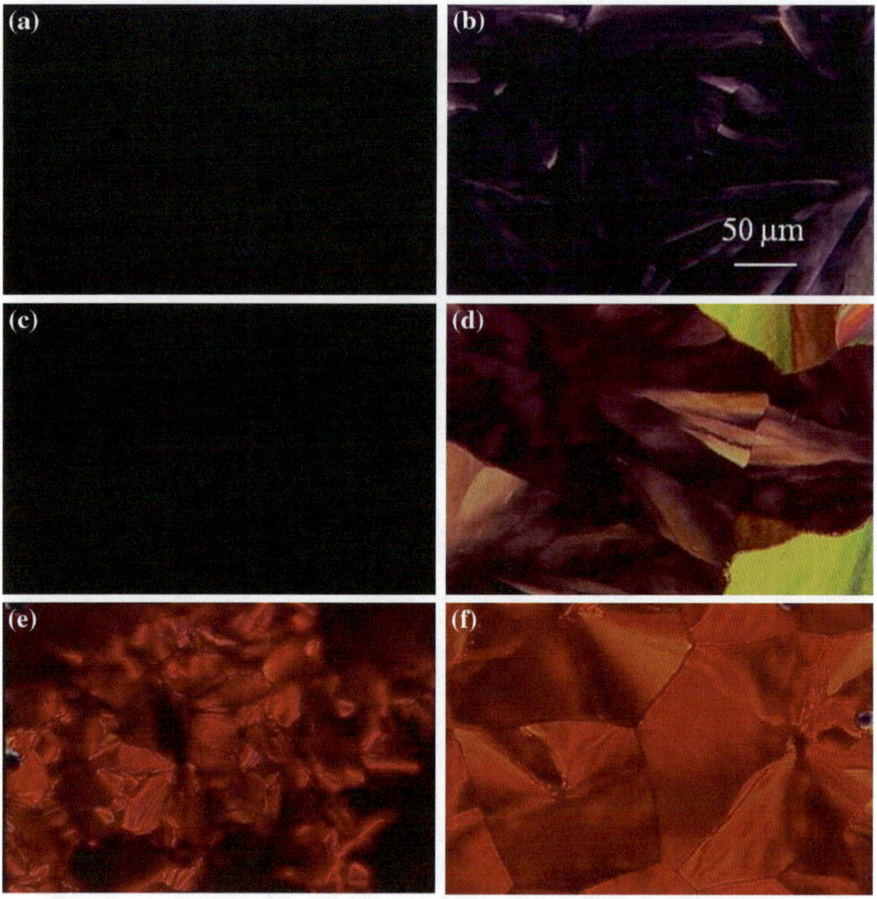

Fig. 7.4 Crossed polarized optical textures of **14** (**a** 5 μm, **c** 9 μm, **e** 20 μm thick) and its corresponding non-fluorinated counterpart **13** (**b** 5 μm, **d** 9 μm, **f** 20 μm thick) at room temperature. The *dark areas* represent homeotropic alignment, and the *bright* domains appear where the porphyrin planes are oblique to the substrate. Reproduced with permission from [107]. Copyright 2009 Taylor & Francis

exceptionally high correlation length as revealed by X-ray analysis. Their spontaneous alignment propensity has been attributed to the dipolar interaction between the columns. The hexacarboxylates displayed dark field of view in POM on quartz, sapphire, mica, indium tin oxide (ITO), silicon wafer, Au, Polyethylene terephthalate (PET), polyvinylidene chloride (PVDC) and polyether ether ketone (PEEK), together with rubbed and unrubbed polyimide films. The homeotropic alignment has been corroborated by the observation of highly anisotropic conductivities during the flash-photolysis time-resolved microwave conductivity measurements of the aligned films.

7.2.3 Surface Modification of Substrates

By modifying the physical and chemical nature of the surfaces of solid substrates it is possible to tune the interaction of DLC molecules with surfaces. Therefore by modulating the surface-molecule interaction desired homeotropic orientation of DLCs can be achieved. Such a simple method for the construction of a tunable self-assembled "command layer" for DLC alignment has been reported by Hoogboom et al. [109]. A pyridine-functionalized oligosiloxane spontaneously forms an anisotropic surface on indium-tin-oxide enabling the alignment of metal phthalocyanines **16** (Scheme 7.5). The pyridine functions as seeds for the epitaxial growth of stacks of highly ordered zinc phthalocyanines. The height of the aligned films can be controlled by varying the immersion time of the substrate. By varying the height of the stacks, the direct control over the properties of the LC domains can be obtained. Interestingly some of the aligned stacks can be dissolved by adding nanomolar concentrations of nitrogen containing compounds.

Recently, the compound **17** was aligned on chemically modified substrate surfaces [110]. Substrates were processed using hydrogen peroxide and coated with silane self-assembly monolayers. Face-on anchoring of **17** was obtained on the peroxidized ITO surface whereas edge-on anchoring was observed on the surface covered by octadecyltrichlorosilane (OTS). They showed that the surface energetic state of the substrates is one of vital factors that determine the way the discotic molecules anchoring at the surface. At a substrate with a high surface free energy, the discotic molecules will reside with disc-surfaces touching the surface and form the face-on anchoring, whereas a low free energy surface can support an edge-on anchoring of the discotic molecules. A surface template comprising domains with different surface free energy levels can be produced by peroxidization process or coating the substrate with proper silanes. This provides the possibility of the multidomain hybrid molecular alignment for discotic molecular materials on substrates with modulated surface energy. Effects of surface free energy on orientational assembly of a hexaazatrinaphthalene-based DLC **18** in sandwich-type cells were also studied by Zheng et al. [111, 112]. The surface free energy of the substrates was modulated by means of oxygen plasma buffing. It was found that the discotic molecules tend to anchor with their disc-face towards the substrates with high surface free energy resulting in the homeotropically aligned columnar phase. However, a decrease in the surface free energy of the substrates leads to the columns tilting away from the normal to the substrate and the tilt angle increases with the decrease in surface free energy.

Geerts et al. have undertaken a study of the alignment of phthalocyanine **19** columns in thick (a few micrometer) and thin (a few tens of nanometers) films and on the influence of confinement on the occurrence of homeotropic alignment [113]. The orientation of the columns in those films is characterized by polarized optical microscopy and atomic force microscopy as a function of the preparation method, the polarity and roughness of the substrate and the number of DLC-solid interfaces, i.e. one or two. The control of columnar alignment is determined by the

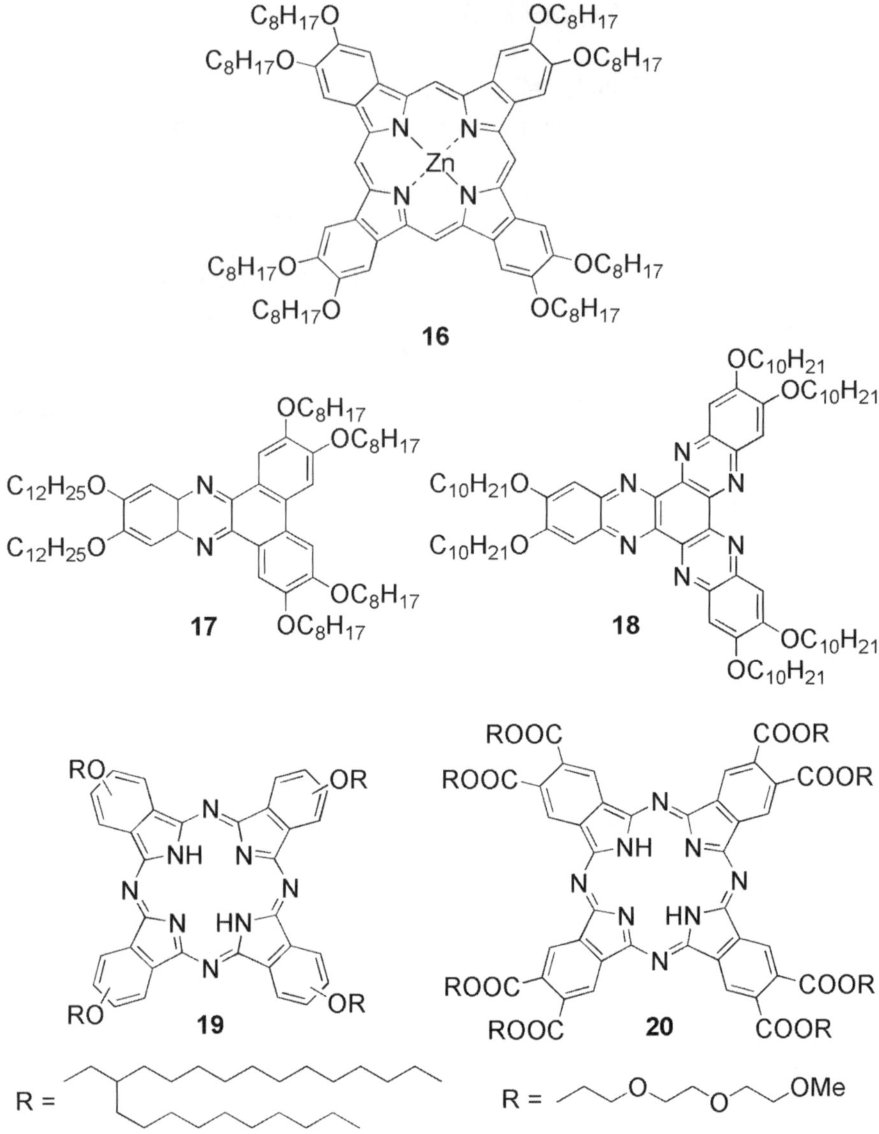

Scheme 7.5 Chemical structures of some discotic molecules aligned on modified surfaces

confinement induced by solid substrates rather than by the nature of those substrates. A planar alignment is always observed on a single substrate regardless of surface energy and roughness; however between two identical substrates homeotropic alignment is observed in most cases.

Evans et al. studied the alignment behavior of discotic columnar phase on surfaces modified by self-assembled monolayer (SAM) [114]. They studied a

COOH-terminated high-energy surface, a CH_3-terminated low energy surface and a SAM composed of DLC-like triphenylene moieties in which triphenylenes are known to lie edge-on to the surface. The columnar phase of a triphenylene-derivative aligns in a homeotropic manner on both the high-energy and low-energy surfaces. In case of the calamitic LCs, it has been shown that a SAM comprised of LC look-alike moieties can be used to dictate the alignment of bulk liquid crystalline phases. So it was expected that the SAM derived triphenylene-thiol would promote the planar anchoring of the columnar phase of triphenylene based LC. Surprisingly, this was not the case rather the compound aligns in a homeotropic manner. This result shows that the surfaces act purely as mechanical barrier although specific interactions between the surface and the discogen can be used to enhance the ease or quality of the alignment.

LC phthalocyanine **20** bearing eight oligo(ethyleneoxy) peripheral substituents forms homeotropically aligned films on the hydrophilic surface, whereas planar alignment with a random distribution of the column directors occurs on the hydrophobic surface [115]. This study shows that the homeotropic alignment of the phthalocyanine depends on the nature of interaction with the surface. Therefore the alignment of the material can be switched by simply changing the nature of a solid substrate from hydrophilic to hydrophobic.

Aida et al. have shown that *o*-phenylene octamers can be used as surface modifiers to obtain homeotropic ordering of a variety of DLCs [116]. The octameric *o*-phenylenes are known to fold helically into a cylindrical architecture that resembles to a pi-stacked column of DLCs. It has been found that the oligomers adhere to the glass substrate with its cylindrical axis orthogonal to the surface. Therefore it is speculated that this face-on orientation of the octamer likely nucleates the homeotropic ordering of the liquid crystalline discotic materials.

7.2.4 Electric Field

It is well known that rod-like LC molecules align uniformly under the influence of an electric field. The direction of molecular orientations can be controlled as desired by electric field and forms the basis for their successful application in flat panel displays. However, the use of electric fields to align discotic liquid crystalline materials has not been explored as much as other alignment techniques. The possibility of using electric fields as a new tool for creating long-range ordered films of discotic semiconducting molecules have been recently demonstrated. If columnar DLCs can be aligned by an electric field, electronic devices with suitably oriented pathways for high carrier transport could be easily developed. Therefore the strategies to control the alignment of DLCs under an electric field are attractive. Electric alignment of anisotropic molecules originates from the anisotropy in their electric susceptibility and the dielectric anisotropy of DLC molecules could be positive or negative depending on their detailed chemical structures. Aida et al. developed discotic liquid crystalline molecules with electric field responsive handles which exhibit columnar phases [117, 118]. The electric field responsive handles enable large-area unidirectional orientation of discotic liquid crystalline molecules. It has been observed that the corannulene and triphenylene derivatives **21** and **22** (Scheme 7.6) with large dipoles can be oriented in such a way that their columnar axes are parallel to the direction of the applied electric field and hence gives rise to a homeotropic alignment of hexagonal columns with respect to the electrode surface. Interestingly the resulting macroscopic orientation can be maintained even after the electric field is switched off. Moreover it is possible to restrict the molecular alignment to only desired regions of the sample by selecting the placement and the size of the electrodes. Self-assembled fibers of discotic phthalocyanine derivative **23** have been shown to lie in a planar way onto an ITO substrate in the absence of an electric field [119]. However when the electric field is turned on the planar fibers stand up with their long axes oriented along the direction of the electric field and hence perpendicular to the surface of ITO substrate. Strong interaction of the self-assembled fibers with the direct current electric field allows for their homeotropic alignment on the conductive substrates for controlling the orientation. Recently the influence of electric fields on the homeotropic alignment of a HBC derivative **24** bearing six branched alkyl side chains containing ether linkages have been reported [120]. When an alternating current (AC) electric field was applied, the original molecular ordering was disturbed by the time-varying electric field indicating that alternating electric field is detrimental to the order of its homeotropic alignment relative to the electrodes. On the other hand, homeotropic orientation of the columns was reinforced to form a large-area unidirectional orientation by application of a direct current (DC) electric field. It is expected that the method of electric field alignment is of more general use and can be applied to a variety of rigid conjugated pi-systems. Therefore demonstrating the feasibility of using an electric field for aligning discotic molecules may be of future value for creating films with new and more precise geometries for improved electronic devices.

7.2.5 Infrared Irradiation

Photoswitching techniques for LCs have been extensively studied in order to obtain a photoinduced change of molecular alignment in calamitic nematic films. Photoinduced control of molecular orientations is quite interesting from the viewpoint of device fabrication technology using organic thin films. Most of the studies of the photoalignment of LCs have been focused on the application of ultraviolet and/or visible light, since this could provide a clean and fast fabrication process for organic thin film devices. The possibility of controlling the molecular alignment and domains of LCs using infrared light relating to the selective excitation of vibrational processes has been explored by Shimizu et al. in a series of studies involving a variety of columnar mesophases [121–127]. This technique has the unique advantage of remote control. The studies have been aided by the

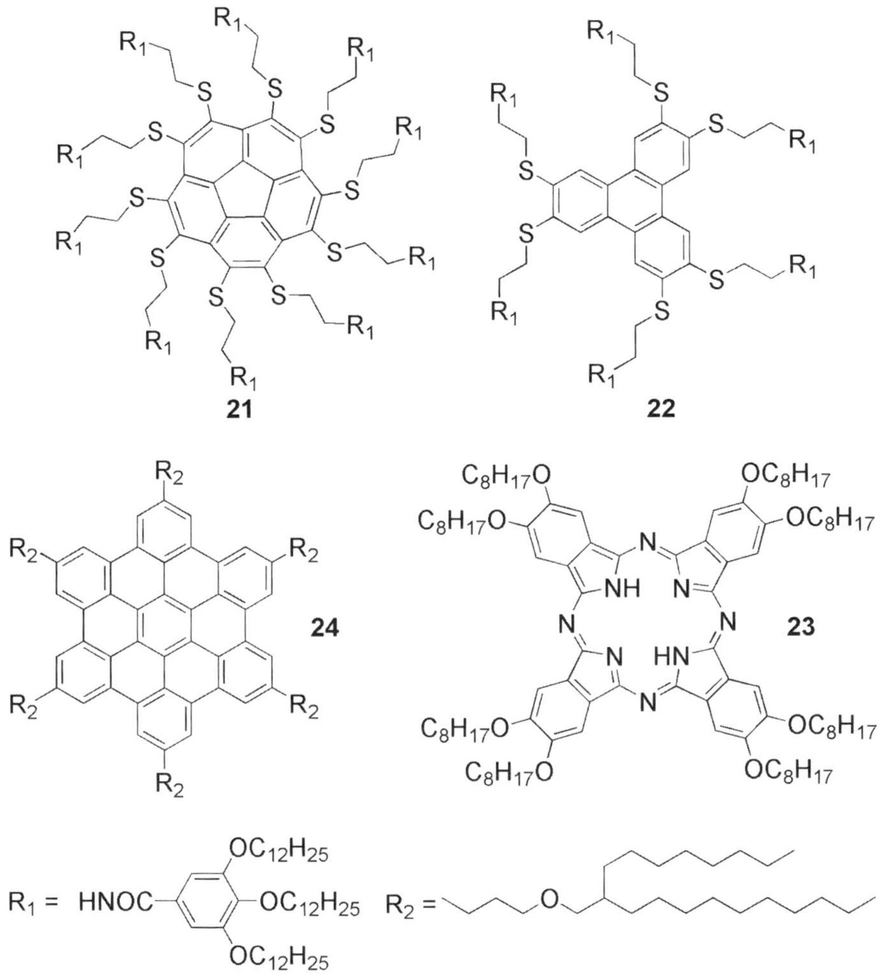

Scheme 7.6 Chemical structures of some discotic compounds aligned by electric field

advancement of highly tunable free-electron laser beams with a variety of infrared wavelengths. The reorientation of discotic triphenylenes (Scheme 7.7) has been achieved by vibrational excitation with pulsed infrared radiations on aligned films confined between two BaF_2 substrates. The triphenylene films were prepared between two BaF_2 substrates which have high transparency in infrared region. Irradiation of the film with infrared radiation corresponding to the aromatic absorption band causes change of orientations. It was found that the homeotropic domain could be changed to a domain with columns aligned in a plane by the excitation of the C-C stretching vibration of triphenylene cores.

It was observed that the irradiation with circularly polarized infrared light of the columnar hexagonal phase yields uniform homeotropically aligned LC domains.

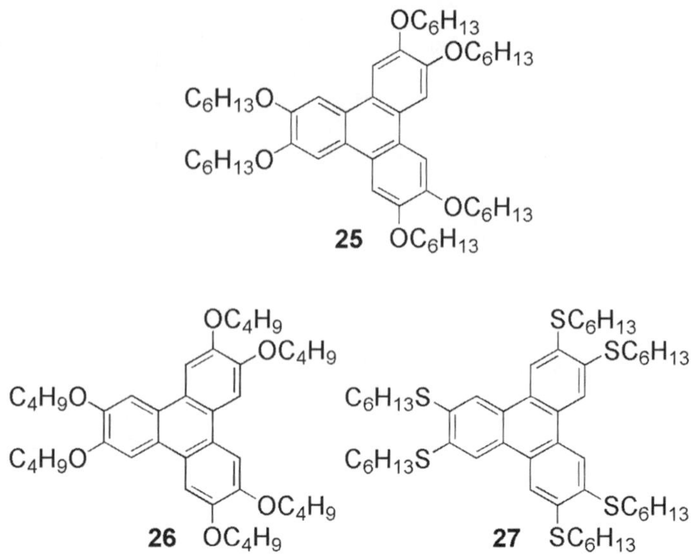

Scheme 7.7 Chemical structures of some discotic compounds aligned by infrared light irradiation

Hence a linearly polarized infrared irradiation results in the alignment change from homeotropic to planar; however subsequent irradiation with circular polarized light yields the homeotropic alignment back. Thus by sequentially irradiating with linearly polarized and circularly polarized infrared light, the alignment of the columnar hexagonal phase could be dynamically changed (Fig. 7.5) [122]. Alternative homeotropic and planar alignment areas of a sample can be obtained by this technique.

7.2.6 Other Methods

Homeotropic alignment of DLCs can be obtained by cooling its isotropic phase between two confining solid substrates. Accordingly local modification of the alignment of DLC columns has been achieved by patterning the DLC film using lithographically controlled wetting which exploits the confinement effect [128]. Phthalocyanine films of few hundred nanometers have been patterned by gently placing a stamp with recesses and protrusions on top of the film. The temperature of the film was increased above the isotropic transition temperature. In these conditions, when the sample is cooled down to below the isotropic temperature the columns retain their original planar alignment in regions under the recesses of the stamp. In the places where the stamp protrusions are in contact with the film, cooling down to the isotropic temperature leads the columns to achieve

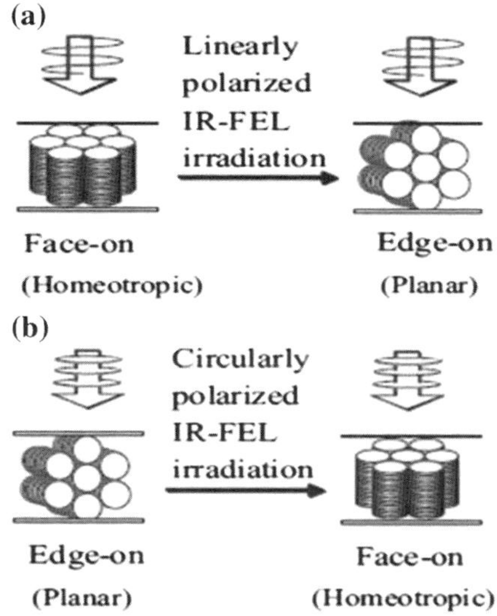

Fig. 7.5 Alignment of DLCs by infrared irradiation. **a** Infrared irradiation with linearly polarized light and **b** irradiation with circular polarized light. Reproduced with permission from [122]. Copyright 2006 Wiley-VCH

homeotropic alignment. Patterned films with homeotropic and planar alignment are very stable and no effects of aging were observed while keeping the sample at ambient conditions for a few months. The process can be applied for the formation of a specific column orientation at pre-set regions of a discotic film.

Obtaining large area homeotropic alignment of an open film on a substrate is very challenging due to the presence of free air interface which imposes planar alignment so that the alkyl side chains are exposed at the outer film surface. This can be overcome by using a removable sacrificial layer on top of the discotic layer [129]. By cooling down the material between the sacrificial layer and the substrate, the desired homeotropic alignment have been achieved over a large area. Subsequently the sacrificial layer was removed by using a proper solvent which does not dissolve and disturb the alignment. This is a promising method for device fabrication when subsequent layer has to be deposited onto an already aligned discotic film.

Immobilizing organic molecules on solid surfaces is of great importance both toward the goal of high-resolution structural characterization and for their potential applications. Considerable efforts have been made toward the goal of immobilizing individual molecules on substrates, with minimal impact on the chemical and physical properties of the adsorbed molecules. During solution processing of large discotic cores such as triphenylene, porphyrin, phthalocyanine and hexabenzocoronene on metallic and non metallic surfaces, it has been observed that the molecules form a monolayer lying flat on the substrate. Such substrates can potentially act as patterned surfaces for further alignment of DLCs on them [130–136]. The immobilization and study of discotics should also be considered in the light of the interaction of discotic molecules with the electrode surface. Owing to the electrode-discotic heterojunction, the interface-resistance determines to a great extent the performance of the organic electronic devices.

7.3 Homogeneous (Planar) Columnar Orientation

In the uniaxial planar alignment, the column axes lie on the substrate surface orienting along one direction. For uniaxial planar aligned samples, rotating the microscope stage by 45° produces alternating bright and dark domains under crossed polarizer conditions. The birefringence of the domains is extinguished when the column axes are aligned either parallel with the polarizer or analyzer and is maximally bright when the stage is rotated by 45°. Besides the above optical test, the quality of planar alignment of discotic columnar phases on a surface can be characterized by grazing-incidence X-ray diffraction and by scanning probe microscopy techniques such as atomic force microscope (AFM), scanning electron microscope (SEM), and transmission electron microscope (TEM).

The planar alignment can be obtained by mechanical shearing the viscous columnar phase along one direction between two substrates. Processing on a single substrate often yields planar alignment of DLCs due to the presence of air interface. When a DLC is put on a solid substrate without cover, a planar alignment of the columnar phase is usually obtained as a result of the balance between the DLC-solid and DLC-air surface tensions respectively. The orientational ordering of the molecules in the first monolayer at the surface gets duplicated and epitaxially grows into bulk through self-assembly. However the quality of the alignment, i.e. degenerate planar or uniaxial planar depends on the controllability of the processing technique used. The following are the different methods used to obtain the planar alignment of discotic liquid crystals.

7.3.1 Mechanical Shearing

As mentioned earlier, the planar alignment of discotic columnar phase can be obtained by mechanical shearing the material in the mesophase. Our group has demonstrated that it is possible to achieve uniaxial planar alignment by unidirectional shearing homeotropically aligned columnar phase (Fig. 7.6) [107]. Ozaki et al. have reported that high-quality planar alignment could be achieved in the hexagonal columnar phase of hexahexyloxytriphenylene by applying a bidirectionally oscillating shear generated by a piezoelectric actuator [137]. DLCs with thickness up to 38 μm could be homogeneously aligned by using this technique where the molecular columns align parallel to the direction of the shear. The quality of the alignment was assessed by polarizing optical microscopy and Fourier transform infrared spectroscopy. Figure 7.7 shows a schematic illustration of the shear device and microscopy photographs of aligned samples obtained after shearing. Before applying shear, a spontaneous homeotropic alignment was achieved in the columnar hexagonal mesophase between two CaF_2 substrates by

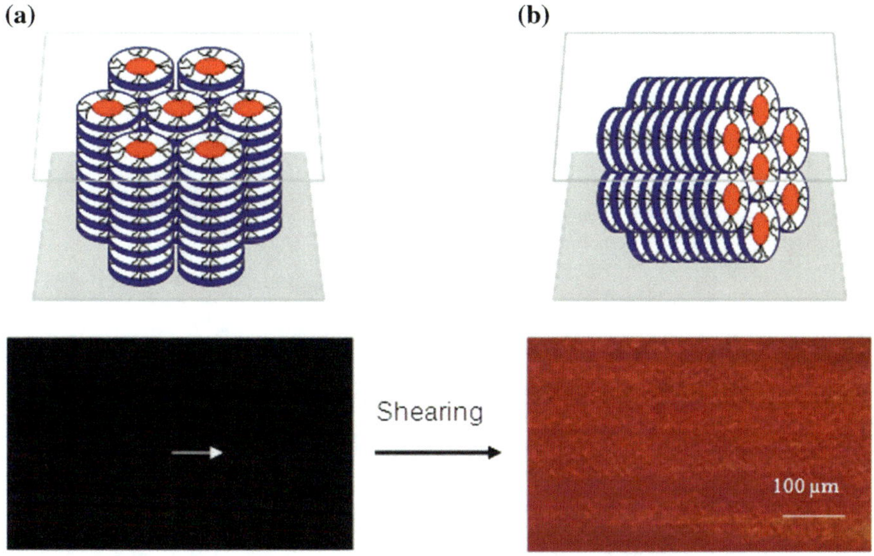

Fig. 7.6 Alignment of discotic liquid crystals by mechanical shearing. **a** Homeotropic alignment before shearing and **b** homogeneous alignment after shearing, arrow indicates shearing direction. Reproduced with permission from [107]. Copyright 2009 Taylor & Francis

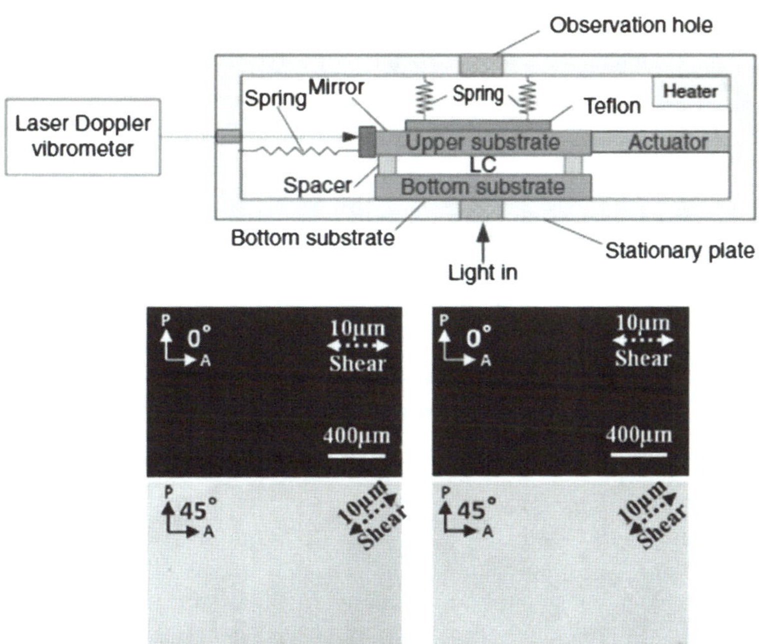

Fig. 7.7 Schematic device configuration for alignment of liquid crystals by oscillating shear. POM photographs of aligned sample are shown below, the bidirectional shear direction is indicated by the *dotted arrow*. Reproduced with permission from [137]. Copyright 2013 The Japan Society of Applied Physics

slow cooling from the isotropic phase. Rotating the microscope stage by 45° following the application of shear produced alternating bright and dark domains between crossed polarizers. This textural observation indicates a planar alignment of the liquid crystal. The order parameter of the discotic core was estimated to be 0.71–0.74 which shows higher degree of ordering in the samples.

7.3.2 Magnetic Field

Magnetic alignment, which utilizes the magnetic anisotropy of materials, is one of the most straightforward and efficient method to fabricate oriented molecular materials over the macroscopic length scale. Magnetic fields have been successfully utilized to align a wide variety of self-assembled system. Magnetic fields are effective for producing oriented films of DLCs which allows the investigation of structure-property relationships in such magnetically aligned films. Goldfarb et al. have applied magnetic field to prepare samples for the study of mesophase using nuclear magnetic resonance (NMR) spectroscopy [138]. They prepared sample by slow cooling of the isotropic liquid in a sufficiently strong magnetic field. Monodomain of the columnar mesophase was achieved by spinning the sample about an axis perpendicular to the field direction. Dueterium NMR measurements on magnetically aligned samples have allowed the study of side-chain fluctuations and motions of rigid cores in the mesophase separately [139]. Mann et al. reported the application of a magnetic field for producing highly oriented films of a substituted HBC [140]. Optical microscopy studies revealed large area HBC monodomains that covered the entire film, while wide angle X-ray measurements showed that the HBC molecules are aligned with their planes along the applied field. On the basis of magnetic alignment, solution-processed field-effect transistors were constructed with very high charge carrier mobilities, which are greatly enhanced with respect to the unaligned material. Significantly high mobility anisotropies for current flow parallel and perpendicular to the alignment direction have been measured as a function of the channel length. The uniaxial alignments of the columnar superstructure of discotic metallomesogens over the centimeter length scale have been achieved by spinning samples under a static magnetic field [141–143]. The orientations of the columnar superstructures have been revealed by small angle neutron scattering and cryogenic transmission electron microscopy. When the sample was continuosly spun during cooling from the isotropic phase to the columnar mesophase in the presence of a static magnetic field, it was found to form uniaxially aligned columnar superstructures with the director normal to the external field (Fig. 7.8). Choi et al. further demonstrated the simultaneous use of magnetic fields and surface interactions to produce uniaxially oriented and highly ordered discotic metallomesogens thin film on substrates over a macroscopic length scale [144]. The substrate was functionalized with octadecyl trichlorosilane (OTS) to increase the

(a) (b)

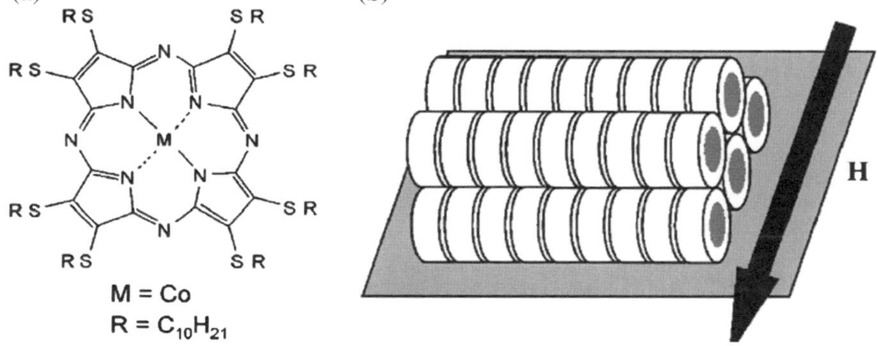

Fig. 7.8 Magnetic alignment of discotic liquid crystals. **a** Molecular structure of a discotic compound and **b** an arrangement of columnar stacks perpendicular to the magnetic field on a substrate. Reproduced with permission from [142]. Copyright 2006 Elsevier

affinity of the surface with the alkyl side chains of the discotics, resulting in an edge-on arrangement of DLC columns. When a spin-coated DLC thin film on an OTS-functionalized silicon wafer is cooled from its isotropic phase to the liquid crystalline phase in the presence of an external magnetic field, it forms highly oriented edge-on columnar superstructures. The detailed structures of the ordered and oriented columnar DLCs on substrates were investigated using the grazing incidence small angle X-ray scattering technique. Effect of film thickness on magnetic alignment of discotic columns shows that the strong interfacial interaction at the film-substrate interface propagates up to 50–100 nm from the substrate, maintaining the orientation of columnar packing in the plane perpendicular to the applied field. When the distance from the film-substrate interface becomes larger than about 100 nm, the tilting of columnar layer occurs due to longitudinal edge dislocation induced by elastic deformation in the film. Subsequently the same group achieved the alignment of columnar phase on a substrate by solvent-evaporation under a magnetic field. Here also the columnar director aligns perpendicular to the applied magnetic field as expected. This method represents a simple and easy solution process to fabricate aligned columnar DLC on substrates. Recently the planar alignment of the columnar phase of triphenylamine-based DLC has been reported by applying magnetic field [145]. Takami et al. investigated the influence of magnetic field on the electrical characteristics of phthalocyanine nanowires produced in porous alumina templates [146]. It was found that the morphologies of the phthalocyanine nanowires such as domain size, columnar alignment and face-to-face spacing etc. are improved by using a high magnetic field which has been attributed to the anisotropic magnetic susceptibility of the phthalocyanine macrocycles. Electrical measurements of individual nanowires show that such highly ordered molecular packing results in a 17-fold increase in conductivity.

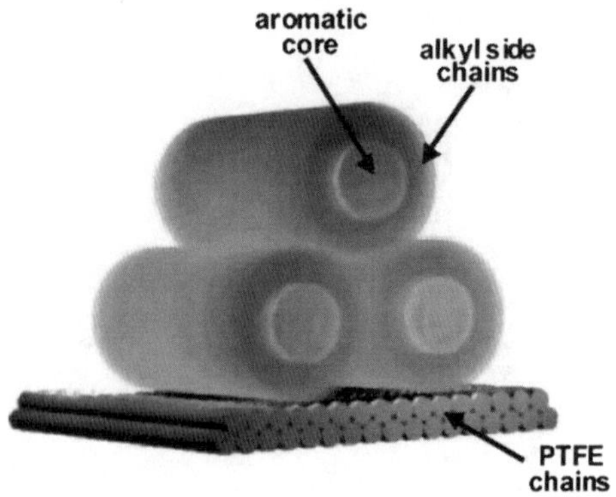

Fig. 7.9 Alignment of discotic liquid crystals on PTFE layers. Columnar HBC stacks are lying edge-on and parallel with respect to the underlying PTFE chain. Reproduced with permission from [43]. Copyright 2007 American Chemical Society

7.3.3 Polytetrafluoroethylene Alignment Layer

Upon mechanically stretching a blend of a host polymer and a discotic guest, it was observed that the discotic columns show uniaxial orientation. Encouraged by these findings, Zimmermann et al. explored the possibility of orienting columnar liquid-crystalline materials on a highly oriented poly(tetrafluoroethylene) (PTFE) layer [147]. Uniaxial in-plane alignment of the columnar phase of a triphenylene derivative **28** has been readily achieved by spin-coating the material onto a friction transfer poly(tetrafluoroethylene) (PTFE) orientation layer. Annealing of the films at temperatures within the mesophase temperature was found to enhance the degree of molecular orientation which was characterized by optical microscopy. The optical absorption and photoluminescence spectra of oriented films were found to be significantly polarized.

Friend et al. reported uniaxially oriented columnar phase forming hexabenzocoronene derivatives **29** and **30** parallel to a substrate coated with poly(tetrafluoroethylene) alignment layer [148]. Both the compounds are processed by casting from solution and slow evaporation of the solvent. It was observed that the alignment layer promoted the direct formation of highly oriented supramolecular architectures for both the materials without the need of an additional annealing step in the mesophase. Electron diffraction studies revealed that the columnar stacks are oriented parallel to the underlying PTFE chains (Fig. 7.9). Evidence shows *meso*-epitaxial growth of the columns, *i.e.* the alignment behavior seems to be dictated by a combination of molecular epitaxy and surface-topology-induced epitaxy. Grelet et al. reported about the unidirectional planar orientation of DLCs exhibiting ambient mesomorphism by using a rubbed Teflon coating [149]. Interestingly, the columnar anchoring is found to depend on the nature of the compound and is either parallel or perpendicular to the Teflon coating. The pyrene derivative aligns parallel to the Teflon layers whereas the benzoperylene aligns perpendicular to the PTFE chains. The different behavior has been attributed to the difference in polarity of the materials. Columnar phase of LC perylene tetracarboxylic bisimide **31** (Scheme 7.8) exhibiting high electron mobility has also been uniaxially aligned on

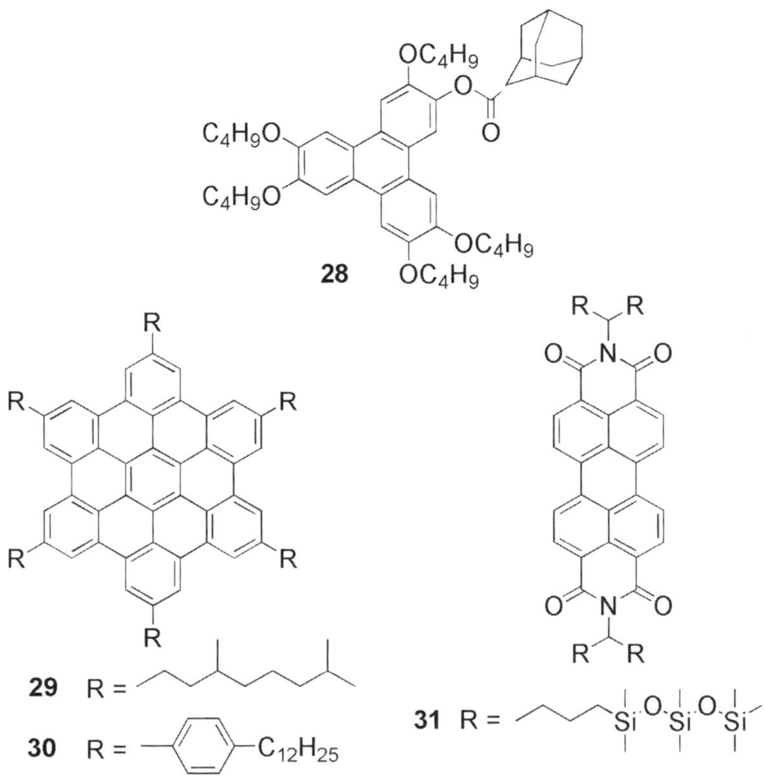

Scheme 7.8 Chemical structures of some discotic compounds aligned on PTFE surface layer

friction transferred PTFE layers [150]. Ivanov et al. reported the spontaneous formation of macroscopic homeotropically aligned LC monodomains on nano-patterned surfaces fabricated by means of friction transfer of PTFE [151]. The columnar alignment in the films was studied as a function of temperature by POM and synchrotron X-ray diffraction. It is noteworthy that in most instances PTFE surfaces were reported to induce homogeneous rather than homeotropic alignment. Though a broad variety of discotic materials exhibit highly oriented growth on aligned PTFE orientation layers, the mechanism of alignment is still not clear and have been speculated that the compound may exhibit intermediate lyotropic phase upon evaporation of the solvent or occurs through epitaxial growth of columns.

7.3.4 Langmuir-Blodgett Technique

Langmuir-Blodgett (LB) technique is one of the processing techniques for the fabrication of thin films of functional materials with controlled film thickness. For the fabrication of the thin films with nano-level control of film thickness, number

Scheme 7.9 Chemical structures of some discotic compounds aligned by LB technique

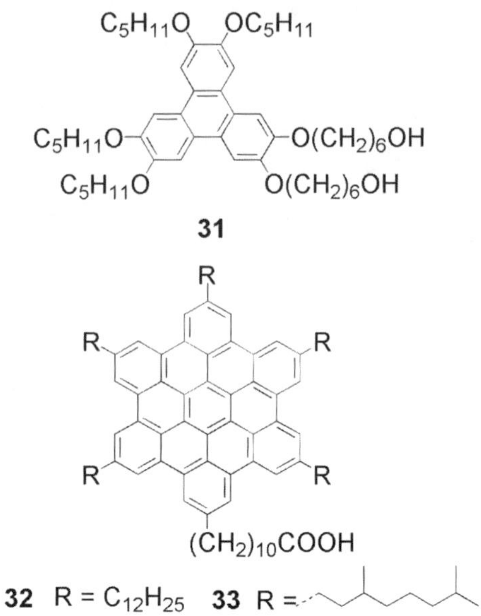

of layers and precise layered sequences, LB technique is preferred for amphiphilic molecules. A similar technique is self-assembled monolayer (SAM) method, which is usually used for formation of single layer structures on surfaces. This method is important for surface modifications of various materials and devices but is not effective for multilayer formation. The LB method is not a technique used simply to fabricate layered structures of lipid-like amphiphilic molecules. It can be applied to a variety of substances including biomolecules and inorganic nanomaterials.

It has been demonstrated that uniaxially planar aligned discotic columnar films can be fabricated by the Langmuir-Blodgett (LB) technique [152–154]. This approach can be applied to unsymmetrically substituted DLCs possessing at least one polar terminal group in their side chains. Accordingly some amphiphilic discotic compounds have been synthesized (Scheme 7.9) and uniaxially aligned by using the LB technique. The Langmuir-Blodgett has been frequently applied for the alignment of triphenylenes, phthalocyanines, hexabenzocoronenes and discotic multiynes etc. [155–161]. Langmuir films of disk-shaped molecules provide an interesting model for a 2D system with anisotropic interactions and properties. The first step in the study of Langmuir-Blodgett films is usually to measure a pressure-area isotherm. Depending on the detail molecular structure, these measurements of disk-shaped molecules usually indicate one of the two molecular configurations at the air-water interface. Molecules with relatively strong core-water interactions and weak core-core interactions generally show a face-on arrangement with the cores co-facial to the water surface and the aliphatic chains extending away from the surface. By contrast, molecules with stronger core-core attractions show an edge-on arrangement. Those films which are edge-on at the air-water interface are

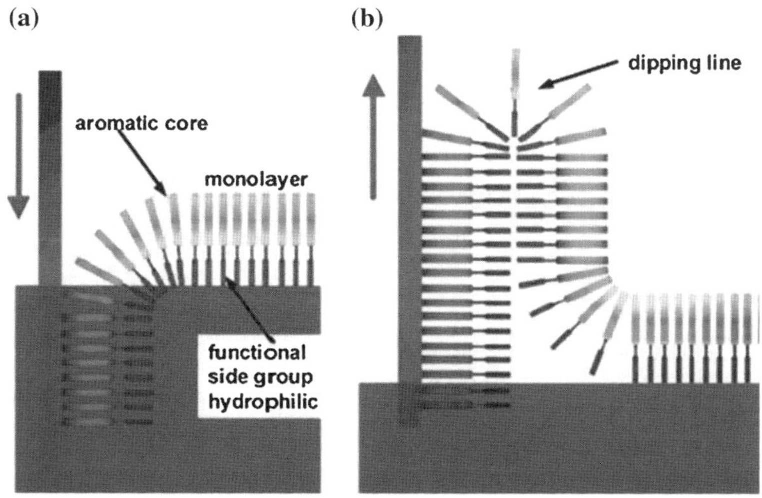

Fig. 7.10 Schematic illustration of the Langmuir-Blodgett technique for the edge-on alignment of discotic liquid crystals onto a substrate. Reproduced with permission from [43]. Copyright 2007 American Chemical Society

most often found to form edge-on columnar structures after transfer to the solid substrate. Macroscopic in-plane alignment is obtained during the monolayer transfer from the air-water interface (Fig. 7.10) onto a solid substrate. After transfer to solid substrates via LB techniques, the discotic monolayer film could be studied by various structural probes available. Non-symmetrically substituted DLC molecules possessing strong pi-pi interaction seem to retain the bulk columnar organization upon transfer to solid substrate. For LB studies, the HBC derivatives **32** and **33** were designed which are asymmetrically substituted and terminated by a carboxylic acid group to provide the desired amphiphilic character [156, 157]. The molecules formed well defined monolayers when spread from a solution at the air-water interface. Efficient transfer of the monolayer onto the substrate by vertical dipping gave well-defined multilayer films. Columns were oriented along the dipping direction with disc planes perpendicular to the column axes and stacked with a cofacial manner. The amphiphilic HBC disks formed a macroscopic in-plane orientation of the columns with their axes parallel to the dipping direction. The pentaynes form discotic nematic phase in the bulk but in the LB films they form columnar stacks with edge-on orientation. Face- and edge-on orientations of octa-acid and —alcohol substituted discotic tetraazaporphyrins in LB monolayers has been recently reported by Eichhorn and co-workers [162].

Ordering of terminally functionalized DLCs as self-assembled monolayer (SAM) on different substrate surfaces has been investigated. Triphenylene derivatives **34-36** (Scheme 7.10) with one or two thiol end groups align homogeneously

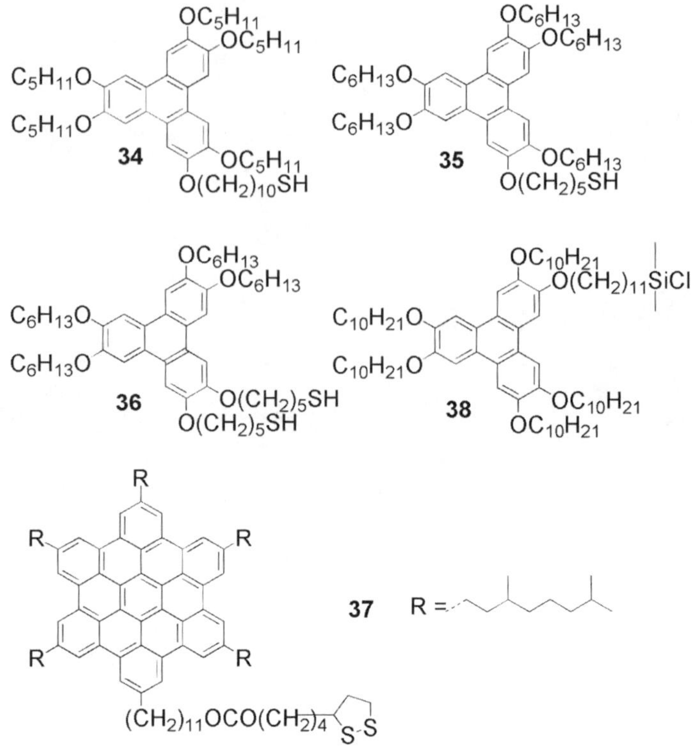

Scheme 7.10 Structure of discotic compounds which form self-assembled monolayers

with columnar stacks on gold surfaces [163]. Decreasing spacer length between the thiol end groups and the aromatic core was found to yield larger domains. Homeotropic alignment was found for triphenylene derivatives with thioethers attached to the core [164]. Planar alignment has also been observed in a metal-SAM-metal junction with an HBC derivative **37** bearing a dithiolane functionality [165–168]. Electron tunneling across the self-assembled columns indicates that HBC cores are transparent to electrons. Inspired by the SAM of alkane silanes, Mansueto et al. reported chloro silane terminated liquid crystalline triphenylene derivatives **38** for direct surface anchoring in self-assembled monolayers, which exhibit planar alignment [169]. SAMs were obtained by adsorption from solution onto SiO_2 and characterized by the combination of water contact angle measurement, null ellipsometry and atomic force microscopy. To study the effect of concentration binary SAMs were coassembled with different alkane chlorosilanes. The self-assembled monolayers of triphenylene based thiols and silanes could potentially act as alignment layers for structurally related DLCs.

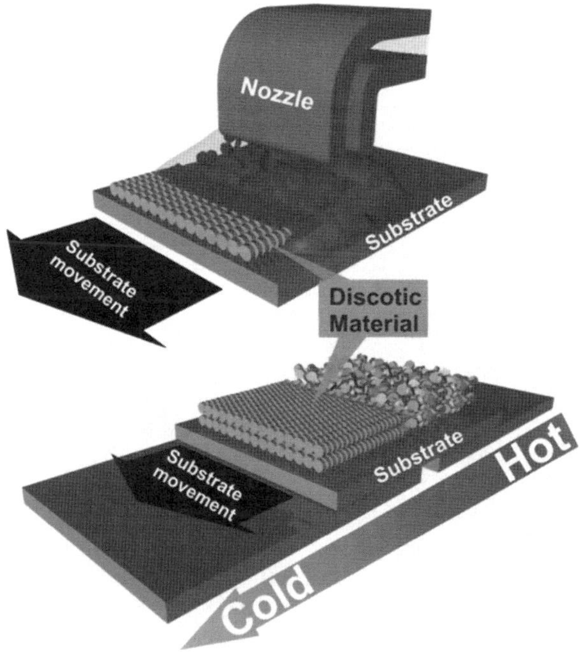

Fig. 7.11 Schematic illustration of zone casting (*top*) and zone melting (*bottom*) methods for the alignment control of discotic liquid crystals. Reproduced with permission from [45]. Copyright 2007 The Royal Society of Chemistry

7.3.5 Zone Casting

Zone casting technique is a facile way to produce uniaxially aligned thin films without the use of an alignment layer. This solution-based processing technique is based on the strong aggregation tendency of the discotic molecules in solution. Controlling the interaction between the molecules and therefore their self-association by changing the chemical composition is the key factor tailoring the solubility and thermal properties. The degree of self-association dominates the processing from solution. However it has been observed that discotic molecules with significantly higher solubility which is related to a lower self-aggregation in solution form ordered surface layers when processed by zone-casting technique. During the zone casting process, there is a change of solution concentration within a defined zone which is the driving force for the directed growth of DLCs. The principle of zone processing is schematically shown in Fig. 7.11. For zone casting, a solution is spread by means of a nozzle onto a moving support, creating a concentration gradient within a meniscus formed between the nozzle and the support. At the critical concentration, the material nucleates from the solution onto the moving substrate to form a uniaxially aligned thin layer. The film morphology is controlled by a number of processing parameters such as evaporation temperature and polarity of the solvent, solution concentration, solution flow rate, and substrate velocity which are related to the nucleation from solution of the corresponding material. This alignment technique has been successfully applied with

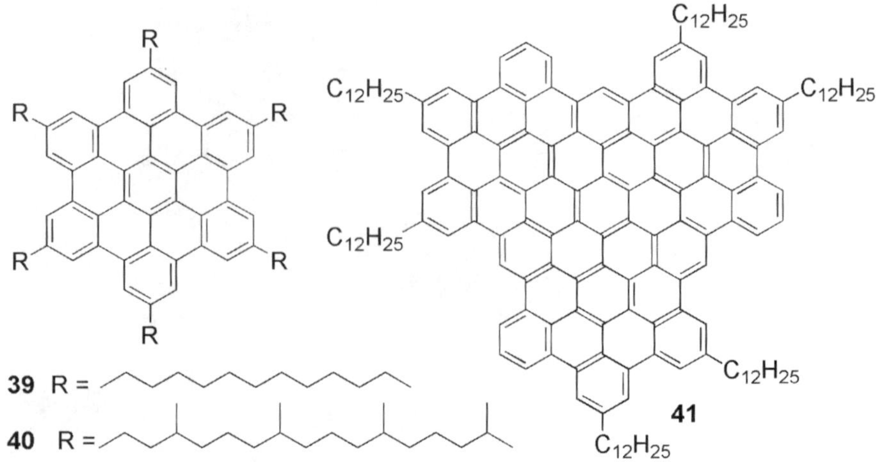

Scheme 7.11 Chemical structures of some discotic compounds aligned by zone casting

extended aromatic cores **39–41** (Scheme 7.11) that possess pronounced self-aggregation in solution [170–172]. The study of the zone cast films by the combination of atomic force microscope, electron microscope and X-ray diffraction has revealed a perfect uniaxial long-range edge-on orientation of the columnar stacks along the processing direction. Moreover, heating the zone-cast films to the mesophase temperature significantly enhanced the co-facial stacking of the molecules in the columns. The zone cast technique has been successfully applied to fabricate field effect transistors by using a HBC derivative which exhibits good device characteristics [173]. Recently, a room temperature liquid crystalline phthalocyanine has been uniaxially aligned on glass support by zone casting technique. The discotic compound forms uniform films with planar columnar orientation over several square centimeters [174].These observations are essential for the film fabrication of other soluble organic semiconductors for the application in electronic devices.

7.3.6 Zone Melting

DLCs with good thermal stability and accessible isotropic temperatures could be aligned by melt processing either from the isotropic phase or in the columnar phase. It is essential for the melt processing of discotics of large aromatic cores to decrease their isotropization temperatures by substitution with suitable side chains without sacrificing their thermal stability. The uniaxial planar alignment of porphyrins based DLCs along a temperature gradient from their melt has been achieved by Bard and coworkers [175]. They could successfully achieve the alignment of a DLC **42** by zone melting technique. In this technique,

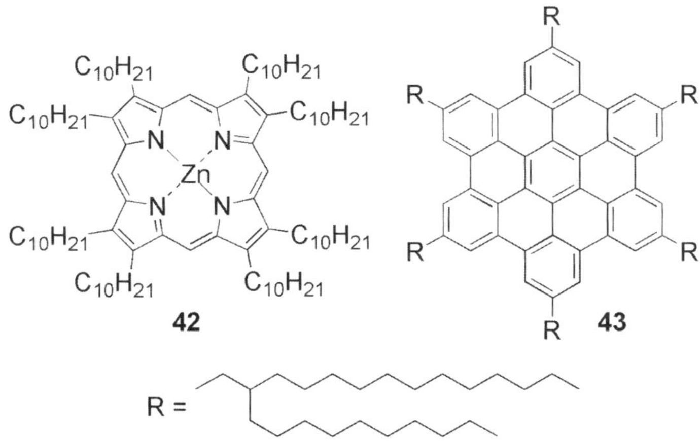

Scheme 7.12 Chemical structures of some DLCs aligned by zone melting

an electrically heated wire was slowly moved across the sample sandwiched between two glass slides. After a single pass of the narrow molten zone, a single-crystal like film was produced. The procedure is depicted in Fig. 7.11. Interestingly, it was found that there was significant improvement of the purity of the materials due to the zone-refinement effect. Subsequently the zone crystallization technique was adopted to align a HBC derivative by processing from its melt [176]. Sterically demanding alkyl chains were substituted to reduce the self-aggregation of the molecules which led to the formation of stable isotropic state at low temperature (Scheme 7.12). Upon moving a sample along the temperature gradient with a constant speed, the material crystallized in the oriented film. 2D-WAXS revealed a columnar growth along the temperature gradient with edge-on arranged discs, and POM displayed high optical anisotropy characteristic for macroscopically oriented planar discotic columnar layer.

7.3.7 Other Methods

A new class of hexabenzocoronene derivatives, e.g. compounds (Scheme 7.13) can self-organize into liquid crystalline phases composed of molecular stacks that orient spontaneously parallel to the surface [177, 178]. Field effect transistors based on these materials show high charge-carrier mobilities, high on/off ratios, and low turn-on voltages. So far compound **44** exhibits the best field effect transistor properties achieved for a columnar discotic material. Polarized light microscopy revealed that these materials tend to orient their columns parallel to the surface upon thermal annealing.

Dip coating is a technique which provides uniaxial planar aligned columnar films onto substrate surfaces from solutions of compound **45** [179]. During the dip

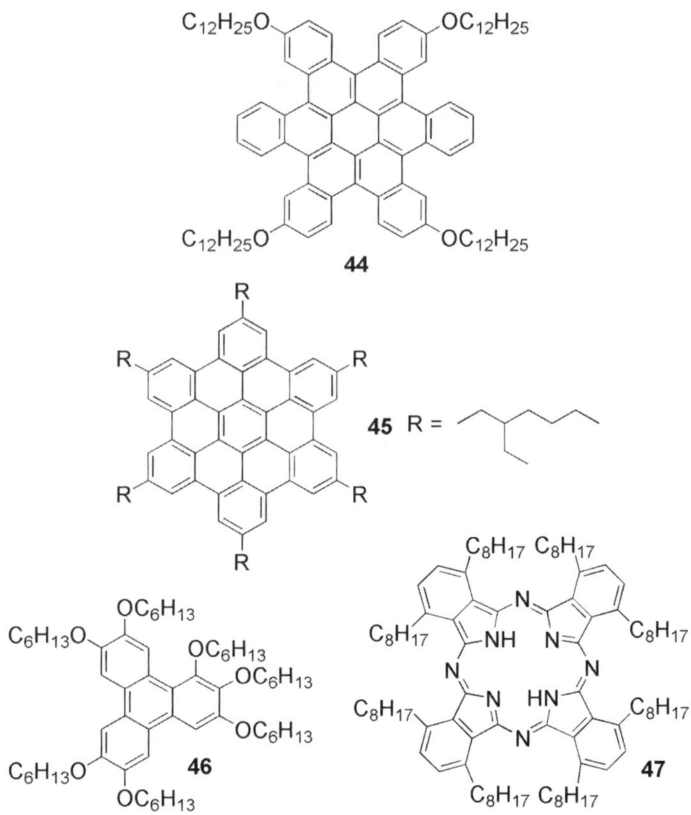

Scheme 7.13 Chemical structures of some discotic compounds exhibiting planar alignment

coating, a substrate is moved out of a solution and the velocity of the substrate from solution leads to the formation of a meniscus at the substrate/solution interface, where the molecules can align due to the concentration gradient. However this method is not suitable to process large areas of samples. Solvent vapor annealing (SVA) is a post-processing technique which has been used to improve the self-organization of molecules within the aligned columnar architectures [180]. SVA is performed by simply placing the sample into a vessel where a vapour-liquid equilibrium is attainable for the chosen solvent which converts poorly assembled deposited materials into highly organized functional structures.

Mullen et al. have demonstrated the use of electric field as a tool to align nonpolar discotic liquid crystalline molecules [181]. They deposited the solution of the compound onto a glass surface by drop-casting and the molecules were oriented into highly ordered structures by an electric field during solvent evaporation. Atomic force microscopy (AFM), scanning electron microscopy (SEM) and transmission electron microscopy (TEM) showed a long-range alignment where the disc-like molecules were organized in columns perpendicular to the direction of the imposed electric field and in the edge-on arrangement on the glass surface. Crystallization in an atmosphere saturated with vapor of the organic solvent was employed to improve the alignment of HBC molecules under electric field influence. Ultrathin films of a phthalocyanine DLC can be organized to form a conductive channel tens of microns long between Au electrodes with thickness control over a single monolayer [182]. This approach exploits the electromigration of the isotropic phase formed starting from the pretransitional region of the columnar-isotropic phase transition. Dewetting and subsequent electromigration expose an ultrathin film, a few monolayers thick, exhibiting columnar LC order. The layers of this ultrathin film melt progressively above clearing temperature and can be individually exfoliated by electromigration, starting from the ninth down to the first monolayer. Thus the combined phenomena of dewetting and electromigration produce discotic functional films on a device structure. This approach represents a powerful tool for aligning DLC in a bottom-electrode device channel.

Uniaxial planar alignment of columnar DLCs on chemically patterned surfaces has been obtained by isotropic phase dewetting of the materials [183]. Several triphenylene and phthalocyanine derivatives have been successfully aligned by this technique. Surfaces with chemically patterned stripes have been fabricated either by microcontact printing of organothiol self-assembled monolayers on gold or by deep UV patterning of organosilane SAMs on silicon. Upon cooling from the isotropic into the hexagonal columnar phase, the DLCs align in a planar orientation. Some of the triphenylene derivatives **46** are found to align with the director perpendicular to the stripe direction but the phthalocyanine derivative **47** align parallel to the stripe direction. A soft-landing methodology which exploits matrix-assisted laser desorption/ionization (MALDI) mass spectrometry has been used to produce ordered structures of large aromatic molecules at surfaces [184]. This technique has the potential to align columnar phases on surfaces with the help of the deposited nanographene monolayer.

7.4 Alignment of DLCs in Micro- and Nanopores and Channels

Investigation of the properties of atoms, molecules, and materials confined within nanometer-sized spaces improves our fundamental understanding of the physics, chemistry and materials science occurring in nanospace. Advanced sciences and technologies depend strongly on the methods used to fabricate nanometer sized structures from functional materials. Geometric confinement of discotic columnar LCs in micro- and nanoporous templates is a promising strategy to fabricate nanofibers with potential applications. However, for many materials, nanoconfinement has been shown to significantly affect the physical properties such morphology and phase behavior. It is recently revealed by Geerts et al. that there occurs substrate induced phase transitions at this length scale [185]. Discotic nanowire arrays prepared by employing nanoporous materials such as porous alumina templates may exhibit high regularity and specific macroscopic orientation over large areas. Wendorff et al. recently demonstrated the use of porous alumina templates to produce aligned nanowires from liquid crystalline triphenylene derivative [186]. When the pores with a diameter of a few hundred nanometers were filled with the molten triphenylene derivative, only the pore walls were found to be wetted by the material. Various parameters including pore geometry, interfacial phenomena and the thermal history influence the order of the discs within the pores. Post-templating, the template was chemically removed which provide aligned nanowires of the liquid crystal compound as revealed by SEM. An interesting extension of the templating method was the preparation of nanotubes consisting of a polymer layer outside and a discotic triphenylene layer inside, by wetting porous alumina templates [187]. The discs arrange to produce a columnar alignment along the axes of the tubes under controlled thermal annealing. Such phenomena should allow the fabrication of materials combining the functionality of the LC with the structural stability of a polymer. These materials can exhibit the mechanical stability of the polymeric component along with the switchable optical and electronic properties depending on the orientation and order of the LCs. A HBC derivative was successfully templated in nanoscopic pores as well as in macroscopic glass capillaries by melt processing [188]. In both the cases, the columnar structures were long range aligned along the template axis. This behavior was explained by the pronounced directional self-assembly of the molecules, while the influence of template curvature was negligible. Duran et al. studied the structure formation, phase behavior and dynamics of a dipole-functionalized HBC derivative in anodic aluminium oxide (AAO) nanopores with narrow size distribution as a confining matrix [189]. Alignment of the HBC columns along the nanopore axes was found to be independent of the pore diameter. Thus, arrays of 1D supramolecular wires with uniform columnar orientation on a macroscopic scale were realized (Fig. 7.12). The cylindrical confinement affects the disc axial dynamics as well as the distribution of relaxation times.

Uniaxial alignment of columnar mesophase of a phthalocyanine derivative due to nanoconfinement effects has been achieved and studied by Mouthuy et al. in a network of crisscrossed nanogrooves [190, 191]. The uniaxial alignment of the columnar mesophase was obtained in nano-grooves that were etched into the surface of an oxidized silicon wafer. Preferential alignment results from the combination of the anisotropy of interface tension between the mesophase and the nanogrooves' walls and the anisotropy of the network cell dimensions. Atomic force microscopy in tapping mode was used to show that the columns adopt a planar configuration with their axis perpendicular to the vertical walls of the grooves (Fig. 7.13). The nonvanishing interfacial tension anisotropy makes the confined alignment permanent in nature. Hence rational engineering of arrays of nanowires of liquid-crystalline organic semiconductors can be accomplished. Aligned nanowires of various shapes could be fabricated by allying the control on the local alignment of columns with an efficient confinement of the matter. Besides the

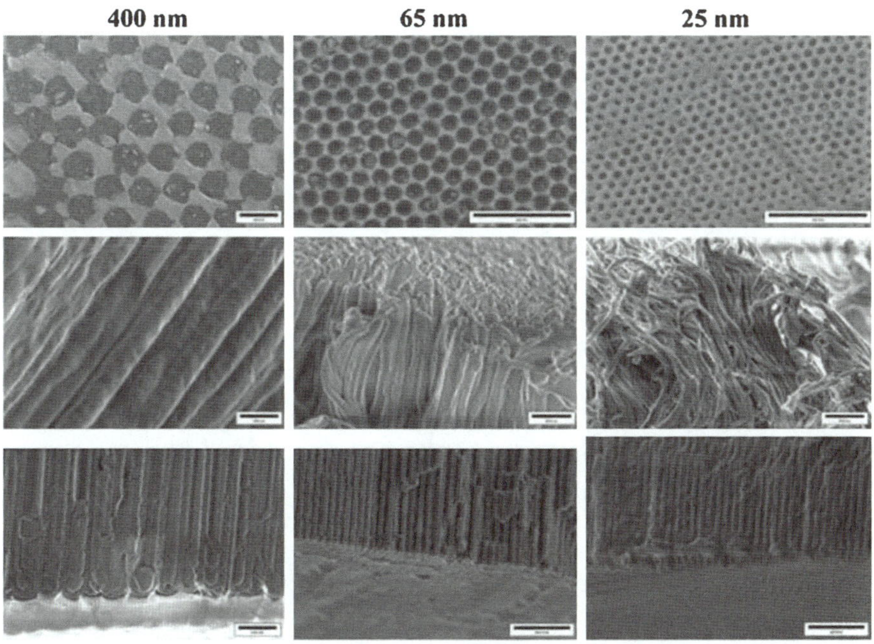

Fig. 7.12 Alignment of DLCs in nanopores. *Top* and *bottom rows* show *top* and *bottom* view of HBC embedded membranes respectively imaged by SEM. SEM images in the *middle row* refer to free-standing HBC nanorods obtained by selective dissolution of the membranes. Reproduced with permission from [189]. Copyright 2012 American Chemical Society

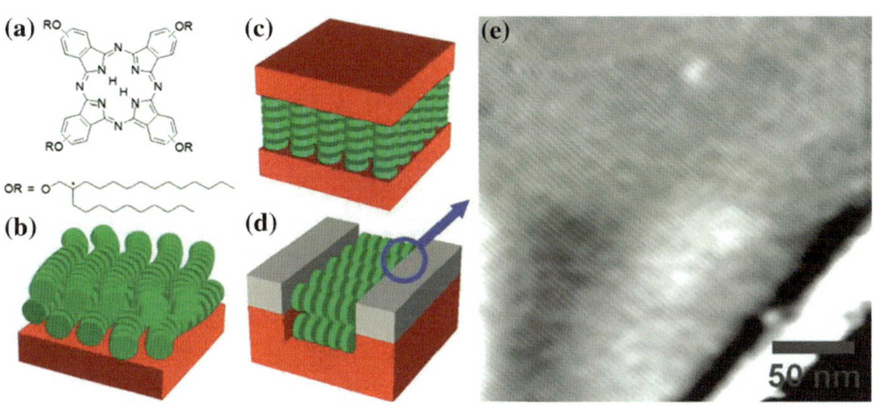

Fig. 7.13 Alignment of discotic phthalocyanine in nanogrooves. **a** Chemical structure of Pc, schematic drawing of the alignment of Pc columns exposed to air (**b**), sandwiched between two solid substrates (**c**), in a nanogroove (**d**), AFM image taken at the edge of the nanogroove (**e**). Reproduced with permission from [190]. Copyright 2007 American Chemical Society

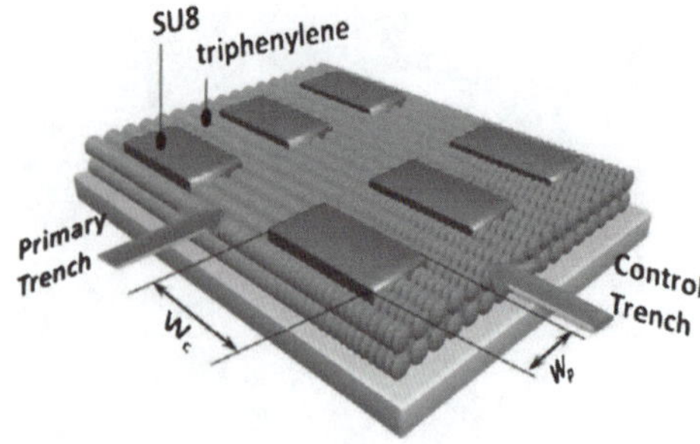

Fig. 7.14 Schematic illustration of alignment of triphenylene in microtrench. Reproduced with permission from [193]. Copyright 2012 Taylor & Francis Group, LLC

Fig. 7.15 POM of planar aligned columnar phase in microchannels on silicon. Channel widths are (from *left* to *right*) 10, 5, 2 μm respectively. The microscope is in the reflection mode. Reproduced with permission from [194]. Copyright 2013 Wiley-VCH

expected advantages in device fabrication, it is possible to extract quantitative information regarding mechanical properties of molecular systems.

Cerclier et al. have studied the structure and phase behavior of the discotic columnar phase confined in various templates [192]. The influence of size, the roughness and chemical nature of the pores was investigated by a combination of different techniques. A homeotropic anchoring is favoured in smooth cylindrical nanochannels of porous alumina while surface roughness of porous silicon promotes more disordered structures. The hexagonal columnar to isotropic phase transition is modified as a result of geometrical constraints and interfacial interactions. The results demonstrate the importance of surface anchoring and topological constraints, which can occur in the nucleation process and the relative stability of confined phases in the case of DLCs.

Recently, Chiang et al. produced microtrenches on substrates and studied the molecular stacking of triphenylene discotic molecules on the substrate [193]. They demonstrated that the discotic molecules in the trenches can assemble into uniaxially aligned columns (Fig. 7.14). The way in which the discotic molecules anchor on the walls of trenches and the over all orientation of the columns are determined by the energetic conditions of the walls. They show that the orientation direction of the columns can be controlled to align either parallel or perpendicular to the wall of the primary trenches using properly designed controlling trenches. Such a technique would be particularly useful for organic semiconductor devices that require a local-planar-alignment of discotic columns.

A variety of triphenylene and phthalocyanine DLCs has been aligned in micron scale channels created using SU8 photoresist and filled by capillary action [194]. It has been shown that the alignment depends on the nature of the LC and on the width of the channel. In all cases, the columns director lies in plane and across the channels as revealed by POM studies (Fig. 7.15).

7.5 Conclusions and Outlook

The alignment control with suitable molecular orientations in the viscous columnar phases formed by DLCs has been obtained by applying different external stimuli such as electric field, magnetic field, light, and thermal treatment. Alignment has also been achieved by a combination of alignment layers and field forces resulting in highly ordered thin films with tailored alignment. Alignment studies enlighten the fundamental understanding of subtle molecular interactions involved at the substrate-LC interface governing different molecular orientations. The columnar nanostructures of DLCs have been aligned in the desired orientation and successfully integrated into the device configurations of OPV and OFET. Different techniques for the alignment in cells, on single open substrates as well as in nanopores have been developed and demonstrated. It is highly preferable for a material to exhibit a single mesophase structure over a wide window of temperature starting from room temperature so that the achieved desired alignment would stay intact under ambient conditions. Although the most promising way is to design columnar discotics that self-organize spontaneously with planar or homeotropic alignment on substrate surfaces, general molecular design principles that allow this controllability on the molecular orientation have not yet been established. Moreover, translating molecular properties to materials properties is not very straight forward which has been the main hindrance for the potential applications of DLCs. It is essential to assess the environmental stability of the alignment layers obtained by different processing techniques as a function of time before they could be integrated into devices. It is time to test the alignment of one type of DLC on different substrates and under different fields. Alignment techniques reveal the molecular orientation and device performance relationships in DLCs and some of the concepts can be extended to conjugated organic materials. The alignment of the columnar phases can be translated to organize and reorient dispersed nanomaterials in predetermined directions. The knowledge gained over the years should aid in reaching the ultimate goal of making a rational connection between molecular structure, bulk organization, the forces that control self-assembly, and device performance. Owing to their promising properties like optical transparency, electrical conductivity, mechanical strength, and stability, graphene, graphene oxide and reduced graphene oxide should be explored as alignment layers for DLCs. Due to the development of liquid phase processing of graphene and graphene oxide, they can be deposited onto substrate surfaces. Moreover, they can also be grown on substrates by different techniques which can act as surface alignment layers.

Acknowledgments The preparation of this chapter benefited from the support to Quan Li by the Department of Defense Multidisciplinary University Research Initiative (AFOSR MURI FA9550-12-1-00370 and FA9550-06-1-0337), the Air Force Office of Scientific Research (AFOSR FA9550-09-1-0193 and FA9550-09-1-0254), the National Science Foundation (NSF IIP 0750379), the National Aeronautics and Space Adminstration (NASA), the Department of Energy (DOE DE-SC0001412), Ohio Third Frontier, and the Ohio Board of Regents under its Research Challenge program.

References

1. M. Bremer, P. Kirsch, M. Klasen-Memmer, K. Tarumi, The TV in your pocket: development of liquid-crystal materials for the new millennium. Angew. Chem. Int. Ed. **52**, 8880–8896 (2013)
2. T. Geelhaar, K. Griesar, B. Reckmann, 125 Years of liquid crystals—a scientific revolution in the home. Angew. Chem. Int. Ed. **52**, 8798–8809 (2013)
3. S.H. Lee, S.S. Bhattacharyya, H.S. Jin, K.-U. Jeong, Devices and materials for high-performance mobile liquid crystal displays. J. Mater. Chem. **22**, 11893–11903 (2012)
4. J.W. Goodby, The nanoscale engineering of nematic liquid crystals for displays. Liq. Cryst. **38**, 1363–1387 (2011)
5. D. Demus, J.W. Goodby, G.W. Gray, H.-W. Spiess, V. Vill (eds.), in *Handbook of Liquid Crystals*, vol. 1–3 (Wiley-VCH, New York, 1998)
6. B. Bahadur (ed.), *Liquid Crystals Applications and Uses* (World Scientific, Singapore, 1992)
7. J.W. Goodby, I.M. Saez, S.J. Cowling, V. Gortz, M. Draper, A.W. Hall, S. Sia, G. Cosquer, S.-E. Lee, E.P. Raynes, Transmission and amplification of information and properties in nanostructured liquid crystals. Angew. Chem. Int. Ed. **47**, 2754–2787 (2008)
8. T. Kato, N. Mizoshita, K. Kishimoto, Functional liquid crystalline assemblies: self-organized soft materials. Angew. Chem. Int. Ed. **45**, 38–68 (2006)
9. C. Tschierske, Liquid crystal engineering-new complex mesophase structures and their relation to polymer morphologies, nanoscale patterning and crystal engineering. Chem. Soc. Rev. **36**, 1930–1970 (2007)
10. C. Tschierske, Development of structural complexity by liquid-crystal self-assembly. Angew. Chem. Int. Ed. **52**, 8828–8878 (2013)
11. E.-K. Fleischmann, R. Zentel, Liquid-crystalline ordering as a concept in materials science: from semiconductors to stimuli-responsive devices. Angew. Chem. Int. Ed. **52**, 8810–8827 (2013)
12. Q. Li (ed.), *Liquid Crystals Beyond Displays: Physics, Chemistry, and Applications* (Wiley, Hoboken, 2012)
13. Q. Li (ed.), *Self-Organized Organic Semiconductors: From Materials to Device Applications* (Wiley, Hoboken, 2011)
14. Q. Li (ed.), in *Intelligent Stimuli Responsive Materials: From Well-Defined Nanostructures to Applications* (Wiley, New York, 2013)
15. Y. Wang, Q. Li, Light-driven chiral molecular switches or motors in liquid crystal media. Adv. Mater. **24**, 1926–1945 (2012)

16. T. Ikeda, J. Mamiya, Y. Yu, Photomechanics of liquid-crystalline elastomers and other polymers. Angew. Chem. Int. Ed. **46**, 506–528 (2007)
17. C. Ohm, M. Brehmer, R. Zentel, Liquid crystalline elastomers as actuators and sensors. Adv. Mater. **22**, 3366–3387 (2010)
18. H.N.W. Lekkerkerker, G.J. Vroege, Liquid crystal phase transitions in suspensions of mineral colloids: new life from old roots. Phil. Trans. R. Soc. A **371**, 20120263 (2013)
19. F.M. van der Kooij, K. Kassapidou, H.N.W. Lekkerkerker, Liquid crystal phase transitions in suspensions of polydisperse plate-like particles. Nature **406**, 868–871 (2000)
20. J.-C.P. Gabriel, F. Camerel, B.J. Lemaire, H. Desvaux, P. Davidson, P. Betail, Swollen liquid-crystalline lamellar phase based on extended solid-like sheets. Nature **413**, 504–508 (2001)
21. S.J. Woltman, G.D. Jay, G.P. Crawford, Liquid-crystal materials find a new order in biomedical applications. Nat. Mater. **6**, 929–938 (2007)
22. I.W. Hamley, Liquid crystal phase formation by biopolymers. Soft Matter **6**, 1863–1871 (2010)
23. A. Angelova, B. Angelov, R. Mutafchieva, S. Leieur, P. Couvreur, Self-assembled multicompartment liquid crystalline lipid carrier for protein, peptide, and nucleic acid drug delivery. Acc. Chem. Res. **44**, 147–156 (2011)
24. A.M. Lowe, N.L. Abbott, Liquid crystalline materials for biological applications. Chem. Mater. **24**, 746–758 (2012)
25. H.K. Bisoyi, S. Kumar, Liquid-crystal nanoscience: an emerging avenue of soft self-assembly. Chem. Soc. Rev. **40**, 306–319 (2011)
26. T. Hegmann, H. Qi, V.M. Marx, Nanoparticles in liquid crystals: synthesis, self-assembly, defect formation and potential applications. J. Inorg. Organomet. Polym Mater. **17**, 483–508 (2007)
27. J.P.F. Lagerwall, G. Scalia, A new era for liquid crystal research: applications of liquid crystals in soft matter nano-, bio- and microtechnology. Curr. Appl. Phys. **12**, 1387–1412 (2012)
28. G.L. Nealon, R. Greget, C. Dominguez, Z.T. Nagy, D. Guillon, J.-L. Gallani, B. Donnio, Liquid-crystalline nanoparticles: hybrid design and mesophase structures. Beilstein J. Org. Chem. **8**, 349–370 (2012)
29. H.K. Bisoyi, S. Kumar, Carbon-based liquid crystals: art and science. Liq. Cryst. **38**, 1427–1449 (2011)
30. S. Kumar, *Chemistry of Discotic Liquid Crystals: From Monomers to Polymers* (CRC Press, Boca Raton, 2011)
31. L. Schmidt-Mende, A. Fechtenkotter, K. Mullen, E. Moons, R.H. Friend, J.D. MacKenzie, Self-organized discotic liquid crystals for high-efficiency organic photovoltaics. Science **293**, 1119–1122 (2001)
32. D. Adam, P. Schumacher, J. Simmerer, L. Haussling, K. Siemensmeyer, K.H. Etzbachi, H. Ringsdorf, D. Haarer, Fast photoconduction in the highly ordered columnar phase of a discotic liquid crystal. Nature **371**, 141–143 (1994)
33. X. Feng, V. Marcon, W. Pisula, M.R. Hansen, J. Kirkpatrick, F. Grozema, D. Andrienko, K. Kremer, K. Mullen, Towards high charge-carrier mobilities by rational design of the shape and periphery of discotics. Nat. Mater. **8**, 421–426 (2009)
34. K. Ohta, K. Hatsusaka, M. Sugibayashi, M. Ariyoshi, K. Ban, F. Maeda, R. Naito, K. Nishizawa, A.M. Van de Craats, J.M. Warman, Discotic liquid crystalline semiconductors. Mol. Cryst. Liq. Cryst. **397**, 25–45 (2003)
35. S. Laschat, A. Baro, N. Steinke, F. Giesselmann, C. Hagele, G. Scalia, R. Judele, E. Kapatsina, S. Sauer, A. Schreivogel, M. Tosoni, Discotic liquid crystals: from tailor-made synthesis to plastic electronics. Angew. Chem. Int. Ed. **46**, 4832–4887 (2007)
36. S. Kumar, Self-organization of disc-like molecules: chemical aspects. Chem. Soc. Rev. **35**, 83–109 (2006)
37. M. Funahasi, Development of liquid-crystalline semiconductors with high carrier mobilities and their application to thin-film transistors. Polym. J. **41**, 459–469 (2009)

38. R.J. Bushby, O.R. Lozman, Discotic liquid crystals 25 years on. Curr. Opin. Colloid Interface Sci. **7**, 343–354 (2002)
39. S. Kumar, Recent developments in the chemistry of triphenylene-based discotic liquid crystals. Liq. Cryst. **31**, 1037–1059 (2004)
40. R.J. Bushby, K. Kawata, Liquid crystals that affected the world: discotic liquid crystals. Liq. Cryst. **38**, 1415–1426 (2011)
41. B.R. Kaafarani, Discotic liquid crystals for opto-electronic applications. Chem. Mater. **23**, 378–396 (2011)
42. M. O'Neill, S.M. Kelly, Ordered materials for organic electronics and photonics. Adv. Mater. **23**, 566–584 (2011)
43. J. Wu, W. Pisula, K. Mullen, Graphene as potential material for electronics. Chem. Rev. **107**, 718–747 (2007)
44. S. Chandrasekhar, S.K. Prasad, Recent developments in discotic liquid crystals. Contemp. Phys. **40**, 237–245 (1999)
45. S. Sergeyev, W. Pisula, Y.H. Geerts, Discotic liquid crystals: a new generation of organic semiconductors. Chem. Soc. Rev. **36**, 1902–1929 (2007)
46. D. Janietz, Structure formation control of disc-shaped molecules. Mol. Cryst. Liq. Cryst. **396**, 251–264 (2003)
47. C.D. Simpson, J. Wu, M.D. Watson, K. Mullen, From graphite molecules to columnar superstructures—an exercise in nanoscience. J. Mater. Chem. **14**, 494–504 (2004)
48. D. Janietz, Structure formation of functional sheet-shaped mesogens. J. Mater. Chem. **8**, 265–274 (1998)
49. W. Pisula, X. Feng, K. Mullen, Tuning the columnar organization of discotic polycyclic aromatic hydrocarbons. Adv. Mater. **22**, 3634–3649 (2010)
50. W. Pisula, M. Zorn, J.Y. Chang, K. Mullen, R. Zentel, Liquid crystalline ordering and charge transport in semiconducting materials. Macromol. Rapid Commun. **30**, 1179–1202 (2009)
51. J.K. Vij, A. Kocot, T.S. Perova, Order parameter, alignment and anchoring transition in discotic liquid crystals. Mol. Cryst. Liq. Cryst. **397**, 231–244 (2003)
52. K. Takatoh, M. Hasegawa, M. Koden, N. Itoh, R. Hasegawa, M. Sakamoto, *Alignment Technologies and Applications of Liquid Crystal Devices* (Taylor & Francis, New York, 2005)
53. S.H. Eichhorn, A. Adavelli, H.S. Li, N. Fox, Alignment of discotic liquid crystals. Mol. Cryst. Liq. Cryst. **397**, 47–58 (2003)
54. V. Percec, M. Glodde, T.K. Bera, Y. Miura, I. Shiyanovskaya, K.D. Singer, V.S.K. Balagurusamy, P.A. Heiney, I. Schnell, A. Rapp, H.-W. Spiess, S.D. Hudson, H. Duan, Self-organization of supramolecular helical dendrimers into complex electronic materials. Nature **419**, 384–387 (2002)
55. M. Yoshio, T. Kagata, K. Hoshino, T. Mukai, H. Ohno, T. Kato, One-dimensional ion-conductive polymer films: alignment and fixation of ionic channels formed by self-organization of polymerizable columnar liquid crystals. J. Am. Chem. Soc. **128**, 5570–5577 (2006)
56. T.-Q. Nguyen, R. Martel, M. Bushey, P. Avouris, A. Carlsen, C. Nuckolls, L. Brus, Self-assembly of 1D organic semiconductor nanostructures. Phys. Chem. Chem. Phys. **9**, 1515–1532 (2007)
57. J. Hoogboom, J.A.A.W. Elemans, A.E. Rowan, T.H.M. Rasing, R.J.M. Nolte, The development of self-assembled liquid crystal display alignment layers. Phil. Trans. R. Soc. A **365**, 1553–1576 (2007)
58. G.D. Luca, W. Pisula, D. Credgington, E. Treossi, O. Fenwick, G.M. Lazzerini, R. Dabirian, E. Orgiu, A. Liscio, V. Palermo, K. Mullen, F. Cacialli, P. Samori, Non-conventional processing and post-processing methods for the nanostructuring of conjugated materials for organic electronics. Adv. Funct. Mater. **21**, 1279–1295 (2011)
59. O. Yaroshchuk, Y. Reznikov, Photoalignment of liquid crystals: basics and current trends. J. Mater. Chem. **22**, 286–300 (2012)

60. K. Ichimura, Photoalignment of liquid-crystals systems. Chem. Rev. **100**, 1847–1873 (2000)
61. H.N. Tsao, H.J. Rader, W. Pisula, A. Rouhanipour, K. Mullen, Novel organic semiconductors and processing techniques for organic field-effect transistors. Phys. Stat. Sol. (a) **205**, 421–429 (2008)
62. L. Li, S. Kang, J. Harden, Q. Sun, X. Zhou, L. Dai, A. Jakli, S. Kumar, Q. Li, Nature-inspired light-harvesting liquid crystalline porphyrins for organic photovoltaics. Liq. Cryst. **35**, 233–239 (2008)
63. H.K. Bisoyi, S. Kumar, Discotic nematic liquid crystals: science and technology. Chem. Soc. Rev. **39**, 264–285 (2010)
64. K. Kawata, Orientation control and fixation of discotic liquid crystal. Chem. Rec. **2**, 59–80 (2002)
65. H. Mori, The wide view (WV) film for enhancing the field of view of LCDs. J. Disp. Technol. **1**, 179–186 (2005)
66. C. Vauchier, A. Zann, P. Le Barny, J.C. Dubois, J. Billard, Orientation of discotic mesophases. Mol. Cryst. Liq. Cryst. **66**, 103–114 (1981)
67. Y. Shimizu, H. Monobe, M. Heya, K. Awazu, A novel technique for the alignment control of highly ordered liquid crystals based on vibrational excitation of chemical bond by polarized infrared laser. Mol. Cryst. Liq. Cryst. **441**, 287–295 (2005)
68. H. Monobe, N. Terasawa, K. Kiyohara, Y. Shimizu, H. Azehara, A. Nakasa, M. Fujihara, Alignment behavior of discotic nematic and rectangular columnar mesophases on self-assembled monolares of alkanethiols and asymmetrical disulphides. Mol. Cryst. Liq. Cryst. **412**, 229–236 (2004)
69. H. Monobe, S. Mima, T. Sugino, Y. Shimizu, M. Ukon, Alignment behavior of the discotic nematic phase of 2,3,6,7,10,11-hexa(4-*n*-octyloxybenzoyloxy)triphenylene on polyimide and cetyltrimethylammonium bromide coated substrates. Liq. Cryst. **28**, 1253–1258 (2001)
70. T. Sergan, M. Sonpatki, J. Kelly, L.C. Chien, Mol. Cryst. Liq. Cryst. **359**, 245–257 (2001)
71. T. Sergan, M. Sonpatki, J. Kelly, L.C. Chien, Polarizing microscopy of a chiral discotic. Mol. Cryst. Liq. Cryst. **359**, 259–267 (2001)
72. K. Ichimura, S. Furumi, S. Morino, M. Kidowaki, M. Nakagawa, M. Ogawa, Y. Nishiura, Photocontrolled orientation of discotic liquid crystals. Adv. Mater. **12**, 950–953 (2000)
73. S. Furumi, M. Kidowaki, M. Ogawa, Y. Nishiura, K. Ichimura, Surface-mediated photoalignment of discotic liquid crystals on azobenzene polymer films. J. Phys. Chem. B **109**, 9245–9254 (2005)
74. S. Furumi, K. Ichimura, Surface-assisted photoalignment of discotic liquid crystals by nonpolarized light irradiation of photo-cross-linkable polymer thin films. J. Phys. Chem. B **111**, 1277–1287 (2007)
75. S. Furumi, K. Ichimura, Highly sensitive photoalignment of calamitic and discotic liquid crystals assisted by axis-selective triplet energy transfer. Phys. Chem. Chem. Phys. **13**, 4919–4927 (2011)
76. S. Furumi, D. Janietz, M. Kidowaki, M. Nakagawa, S. Morino, J. Stumpe, K. Ichimura, Polarized photoluminescence from photopatterned discotic liquid crystal films. Chem. Mater. **13**, 1434–1437 (2001)
77. S. Furumi, D. Janietz, M. Kidowaki, M. Nakagawa, S. Morino, J. Stumpe, K. Ichimura, Surface-assisted orientational control of discotic liquid crystals by light. Mol. Cryst. Liq. Cryst. **368**, 517–524 (2001)
78. S. Ikeda, Y. Takanishi, K. Ishikawa, H. Takezoe, Magnetic field effect on the alignment of a discotic liquid crystal. Mol. Cryst. Liq. Cryst. **329**, 589–595 (1999)
79. X. Zhou, S. Kang, S. Kumar, R.R. Kulkarni, S.Z.D. Cheng, Q. Li, Self-assembly of porphyrin and fullerene supramolecular complex into highly ordered nanostructure by simple thermal annealing. Chem. Mater. **20**, 3551–3553 (2008)
80. S. Kumar, H.K. Bisoyi, Aligned carbon nanotubes in the supramolecular order of discotic liquid crystals. Angew. Chem. Int. Ed. **46**, 1501–1503 (2007)

81. H.K. Bisoyi, S. Kumar, Carbon nanotubes in triphenylene and rufigallol-based room temperature monomeric and polymeric discotic liquid crystals. J. Mater. Chem. **18**, 3032–3039 (2008)
82. J.J. Lee, A. Yamaguchi, M.A. Alam, Y. Yamamoto, T. Fukushima, K. Kato, M. Takata, T. Aida, Discotic ionic liquid crystals of triphenylene as dispersants for orientating single-walled carbon nanotubes. Angew. Chem. Int. Ed. **51**, 8490–8494 (2012)
83. T.S. Perova, J.K. Vij, The influence of surface structure on the discotic liquid crystalline alignment: an infrared spectroscopy study. Adv. Mater. **7**, 919–922 (1995)
84. T.S. Perova, J.K. Vij, A. Kocot, Observation of an anchoring transition in a discotic liquid crystal. Europhys. Lett. **44**, 198–204 (1998)
85. T. Perova, S. Tsvetkov, J. Vij, S. Kumar, Observation of the orientational transition in hexa(hexylthio)triphenylene using polarized FTIR study. Mol. Cryst. Liq. Cryst. **351**, 95–102 (2000)
86. K. Hatsusaka, K. Ohta, I. Yamamoto, H. Shirai, Discotic liquid crystals of transition metal complexes, part 30: spontaneous uniform homeotropic alignment of octakis(dialkoxyphenoxy)phthalocyaninatocopper(II) complexes. J. Mater. Chem. **11**, 423–433 (2001)
87. T. Kamei, T. Kato, E. Itoh, K. Ohta, Discotic liquid crystals of transition metal complexes 47: synthesis of phthalocyanine-fullerene dyads showing spontaneous homeotropic alignment. J. Porphyrins Phthalocyanines **16**, 1261–1275 (2012)
88. M. Ariyoshi, M. Sugibayashi-Kajita, A. Suzuki-Ichihara, T. Kato, T. Kamei, E. Itoh, K. Ohta, Discotic liquid crystals of transition metal complexes 44: synthesis of hexaphenoxy-substituted phthalocyanine derivatives showing spontaneous perfect homeotropic alignment. J. Porphyrins Phthalocyanines **16**, 1114–1123 (2012)
89. S. Tunel, H.A.J. Banimuslem, M. Durmus, A.G. Gurek, V. Ahsen, T.V. Basova, A.K. Hassan, Liquid crystalline octasubstiuted lead (II) phthalocyanines: effects of alkoxy and alkylthio substituents on film alignment and electrical properties. New J. Chem. **36**, 1665–1672 (2012)
90. W. Pisula, Z. Tomovic, B.E. Hamaoui, M.D. Watson, T. Pakula, K. Mullen, Control of the homeotropic order of discotic hexa-peri-hexabenzocoronene. Adv. Funct. Mater. **15**, 893–904 (2005)
91. C. Liu, A. Fechtenkotter, M.D. Watson, K. Mullen, A.J. Bard, Room temperature discotic liquid crystalline thin films of hexa-peri-hexabenzocoronene: synthesis and optoelectronic properties. Chem. Mater. **15**, 124–130 (2003)
92. E. Grelet, H. Bock, Control of the orientation of thin open supported columnar liquid crystal films by the kinetics of growth. Europhys. Lett. **73**, 712–718 (2006)
93. E. Charlet, E. Grelet, P. Brettes, H. Bock, H. Saadaoui, L. Cisse, P. Destruel, N. Gheradi, I. Seguy, Ultrathin films of homeotropically aligned columnar liquid crystals on indium tin oxide electrodes. Appl. Phys. Lett. **92**(024107), 1–3 (2008)
94. T. Brunet, O. Thiebaut, E. Charlet, H. Bock, J. Kelber, E. Grelet, Anchoring transition in confined discotic columnar liquid crystal films. Eur. Phys. Lett. **93**(16004), 1–6 (2011)
95. L. Cisse, P. Destruel, S. Archambeau, I. Seguy, P. Jolinat, H. Bock, E. Grelet, Measurement of the exciton diffusion length in discotic columnar liquid crystals: comparison between homeotropically oriented and non-oriented samples. Chem. Phys. Lett. **476**, 89–91 (2009)
96. E. Grelet, S. Dardel, H. Bock, M. Goldmann, E. Lacaze, F. Nallet, Morphology of open films of discotic hexagonal columnar liquid crystals as probed by grazing incidence X-ray diffraction. Eur. Phys. J. E **31**, 343–349 (2010)
97. S. Archambeau, I. Seguy, P. Jolinat, J. farenc, P. Destruel, T.P. Nguyen, H. Bock, E. Grelet, Stabilization of discotic liquid organic thin films by ITO surface treatment. Appl. Surf. Sci. **253**, 2078–2086 (2006)
98. O. Thiebaut, H. Bock, E. Grelet, Face-on oriented bilayer of two discotic columnar liquid crystals for organic donor-acceptor heterojunction. J. Am. Chem. Soc. **132**, 6886–6887 (2010)

99. J. Wang, Z. He, Y. Zhang, H. Zhao, C. Zhang, X. Kong, L. Mu, C. Liang, The driving force for homeotropic alignment of a triphenylene derivative in a hexagonal columnar mesophase on single substrate. Thin Solid Films **518**, 1973–1979 (2010)
100. J. Li, Z. He, H. Zhao, H. Gopee, X. Kong, M. Xu, X. An, X. Jing, A.N. Cammidge, Homeotropic alignment through charge transfer-induced columnar mesophase formation in an unsymmetricallly substituted triphenylene derivative. Pure Appl. Chem. **11**, 1993–2003 (2010)
101. G. Zucci, P. Viville, B. Donnio, A. Vlad, S. Melinte, M. Mondeshki, R. Graf, H.W. Spiess, Y.H. Geerts, R. Lazzaroni, Miscibility between differently shaped mesogens: structural and morphological study of a phthalocyanine-perylene binary system. J. Phys. Chem. B **113**, 5448–5457 (2009)
102. G. Schweicher, G. Gbabode, F. Quist, O. Debever, N. Dumont, S. Sergeyev, Y.H. Geerts, Homeotropic and planar alignment of discotic liquid crystals: the role of the columnar phase. Chem. Mater. **21**, 5867–5874 (2009)
103. H.K. Bisoyi, S. Kumar, First examples of monodisperse discotic liquid crystal pentamers: synthesis and mesomorphism. Tetrahedron Lett. **49**, 3628–3631 (2008)
104. I. Cour, Z. Pan, L.T. Lebruin, M.A. Case, M. Furis, R.L. Headrick, Selective orientation of discotic films by interface nucleation. Org. Electron. **13**, 419–424 (2012)
105. N. Terasawa, H. Monobe, K. Kiyohara, Y. Shimizu, Strong tendency towards homeotropic alignment in a hexagonal columnar mesophase of fluoroalkylated triphenylenes. Chem. Commun. 1678–1679 (2003)
106. N. Terasawa, N. Tanigaki, H. Monobe, K. Kiyohara, Alignment behavior for novel triphenylene compounds possessing fluroalkylated side chains on modified substrates. J. Fluorine Chem. **127**, 1096–1104 (2006)
107. X. Zhou, S. Kang, S. Kumar, Q. Li, Self-assembly of discotic liquid crystal porphyrin into more controllable ordered nanostructure mediated by fluorophobic effect. Liq. Cryst. **36**, 267–274 (2009)
108. Q. Sun, L. Dai, X. Xiaoli, L. Li, Q. Li, Bilayer- and bulk-heterojunction solar cells using liquid crystalline porphyrins as donors by solution processing. Appl. Phys. Lett. **91**, 253505 (2007)
109. J. Hoogboom, P.M.L. Garcia, M.B.J. Otten, J.A.A.W. Elemans, J. Sly, S.V. Lazarenko, T. Rasing, A.E. Rowan, R.J.M. Nolte, Tunable command layers for liquid crystal alignment. Angew. Chem. Int. Ed. **51**, 7990–7993 (2012)
110. W. Zheng, C.-Y. Chiang, I. Underwood, Hybrid molecular orientation of sandwich-type structure discogen hexaalkoxydibenzo[a, c]phenazine on the surfaces modified using silane self-assembled monolayers. Mol. Cryst. Liq. Cryst. **540**, 94–101 (2011)
111. W. Zheng, C.-Y. Chiang, Y.-T. Hu, C.W. Ong, Effect of surface free energy on orientational assembly of hexaazatrinaphthalene-based discotic mesogens in sandwich-type structure. Jpn. J. Appl. Phys. **50**, 021701-1-5 (2011)
112. W. Zheng, Y.-T. Hu, C.-Y. Chiang, C.W. Ong, Orientational packing of a confined discotic mesogens in the columnar phase. Int. J. Mol. Sci. **11**, 943–955 (2010)
113. V.D. Cupere, J. Tant, P. Viville, R. Lazzaroni, W. Osikowicz, W.R. Salaneck, Y.H. Geerts, Effect of interface on the alignment of a discotic liquid crystalline phthalocyanine. Langmuir **22**, 7798–7806 (2006)
114. Z.H. Al-Lawati, R.J. Bushby, S.D. Evans, Alignment of a columnar hexagonal discotic liquid crystal on self-assembled monolayers. J. Phys. Chem. C **117**, 7533–7539 (2013)
115. S. Sergeyev, J. Levin, J.-Y. Balandier, E. Pouzet, Y.H. Geerts, Homeotropic alignment of a mesogenic phthalocyanine depends on the nature of interactions with the surface. Mendeleev Commun. **19**, 185–186 (2009)
116. T. Kajitani, Y. Suna, A. Kosaka, T. Osawa, S. Fujikawa, M. Takata, T. Fukushima, T. Aida, o-Phenylene octamers as surface modifiers for homeotropic columnar ordering of discotic liquid crystals. J. Am. Chem. Soc. **135**, 14564–14567 (2013)

117. D. Miyajima, K. Tashiro, F. Araoka, H. Takezoe, J. Kim, K. Kato, M. Takata, T. Aida, Liquid crystalline corannulene responsive to electric field. J. Am. Chem. Soc. **131**, 44–45 (2009)
118. D. Miyajima, F. Araoka, H. Takezoe, J. Kim, K. Kato, M. Takata, T. Aida, Electric-field-responsive handle for large-area orientation of discotic liquid-crystalline molecules in millimeter-thick films. Angew. Chem. Int. Ed. **50**, 7865–7869 (2011)
119. V. Duzhko, K.D. Singer, Self-assembled fibers of a discotic phthalocyanine derivative: internal structure, tailoring of geometry and alignment by a direct current electric field. J. Phys. Chem. C **111**, 27–31 (2007)
120. W. Wang, X. Liu, J. Pu, Electric-field response of discotic hexabenzocoronene(HBC) liquid crystals. Molecules **16**, 9101–9108 (2011)
121. H. Monobe, K. Awazu, Y. Shimizu, Change of liquid-crystal domains by vibrational excitation for a columnar mesophase. Adv. Mater. **12**, 1495–1499 (2000)
122. H. Monobe, K. Awazu, Y. Shimizu, Alignment control of a columnar liquid crystal for a uniformly homeotropic domain using circularly polarized infrared irradiation. Adv. Mater. **18**, 607–610 (2006)
123. H. Monobe, Y. Shimizu, Anisotropic photoconduction of triphenylene-based DLC in aligned domains by wavelength tunable CO_2 laser irradiation. Mol. Cryst. Liq. Cryst. **542**, 151–157 (2011)
124. H. Monobe, K. Kiyohara, M. Heya, K. Awazu, Y. Shimizu, An infrared technique for alignment control of discotic liquid crystals: a possible fabrication technology for organic micro/nano electronic devices. Mol. Cryst. Liq. Cryst. **397**, 59–65 (2003)
125. H. Monobe, K. Kiyohara, N. Terasawa, M. Heya, K. Awazu, Y. Shimizu, Infrared photoinduced alignment change for triphenylene-based columnar liquid crystals by using free electron laser. Thin Solid Films **438–439**, 418–422 (2003)
126. H. Monobe, H. Hori, M. Heya, K. Awazu, Y. Shimizu, Homeotropic alignment change for discotics in plastic columnar mesophase by infrared irradiation. Thin Solid Films **499**, 259–262 (2006)
127. H. Monobe, K. Awazu, Y. Shimizu, Alignment change of hexahexyloxythiotriphenylene in the helical columnar phase by infrared laser irradiation. Thin Solid Films **518**, 762–766 (2009)
128. M. Cavallini, A. Calo, P. Stoliar, J.C. Kengne, S. Martins, F.C. Matacotta, F. Quist, G. Gbabode, N. Dumont, Y.H. Geerts, F. Biscarini, Lithographic alignment of discotc liquid crystals: a new time-temperature integrating framework. Adv. Mater. **21**, 4688–4691 (2009)
129. E. Pouzet, V.D. Cupere, C. Heintz, J.W. Andereasen, D.W. Breiby, M.M. Nielsen, P. Viville, R. Lazzaroni, G. Gbabode, Y.H. Geerts, Homeotropic alignment of a discotic liquid crystal induced by a sacrificial layer. J. Phys. Chem. C **113**, 14398–14406 (2009)
130. P. Samori, X. Yin, N. Tchebotareva, Z. Wang, T. Pakula, F. Jackel, M.D. Watson, A. Venturini, K. Mullen, J.P. Rabe, Self-assembly of electron donor-acceptor dyads into ordered architectures in two and three dimensions: surface patterning and columnar "double cables". J. Am. Chem. Soc. **126**, 3567–3575 (2004)
131. X. Qiu, C. Wang, Q. Zeng, B. Xu, S. Yin, H. Wang, S. Xu, C. Bai, Alkane-assisted adsorption and assembly of phthalocyanines and porphyrins. J. Am. Chem. Soc. **122**, 5550–5556 (2000)
132. N. Katsonis, A. Marchenko, D. Fichou, Substrate-induced pairing in 2,3,6,7,10,11-hexakis-undecalkoxy-triphenylene self-assembled monolayers on Au (111). J. Am. Chem. Soc. **125**, 13682–13683 (2003)
133. J.-C. Gabriel, N.B. Larsen, M. Larsen, N. Harrit, J.S. Pedersen, K. Schaumburg, K. Bechgaard, Ordering of the disk-like 2,3,6,7,10,11-hexakis(hexylthio)triphenylene in solution and at a liquid-solid interface. Langmuir **12**, 1690–1692 (1996)
134. F. Charra, J. Cousty, Surface-induced chirality in a self-assembled monolayer of discotic liquid crystal. Phys. Rev. Lett. **80**, 1682–1685 (1998)

135. P. Samori, N. Severin, C.D. Simpson, K. Mullen, J.P. Rabe, Epitaxial composite layers of electron donors and acceptors from very large polycyclic aromatic hydrocarbon. J. Am. Chem. Soc. **124**, 9454–9457 (2002)
136. R. Friellein, X. Crispin, C.D. Simpson, M.D. Watson, F. Jackel, W. Osikowicz, S. Marciniak, M.P. de Jong, P. Samori, S.K.M. Jonsson, M. Fahlman, K. Mullen, J.P. Rabe, W.R. Salaneck, Electronic structure of highly ordered films of self-assembled graphitic nanocolumns. Phys. Rev. B **68**, 195414 1-7 (2003)
137. J. Kim, N. Yamasaki, T. Hayashi, H. Yoshida, H. Moritake, A. Fujii, Y. Shimizu, M. Ozaki, High-quality planar alignment of discotic liquid crystals using oscillating shear. Appl. Phys. Express **6**(061702), 1–3 (2013)
138. D. Goldfarb, Z. Luz, H. Zimmermann, A deuterium NMR study of the discotic mesophase of hexa-hexyloxytriphenylene. J. Phys. **42**, 1303–1311 (1981)
139. W. Kranig, C. Boeffel, H.W. Spiess, Deuterium nuclear magnetic resonance studies of molecular motions and alignment process of discotic liquid-crystalline compounds based on substituted triphenylenes. Macromolecules **23**, 4061–4067 (1990)
140. I.O. Shklyarevskiy, P. Jonkeejim, N. Stutzmann, D. Wasserberg, H.J. Wondergem, P.C.M. Christianen, A.P.H.J. Schenning, D.M. de Leeuw, Z. Tomovic, J. Wu, K. Mullen, J.C. Mann, High anisotropy of the field effect transistor mobility in magnetically aligned discotic liquid crystalline semiconductors. J. Am. Chem. Soc. **127**, 16233–16237 (2005)
141. J.-H. Lee, S.-M. Choi, B.D. Pate, M.H. Chisholm, Y.-S. Han, Magnetic uniaxial alignment of the columnar superstructure of discotic metallomesogens over the centimeter length scale. J. Mater. Chem. **16**, 2785–2791 (2006)
142. J.-H. Lee, H.-S. Kim, B.D. Pate, S.-M. Choi, Magnetic alignment of discotic liquid crystals on substrates. Phys. B **385–386**, 798–800 (2006)
143. H.-S. Kim, S.-M. Choi, B.D. Pate, P.G. Park, Effect of film thickness on the columnar packing structures of discotic supramolecules in thin films. ChemPhysChem **10**, 2642–2646 (2009)
144. H.-S. Kim, S.-M. Choi, J.-H. Lee, P. Busch, S.J. Koza, E.A. Verploegen, B.D. Pate, Uniaxially oriented, highly ordered, large area columnar superstructures of discotic supramolecules using magnetic field and surface interactions. Adv. Mater. **20**, 1105–1109 (2008)
145. T.D. Choudhury, N.V.S. Rao, R. Tenent, J. Blackburn, B. Gregg, I.I. Smalyukh, Homeotropic alignment and director structure in thin films of triphenylamine-based discotic liquid crystals controlled by supporting nanostructured substrates and surface confinement. J. Phys. Chem. B **115**, 609–617 (2011)
146. S. Takami, S. Furumi, Y. Shirai, Y. Sakka, Y. Wakayama, Impact of magnetic field on molecular alignment and electrical conductivity in phthalocyanine nanowires. J. Mater. Chem. **22**, 8629–8633 (2012)
147. S. Zimmermann, J.H. Wendorff, C. Weder, Uniaxial orientation of columnar discotic liquid crystals. Chem. Mater. **14**, 2218–2223 (2002)
148. A.M. van de Craats, N. Stutzmann, O. Bunk, M.M. Nielsen, M. Watson, K. Mullen, H.D. Chanzy, H. Sirringhaus, R.H. Friend, Meso-epitaxial solution-growth of self-organizing discotic liquid-crystalline semiconductors. Adv. Mater. **15**, 495–499 (2003)
149. E. Charlet, E. Grelet, Anisotropic light absorption, refractive indices and orientational order parameter of unidirectionally aligned columnar liquid crystal films. Phys. Rev. E **78**, 041707-1-8 (2008)
150. M. Funahashi, A. Sonoda, High electron mobility in a columnar phase of liquid-crystalline perylene tetracarboxylic bisimide bearing oligosiloxane chains. J. Mater. Chem. **22**, 25190–25197 (2012)
151. R.I. Gearba, D.V. Anokhin, A.I. Bondar, W. Bras, M. Jahr, M. Lehmann, D.A. Ivanov, Homeotropic alignment of columnar liquid crystals in open films by means of surface nanopatterning. Adv. Mater. **19**, 815–820 (2007)
152. Y. Mindyuk, P.A. Heiney, Structural studies of Langmuir films of disc-shaped molecules. Adv. Mater. **11**, 341–344 (1999)

153. T. Bjornholm, T. Hassenkam, N. Reitzel, Supramolecular organization of highly conducting organic thin films by the Langmuir-Blodgett technique. J. Mater. Chem. **9**, 1975–1990 (1999)
154. H. Eichhorn, D.W. Bruce, D. Wohrle, Amphitropic mesomorphic phthalocyanines—a new approach to highly ordered layers. Adv. Mater. **10**, 419–422 (1998)
155. O. Karthaus, H. Ringsdorf, V.V. Tsukruk, J.H. Wendorff, Columnar ordering of liquid-crystalline discotics in Langmuir-Blodget films. Langmuir **8**, 2279–2283 (1992)
156. S. Kubowicz, U. Pietsch, M.D. Watson, N. Tchebotareva, K. Mullen, A.F. Thunemann, Thin layers of columns of an amphiphilic hexa-peri-hexabenzocoronene at silicon wafer surface. Langmuir **19**, 5036–5041 (2003)
157. N. Reitzel, T. Hassenkam, K. Balashev, T.R. Jensen, P.B. Howes, K. Kjaer, A. Fechtenkotter, N. Tchebottareva, S. Ito, K. Mullen, T. Bjornholm, Langmuir and Langmuir-Blodget films of amphiphilic hexa-peri-hexabenzocoronene: new phase transitions and electronic properties by pressure. Chem. Eur. J. **7**, 4894–4901 (2001)
158. D. Janietz, R.C. Ahuja, D. Mobius, Langmuir monolayers of sheet-shaped multialkyne amphiphiles. Langmuir **13**, 305–309 (1997)
159. A. Julita, D. Janietz, J. Reiche, H. Lemmetyinen, Photophysical investigation of Langmuir-Blodget films of amphiphilic discotic penta-alkynes. Thin Solid Films **268**, 121–129 (1995)
160. J. Reiche, R. Dietel, D. Janietz, H. Lemmetyinen, L. Brehmer, Edge-on Langmuir-Blodget multilayers derived from disc-shaped multiyne mesogens. Thin Solid Films **226**, 265–269 (1992)
161. A. Angelova, J. Reiche, R. Ionov, D. Janietz, L. Brehmer, Control of the structure of Langmuir-Blodget films of a discotic liquid crystalline compound via the subphase composition and the adjacent molecular environment. Thin Solid Films **242**, 289–294 (1994)
162. M. Ahmida, S. Dufour, H.-S. Li, H. Kayal, R. Schmidt, C.E. DeWolf, S. Holger Eichhorn, Face- and edge-on orientations of octa-acid and –alcohol substituted tetraazaporphyrins in Langmuir and Langmuir-Blodgett monolayers. Soft Matter **9**, 811–819 (2013)
163. N. Boden, R.J. Bushby, P.S. Martin, S.D. Evans, R.W. Owens, D.A. Smith, Triphenylene-based discotic liquid crystals as self-assembled monolayers. Langmuir **15**, 3790–3797 (1999)
164. H. Schonherr, F.J.B. Kremer, S. Kumar, J.A. Rego, H. Wolf, H. Ringsdorf, M. Jaschke, H.-J. Butt, E. Bamberg, Self-assembled monolayers of discotic liquid crystalline thioethers, discoid disulphides and thiols on gold: molecular engineering of ordered surfaces. J. Am. Chem. Soc. **118**, 13051–13057 (1996)
165. M. Duati, C. Grave, N. Tcbeborateva, J. Wu, K. Mullen, A. Shaporenko, M. Zharnikov, J.K. Kriebel, G.M. Whitesides, M.A. Rampi, Electron transport across hexa-*peri*-hexabenzocoronene units in a metal-self-assembled monolayer-metal junction. Adv. Mater. **18**, 329–333 (2006)
166. D. Kafer, A. Basir, X. Dou, G. Witte, K. Mullen, C. Woll, Evidence for band-like transport in graphene-based organic monolayers. Adv. Mater. **22**, 384–388 (2010)
167. L. Piot, C. Marie, X. Dou, X. Feng, K. Mullen, D. Fichou, Growth of long, highly stable, and densely packed worm-like nanocolumns of hexa-peri-hexabenzocoronenes via chemisorptions on Au(111). J. Am. Chem. Soc. **131**, 1378–1379 (2009)
168. L. Piot, C. Marie, X. Feng, K. Mullen, D. Fichou, Hierarchical self-assembly of edge-on nanocolumnar superstructures of large disc-like molecules. Adv. Mater. **20**, 3854–3858 (2008)
169. M. Mansueto, S. Sauer, M. Butschies, M. Kaller, A. Baro, R. Woerner, N.H. Hansen, G. Tovar, J. Pflaum, S. Laschat, Triphenylene silanes for direct surface anchoring in binary mixed self-assembled monolayers. Langmuir **28**, 8399–8407 (2012)
170. A. Tracz, J.K. Jeszka, M.D. Watson, W. Pisula, K. Mullen, T. Pakula, Uniaxial alignment of the columnar super-structure of a hexa(alkyl) hexa-peri-hexabenzocoronene on untreated glass by simple solution processing. J. Am. Chem. Soc. **125**, 1682–1683 (2003)

171. W. Pisula, Z. Tomovic, M. Stepputat, U. Kolb, T. Pakula, K. Mullen, Uniaxial alignment of polycyclic aromatic hydrocarbons by solution processing. Chem. Mater. **17**, 2641–2647 (2005)
172. D.W. Breiby, O. Bunk, W. Pisula, T.I. Solling, A. Tracz, T. Pakula, K. Mullen, M.M. Nielsen, Structure of zone-cast HBC-$C_{12}H_{25}$ films. J. Am. Chem. Soc. **127**, 11288–11293 (2005)
173. W. Pisula, A. Menon, M. Stepputat, I. Lieberwirth, U. Kolb, A. Tracz, H. Sirringhaus, T. Pakula, K. Mullen, A zone-casting technique for device fabrication of field-effect transistors based discotic hexa-per-hexabenzocoronene. Adv. Mater. **17**, 684–689 (2005)
174. A. Tracz, T. Makowski, S. Masirek, W. Pisula, Y.H. Geerts, Macroscopically aligned films of discotic phthalocyanine by zone casting. Nanotechnology **18**(485303), 1–5 (2007)
175. C.-Y. Liu, A.J. Bard, In-situ regrowth and purification by zone melting of organic single crystal thin films yielding significantly enhanced optoelectronic properties. Chem. Mater. **12**, 2353–2362 (2000)
176. W. Pisula, M. Kastler, D. Wasserfallen, T. Pakula, K. Mullen, Exceptionally long-range self-assembly of hexa-peri-hexabenzocoronene with dove tailed alkyl substituents. J. Am. Chem. Soc. **126**, 8074–8075 (2004)
177. S. Xiao, M. Myers, Q. Miao, S. Sanaur, K. Pang, M.L. Steigerwald, C. Nuckolls, Molecular wires from contorted aromatic compounds. Angew. Chem. Int. Ed. **44**, 7390–7394 (2005)
178. S. Xiao, J. Tang, T. Beetz, X. Guo, N. Tremblay, T. Siegrist, Y. Zhu, M. Stegerwald, C. Nuckolls, Transferring self-assembled, nanoscale cables into electrical devices. J. Am. Chem. Soc. **128**, 10700–107001 (2006)
179. M. Kastler, W. Pisula, D. Wasserfallen, T. Pakula, K. Mullen, Influence of alkyl substituents on the solution- and surface-organization of hexa-*peri*-hexabenzocoronenes. J. Am. Chem. Soc. **127**, 4286–4296 (2005)
180. G.D. Luca, A. Liscio, F. Nolde, L.M. Scolaro, V. Palermo, K. Mullen, P. Samori, Self-assembly of discotic molecules into mesoscopic crystals by solvent-vapour annealing. Soft Matter **4**, 2064–2070 (2008)
181. A. Cristadoro, G. Lieser, H.J. Rader, K. Mullen, Field-force alignment of disc-type pi systems. ChemPhysChem **8**, 586–591 (2007)
182. A. Calo, P. Stoliar, M. Cavallini, S. Sergeyev, Y.H. Geerts, F. Biscarini, Monolayer control of discotic liquid crystal by electromigration of dewetted layers in thin film devices. J. Am. Chem. Soc. **130**, 11953–11958 (2008)
183. J.P. Bramble, D.J. Tate, D.J. Revill, K.H. Sheikh, J.R. Henderson, F. Liu, X. Zeng, G. Ungar, R.J. Bushby, S.D. Evans, Planar alignment of columnar discotic liquid crystals by isotropic phase dewetting on chemically patterned surfaces. Adv. Funct. Mater. **20**, 914–920 (2010)
184. H.J. Rader, A. Rouhanipour, A.M. Talarico, V. Palermo, P. Samori, K. Mullen, Processing of giant graphene molecules by soft-landing mass spectrometry. Nat. Mater. **5**, 276–280 (2006)
185. G. Gbabode, N. Dumont, F. Quist, G. Schweicher, A. Moser, P. Viville, R. Lazzaroni, Y.H. Geerts, Substrate-induced crystal plastic phase of a discotic liquid crystals. Adv. Mater. **24**, 658–662 (2012)
186. M. Steinhart, S. Zimmermann, P. Goring, A.K. Schaper, U. Gosele, C. Weder, J.H. Wendorff, Liquid crystalline nanowires in porous alumina: geometric confinement versus influence of pore walls. Nano Lett. **5**, 429–434 (2005)
187. M. Steinhart, S. Murano, A.K. Schaper, T. Ogawa, M. Tsuji, U. Gosele, C. Weder, J.H. Wendorff, Morphology of polymer/liquid-crystal nanotubes: influence of confinement. Adv. Funct. Mater. **15**, 1656–1664 (2005)
188. W. Pisula, M. Kastler, D. Wasserfallen, R.J. Davies, M.-C. Garcia-Gutierrez, K. Mullen, From macro- to nanoscopic templating with nanographene. J. Am. Chem. Soc. **128**, 14424–14425 (2006)
189. H. Duran, B. Hartmann-Azanza, M. Steinhart, D. Gehrig, F. Laquai, X. Feng, K. Mullen, H.-J. Butt, G. Floudas, Arrays of aligned supramolecular wires by macroscopic orientation of columnar discotic mesophases. ACS Nano **6**, 9359–9365 (2012)

190. P.-O. Mouthuy, S. Melinte, Y.H. Geerts, A.M. Jonas, Uniaxial alignment of nanoconfined columnar mesophase. Nano Lett. **7**, 2627–2632 (2007)
191. P.-O. Mouthuy, S. Melinte, Y.H. Geerts, B. Nysten, A.M. Jonas, Nanocontrolled bending of discotic columns by spiral networks. Small **4**, 728–732 (2008)
192. C.V. Cerclier, M. Ndao, R. Busselez, R. Lefort, E. Grelet, P. Huber, A.V. Kityk, L. Noirez, A. Schonhals, D. Morineau, Structure and phase behavior of a discotic columnar liquid crystal confined in nanochannels. J. Phys. Chem. C **116**, 18990–18998 (2012)
193. C.-Y. Chiang, I. Underwood, W. Zheng, Controlling orientation direction of discotic columns assembled in microtrenches. Mol. Cryst. Liq. Cryst. **553**, 185–192 (2012)
194. J. Cattle, P. Bao, J.P. Bramble, R.J. Bushby, S.D. Evans, J.E. Lydon, D.J. Tate, Controlled planar alignment of discotic liquid crystals in microchannels made using SU8 photoresist. Adv. Funct. Mater. **23**, 5997–6006 (2013)

Chapter 8
Discotic Liquid Crystalline Blends for Nano-Structure Formation Toward Bulk Heterojunction Active Layer in Organic Photovoltaics

Yo Shimizu

Abstract Liquid crystalline semiconductor is a new category of organic semiconductors and lots of studies on the materials and application to electronic devices have been carried out. In particular, organic thin film photovoltaics is an interesting research field by use of liquid crystalline semiconductors because of the characteristic properties of liquid crystals. In this chapter, the miscibility in liquid crystals and the related issues on the application of liquid crystalline semiconductors are overviewed to give an insight for their possible contribution to the drastic improvement of organic photovoltaics technology including the fabrication by printed electronics.

8.1 Introduction

Research in organic electronics is now directing to a new era of electronic devices aiming the production of flexible thin film devices that could be fabricated by roll-to-roll process on polymer substrates. In this way, making electronically active parts on polymer substrate could be realized by use of printing techniques to give functionally monolithic devices. This means that all materials for the active-part making should be resolved or dispersed into solvents and/or binders for the preparation of "inks". If one considers a possibility of semiconducting inks of which semiconducting solute is an organic compound, it should be facilitated with some important characteristic properties such as solubility into solvents and self-organizing nature for the spontaneous processing of high quality of thin films in addition to the charged carrier mobility as semiconductor. In the self-organizing nature, phase separation is included as emerging in spontaneous ways.

Y. Shimizu (✉)
Research Institute for Ubiquitous Energy Devices, National Institute of Advanced Industrial Science and Technology, Kansai Center (AIST-Kansai), Osaka 563-8577, Japan
e-mail: yo-shimizu@aist.go.jp

Q. Li (ed.), *Nanoscience with Liquid Crystals*, NanoScience and Technology,
DOI: 10.1007/978-3-319-04867-3_8,

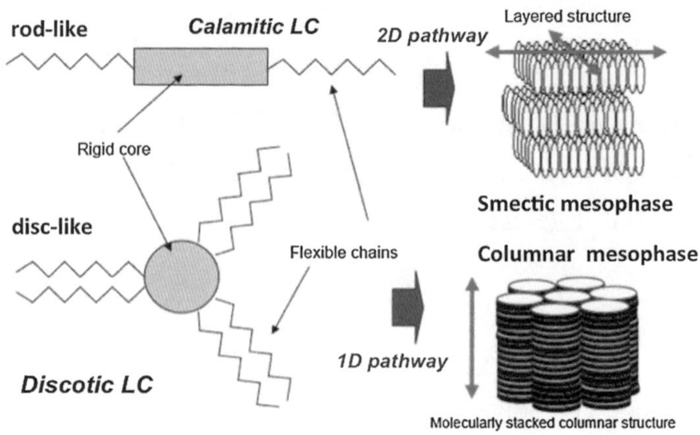

Fig. 8.1 Typical structures of mesogenic molecules for calamitic and discotic liquid crystals. A central core is made of aromatics and the flexible parts are of, typically, alkyl chains

In these points of view, characteristic properties of liquid crystalline materials are certainly related to the specifications as a solute for semiconductor inks. The relatively high solubility into common organic solvents is a gain coming from the characteristic chemical structure of mesogenic molecules represented by the combination of rigid π-electronic conjugation systems as a core part and insulating chains as a flexible part (Fig. 8.1). In particular, it is noteworthy to remind that on a chemical modification of chain parts, one can introduce a certain level of solubility even into aqueous solvents. Moreover, this flexible chain is an origin of self-organizing nature in liquid crystals. The strong self-organizing nature of liquid crystalline molecules would lead to easy formation of the relatively larger size of domains where the molecules uniformly align in one way and this could be controlled by surface modifications on substrate and by some external fields such as electric and magnetic field. These imply the mesogenic characters in organic semiconductors could contribute to the lower cost fabrication of thin film devices in electronics and photonics [1–3]. In recent studies on the formation of molecularly aligned films by way of the phase transition from mesophase to crystalline solid phase, it has been revealed that relatively larger size of mono-domain films could be more easily formed at the mesophase temperature range and even after the crystallization on cooling, the domain tends to be maintained to give a good performance as field effect transistor (FET) [4–6].

Liquid crystal is a state of matter where the molecules are so mobile and a certain anisotropy in motion is involved depending on the inter- and intramolecular forces. This is an origin of softness and sometimes of fluidity of materials. As electronic materials such as semiconductors, one has to consider that the efficient charge transport in ordered molecular systems is ascribed to periodical arrangement of molecules giving a certain intermolecular interaction for strongly coupled electronic states. When the transfer integral of HOMO-HOMO and/or

LUMO/LUMO gets larger to some extent, even such periodic systems of dynamic molecular aggregation could provide a high mobility of charged carriers in electronic manner and it could be even though thermal motions of molecules surely disturb such periodical electronic chain for charge migration. In fact, some experimental evidences surely let us know it could be comparable to that of amorphous silicon ($\sim 10^{-1}$ cm^2 V^{-1} s^{-1}) [7, 8]. In macroscopic point of view, such a molecular arrangement often has defect and/or dislocation as a definite boundary at which electronic pathways mismatch and this is typically seen in ordered solid. Elastic properties of liquid crystals, however, enable us to consider that a "self-healing" of domain boundaries is attained with a gradient change of the director and it would be no longer a definite boundary for charge migration, indicating it could give a high reliability as charge transport materials [9].

On a strong self-organizing nature of liquid crystals, spontaneous formation of nano-structure in molecular aggregation is also an important issue in research as an origin of superfunctionality in molecular materials. Recent studies on how to build-up such nano-structure in organic materials have shown the importance of application of liquid crystalline field. Molecular dynamics (MD) simulation has been applied to this issue [10] and the wider variation in experimental approaches has been seen [11] in recent years.

On the other hand, "*miscibility*" is, in one sense, one of the characteristic properties in liquid crystals. "*Miscibility*" in liquid crystals is a well-known macroscopic property where one can see two mesogens exhibiting a thermodynamically identical liquid crystalline phase are mixed to show its phase at arbitrary component ratio. This has already been applied to liquid crystals for LCDs to control some properties such as temperature range of nematic phase. A diversity of functional properties such as temperature range, anisotropic electrical permittivity, viscosity etc. can be controlled in nematic blends and non-mesogenic molecules also can be a component which contributes to the resultant properties as they behave like a solute in liquid solution. However, charge transport property has not yet been well studied in terms of molecular blends with liquid crystalline materials, while thin film organic photovoltaics have been so extensively studied in recent years as molecular blends.

In this research field, polymeric semiconductors have been so often used because of easy fabrication of thin film as well as high efficiency in charge transport. However, small molecules have drawn much more attention in recent years and one can study periodical nano-structures in relation to the active layer performance. In this sense, such nano-structures should be well controlled and also formed in spontaneous ways. Therefore, self-organization ability would surely be an important character that is furnished in liquid crystals.

Here in this chapter, studies on molecular blends of liquid crystalline materials, especially discotic liquid crystalline blends as semiconductor are overviewed in terms of both macroscopic miscibility and nanoscopic phase separation. Discotic liquid crystal has been extensively studied as one-dimentional semiconducting materials in recent years [12–14]. In particular, organic thin film photovoltaics is an interesting research field of its application. We see blended molecular systems

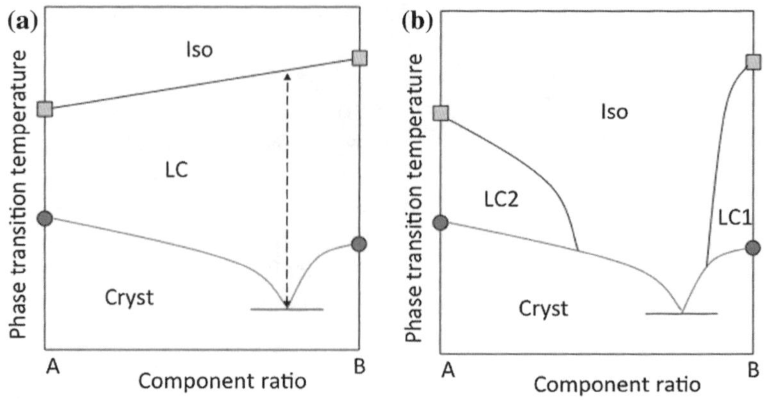

Fig. 8.2 Typical phase diagrams of liquid crystalline (LC) binary mixtures without any additional interaction induced by mixing. **a** Compounds, A and B are miscible in the LC phase when these exhibit the identical LC phase which comes up at any component ratio of two mesogens and the linear correlation is seen for the isotropization temperature. The dotted line means the extended range of temperature for the LC phase at the eutectic component. **b** The case of "immiscible" is shown. This can take place when two mesogens do not have the identical LC phase to each (LC1 and LC2)

of discotic liquid crystal as a good candidature for building-up well-controlled bulk heterojunction (BHJ) active layer where nano-sized structures are furnished to give the better performance in photo-carrier generation and exciton diffusion as well as charge transport.

8.2 Miscibility and Phase Separation in Liquid Crystals

Miscibility of liquid crystal is a well-known phenomenon and this is sometimes a good tool for tuning the properties as functional materials. When one has two mesogenic compounds that exhibit an identical mesophase to each, but in the different range of temperature, sometimes one can see the identical mesophase in the mixture at any component ratio. The temperature range of liquid crystalline phase is variant depending on the difference of phase transition temperatures of those and eutectic behavior of the melting point. A typical and simple binary phase diagram is shown in Fig. 8.2a. One can attain widening temperature range and the maximum is at the eutectic mixture. Thus, on mixing of mesogenic compounds toward functional solid, we may have a solid solution if no sign of eutectic drop of melting point.

However, some limitations are considered to realize this miscibility. The molecular size including molecular anisotropy in shape and intermolecular specific interaction that is weak in one compound system, but may be induced and/or enhanced by mixing, are typical factors for affecting miscibility. When two

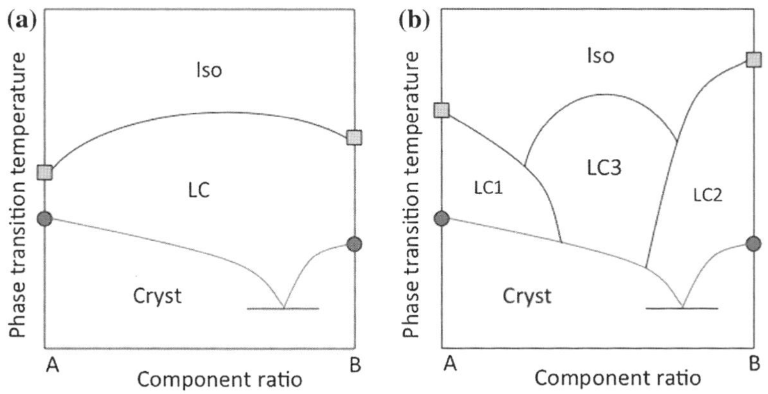

Fig. 8.3 Typical phase diagrams of LC binary mixtures with a certain intermolecular interaction induced by mixing. **a** The case that two mesogens have the identical LC phase to each with thermal enhancement of LC phase which is due to a stoichiometric specific intermolecular interaction. **b** The case that two mesogens have the different type of LC phase (LC1 and LC2) with generation of a new LC phase (LC3)

mesogenic compounds which show the different type of mesophases to each, "miscibility" does not come up (Fig. 8.2b). The phase transition temperature of isotropization goes down as the counterpart compound is added. Addition of the excess amount of one compound leads to disappearance of the mesophase. On the other hand, when two mesogenic molecules have a strong intermolecular interaction such as charge transfer interaction, for the mesogens of which liquid crystalline phase is identical to each, the thermal enhancement is expected as simply drawn in Fig. 8.3a and for the mesogens with different mesophase, occasionally new phase arises as shown in Fig. 8.3b.

In the case of discotic liquid crystal, we have more difficulty in emerging "miscibility" in comparison to the case of calamitic ones. In the homologues of discotic liquid crystals, a difference in alkyl chain length is sometimes a serious problem. In the case where the difference in chain length is, for example, only four methylene groups and the columnar arrangement is not the same type, miscibility of mesophase is reasonably lost. However, when the difference in alkyl chain length gets larger, the identical columnar mesophase is exhibited by two alkyl homologues, it is not always miscible, while miscibility is observed in the case that the smaller difference in chain length. Furthermore, the immiscibility is sometimes more remarkable in the difference in the linkage groups between the core and chains and sometimes a thermal enhancement of mesophase is observed due to intermolecular specific interaction [15]. For the binary system of two discogens with a different type of core part, immiscibility is essentially promoted.

However, "immiscibility" in these facts does not always mean that phase separation takes place. As seen in Fig. 8.2b, one can see the melting point of binary mixture even if immiscible for the mesophase. When two compounds are immiscible and thus, show phase separation in the isotropic liquid phase, for

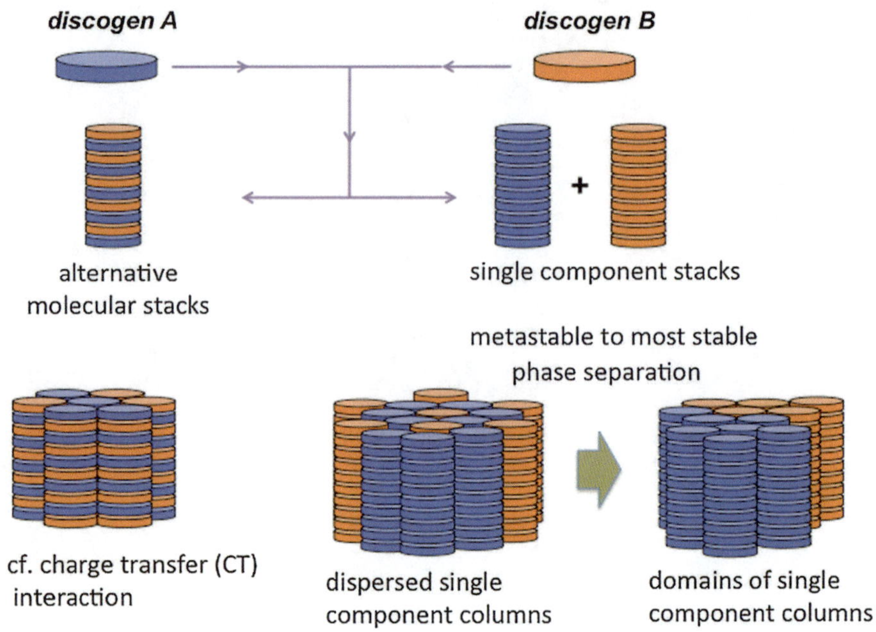

Fig. 8.4 Two cases of miscible columnar blends. The segregation of single component columns is a metastable state and the macroscopic domain formation is promoted as phase separation

crystals formed on cooling, we should observe two melting points corresponding to the melt of single component domains. In a typical case of melting, we find a eutectic point at which eutectic crystals melt to the fluid state of mixture. At other ratio of the mixture, one observes a melting of the eutectic crystal and the following melting of the excess amount of one component. Therefore, a biphasic region in temperature is seen between these two melting points. The complete phase separation to give macroscopic domains results in independent phase transitions derived from the domains with a single compound.

Therefore, miscibility in discotic columnar mesophase is recognized as a result of molecularly dispersed systems. However, we should pay some attention to a state of mixture for two discogens to form the identical columnar mesophase. As a more realistic system, is it a molecularly dispersed or a single-component column dispersed system? This is an essentially important question when we consider the application of discotic liquid crystalline blends as a new tool for organic electronic materials (Fig. 8.4).

Unfortunately, not so many experimental evidences and theoretical studies have been reported so far. Especially, compounds provided for the studies on binary systems as a semiconductor and its related devices do not show so much variation. In the next sections, we overview the research results on discotic liquid crystalline blends for semiconducting materials and for organic thin film photovoltaics.

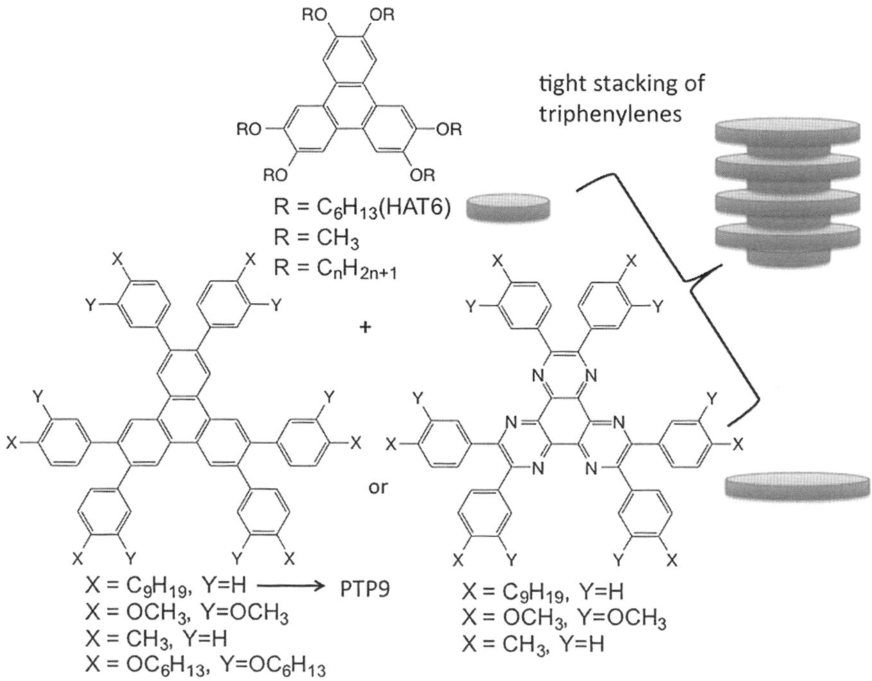

Fig. 8.5 Chemical structures of two types of triphenylene derivatives for Complimentary Polytopic Interaction (CPI) studies toward highly efficient charge transport

8.3 Liquid Crystalline Blends Toward Semiconductors

A well-studied binary system of discotic liquid crystal toward high carrier mobility systems is a blend of two types of triphenylene-based compounds (Fig. 8.5). These two compounds were mixed in expectation to give a new semiconducting system that exhibits enhancement of charge transport efficiency. The concept is based on Complimentary Polytopic Interaction (CPI) [16–18].

These two types of mesogenic compounds were mixed at 1:1 molar ratio to give an enhancement of clearing temperature, resulting in the extended range of temperature for the mesophase [16]. For the binary system of HAT6/PTP9 and HAT11/PTP9, the thermal stability of Col_{ho} mesphase is increased. The carrier mobility was measured by Time-Of-Flight (TOF) method to reveal a certain increase of hole mobility [19]. HAT6 and HAT11 exhibit hole mobility of 7.1×10^{-4} and 1.1×10^{-4} cm^2 V^{-1} s^{-1} in the hexagonal columnar (Col_h) mesophase, respectively, as mostly temperature-independent nature [20]. These mixtures still keep a mesomorphic state at 1:1 component ratio for a certain range of temperature and the mesophase does not show crystallization on cooling to give a glassy state with the liquid crystalline order. The hole mobility was enhanced up to 2.1×10^{-2} cm^2 V^{-1} s^{-1} in the mesophase and could be determined for the glassy state to be 1.6×10^{-2} cm^2 V^{-1} s^{-1}. In this binary system, homeotropic alignment is easily attained as the single compound does show in the Col_h phase and the resultant electrical conductivity is also increased [21]. The larger phase transition enthalpy (ΔH) for the clearing and a more number of diffraction peaks in XRD studies indicate the relatively higher ordering takes place in the newly formed columnar mesophase. Consequently, it is shown that a complimentary stacking system for column formation is a good strategy to achieve the enhancement of carrier mobility in mesophase. Probably, this would be good for design of solid-state columns. More interesting thing reported is a possibility to give "alloy band formation" in this binary system [22]. The origin of mobility enhancement was discussed to give a conclusion that it is partly due to more stable (more rigid) columnar structure formed by mixing and partly due to efficient band formation for electronic transport by mixing to give an alloy-like state in solid. This is a view as columns behave as solidified linear chains. In addition, the hole mobility higher than 1 cm^2 V^{-1} s^1 was suggested as an attainable value based on theoretical treatments for molecularly tight stacking in columns. The enhancement of carrier mobility was also theoretically analyzed to give an interpretation on a homogeneously distributed single-trap model [23], and this may imply two types of molecules are homogeneously distributed along the columnar axis. No experimental data was reported on the HOMO and LUMO levels for the mixed mesogenic compounds. However, those seem to be in the comparable levels considering their chemical structures.

It was also shown that CPI is a useful concept for controlling mesogenic temperature range as a result of increasing clearing temperature, inducing a new mesophase, and improved alignment behavior in this combination of mesogens [24]. The fluorine atom-introduced HAT6 resulting in induced lateral dipole shows a columnar mesophase below room temperature and the mixture with PTP9 also such a wide range of mesophase temperature with increase of mobility [25]. This also indicates that the lateral dipole could be considered for improving charge transport efficiency even for such a binary system.

Although the HATn/PTP9 system exhibits the highest thermal stability of Col_{ho} mesophase as a composite at 1:1 molar ratio, this composite is not miscible with each compound in the mesophase. This indicates the columnar structure with an alternative stacking of two types of molecule is so stable (rigid) and then such rigid columns align in an ordered way to give, probably a plastic columnar mesophase. However, there must be a eutectic mixture showing the lower limitation of melting temperature, when one considers a binary system of columnar discogens without a CPI concept.

When one considers use of polymerizable mesogens to fix the liquid crystalline order as a solid film, miscibility between the semiconducting and polymerizable mesogens is also important. Unfortunately, we do not know any results that polymerization of liquid crystals for solid state semiconductor is successful to enhance the carrier mobility, except for the case of discotic nematic phase [26]. In the case of a mixture of columnar mesogen, HAT5 (melting at 65 °C and clearing at 121 °C) and a photo-polymerizable calamitic mesogen showing nematic phase

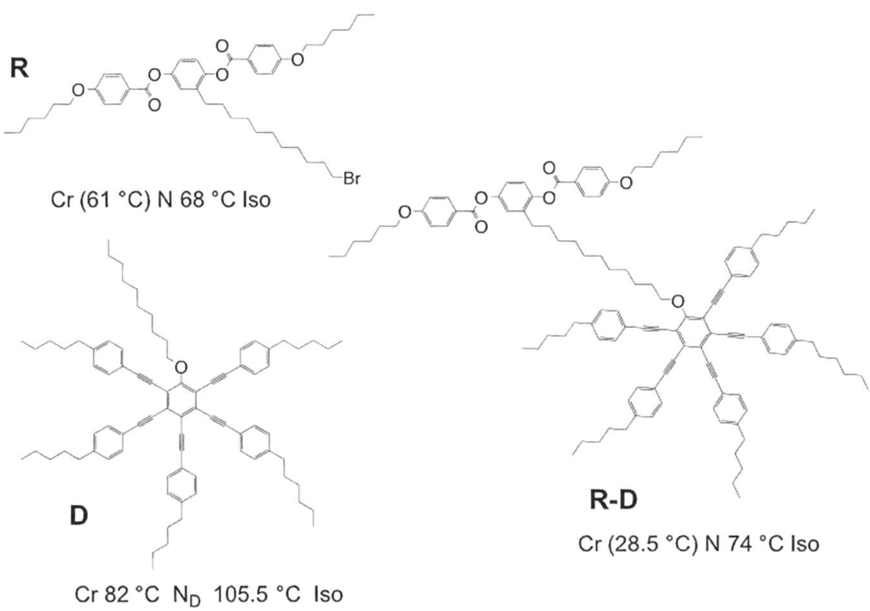

Fig. 8.6 A rod-like and disc-like molecules (R and D, respectively) show a miscibility in nematic phase via buffering of order by a rod-disc connected molecule (R-D)

between 65 and 127 °C, a eutectic behavior is seen at the 3:7 composition. These are reasonably immiscible [27]. This is a normal behavior as a binary mesogenic system and indicates that phase separation would be preferred in the mixture of discotic and calamitic mesogens,. However, when one adopts a specified concept to attain miscibility between discotic (disc-shaped) and calamitic (rod-shaped) mesogens, it is realized.

Two nematogens, one of which shows discotic nematic phase as a disc-shaped mesogenic molecule and the other nematic phase as a rod-shaped mesogenic molecule were studied on the miscibility. In a moment, at least two concepts seem to be good for realizing the miscibility in mesophase between discotic and calamitic liquid crystals. One is to add the third mesogen of which chemical structure is a combination of rod-shaped and disc-shaped moieties (Fig. 8.6) [28]. In this case, the compound R-D with the combination of disc- and rod-shaped moieties exhibits a monotropic nematic phase and the rod-like and disc-like mesogens (R and D, respectively) show a monotropic nematic and an enantiotropic discotic nematic phases, respectively. It is reasonable to see that R and D are never miscible. However, addition of R-D to the 1:1 mixture of R and D (with no mesophase) leads to the appearance of nematic phase in a wide range of component ratio and it was interpreted as shape *amphiphilicity*. Nematic phase is the lowest ordered liquid crystalline phase. Therefore, it is reasonably imagined that columnar mesophase should have more difficulty to realize miscibility in disc- and rod-shaped mixtures.

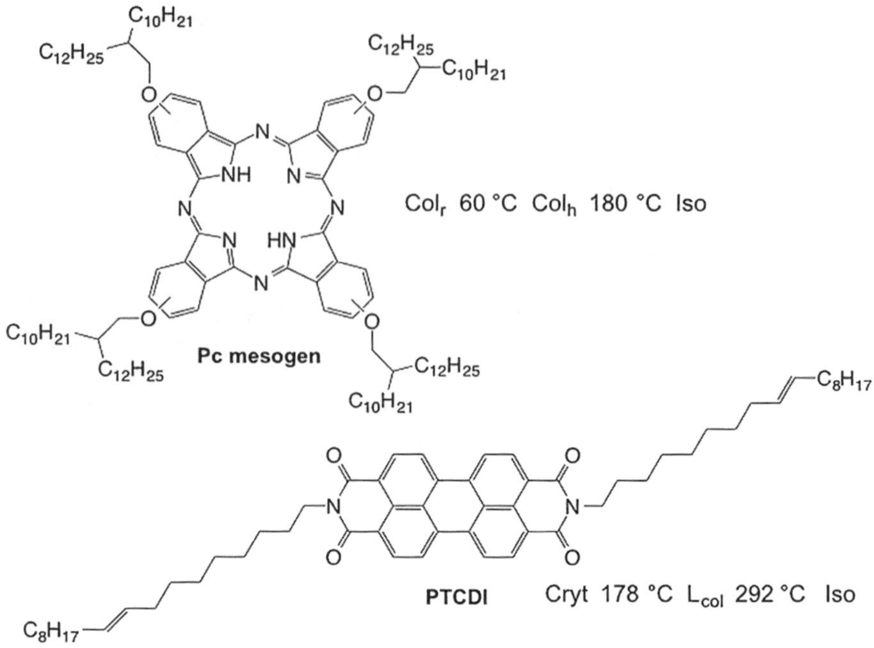

Fig. 8.7 Two mesogens exhibiting remarkable miscibility in a columnar mesophase. The mixture shows enhanced Col_h mesophase

Another concept describes, however, that it would be possible in the combination of disc- and lath-shaped mesogens (Fig. 8.7) [29]. The mixture of a phthalocyanine (Pc) discogen and lath-shaped mesogenic perylenetetracarboxidiimide (PTCDI) exhibits remarkable miscibility in the Col_h mesophase. The Col_h mesophase was stabilized with an increase of clearing temperature by about 100 K at 75 % of the Pc mesogen. It is noteworthy to remind that phthalocyanine is a well-known p-type of semiconductor and PTCDI is of typical n-type. This relation of electronic property for these compounds makes us imagine a possibility of charge transfer interaction. If two types of molecules are involved into the formation of columnar structure by molecular stacking, the resultant columns would be thermally stabilized due to its strong interaction along the columnar axis. Indeed, the contaminant PTCDI mesogens affect the dynamics of molecules in the columnar phase and the higher ordering is attained [30]. The experimental results indicate the intercalation of PTCDI molecules into the Pc stacking columns takes place.

The binary systems composed of p- and n-type of semiconductors is just a content of the active layer in organic thin film solar cells and it could work as a bulk heterojunction if those form carrier transport path for hole and electron with phase separation and the charges are generated at the interface of p- and n-types of semiconductor domains. However, when both p- and n-types of molecules are

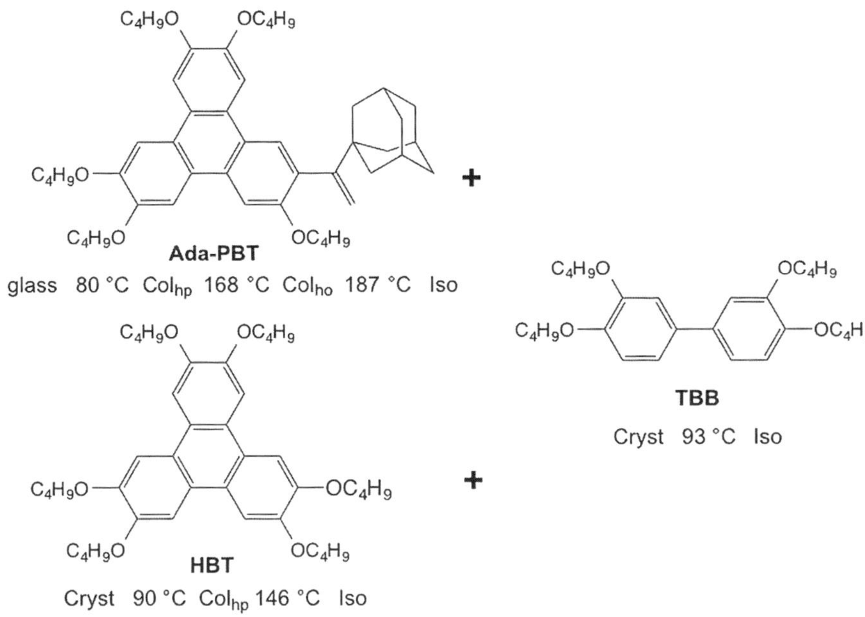

Fig. 8.8 Two binary systems examined for mesomorphism with Ada-PBT+TBB and HAT4+TBB

involved in a columnar structure, a single column would not work as a one-dimensional path for charge migration along the columnar axis, due to the inhomogeneity of electronic state along the columnar axis (act as trap sites for hopping charges). Therefore, the carrier mobility, especially the drift mobility that can be determined by TOF method, would be smaller than those of single component columnar phase. In order to obtain the more efficient transport of charged carriers, the p- and n-types of semiconductors have to molecularly stack independently for making columns as segregated between p- and n-types of molecules.

On the other hand, eutectic crystal is also an interesting state of binary system as electronic materials. Especially it is important when one considers a crystalline thin film is used as active layer which is fabricated by way of liquid crystalline phases. A binary system with a discotic mesogen exhibiting a plastic columnar mesophase and non-disc-shaped tetrabutyloxybiphenyl (TBB) were studied on their phase morphology and phase separation (Fig. 8.8) [31]. Asymmetrical triphenylene mesogen with an adamantyl group at a peripheral position (Ada-PBT) exhibits a plastic hexagonal columnar (Col_{hp}) and hexagonal ordered columnar (Col_{ho}) mesophases. The addition of TBB into Ada-PBT destabilizes the Col mesophase and phase separation of TBB takes place as its concentration is increased. On the other hand, the mixture of HAT4 and TBB shows a eutectic mixture meaning miscibility in the isotropic phase, though phase separation is caused in the Col mesophases. These indicate a limitation of miscibility relates to solubility in isotropic phase.

These examples of study on binary mixtures with discotic systems show an importance of solubility and molecular orders in mesophase for the resultant structure of mesogenic binary systems.

8.4 Liquid Crystalline Blends in Organic Photovoltaics

Bulk heterojunction active layer in organic thin film solar cells is a typical example of mixture for functional blends. The mixture of p- and n-types of semiconductors have to form an appropriate geometry of two compounds to make effective charge transport paths for hole and electron as well as effective interfaces at which charges are generated with less efficient recombination. In addition, excitons generated by photon abosorption have to migrate for a certain distance to reach to the interface and it also needs an appropriate path.

Most of studies on organic thin film photovoltaics have been focused so far on the combination of semiconducting polymers as p-type and fullerene derivatives as n-type of compounds [32]. New conjugated polymers have been developed so far for aiming the better light harvesting and charge transport properties [33]. Recently, small molecule is also drawn much attention as the material for the active layer [34, 35] and wide variation of chemical structures have been extensively studied [36]. Most of such small molecules, just like the case of polymeric systems, tend to have, more or less, alkyl chains to obtain good solubility into common organic solvents and strong self-organization of aromatic moiety. Nowadays, over 6 % of conversion efficiency was reported for solution-processed small-molecule solar cells [37, 38].

Actually, small molecular system is good for pursuing the best geometry in active layer by controlling nano-structures as the geometry of p- and n-types of semiconductors. The introduction of long chains into the aromatic core part in such a molecule leads to liquid crystallinity as well, though essentially depending on the anisotropic shape of molecules. Strong self-organizing nature of small molecules is significantly important, when one considers that the processing of such nanostructures should be realized on spontaneous formation. In the following, we overview the research results on structural studies of BHJ active layer in organic thin film photovoltaics.

Most of BHJ active layers studied so far for conjugated polymers are not of well-controlled geometry for p-n junction in a layer. Some results have indicated for the systems of alkylated conjugated polymers with fullerene derivatives that fullerene molecules are positioned in forests of alkyl chains as schematically drawn in Fig. 8.9 [39, 40]. In small molecule system, it is reasonably expected that one could obtain the better clearance of nano-scale periodical structures and more detailed information of structure-property relationship [41]. In particular, by use of mesomorphic state, one could control such structures and thus, could obtain more useful information to improve the performance.

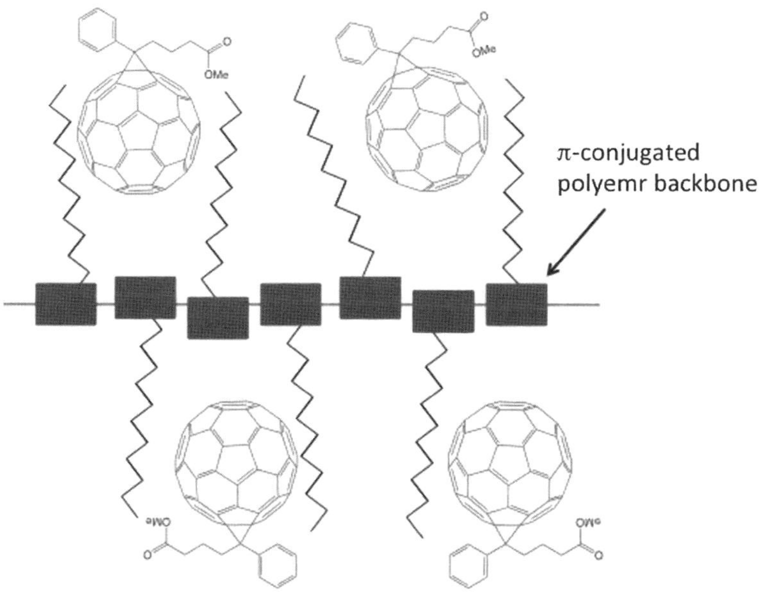

Fig. 8.9 An image of relative positioning of C60 derivatives in the mixture with an alkylated conjugated polymer semiconductor

The better performance requires an appropriate geometry of p-n junction in a layer and it is reported that interdegitated p-n junction is most promising in performance (Fig. 8.10) [42] and further, bicontinuous network is proposed for extracting maximum performance [43]. Fortunately, these two conditions of well-controlled BHJ active layer can be seen in mesopmorphism. The former is an inspiration from columnar mesophase of discotics that tend to spontaneously align in a homeotropic manner best fit to connect two counter electrodes in sandwich-type cells and the latter is bicontinuous cubic mesophase. Recent progress of morphological studies on BHJ active layers has revealed the importance of phase behavior, domain state and size, miscibility and purity, processing conditions for controlled nano-structured active layer, and thin film morphology such as depth profile and 3D-molecular ordering [44].

The first application of discotic liquid crystal to organic thin film solar cells is of a binary blend of liquid crystalline phthalocyanine (Pc) as p-type of semiconductor and non- mesogenic PTCDI derivative as n-type [45], where the active layer was not a BHJ type, but had a p-i-n junction (Fig. 8.11). The Pc layer was fabricated by spin-coating and heated up to the isotropic phase followed by cooling to obtain a spontaneously formed homeotoropic alignment on ITO-coated substrate. The n-type layer was deposited by sublimation on Pc layer and finally the counter Al electrode was deposited in vacuo. The performance was shown simply as short circuit current, $I_{SC} = 0.4$ mA cm^{-2}, open circuit voltage, $V_{OC} = 0.3$ V and external quantum efficiency, EQE ~0.5 %.

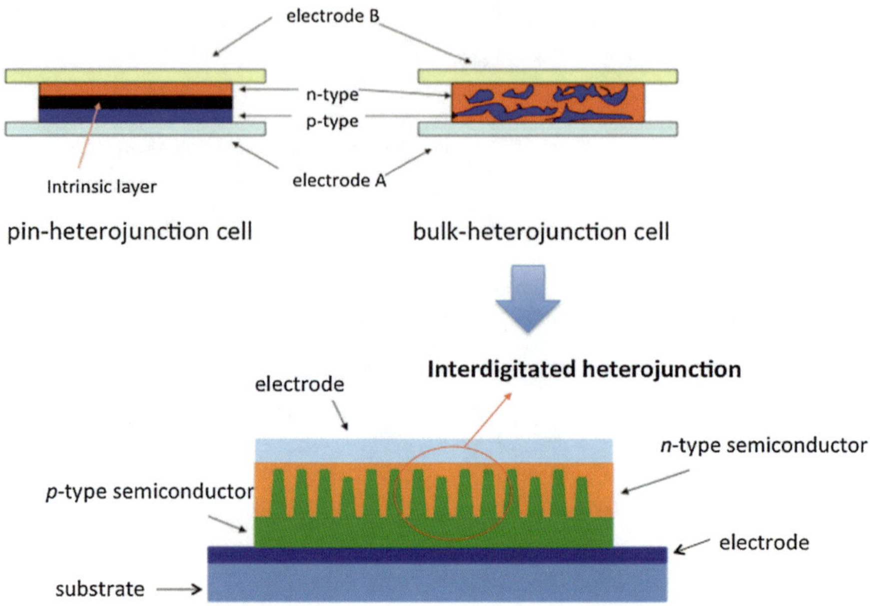

Fig. 8.10 Schematic drawings of interdigitated p-n junction in an active layer which is evolved from non-controlled BHJ layer to gain the better performance

As a fully studied system of which active layer is consist of liquid crystalline semiconductors, a combination of hexabenzocolonene (HBC) mesogen as p-type and non-mesogenic PTCDI as n-type was applied for the first BHJ solar cells (Fig. 8.12a) [46]. The BHJ active layer was fabricated by spin-coating technique to give ~2 % of power conversion efficiency and >34 % of EQE for the 40:60 blend (V_{OC}:0.69 V, I_{SC}:33.5 mA cm^{-2}). The p-n junction geometry in the active layer was not controlled. The molecular alignment in the active layer is also not controlled and the homeotropic alignment of HBC mesogens probably is a factor that can improve the performance.

In the studies on the combination of HBC mesogens and PTCDI derivatives for the BHJ active layer, the external quantum efficiency (EQE) were investigated to clarify the solvent-dependent nature of thin film morphology and of the EQE performance. However, the structural studies of active layer were not shown [47, 48]. Liquid crystallinity is, generally, strongly dependent on the type and length of flexible chains. A series of HBC mesogens was studied on alkyl chain effect on photovoltaic behavior with a PTCDI derivative (Fig. 8.12b). Replacement of the alkyl groups from C_2C_6 to $C_{12}C_{14}$ leads to the drastic decrease of clearing point and get the melting lowered below room temperature. It was revealed that the elongation of chain negatively affects the solar cell performance. It was found that the shortest side chain is better and the HBC with the shortest chain (C_2C_6) shows the best with highest EQE of 12 % [49]. Also for HBC-PhCn,

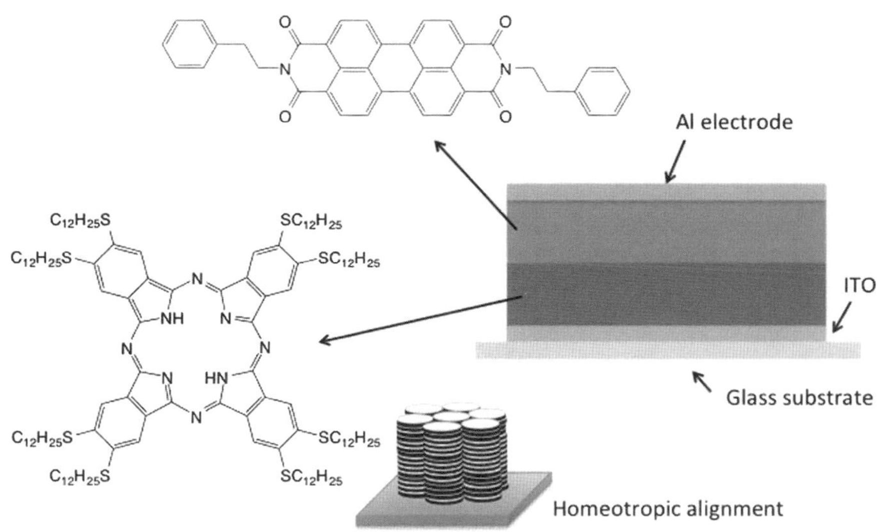

Fig. 8.11 The early study on organic solar cell where active layer contains liquid crystalline phthalocyanine as p-type of semiconductor. This is not a BHJ active layer, but of p-i-n junction type

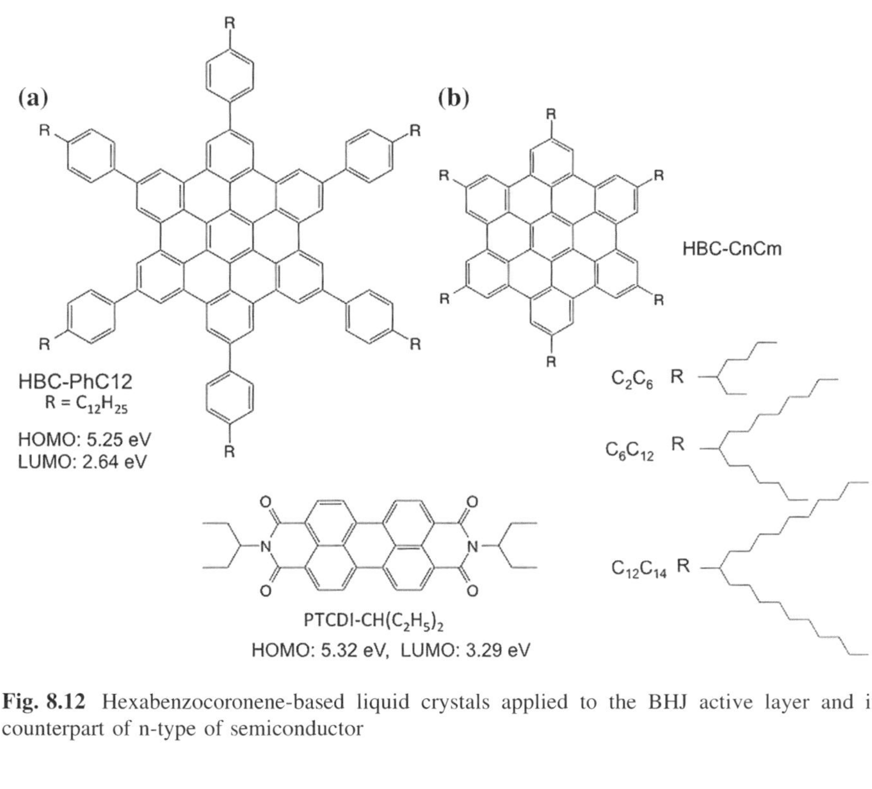

Fig. 8.12 Hexabenzocoronene-based liquid crystals applied to the BHJ active layer and its counterpart of n-type of semiconductor

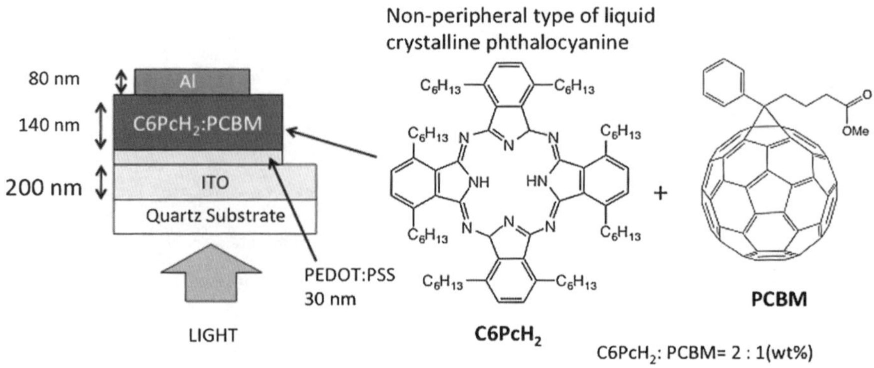

Fig. 8.13 A simple drawing of the BHJ solar cell of which active layer is composed of non-peripheral mesogenic Pc and PCBM and the 60:40 wt% mixture exhibiting the best performance (PCE = 3.1 %)

the effect of alkyl chain length on photovoltaic performance (with a PTCDI derivative) was studied. Photoluminescence quenching, transient photovoltage and photocurrent decay measurements indicate that the shorter homologue allows the better contact to PTCDI molecules and thus gives the better charge separation. AFM analyses of thin film morphology reveal that the larger interfacial area is formed and the higher mobility is attained, while the longer alkyl chains could lead to the less charge recombination in the homeotropic domains [50].

The HBC molecule, however, have its photo-absorption onset, more or less, up to 500 nm and only a part of sunlight in wavelength is covered. Phtalocyanine (Pc), on the other hand, is well known dye molecule and the photo-absorption onset goes to near infrared region depending on the chemical modification. In particular, strong Q band absorption at around 700 nm and strong electron donating property is quite attractive for organic photovoltaic studies. However, not so many Pc mesogens have been applied to photovoltaic study, though a wide variation of mesogenic Pc have been reported and mostly high carrier mobility is observed in their columnar mesophases. In particular, for non-peripheral Pc mesogens (Fig. 8.13), high carrier mobility obtained by TOF method (meaning it is a drift mobility as a macroscopic property) was reported with its ambipolar nature [51–53] and especially the hexyl homologue (C6PcH$_2$) exhibits high hole mobility of >1 cm^2 V^{-1} s^{-1} for the crystalline polydomain films.

A simple BHJ solar cell was fabricated with C6PcH$_2$ and PCBM by spin coating. For a 60:40 wt% blend of C6PcH$_2$ and PCBM, power conversion efficiency (PCE) showed 3.1 % and >70 % of EQE was observed at Q-band region. Studies on the binary mixtures of C6PcH$_2$ and PCBM indicate that the addition of PCBM into C6PcH$_2$ leads to deformation of columnar lattice in the Col mesophase with decrease of isotropization temperature. At ca. 25 mol % of PCBM, one can see the eutectic point where the Col mesophase disappears. This molar ratio implies in the crystal blend that chain-like arrangement of PCBM is positioned

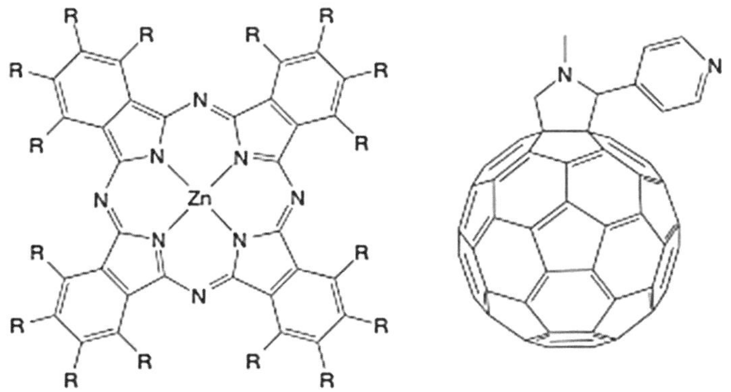

Fig. 8.14 Phthalocyanine derivatives with the different absorption band to examine the multi-component system of BHJ active layer. The n-type semiconductor is a fullerene derivative (Py-C60)

among the columns of C6PcH_2. Interestingly, the hole mobility determined by TOF technique is the highest one for the eutectic solid [54].

When C6PcH_2 was added into the well-studied BHJ photovoltaic system, P3HT/PCBM binary blends, the observed performance is improved [55], indicating the better light harvesting is attained and C6PcH_2 domains form charge separation interfaces and resultant charges efficiently move to P3HT and PCBM. This effect of additives on solar cell performance was reported also for a triphenylene discogen to P3HT/PCBM binary blends [56]. Another alkyl homologues were examined as additives to reveal that the size of micron domains of added Pc mesogens is changed depending on the chain length [57]. The alkyl chain dependence (C6, C7, C9, C10) was also studied to give slight change of the performance [58]. Small amount of dihalogenoalkanes was added to promote the micro-scale phase separation and the performance is raised up to 4.2 % [59, 60]. It was also reported that the added HAT4 to P3HT/PCBM BHJ layer could contribute to the performance as an improver for charge transport efficiency [61].

In order to improve the light harvesting ability of BHJ active layer, three Pc compounds were mixed with a fullerene derivative (Py-C60) [62]. The chemical structures of these Pcs show variant combination of the attached groups such as *t*-Butyl, alkylthio, H and F (Fig. 8.14) to give a variation of wavelength region for photo-absorption. A certain improvement could be attained by such a mixing, though the PCE is still in the lower level around 0.1 %. A concept of multi-blended semiconductors in organics is quite interesting, but at the same time, it is so challenging works to set up the methodology to control the crystal morphology. Mesogenic compounds, however, might be useful in this case.

On the other hand, porphyrin is an analogous compound to Pcs. Therefore, porphyrins have also strong absorption bands in both UV and visible region such as Soret and Q bands. Recently, mesogenic porphyrins were synthesized for thin

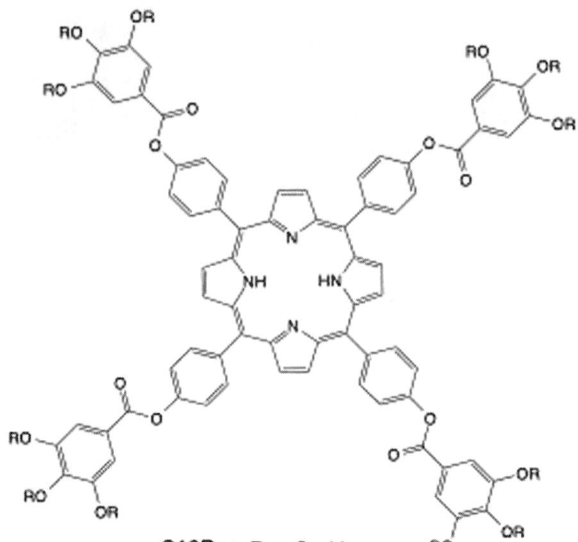

Fig. 8.15 Tetraphenylporphyrin derivatives showing Col_h mesohase

film photovoltaic studies and their photovoltaic characteristics were reported (Fig. 8.15) [63]. The cells with BHJ type active layer were investigated on their performance. Two alkyl homologues C12Por and C14Por were used and PCBM (or C60) was mixed as the counter n-type of semiconductor (1:1 ratio). Chain length dependence on the performance was seen and the cell with C12Por shows higher conversion efficiency (0.222 %) than C14Por (0.155 %). Thermal annealing of the cells surely leads to improvement of cell performance. In comparison to the as-prepared cells with C12Por and C14Por, the PCE after thermal annealing goes up to 0.712 and 0.775 %, respectively. Both porphyrin compounds in the active layers are in a mesophase at room temperature (Col_{ho} phase) and thermal annealing (sequentially carried out) directly affect the I_{SC}, indicating the better phase separation takes place during the thermal treatment accompanied with re-alignment of columns. In this binary system, strong tendency for homeotropic alignment is also observed [64].

8.5 Summary

Simple overviews of organic thin film photovoltaics involving liquid crystalline semiconductors (not polymeric systems) have shown that some results are derived from unique properties of liquid crystals. Thermal treatment to make the as-prepared molecular aggregation relaxed to the more stable state is, in particular, effective by way of mesophase. Most of publications show evidences of surface morphology of the active layer, for example, by AFM measurements. However, there we can see almost no discussion about molecular level issues. If one can study on thin film of organics, it is reasonably expected that small molecular systems enable us to see nano-structure of the film that depends on stoichiometry of the system in addition to its surface morphology. Moreover, one can pursue a thermodynamically most stable state of the active layer. Therefore, studies on small molecular systems would make it possible to clearly see the structure-property relation in organic thin film photovoltaics. In this point of view, mixing molecules could be a methodology for obtaining a blend to exhibit the best performance as desired to the materials. Mixing of molecules can be designed by use of self-organizing nature of molecules, typically liquid crystallinity. One has to learn more about miscibility and phase separation in nano-to-macro scale of matters.

In liquid crystalline state, miscibility is sometimes observed and this is a technique to identify the mesophase of one compound in the combination with compounds of which mesomorphism is established. Molecularly dispersed mixture of liquid crystals is a result of miscibility. However, this macroscopic phenomenon might include short-range order in a dynamical state. For discotic columnar liquid crystals, a binary system might be able to form 2-dimentional columnar arrangement with columns made of one type of molecules in the mixture of the different types of molecules. Even if this is a thermodynamically metastable state, one might be able to control the size and ordering of columns for the better geometry in functions. As shown in miscibility of discotics, we need more information to make a new concept for generating a novel molecular system for electronic applications.

Recent studies on binary mixtures of discotic liquid crystals have shown that a binary mixture of a non-peripheral hexaalkylphthalocyanine ($C10PcH_2$), and its Zn complex exhibits miscibility for the Col_h mesophase. The Col_h mesophase of both mesogens is identical because the clearing temperature shows a linear correlation on component ratio and no sign of eutectic point for the melting may indicate the identical crystalline structure [65] (Fig. 8.16). Furthermore, preliminary results indicate that the phase separation takes place to give separated domains of two mesogens because the observed carrier mobility gets recovered to the original level ($\sim$0.1 $cm^2\ V^{-1}\ s^{-1}$) by thermal treatment, though the as-prepared mixture shows the decreased mobility [66]. This would be an implication to the interdegitated p-n junction in a BHJ active layer that has the better light harvesting ability

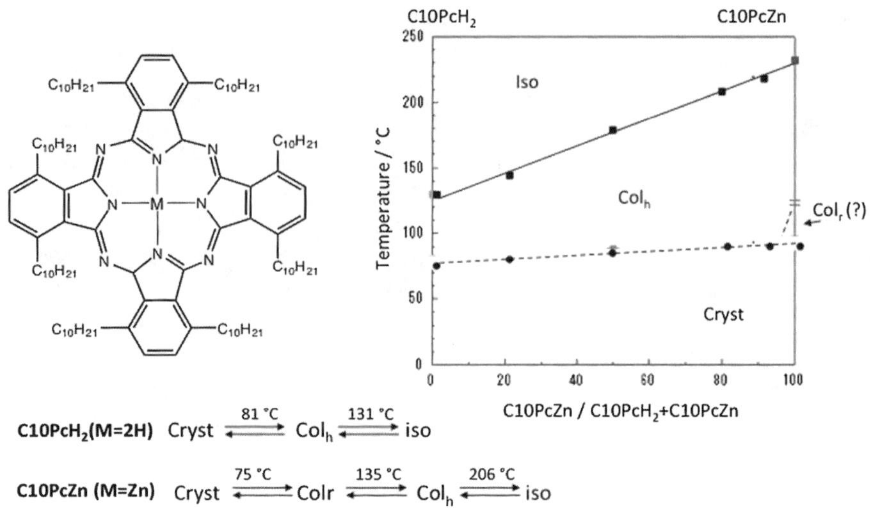

Fig. 8.16 Two phthalocyanine mesogens for the binary mixtures and the phase diagram where one can see a complete miscibility in the Col_h mesophase. However, this is a macroscopic property

which is realized by mixing of some dye semiconducting molecules in conjunction with their liquid crystallinity.

In recent years, synthetic strategies of p-n dyad molecules have shown attractive systems and some of them exhibit mesomorphism [67–73]. These also would be interesting candidatures for the active layer in organic photovoltaics. However, simple mixing would be the best method to give a highly functional blend, if mixing of variety of molecules by use of liquid crystalline field can successfully produce a variety of electronic materials under the morphology control.

Finally, the application of liquid crystals would not be limited to photovoltaics by new methodologies of molecular mixing [74, 75].

References

1. M.O. O'Neill, S.M. Kelly, Ordered materials for organic electronic and photonics. Adv. Mater. **23**, 566–584 (2011)
2. M. Funahashi, Development of liquid-crystalline semiconductors with high carrier mobilities and their application to thin-film transistors. Polym. J. **41**, 459–469 (2009)
3. Y. Shimizu, K. Oikawa, K. Nakayama, D. Guillon, Mesophase semiconductors in field effect transistors. J. Mater. Chem. **17**, 4223–4229 (2007)
4. H. Iino, J. Hanna, Liquid crystalline thin films as a precursor for polycrystalline thin films aimed at field effect transistors. J. Appl. Phys. **109**, 074505-1-5 (2011)
5. H. Iino, J. Hanna, Availability of liquid crystallinity in solution processing for polycrystalline thin films. Adv. Mater. **23**, 1748–1751 (2011)

6. H. Iino, J. Hanna, Availability of liquid crystalline molecules for polycrystalline organic semiconductor thin films. Jpn. J. Appl. Phys. **45**, L867–L870 (2006)
7. D. Adam, P. Schuhmacher, J. Simmerer, L. Häussling, K. Siemensmeyer, K.H. Etzbacj, H. Ringsdorf, D. Haarer, Fast photoconduction in the highly ordered columnar phase of a discotic liquid crystal. Nature **371**, 141–143 (1994)
8. M. Funahashi, J. Hanna, High carrier mobility up to 0.1 $cm^2\ V^{-1}\ s^{-1}$ at ambient temperatures in thiophene-based smectic liquid crystals. Adv. Mater. **17**, 594–598 (2005)
9. H. Maeda, M. Funahashi, J. Hanna, Effect of domain boundary on carrier transport of calamitic liquid crystalline photoconductive materials. Mol. Cryst. Liq. Cryst. **346**, 183–192 (2000)
10. M. Yoneya, Toward rational design of complex nanostructured liquid crystals. Chem. Rec. **11**, 66–76 (2011)
11. T. Kato, N. Mizoshita, K. Kishimoto, Functional liquid-crystalline assemblies: self-organized soft materials. Angew. Chem. Int. Ed. **45**, 38–68 (2006)
12. W. Pisula, M. Zorn, J.-Y. Chang, K. Müllen, R. Zentel, Liquid crystalline ordering and charge transport in semiconducting materials. Macromol. Rapid Commun. **30**, 1179–1202 (2009)
13. S. Laschat, A. Baro, N. Steinke, F. Giesselmann, C. Hägele, G. Scalia, R. Judele, E. Kapatsina, S. Sauer, A. Scheivogel, M. Tosini, Discotic liquid crystals: from tailor-made synthesis to plastic electronics. Angew. Chem. Int. Ed. **46**, 4832–4887 (2007)
14. S. Sergeyev, W. Pisula, Y.H. Geerts, Discotic liquid crystals: a new generation of organic semiconductors. Chem. Soc. Rev. **36**, 1902–1929 (2007)
15. C. Destrade, N.-H. Tinh, H. Gasparoux, J. Malthete, A.M. Levelut, A. Disk-Like, Mesogens: a classification. Mol. Cryst. Liq. Cryst. **71**, 111–135 (1981)
16. E.O. Arkainen, N. Boden, R.J. Bushby, O.R. Lozman, J.G. Vinter, A. Wood, Complimentary polytopic interaction. Angew. Chem. Int. Ed. **39**, 2333–2336 (2000)
17. J.G. Vinter, Extended electron distributions applied to the molecular mechanics of some intermolecular interactions. J. Comput.-Aided Mol. Des. **8**, 653–668 (1994)
18. C.A. Hunter, Aromatic-aromatic interactions: electrostatic or charge transfer? Angew. Chem. Int. Ed. Engl. **32**, 1584–1586 (1993)
19. T. Kreouzis, K. Scott, K.J. Donnovan, N. Boden, R.J. Bushby, O.R. Lozman, Q. Liu, Enhanced electronic transport properties in complimentary binary discotic liquid crystal systems. Chem. Phys. **262**, 489–497 (2000)
20. N. Boden, R.J. Bushby, J. Clements, B. Movaghar, K.J. Donovan, T. Kreouzis, Mechanism of charge transport in discotic liquid crystals. Phys. Rev. B **52**, 13274–13280 (1995)
21. R.J. Bushby, S.D. Evance, O.R. Lozman, A. McNeill, B. Movaghar, Enhanced charge conduction in discotic liquid crystals. J. Mater. Chem. **11**, 1982–1984 (2001)
22. B.R. Wegewijs, L.D.A. Siebbeles, N. Boden, R.J. Busgby, B. Movaghar, O.R. Lozman, Q. Liu, A. Pecchia, L. A. Mason, Charge-carrier mobilities in binary mixtures of discotic triphenylene derivatives as a function of temperature. Phy. Rev. B **65**, 245112-1-8 (2002)
23. A. Pecchia, O.R. Lozman, B. Movaghar, N. Boden, R.J. Bushby, Photoconductive transients and one-dimentional charge carrier dynamics in discotic liquid crystals. Phys. Rev. B **65**, 104204-1-10 (2002)
24. N. Boden, R.J. Bushby, G. Cooke, O.R. Lozman, Z. Lu, CPI: a recipe for improving applicable properties of discotic liquid crystals. J. Am. Chem. Soc. **123**, 7915–7916 (2001)
25. R.J. Bushby, K.J. Donovan, T. Kreouzis, O.R. Lozman, Molecular engineering of tripheynlene-based disotic liquid crystal conductors. Opto-Electron. Rev. **13**, 269–279 (2005)
26. M. Inoue, H. Monobe, M. Ukon, V.F. Petrov, T. Watanabe, A. Kumano, Y. Shimizu, Fast charged carrier mobility of a triphenylene-based polymer film possessing nematic order. Opto-Electron. Rev. **13**, 303–308 (2005)
27. T.-M. Huang, R.V. Talroze, T. Kyu, Eutectic mesophase transitions and induced crystalline phase in mixtures of hexagonal columnar liquid crystal and mesogenic diacrylate. J. Phys. Chem. B **114**, 13031–13041 (2010)

28. R.W. Date, D.W. Bruce, Shape amphiphiles: mixing rods and disks in liquid crystals. J. Am. Chem. Soc. **125**, 9012–9013 (2003)
29. G. Zucchi, B. Donnio, Y.H. Geerts, Remarkable miscibility between disk- and lathlike mesogens. Chem. Mater. **17**, 4273–4277 (2005)
30. G. Zucchi, P. Viville, B. Donnio, A. Vlad, S. Melinte, M. Mondeshki, R. Graf, H.W. Spiess, Y.H. Geerts, R. Lazzaroni, Miscibility between differently shaped mesogens: structural and morphological study of a phthalocyanine-perylene binary system. J. Phys. Chem. B **113**, 5448–5457 (2009)
31. B. Brandl, J.H. Wendorff, Eutectic mixtures with plastic columnar discotics: molecular structure, phase morphology and kinetics of phase separation. Liq. Cryst. **32**, 553–563 (2005)
32. A. Facchetti, π-Conjugated polymers for organic electronics and photovoltaic cell applications. Chem. Mater. **23**, 733–758 (2011)
33. Y. Liang, L. Yu, A new class of semiconducting polymers for bulk heterojunction solar cells with exceptionally high performance. Acc. Chem. Res. **43**, 1227–1236 (2010)
34. B. Walker, C. Kim, T.-Q. Nguyen, Small molecule solution-processed bulk heterojunction solar cells. Chem. Mater. **23**, 470 (2010)
35. A.W. Hains, Z. Laing, M.A. Woodhouse, B.A. Gregg, Molecular semiconductors in organic photovoltaic cells. Chem. Rev. **110**, 6689–6735 (2010)
36. A. Mishra, P. Bäuerle, Small molecule organic semiconductors on the move: promises for future solar energy technology. Angew. Chem. Int. Ed. **51**, 2020–2067 (2010)
37. Y. Sun, G.C. Welch, W.-L. Leong, C.J. Takacs, G.C. Bazan, A.J. Heeger, Solution-processed small-molecule solar cells with 6.7 % efficiency. Nat. Mater. **11**, 44–48 (2012)
38. Z. Li, G. He, X. Wan, Y. Liu, J. Zhou, G. Long, Y. Zuo, M. Zhang, Y. Chen, Solution processable rhodanine-based small molecule organic photovoltaic cells with a power conversion efficiency of 6.1 %. Adv. Energy Mater. **2**, 74–77 (2012)
39. C. Müller, J. Bergqvist, K. Vandewal, K. Tvingstedt, A.S. Anselmo, R. Magnusson, M.I. Alonso, E. Moons, H. Arwin, M. Campoy-Quiles, O. Inganäs, Phase behaviour of liquid-crystalline polymer/fullerene organic photovoltaic blends: thermal stability and miscibility. J. Mater. Chem. **21**, 10676–10684 (2011)
40. A.C. Mayer, M.F. Toney, S.R. Scully, J. Rivenay, C.J. Brabec, M. Scharber, M. Koppe, M. Heeney, I. McCulloch, M.D. McGehee, Bimolecular crystals of fullerenes in conjugated polymers and the implications of molecular mixing for solar cells. Adv. Funct. Mater. **19**, 1173–1179 (2009)
41. F. Würthner, K. Meerholz, System chemistry approach in organic photovoltaics. Chem. Eur. J. **16**, 9366–9373 (2010)
42. S.R. Scully, M.D. McGehee, in *Flexible Electronics: Materials and Applications*, ed. by W.S. Wong, A. Salleo (Springer, New York, 2009)
43. G. Dennler, M.C. Scharber, C.J. Brabec, Polymer-fullerene bulk heterojunction solar cells. Adv. Mater. **21**, 1232–1338 (2009)
44. M.A. Brady, G.M. Su, M.L. Chabinyc, Recent progress in the morphology of bulk heterojunction photovoltaics. Soft Matter **7**, 11065–11077 (2011)
45. K. Petritsch, R.H. Friend, A. Lux, G. Rozenberg, S.C. Moratti, A.B. Holmes, Liquid crystalline phthalocyanines in organic solar cells. Syn. Met **102**, 1776–1777 (1999)
46. L. Schmidt-Mende, A. Fechtenkötter, K. Müllen, E. Moons, R.H. Friend, J.D. MacKenzie, Self-organized discotic liquid crystals for high-efficiency organic photovoltaics. Science **293**, 1119–1122 (2001)
47. L. Schmidt-Mende, A. Fechtenkötter, K. Müllen, R.H. Friend, J.D. MacKenzie, Efficient organic photovoltaics from soluble discotic liquid crystalline materials. Physica E **14**, 263–267 (2002)
48. L. Schmidt-Mende, M. Watson, K. Müllen, R.H. Friend, Organic thin film photovoltaic devices from discotic materials. Mol. Cryst. Liq. Cryst. **396**, 73–90 (2003)
49. J. Li, M. Kastler, W. Pisula, J.W.F. Robertson, D. Wasserfallen, A. Clive, G. Grimsdale, J. Wu, K. Müllen, Organic bulk-heterojunction photovoltaics based on alkyl substituted discotics. Adv. Funct. Mater. **17**, 2528–2533 (2007)

50. H.C. Hesse, J. Weickert, M. Al-Hussein, L. Dössel, X. Feng, K. Müllen, L. Schmidt-Mende, Discotic materials for organic solar cells: effect of chemical structure on assembly and performance. Solar Energy Mater. Solar Cells **94**, 560–567 (2010)
51. H. Iino, J. Hanna, R.J.Bushby, B. Movaghar, B.J. Whitaker, M.J. Cook, Very high time-of-flight mobility in the columnar phases of a discotic liquid crystals. Appl. Phys Lett. **87**, 132102-1-3 (2005)
52. H. Iino, Y. Takayashiki, J. Hanna, R.J. Bushby, Fast ambipolar carrier transport and easy homeotropic alignment in a metal-free phthalocyanine derivative. Jpn. J. Appl. Phys. **44**, L1310–L1313 (2005)
53. Y. Miyake, Y. Shiraiwa, K. Okada, H. Monobe, T. Hori, N. Yamasaki, H. Yoshida, A. Fujii, M. Ozaki, Y. Shimizu, High carrier mobility up to 1.4 $cm^2\ V^{-1}\ s^{-1}$ in non-peripheral octahexyl phthalocyanine. Appl. Phys. Express **4**, 021604-1-3 (2011)
54. F. Nekelson, Q.D. Dao, T. Hori, T. Nakao, H.Yoshida, A. Fujii, M. Ozaki, Y. Shimizu, A binary mixture of a non-peripheral octahexylphthalocyanine and PCBM: thermotropic behavior and carrier mobility (submitted)
55. T. Hori, T. Masuda, N. Fukuoka, T. Hayashi, Y. Miyake, T. Kamikado, H. Yoshida, A. Fujii, Y. Shimizu, M. Ozaki, Non-peripheral octahexylphthalocyanine doping effects in bulk heterojunction polymer solar cells. Org. Electron. **13**, 335–340 (2012)
56. S. Jeong, Y. Kwon, B.-D. Choi, H. Ade, Y.-S. Han, Improved efficiency of bulk heterojunction poly(3-hexylthiophene):[6,6]-phenyl-c_{61}-butyric acid methyl ester photovoltaic devices using discotic liquid crystal additives. Appl. Phys. Lett. **96**, 183305-1-3 (2010)
57. T. Masuda, T. Hori, K. Fukumura, Y. Miyake, Q.D. Dao, T. Hayashi, T. Kamikado, H. Yoshida, A. Fujii, Y. Shimizu, M. Ozaki, Photovoltaic properties of 1,4,8,11,15,18,22,25-octaalkylphthalocyanine doped polymer bulk heterojunction solar cells. Jpn. J. Appl. Phys. **51**, 02BK15-1-4 (2012)
58. T. Hori, Y. Miyake, T. Masuda, T. Hayashi, K. Fukumura, H. Yoshida, A. Fujii, Y. Shimizu, M. Ozaki, Dependence of alkyl-substituent length for bulk heterojunction solar cells utilizing 1,4,8,11,15,18,22,25-octa-alkylphthalocyanine. J. Photon. Energy **2**, 021004-1-7 (2012)
59. Q.-D. Dao, T. Hori, K. Fukumura, T. Masuda, T. Kamikado, A. Fujii, Y. Shimizu, M. Ozaki, Effects of processing additives on nanoscale phase separation, crystallization and photovoltaic performance of solar cells based on mesogenic phthalocyanine. Org. Electron. **14**, 2628–2634 (2013)
60. Q.D. Dao, T. Hori, K. Fukushima, T. Masuda, T. Kamikado, A. Fujii, Y. Shimizu, M. Ozaki, Efficiency enhancement in mesogenic-phthalocyanine-based solar cells with processing additives. Appl. Phys. Lett. **101**, 263301-1-3 (2012)
61. Q. Zheng, G. Fang, W. Bai, N. Sun, P. Qin, X. Fan, F. Cheng, L. Yuan, X. Zhao, Efficiency improvement in organic solar cells by inserting a discotic liquid crystal. Solar Energy Mater. Solar Cells **95**, 2200–2205 (2011)
62. A. Varotto, C.-Y. Nam, I. Radivojevic, J.P.C. Tomé, J.A.S. Cavaleiro, C.T. Black, C.M. Drain, Phthalocyanine blends improve bulk heterojunction solar cells. J. Am. Chem. Soc. **132**, 2552–2554 (2010)
63. Q. Sun, L. Dai, X. Zhou, L. Li, and Q. Li, Bilayer- and bulk-heterojunction solar cells using liquid crystalline porphyrins as donors by solution processing. Appl. Phys. Lett. **91**, 253505-1-3 (2007)
64. X. Zhou, S.W. Kang, S. Kumar, R.R. Kulkarni, S.Z.D. Cheng, Q. Li, Self-assembly of porphyrin and fullerene supramolecular complex into highly ordered nanostructure by simple thermal annealing. Chem. Mater. **20**, 3551–3553 (2008)
65. Y. Shimizu, Y. Matsuda, F. Nekelson, Y. Miyake, H. Yoshida, A. Fujii, M. Ozaki, Binary systems of discotic liquid crystalline semiconductors toward solution-processing thin film devices. Proc. SPIE **8279**, 82790G-1–82790G-8 (2012)
66. Y. Shimizu, Y. Matsuda, T. Nakao, L. Sosa-Vargas, M. Takahashi, H. Yoshida, A. Fujii, M. Ozaki, Unpublished results
67. M. Shimizu, L. Tauchi, T. Nakagaki, A. Ishikawa, E. Itoh, K. Ohta, Discotic liquid crystals of transition metal complexes 48: synthesis of novel phthalocyanine-fullerene dyads and effect

of a methoxy group on their clearing points. J. Porphyrins Phthalocyanines **17**, 264–282 (2013)
68. C.L. Wang, W.B. Zhang, H.-J. Sun, R.M. Van Horn, R.R. Kulkarni, C.-C. Tsai, C.-S. Hsu, B. Lotz, X. Gong, S.Z.D. Cheng, A supramolecular, "double-cable" structure with a 12944 helix in a cilumnar porphyrin-c60 dyad and its application in polymer solar cells. Adv. Energy Mater. **2**, 1375–1382 (2012)
69. H. Hayashi, W. Nihashi, T. Umeyama, Y. Matano, S. Seki, Y. Shimizu, H. Imahori, Segregated donor-acceptor columns in liquid crystals that exhibit highly efficient ambipolar charge transport. J. Am. Chem. Soc. **133**, 10736–10739 (2011)
70. M. Ince, M.V. Martínez-Díaz, J. Barberá, T. Torres, Liquid crystalline phthalocyanine-fullerene dyads. J. Mater. Chem. **21**, 1531–1536 (2011)
71. Y.H. Geerts, O. Debever, C. Amato, S. Sergeyev, Synthesis of mesogenic phthalocyanine-C60 donor-acceptor dyads designed for molecular heterojunction photovoltaic devices. Beilstein J. Org. Chem. **5**, 1–9 (2009)
72. A. de la Escosura, M.V. Martínez-Díaz, J. Barberá, T. Torres, Self-organization of phthalocyanine-[60]fullerene dyads in liquid crystals. J. Org. Chem. **73**, 1475–1480 (2008)
73. N. Tchebotareva, X. Yin, M.D. Watson, P. Samori, J.P. Rabe, K. Müllen, Ordered architectures of a soluble hexa-peri-hexabenzocolonene-pyrene dyad: thermotropic bulk properties and nanoscale phase segregation at surfaces. J. Am. Chem. Soc. **125**, 9734–9739 (2003)
74. M. Mathews, Q. Li, Chapter 4 in *Self-Organized Organic Semiconductors: From Materials to Device Applications*, ed. by Q. Li (Wiley, Hoboken, 2011)
75. B.R. Kaafarani, Discotic liquid crystals for opto-electronic applications. Chem. Mater. **23**, 378–396 (2011)

Chapter 9
Ion-Based Liquid Crystals: From Well-Defined Self-Organized Nanostructures to Applications

Hiromitsu Maeda

Abstract Recent progress in the chemistry of ion-based liquid crystals and related materials based on anion-responsive π-conjugated molecules is summarized. Thermotropic liquid crystals with highly ordered positively and negatively charged species are promising materials as organic semiconductors that show fascinating properties compared to those of electronically neutral species. The achievement of ion-based liquid crystals requires the preparation of appropriate charged building subunits, in particular, planar anionic species, which can be obtained by the complexation of electronically neutral anion-responsive π-conjugated molecules. The author's group has fabricated a variety of ion-based organized structures as liquid crystals comprising pyrrole-based anion receptor molecules. The charge-carrier transporting properties exhibited by some of the obtained materials highlight their potential utility in future applications.

9.1 Introduction

The geometries and substituents of organic molecules are very important factors that influence the formation of dimension-controlled organized structures [1–5]. The arrangement of appropriately designed molecules in ordered forms can facilitate the formation of soft materials [6], such as liquid crystals [7–12], which are extremely useful owing to their ability to change their bulk structures according to external conditions. π-Electronic molecules are promising building blocks of organized structures with their highly planar structures enabling efficient

H. Maeda (✉)
College of Pharmaceutical Sciences, Ritsumeikan University, Kusatsu 525-8577, Japan
e-mail: maedahir@ph.ritsumei.ac.jp

Q. Li (ed.), *Nanoscience with Liquid Crystals*, NanoScience and Technology,
DOI: 10.1007/978-3-319-04867-3_9,

stacking. In addition, they often exhibit optical absorption in the visible region, resulting in the ability to fabricate functional electronic materials. As one of the noncovalent interactions to afford molecular assemblies, the electrostatic interaction among charged species provides ion pairs both in solution and in the bulk states. Therefore, the geometries and electronic states of the ionic species are highly important for determining the properties of the ion pairs and their resulting assemblies. As is well known, bulky geometries in both the cationic and anionic components can produce ionic liquids by preventing crystallization through weakening of the ionic interactions [13–18], whereas carefully designed ionic species can provide dimension-controlled assemblies in the form of liquid crystals and related materials [8, 19–21]. However, most of the ion-based materials reported thus far have been fabricated using either cationic or anionic components as the main building blocks of the assemblies accompanied by the respective counter ions, or used electronically neutral components possessing charged moieties in their side chains. In order to effectively utilize the great potential of ion-based assemblies, a new strategy is required for fabricating the materials comprising both cations and anions as crucial building blocks.

The achievement of effectively stacking charged species requires planar geometries that can form dimension-controlled assemblies comprising cationic and anionic components in ordered arrangements. Appropriately designed planar charged species could provide a variety of assembly modes through the control of their interactions (Fig. 9.1). In this figure, a *charge-by-charge assembly* is defined as an organized structure comprising alternately stacking positively and negatively charged species, whereas a *charge-segregated assembly* results from the combination of the appropriate cations and anions and gives rise to electrostatic repulsion between species of the same charge. In the obtained materials, partial contributions from charge-by-charge and charge-segregated assemblies would be observed as an intermediate assembly mode. These ordered structures containing charged components (i.e. electron-deficient and electron-rich species) exhibit potential for use as organic semiconductors; in particular, charge-segregated assemblies may enable high charge-carrier densities by decreasing the electrostatic repulsion between identical charged species.

As compared to planar cations, it is not easy to synthesize planar anions for use as building blocks in stacking assemblies because of the excess electrons that encourage electrophilic attack on the anions. One promising strategy for preparing planar anionic structures is to combine electronically neutral planar anion-responsive molecules with inorganic anions such as halides, resulting in the formation of receptor–anion complexes as pseudo planar anions. Therefore, the design and synthesis of π-electronic systems with efficient anion-binding abilities would be extremely useful. Potential anion-responsive π-conjugated molecules [22–28] are dipyrrolyldiketone boron complexes (e.g. **1**–**3**, Fig. 9.2) [29–48], which show anion-binding behavior with inversion of the pyrrole rings to form planar receptor–anion complexes. Such anion receptors have been shown to be suitable motifs for fabricating assemblies and organized structures such as thermotropic liquid crystals in the anion-free form [35, 41, 44–46], owing to their

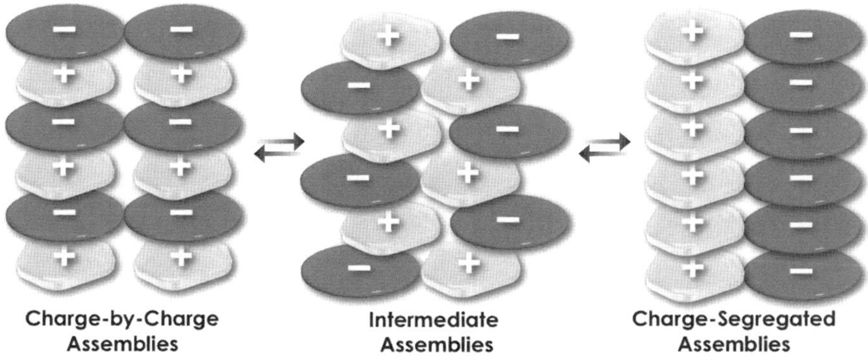

Fig. 9.1 Conceptual diagram of the assembly modes comprising charged species

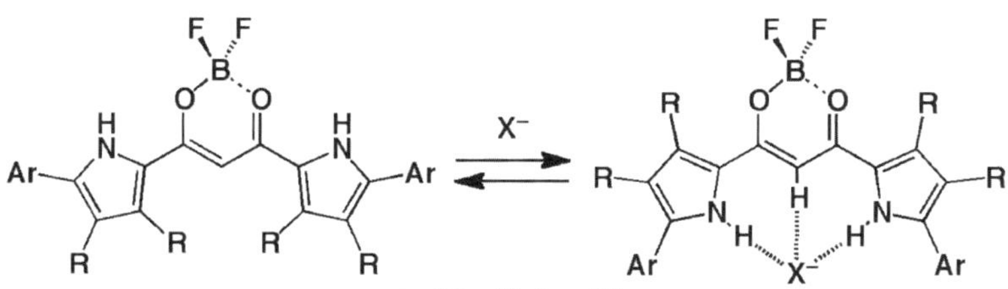

Fig. 9.2 Dipyrrolyldiketone BF_2 complexes and their anion-binding mode

fairly planar conformations and the ability to introduce various substituents that induce additional noncovalent interactions. Assemblies of such molecules exhibit anion-responsive behavior in the bulk state and fascinatingly, the formation of ion-pairing assemblies of the receptor–anion complexes and counter cations.

9.2 Solid-State Ion-Based Assembled Structures

The boron-1,3-propanedione moiety between the two pyrrole rings of the dipyrrolyldiketone boron complexes is effective in affording suitable electronic states that exhibit UV/vis absorption and emission maxima in the visible region, as observed at 432 and 451 nm, respectively, in CH_2Cl_2 for **1a** [32]. Upon the addition of anions as tetrabutylammonium (TBA^+) salts, interactions with the anions through the pyrrole NH and bridging CH can be identified by ^1H NMR chemical shifts, as has been examined for a series of anion receptors. UV/vis

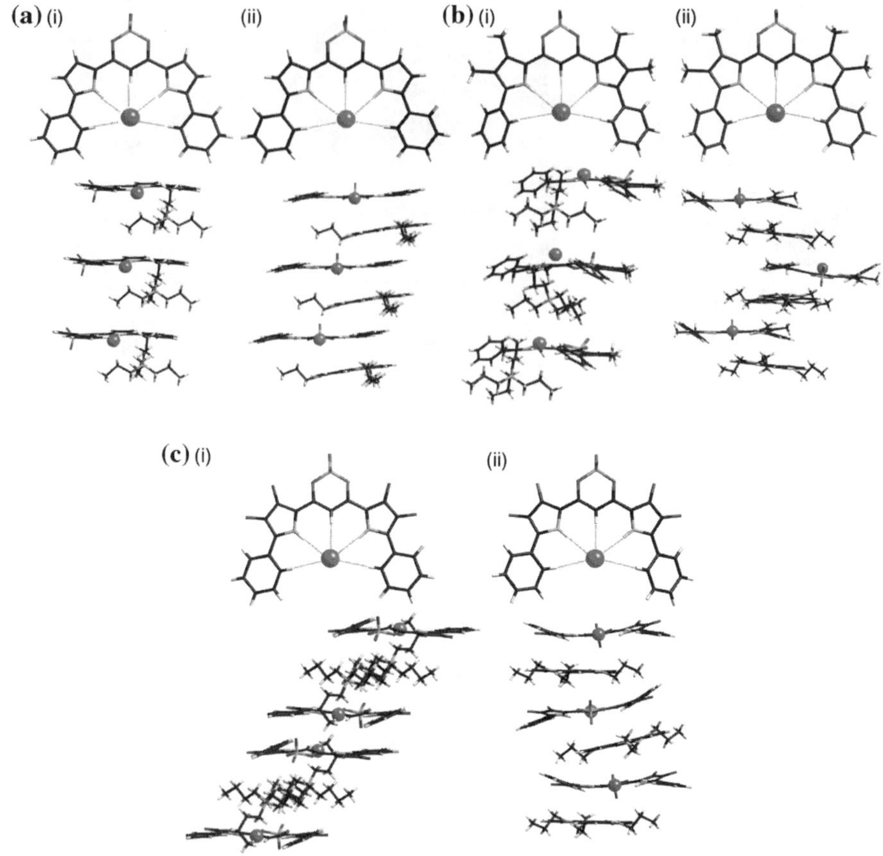

Fig. 9.3 Top and side-packing view from single-crystal X-ray analysis of **a***i* **1b**·Cl$^-$-TPA$^+$ and *ii* **1b**·Cl$^-$-TATA$^+$, **b***i* **2b**·Cl$^-$-TPA$^+$ and *ii* **2b**·Cl$^-$-TATA$^+$, and **c***i* **3b**·Cl$^-$-TBA$^+$ and *ii* **3b**·Cl$^-$-TATA$^+$ (reproduced from cif files: CCDC-646480, 781980, 78198, 745780, 853090, and 853091, respectively)

absorption and emission spectra of the anion receptors were also changed by anion binding, suggesting their potential as colorimetric and fluorescent sensors for anions. The anion-binding constants (K_a) of, for example, α-phenyl-substituted **1b** in CH_2Cl_2 were estimated to be 30,000, 2,800, and 210,000 M^{-1} for Cl^-, Br^-, and $CH_3CO_2^-$, respectively [33].

In the solid state, a series of anion receptors may assemble in various assembling modes in the forms of receptor–anion complexes by combination with counter cations. β-Unsubstituted **1b** afforded a planar [1 + 1]-type pentacoordinated Cl^- complex **1b**·Cl$^-$ that involved the pyrrole NH, the bridging CH, and phenyl *ortho*-CH units (Fig. 9.3a). Planar anion **1b**·Cl$^-$ stacked with tetrapropylammonium cations (TPA$^+$) to form a charge-by-charge columnar structure with a Cl···Cl distance of 8.54 Å, and a separation of 7.29 Å between the **1b**·Cl$^-$ units

(Fig. 9.3a(i)) [33]. Similarly, β-substituted **2b** and **3b** formed [1 + 1]-type complexes **2b**·Cl^- and **3b**·Cl^-; they assembled into columnar structures with the counter cations, with the TPA^+ salt of **2b**·Cl^- showing alternately stacking cationic and anionic components and the TBA^+ salt of **3b**·Cl^- forming a columnar assembly of two **3b**·Cl^- and two TBA^+ in a row (Fig. 9.3b(i), c(i)) [39]. The effects of the cation geometries were observed in the case of the planar 4,8,12-tripropyl-4,8,12-triazatriangulenium cation ($TATA^+$) [49, 50] in place of the tetraalkylammonium cations. In fact, like **1b**·Cl^--TPA^+, the ion pair **1b**·Cl^--$TATA^+$ formed a charge-by-charge columnar structure, but with a smaller distance between the **1b**·Cl^- units (6.85 and 7.29 Å for $TATA^+$ and TPA^+ salts, respectively) (Fig. 9.3a(ii)) [37]. Furthermore, both β-methyl **2b**·Cl^--$TATA^+$ and β-fluorinated **3b**·Cl^--$TATA^+$ also formed columnar assemblies with alternately stacking receptor–Cl^- complexes and $TATA^+$ cations (Fig. 9.3b(ii), c(ii)). It is noteworthy that the Cl···Cl distances of 12.43 and 10.57 Å for **2b**·Cl^--$TATA^+$ and **3b**·Cl^--$TATA^+$, respectively, along the columns were much longer than the 6.85 Å observed for **1b**·Cl^--$TATA^+$. As related to this observation, the smaller overlaps between the receptor–Cl^- complexes and $TATA^+$ in **2b**·Cl^--$TATA^+$ and **3b**·Cl^--$TATA^+$ implied that the β-substituents of the receptor molecules may interfere with the formation of stable charge-by-charge columnar structures [42].

9.3 Thermotropic Liquid Crystals Based on Planar Receptor–Anion Complexes and Appropriate Cations

Modification at the periphery of anion receptor molecules makes it possible to form dimension-controlled organized structures as liquid crystals. In particular, introduction of aliphatic chains to π-conjugated molecules, as seen for **1c**, **2c**, and **3c** (Fig. 9.4a), is an effective strategy for fabricating such assemblies. For example, **1c** appeared as a thermotropic liquid crystal with a mesophase at 36.7–172.5 °C, as revealed by differential scanning calorimetry (DSC). Polarized optical microscopy (POM) showed a ribbon-like texture, and X-ray diffraction (XRD) measurement suggested the formation of a hexagonal columnar (Col_h) phase, in which one unit lattice consisted of a dimer ($Z = 2$ for $\rho = 1$, $a = 3.98$ nm for **1c**) (Fig. 9.4b(i)). Although it is rare to form a stacking columnar structure comprising two rod-like molecules, it can be achieved by N–H···F–B hydrogen-bonding and dipole–dipole interactions. Flash-photolysis time-resolved microwave conductivity (FP-TRMC) measurement [51–54] demonstrated crystal-state charge-carrier mobility in **1c** (0.25 $cm^2\ V^{-1}\ s^{-1}$) at 25 °C, suggesting that **1c** may be useful as an electrically conductive material [35]. Furthermore, β-methyl **2c** and β-fluorinated **3c** showed similar Col_h mesophases ($a = 4.30$ and 4.18 nm, respectively) with transition temperatures (°C) of 89/44 (1st cooling) and 32/101 (2nd heating) for **2c** and 202/42 (1st cooling) and 45/204 (2nd heating) for **3c** (Fig. 9.4b(ii, iii)) [42].

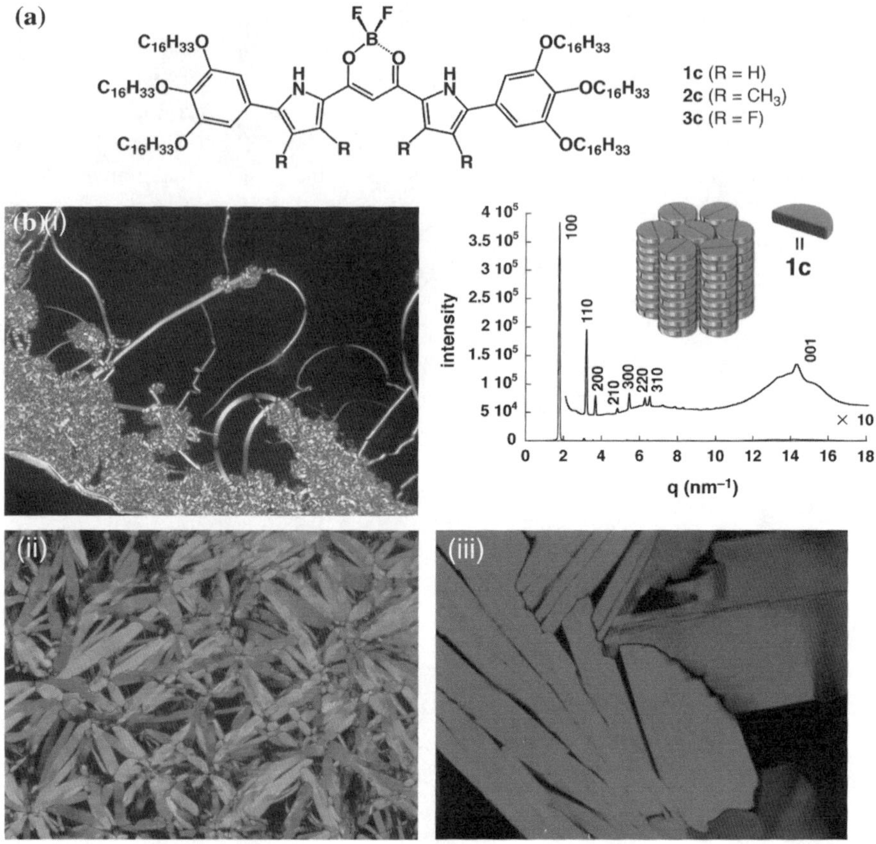

Fig. 9.4 **a** Anion receptors possessing long alkyl chains and **b** POM images of *i* **1c**, *ii* **2c**, and *iii* **3c** as mesophases at 170, 70, and 196 °C, respectively, upon cooling from Iso. Also shown in **b***i* is the synchrotron XRD pattern of **1c** at 120 °C upon cooling from Iso and a proposed Col_h model. The XRD by synchrotron irradiation was performed in [42] and the data was basically consistent with the XRD in [35]

Mesophase behaviors were also observed in ion-based assemblies of **1c**, **2c**, and **3c** as $TATA^+$ salts of receptor–Cl^- complexes. DSC analysis of **1c**·Cl^--$TATA^+$ suggested the formation of a mesophase, as observed in the phase transitions at 88/42 °C (1st cooling) and at 44/96 °C (2nd heating). Cooling from the isotropic liquid (Iso) afforded larger focal conic domains and dark domains in the POM images (Fig. 9.5a(i)), suggesting that discotic columnar structures were well aligned perpendicularly to the substrates. Upon cooling to 70 °C from Iso, synchrotron XRD analysis showed the relatively sharp peaks of a Col_h phase with $a = 4.64$ nm and $c = 0.73$ nm based on a tetrameric assembly ($Z = 3.58$ for $\rho = 1$) (Fig. 9.5a(ii)), with shear-driven alignment of the columnar structure. The c value of 0.73 nm corresponded to the distance ascribed to the alternately stacking ion pairs of **1c**·Cl^- and $TATA^+$, strongly suggesting the contribution of a

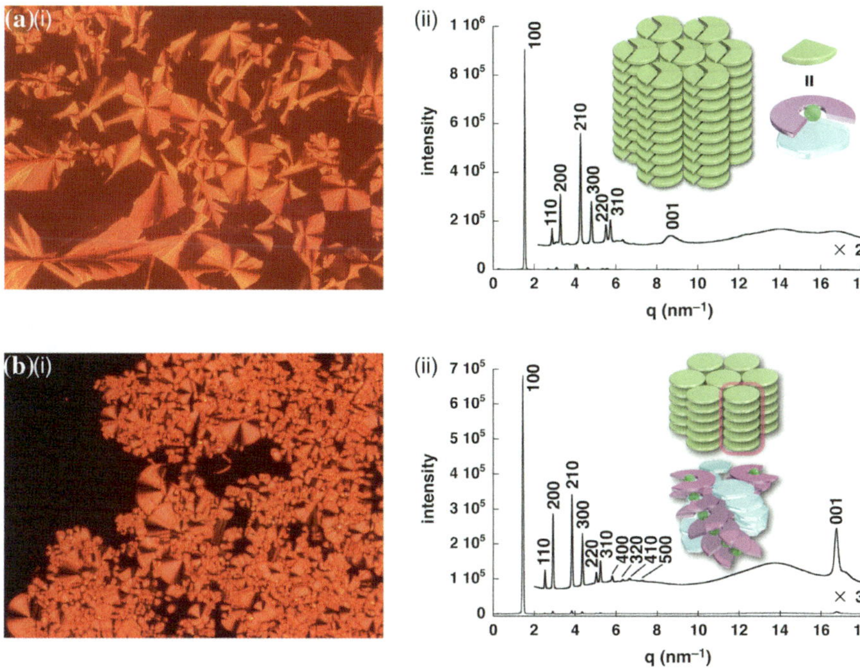

Fig. 9.5 *i* POM images and *ii* synchrotron XRD patterns of (**a**) **1c**·Cl^--$TATA^+$ (70 °C for both cases) and (**b**) **2c**·Cl^--$TATA^+$ (150 and 101 °C, respectively) upon cooling from Iso along with proposed Col_h models based on charge-by-charge and charge-segregated assemblies

charge-by-charge assembly, whereas the circular tetrameric assembly in a single disk unit was also consistent with the fan-like geometry of **1c**·Cl^-. As a preliminary result, the application of an electric field in the mesophase resulted in an optical response, as observed in the POM images [37]. On the other hand, **2c**·Cl^--$TATA^+$ and **3c**·Cl^--$TATA^+$ exhibited POM textures as mesophases (Fig. 9.5b(i)), with transition temperatures (°C) of 149/34 (1st cooling) and 40/153 (2nd heating) and 145/38 (1st cooling) and 45/146 (2nd heating), respectively. Synchrotron XRD analysis of the mesophases of **2c**·Cl^--$TATA^+$ and **3c**·Cl^--$TATA^+$ revealed Col_h structures with $a = 4.99$ nm and $c = 0.37$ nm ($Z = 2.05$ for $\rho = 1$) and $a = 4.92$ nm and $c = 0.37$ nm ($Z = 1.98$ for $\rho = 1$), respectively (Fig. 9.5b(ii)). The periodicity of 0.37 nm comparable to the ordinary π–π stacking distances strongly suggested the local stacking of identically charged planes. The contribution of the charge-segregated assembly may result from the distorted π-conjugated units owing to the β-substituents [42].

Time-of-flight (TOF) electrical conductivity measurements of ion-based materials **1c**·Cl^--$TATA^+$, **2c**·Cl^--$TATA^+$, and **3c**·Cl^--$TATA^+$ indicated ambipolar charge-carrier transport behavior with well-balanced values at high mobilities (10^{-2}–10^{-3} $cm^2\ V^{-1}\ s^{-1}$) for both holes and electrons without special purification procedures. The highest value of the zero-field limit mobility was observed for the

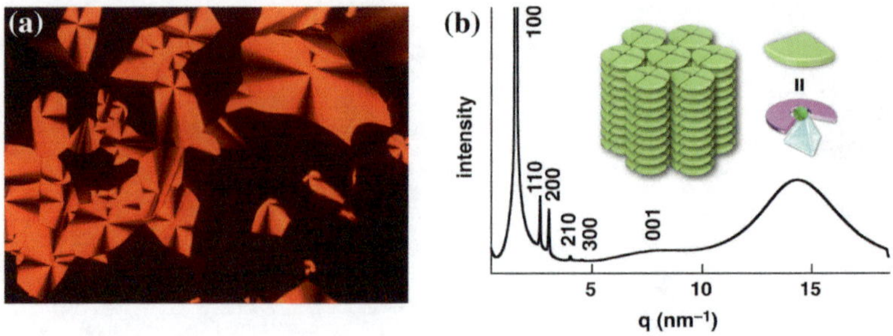

Fig. 9.6 **a** POM texture and **b** synchrotron XRD pattern of **1c**·Cl^--TBA^+ at 62 °C upon cooling from Iso and a proposed Col_h model based on charge-by-charge assembly

positive charge in **3c**·Cl^--$TATA^+$ (0.11 $cm^2\ V^{-1}\ s^{-1}$) probably as a result of the partial contribution of the charge-segregated assembly with distinct arrays of identical charged species. In the case of the negative charge carriers, **3c**·Cl^--$TATA^+$ exhibited almost equivalent values of mobility (5×10^{-3} $cm^2\ V^{-1}\ s^{-1}$ at $E = 4 \times 10^3$ V cm^{-1}) at 100–140 °C, with negligible electric-field dependence. The electron-deficient nature of **3c**·Cl^--$TATA^+$ led to higher stability of the electrons in the Col_h structure, resulting in relatively higher values of electron mobility [42]. Considering the results from these aliphatic receptors, the properties and packing structures of ion-based assemblies can be modulated by further modifications of the anion receptors, such as semifluoroalkyl-substituted [46], β-benzo- and β-corannulene-fused [41, 45], and boron-modified derivatives [44].

Not only planar cations, but bulky cations can also act as building blocks for the formation of ion-based assemblies, even though their steric hindrance interferes with the stacking of the receptor–anion complexes. Similar to the assemblies of planar cations, Col_h mesophases based on charge-by-charge assembly were obtained by the complexation of **1c** with Cl^- as the salts of different tetraalkylammonium cations $n_mMe_{4-m}N^+$ ($(C_nH_{2n+1})_mMe_{4-m}N^+$: $m = 1$–4) (Table 9.1), with broken-fan-like POM textures seen upon cooling from Iso (Fig. 9.6a for **1c**·Cl^--TBA^+ (4_4N^+)). In this case, **1c**·Cl^--TBA^+ as a Col_h mesophase included alternately stacking planar **1c**·Cl^- and bulky TBA^+ via charge-by-charge assembly (Fig. 9.6b). As seen in Table 9.1, ion pairs of a planar anionic component and a bulky tetraalkylammonium cation tune their assembled structures and properties according to the number and length of the aliphatic chains in the employed cations. In particular, as the chain number and length increased, the ability of the cations to assemble with **1c**·Cl^- was reduced owing to the increased steric hindrance, resulting in the formation of ionic liquids at fairly low temperatures [40].

Cationic species as counterparts of the planar receptor–anion complexes can be prepared by *anion binding*. In fact, organized structures based on planar charged species were fabricated by combining positively and negatively charged receptor–anion complexes using dicationic and electronically neutral π-conjugated receptors. Phenylene- or pyrimidine-bridged bis(imidazolium) dicationic anion receptors

Table 9.1 Phase transitions of **1c**·Cl^--$n_m Me_{4-m}N^+$

1c·Cl^--$n_m Me_{4-m}N^+$		Cooling[a]	Heating[a]
1c·Cl^--$n_1 Me_3N^+$	$n = 8$	Cr[c] 36.3 Col_h 72.1 Iso	Cr[c] 43.3 Col_h 74.5 Iso
	$n = 12$	Cr[c] 37.9 Col_h 72.3 Iso	Cr[c] 42.9 Col_h 73.8 Iso
	$n = 16$	Cr[c] 41.9 Col_h 68.1 Iso	Cr[c] 44.9 Col_h 73.0 Iso
1c·Cl^--$n_2 Me_2N^+$	$n = 8$	Cr[c] 33.7 Col_h 61.2 Iso	Cr[c] 39.2 Col_h 62.6 Iso
	$n = 12$	Cr[c] 28.9 Col_h 69.1 Iso	Cr[c] 35.5 Col_h 70.1 Iso
	$n = 18$[b]	Cr[c] 40.2 Col_h 79.8 Iso	Cr[c] 44.5 Col_h 80.8 Iso
1c·Cl^--$n_3 Me_1N^+$	$n = 4$	Cr[c] 41.9 Col_h 54.1 Iso	Cr[c] 46.2 Col_h 61.9 Iso
	$n = 8$	Cr[d] 30.5 Iso	Cr[d] 35.4 Iso
	$n = 12$	Cr[d] 24.5 Iso	Cr[d] 30.8 Iso
1c·Cl^--n_4N^+	$n = 4$	Cr[c] 40.3 Col_h 84.1 Iso	Cr[c] 46.5 Col_h 84.4 Iso
	$n = 8$	Cr[d] 29.2 Iso	Cr[d] 32.7 Cr′[e] 45.6 Iso
	$n = 12$	Cr[d] 23.6 Iso	Cr[d] 27.6 Cr′[e] 42.0 Iso

[a] Transition temperatures (°C, the onset of the peak) from DSC 1st cooling and 2nd heating scans (5 °C min^{-1})
[b] Used because 16_2Me_2NCl is not readily available
[c] Cr with Col_h structures
[d] Cr with unidentified structures
[e] Cr′ with rectangular columnar (Col_r) structures

4a^{2+}–**c**$^{2+}$ (Fig. 9.7a) [55, 56] as precursor cationic anion complexes formed monocationic receptor–Cl^- complexes that were accompanied by a free Cl^-. This free Cl^- was subsequently captured by electronically neutral anion receptors **1a**–**c** to form negatively charged receptor–Cl^- complexes. The ion pairs of the resulting positively and negatively charged planar receptor–Cl^- complexes showed mesophase behaviors. Typical POM textures such as broken-fan-shaped types were observed during heating and cooling processes (Fig. 9.7b(i) for **1c**·Cl^--**4b**$^{2+}$·Cl^-). The phase transition temperatures (°C) of the ion pairs with **1c**·Cl^- were as 103.9/46.7 (**1c**·Cl^--**4a**$^{2+}$·Cl^-), 148.3/33.9 (**1c**·Cl^--**4b**$^{2+}$·Cl^-), and 119.3/28.8 (**1c**·Cl^--**4c**$^{2+}$·Cl^-) upon cooling from Iso. The temperature range of the mesophase in **1c**·Cl^--**4b**$^{2+}$·Cl^- was greater than that in **1c**·Cl^--**4a**$^{2+}$·Cl^- by incorporation of the pyrimidine moieties. The synchrotron XRD pattern of **1c**·Cl^--**4b**$^{2+}$·Cl^- at 115 °C showed the formation of a Col_h mesophase with $a = 4.34$ nm (Fig. 9.7b(ii)). Besides the sharp in-plane diffractions, a broad and weak peak appeared around 0.75 nm, which was derived from the stacking periodicity between the adjacent assembled units with the contribution of charge-by-charge assembly. After cooling to 20 °C, the Col_h structure was maintained, giving $a = 4.73$ nm, and the peak at 0.75 nm became more evident. Accordingly, the average number of ion pairs in an assembled unit (Z) was estimated to be 3 for $\rho = 1$ ($Z = 2.9$ and 3.4 at 115 and 20 °C, respectively). Meanwhile, **1c**·Cl^--**4a**$^{2+}$·Cl^- also provided a Col_h structure ($a = 4.65$ nm, $c = 0.75$ nm, and $Z = 3.3$ for $\rho = 1$ at 80 °C upon cooling from Iso), whereas **1c**·Cl^--**4c**$^{2+}$·Cl^- had complicated XRD patterns, which may be ascribed to the increased rigidity of the dicationic anion receptor with the introduction of aryl moieties. The observations by synchrotron XRD indicated that the

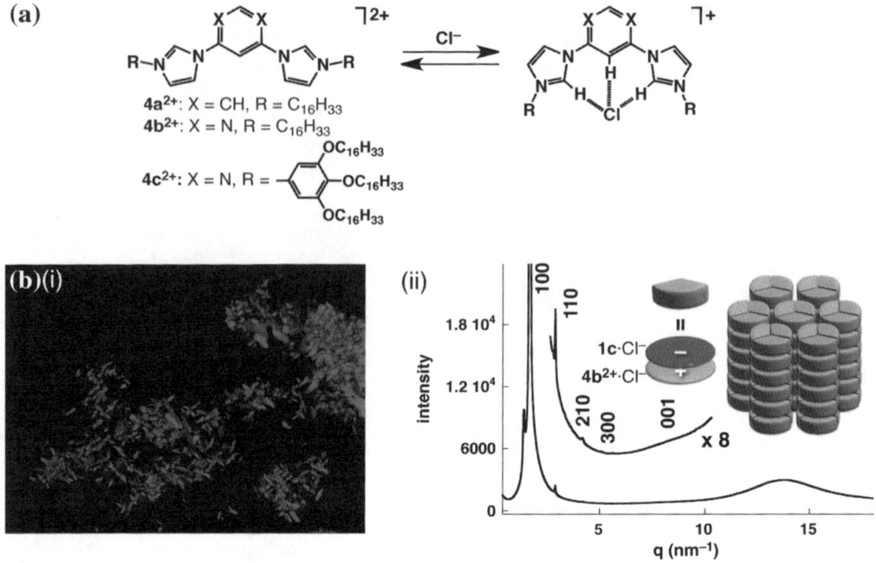

Fig. 9.7 **a** Positively charged anion receptors and their Cl^--binding modes forming cationic receptor–Cl^- complexes and **b***i* POM texture and *ii* synchrotron XRD pattern of **1c**·Cl^--**4b**$^{2+}$·Cl^- at 115 °C upon cooling from Iso, along with a proposed Col_h model wherein the exact molecular orientation cannot be shown

introduction of dicationic anion receptors was effective for constructing ordered assembled structures. It is noteworthy that the geometries in the core π-conjugated parts of the cationic and anionic receptor–Cl^- complexes are quite similar by considering the constituents, one six-membered ring at the center and two neighboring five-membered rings. The similarity in the sizes and shapes of the two ionic components is a very important factor in forming assemblies consisting of multiple components. Furthermore, using the FP-TRMC technique, the one-dimensional charge-carrier transporting properties, with the mobilities of 0.05 and 0.03 $cm^2\ V^{-1}\ s^{-1}$ for **1c**·Cl^--**4a**$^{2+}$·Cl^- and **1c**·Cl^--**4b**$^{2+}$·Cl^-, respectively, were observed for the newly prepared solid-state ion pairs [48].

In the preceding examples, the dimension-controlled organized structures in the ion-based materials were mainly fabricated by van der Waals interactions between the aliphatic chains of the receptor molecules. In contrast, the anion receptors in the absence of aliphatic chains were found to form dimension-controlled structures by combination with cationic species possessing aliphatic units. In fact, the employment of benzyltrialkylammonium chlorides ($16Bn_3NCl$, $n = 2$ and 4) and benzylpyridinium chloride (16BPyCl) (Fig. 9.8a) with β-unsubstituted **1a**, **b** and β-fluorinated **3a**, **b** resulted in the construction of ion-based assemblies. Cl^- complexes of **1a**, **b** and **3a**, **b** as $16Bn_3N^+$ ($n = 2$ and 4) and $16BPy^+$ salts showed different thermal behaviors from the individual components (Table 9.2). Self-organization into mesophases was achieved by the delicate balance between the

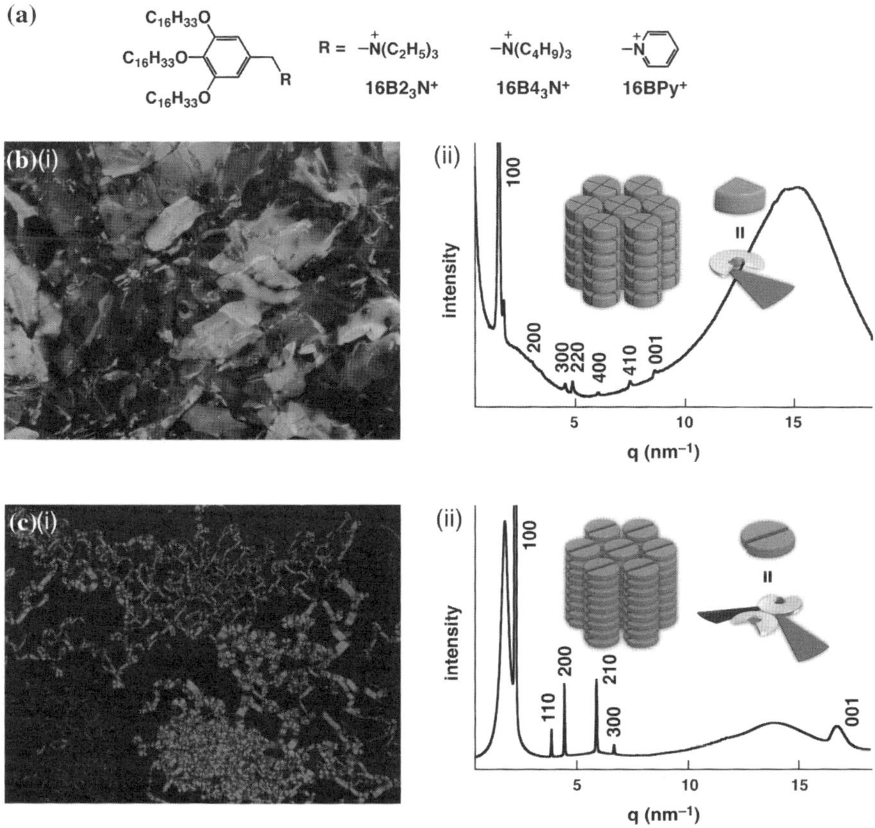

Fig. 9.8 **a** Structures of 16Bn_3NCl ($n = 2$ and 4) and 16BPyCl; *i* POM textures and *ii* synchrotron XRD patterns of **b** **1b**·Cl^--16B4_3N^+ (125 and 90 °C, respectively) and **c** **3a**·Cl^--16B2_3N^+ (85 °C for both cases) upon 2nd heating, along with proposed Col_h models, wherein module cations are shown as fan-like shapes based on the optimized structures and the charge-by-charge and charge-segregated assemblies are represented by thick and thin disk components, respectively

positively and negatively charged components, in terms of the size of the ionic parts in the cation species and the substituents on the anion receptors. For example, **1b**·Cl^--16B4_3N^+ and **3a**·Cl^--16B2_3N^+ showed flake-like and fiber-like POM textures, respectively, upon 2nd heating (Fig. 9.8b, c(i)), with the formation of Col_h mesophases with $a = 4.79$ and 3.28 nm and $c = 0.74$ and 0.38 nm, respectively (Fig. 9.8b, c(ii)). The c values of **1b**·Cl^--16B4_3N^+ and **3a**·Cl^--16B2_3N^+ indicated a charge-by-charge arrangement and the local stacking of identically charged species to produce a column, respectively. The β-substituents of the anion receptor, inductively electron-withdrawing fluorine moieties, were responsible for the contribution of the charge-segregated assembly, making the intermolecular interactions among the anionic complexes sufficiently robust for stacking structure formation. Furthermore, the charge-carrier mobilities $\sum\mu$ of

Table 9.2 Phase transitions[a] of cation-module Cl^- salts and their ion pairs in the presence of **1a**, **b** and **3a**, **b**

	$16B2_3NCl$	$16B4_3NCl$	16BPyCl
–	Cr^b 38.7 Cr'^b 63.7 Col_h 129.4[f] Iso	Cr^b 42.4 Cr'^c 54.0 Cr''^c 76.4 Iso [Cr^b 38.5 Col_h 81.0 Iso]	Cr^b 55.2 Cr'^b 81.1 Col_h 125.9[f] Iso
1a	Cr^b 50.1 Col_h 132.5[f] Iso	Cr^d 43.8 Iso	Cr^b 54.3 Col_h 133.7[f] Iso
1b	Cr^c 37.1 Cr'^c 45.6 Iso [Cr^c 40.4 Iso]	Cr^b 38.6 Col_h 143.1[f] Iso	Cr^c 48.3 Cr'^c 127.3[f] Iso
3a	Cr^c 6.5 Cr'^b 35.7 Col_h 102.4 Iso	Cr^b 39.1 Cr'^b 59.6 Cr''^b 92.1 Iso	Cr^b 39.3 Col_h 53.1[f] Iso
3b	Cr^e 29.1 Cr'^e 51.3 Iso [Cr^e 30.7 Cr'^e 49.4 Cr''^e 54.1 Iso]	Cr^d 30.7 M^e 69.8 Iso [Cr^d 36.6 M'^c 74.2 Iso]	Cr^b 30.2 Cr'^b 45.7 Iso

[a] Transition temperatures (°C, the onset of the peak) from DSC upon 2nd heating (5 °C min^{-1}). The entries in brackets show the transitions upon 1st cooling which exhibit phases different from those upon 2nd heating
[b] Basically as Col_h structures
[c] Unidentified structures
[d] Basically as lamellar structures
[e] Basically as Col_r structures
[f] Values from POM measurements

1b·Cl^--$16B4_3N^+$ and **3a**·Cl^--$16B2_3N^+$ were estimated by FP-TRMC measurements as 0.05 ± 0.01 and 0.22 ± 0.03 $cm^2\ V^{-1}\ s^{-1}$, respectively. The increase in mobility for the assembly with the contribution of the charge-segregated mode offers the possibility of enhanced efficiency of charge-carrier transport in organic electronic devices [43].

Anion receptors that cannot form soft materials by themselves can also form dimension-controlled organized structures by combining with anionic species possessing aliphatic units. The combination of **1a**, **b** with aliphatic-chain-substituted gallic carboxylates ($Ar^{Cn}CO_2^-$, $n = 16$, 18, and 20, Fig. 9.9a) as TBA^+ salts provided mesophases mainly of lamellar structures, with transition temperatures of 42/56/103, 67/73/101, and 67/81/96 °C for **1a**·$Ar^{Cn}CO_2^-$·TBA ($n = 16$, 18, and 20, respectively) and 40/61/81, 53/64/86, and 61/82/89 °C for **1b**·$Ar^{Cn}CO_2^-$·TBA ($n = 16$, 18, and 20, respectively) upon 2nd heating (Fig. 9.9b, c). The diffractions assignable to the charge-by-charge assemblies of receptor–carboxylate moieties and TBA^+ cations could not all be clearly observed, suggesting that fairly disordered structures were produced. On consideration of the crystal-state assembled mode of **1a**·$Ar^{C1}CO_2^-$-TPA^+, the modified anions may be located at distorted angles to the receptor planes and could predominantly control the assembled structures through van der Waals interactions between the aliphatic chains. Furthermore, on increasing the temperature from 28 to 67 °C, the electrical conductivity of **1b**·$Ar^{C16}CO_2^-$-TBA^+ increased from 5×10^{-11} to 3×10^{-8} S m^{-1}, owing to the increased population of thermally activated charge carriers with equivalent mobility in the mesophases. On the other hand, at r.t. upon cooling from

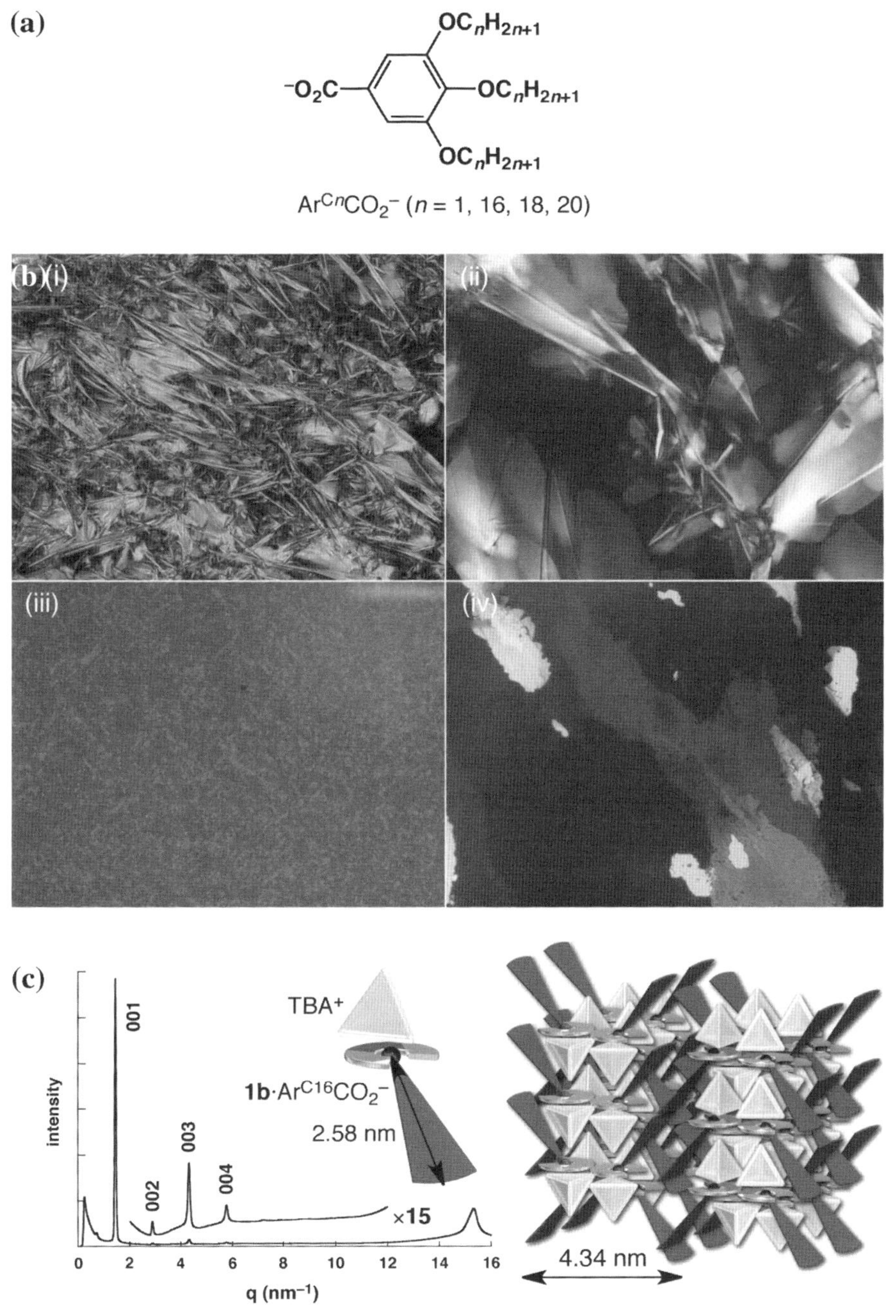

Fig. 9.9 **a** Structures of gallic carboxylates $Ar^{Cn}CO_2^-$ (n = 1, 16, 18, 20); **b** POM images of *i* **1a**·$Ar^{C18}CO_2^-$-TBA^+ at 95 °C, *ii* **1a**·$Ar^{C20}CO_2^-$-TBA^+ at 92 and 80 °C, *iii* **1b**·$Ar^{C16}CO_2^-$-TBA^+ at 80.5 °C, and *iv* **1b**·$Ar^{C18}CO_2^-$-TBA^+ at 83 °C upon cooling from Iso; **c** synchrotron XRD pattern (*left*) and a proposed assembled model (*right*) of **1b**·$Ar^{C16}CO_2^-$-TBA^+ as the solid state at r.t. upon cooling from Iso, suggesting that the observed assembled structure was more disordered than the proposed model, with a value of 2.58 nm estimated by density functional theory (DFT) calculations

Iso, **1a**·$Ar^{C16}CO_2^-$-TBA^+ and **1b**·$Ar^{C16}CO_2^-$-TBA^+ showed charge-carrier mobilities of 0.02 and 0.05 $cm^2\ V^{-1}\ s^{-1}$, respectively. At elevated temperatures, the values for **1a**·$Ar^{C16}CO_2^-$-TBA^+ decreased to 0.007 and $9 \times 10^{-4}\ cm^2\ V^{-1}\ s^{-1}$ at 46 and 70 °C, respectively, whereas those for **1b**·$Ar^{C16}CO_2^-$-TBA^+ were 0.003 and 0.04 $cm^2\ V^{-1}\ s^{-1}$ at 50 and 70 °C, respectively, owing to transitions between the solid state and different mesophases [38].

9.4 Thermotropic Liquid Crystals Based on Planar Ion Pairs

The choice of constituent anion-responsive π-conjugated molecules is very important for controlling the assembly modes of ion-based organized structures. As a building subunit of the π-conjugated molecules, pyrazole can interact electrostatically or via hydrogen bonding with anionic or polar substrates in its partially or fully protonated form. Dipyrrolylpyrazoles (DPPs) (**5a–c** and **6a–c**, Fig. 9.10a(i)), which are small simple structures, are potential candidate precursors for positively charged π-electronic units as they form planar [2 + 2]-type complexes in their protonated forms by combination with trifluoroacetate ($CF_3CO_2^-$) (Fig. 9.10a(i)) [57, 58]. The [2 + 2]-type complexes of **5a**·H^+ and **6a**·H^+ with $CF_3CO_2^-$ formed herringbone-like stacking assemblies in the solid state (Fig. 9.10b) [57]. Based on these observations, it was deduced that an appropriate arrangement of [2 + 2]-type complexes of modified DPPs such as **5c** and **6c** could lead to ordered structures such as stacking columnar assemblies. Ion pairs **5c**·H^+ and **6c**·H^+ as $CF_3CO_2^-$ complexes exhibited enantiotropic mesophases at 66–85 °C (M1) and 85–96 °C (M2) (**5c**·H^+-$CF_3CO_2^-$) and at 68–131 °C (**6c**·H^+-$CF_3CO_2^-$) upon cooling from Iso. POM of **5c**·H^+ as a $CF_3CO_2^-$ complex showed a broken-fan-like texture and no texture for M1 and M2, respectively (Fig. 9.10c(i)), whereas **6c**·H^+-$CF_3CO_2^-$ as a mesophase exhibited a sand-like POM texture (Fig. 9.10d(i)). In contrast to **5c**, which showed no mesophase, **5c**·H^+-$CF_3CO_2^-$ as M2 at 90 °C upon cooling from Iso indicated the formation of a cubic (Cub) ($Pn\bar{3}m$) phase with $a = 10.9$ nm (Fig. 9.10c(ii)). **5c**·H^+-$CF_3CO_2^-$ as M1 at 80 °C showed a slightly complicated phase as a mixture of a rectangular columnar (Col_r) phase ($a = 7.32$ nm, $b = 3.52$ nm, and $Z =$ ca. 2 for $\rho = 1$) and another phase such as Cub. On the other hand, similar to **6c** showing a Col_h phase ($Z =$ ca. 2 for $\rho = 1$) with $a = 4.00$ nm (80 °C), **6c**·H^+-$CF_3CO_2^-$ as a mesophase indicated the formation of a Col_h phase ($a = 3.91$ nm at 120 °C) based on a [2 + 2]-type complex ($Z =$ ca. 2 for $\rho = 1$) (Fig. 9.10d(ii)). In this study, the anion ($CF_3CO_2^-$) served as a bridging unit connecting two cationic π-electronic species, resulting in the formation of a larger planar area as a subunit of the columnar structure [58].

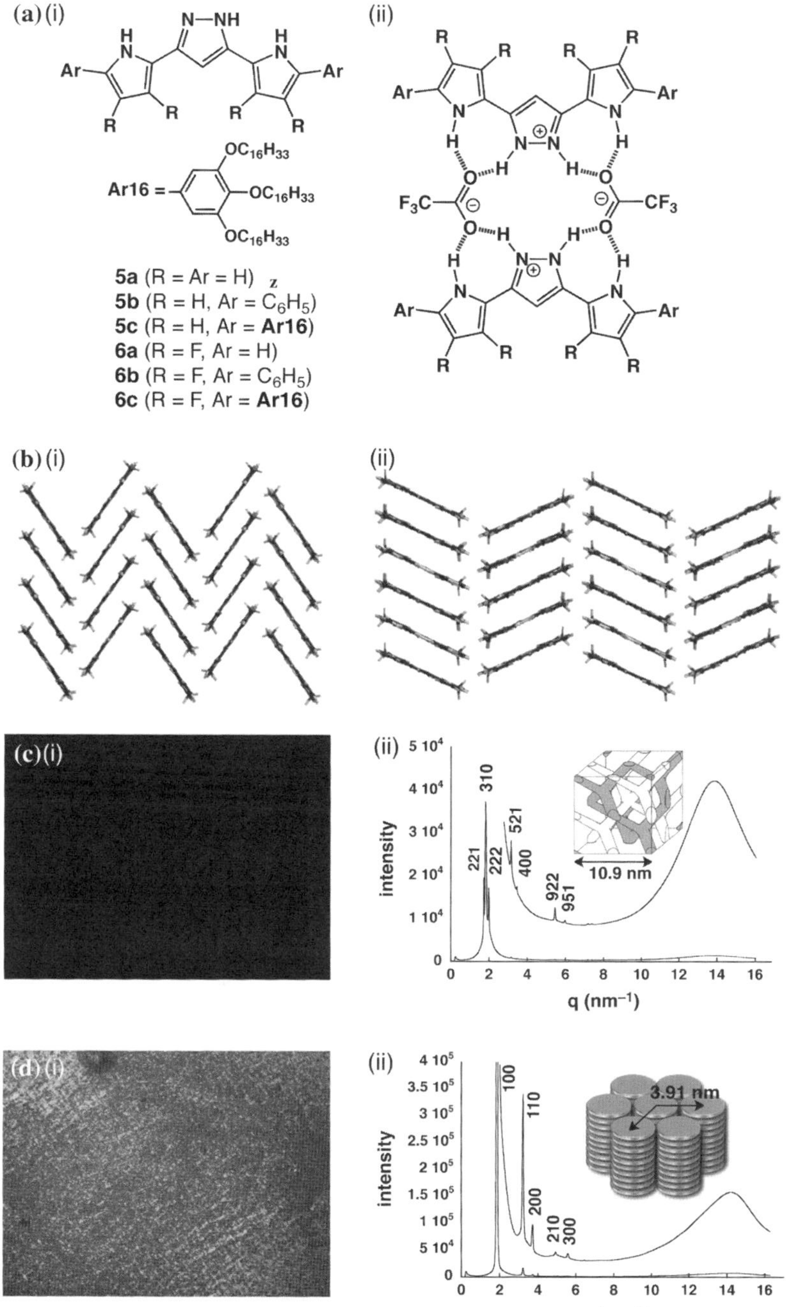

Fig. 9.10 **a** *i* Dipyrrolylpyrazoles (DPPs) and *ii* planar [2 + 2]-type complexes; **b** solid-state stacking diagrams of [2 + 2] assemblies of *i* $\mathbf{5a}_2\cdot TFA_2$ and *ii* $\mathbf{6a}_2\cdot TFA_2$ (reproduced from cif files: CCDC-625142, 625144); *i* POM images and *ii* synchrotron XRD patterns with proposed Cub and Col_h models (insets) of **c** $\mathbf{5c}\cdot H^+$-$CF_3CO_2^-$ and **d** $\mathbf{6c}\cdot H^+$-$CF_3CO_2^-$ as mesophases at 90 and 120 °C, respectively, upon cooling from Iso

9.5 Summary

This chapter discussed selected examples of ion-based thermotropic liquid crystals mainly comprising anion complexes of π-conjugated molecules and counter cations. The ordered arrangement of positively and negatively charged ionic species for the fabrication of liquid crystal materials requires their appropriate geometries, substituents, and electronic properties. Among the candidate components for achieving such materials, the pyrrole-based anion receptors synthesized by our group have been demonstrated to be efficient motifs for the formation of planar anionic complexes by binding anions, resulting in the formation of supramolecular assemblies by combination with cationic species. Modification of the anion receptors and the appropriate choice of anions and cations were found to be effective in forming functional ion-based materials with contributions from charge-by-charge and charge-segregated modes. As demonstrated in this chapter, a variety of ion-based liquid crystal materials have been fabricated to date; however, the exact mechanisms of assembly have not yet been fully elucidated. The innovative design and synthesis of π-electronic systems by considering their geometries, electronic properties, and stabilities would enable the preparation of fascinating ion-based materials for use in improved electronic devices.

Acknowledgments The contributions reported herein have been supported by PRESTO, Japan Science and Technology Agency (JST) ("Structure Control and Function", 2007–2011), Grants-in-Aid for Young Scientists (B) (No. 21750155) and (A) (No. 23685032) and Scientific Research in a Priority Area "Super-Hierarchical Structures" (No. 18039038, 19022036) from the Ministry of Education, Culture, Sports, Science, and Technology (MEXT), the matching fund subsidies for private universities from the MEXT, 2003–2008 and 2009–2014, and the Ritsumeikan Global Innovation Research Organization (R-GIRO) project, 2008–2013. The author thanks all the authors and the collaborators described in the acknowledgement in the previous publications, in particular, Prof. Atsuhiro Osuka and his group members for single-crystal X-ray analysis, Dr. Takashi Nakanishi, NIMS, for his kind help with various analyses of molecular assemblies, Prof. Shu Seki, Osaka University, and his group members for electrical conductivity measurements, Prof. Hitoshi Tamiaki, Ritsumeikan University, for various measurements, and all the group members, especially, Dr. Yohei Haketa and Dr. Bin Dong, for their contributions on ion-based materials.

References

1. G. Tsoucaris (ed.), *Current Challenges on Large Supramolecular Assemblies, NATO Science Series* (Kluwer, South Holland, 1999)
2. A. Ciferri (ed.), *Supramolecular Polymers* (Marcel Dekker, New York, 2000)
3. F. Würthner (ed.), *Supramolecular Dye Chemistry, Topics in Current Chemistry*, vol. 258 (Springer, Berlin, 2005), pp. 1–324
4. J.L. Atwood, J.W. Steed (eds.), *Organic Nanostructures* (Wiley, Weinheim, 2007)
5. P.A. Gale, J.W. Steed (eds.), *Supramolecular Chemistry: From Molecules to Nanomaterials* (Jon Wiley & Sons, Chichester, 2012)
6. W. Hamley, *Introduction to Soft Matter—Polymers, Colloids, Amphiphiles and Liquid Crystals* (John Wiley & Sons, West Sussex, 2000)

7. I. Dierking, *Textures of Liquid Crystals* (Wiley, Weinheim, 2003)
8. T. Kato, N. Mizoshita, K. Kishimoto, Functional liquid-crystalline assemblies: self-organized soft materials. Angew. Chem. Int. Ed. **45**, 38–68 (2006)
9. T. Kato (ed.), *Liquid Crystalline Functional Assemblies and Their Supramolecular Structures, Structure and Bonding*, vol. 128 (Springer, Berlin, 2008), pp. 1–237
10. T. Kato, T. Yasuda, Y. Kamikawa, M. Yoshio, Self-assembly of functional columnar liquid crystals. Chem. Commun. **12**, 729–739 (2009)
11. B.R. Kaafarani, Discotic liquid crystals for opto-electronic applications. Chem. Mater. **23**, 378–396 (2011)
12. S. Kumar, *Chemistry of Discotic Liquid Crystals: From Monomers to Polymers; The Liquid Crystals Book Series* (CRC Press, Boca Raton, 2011)
13. T. Welton, Room-temperature ionic liquids. solvents for synthesis and catalysis. Chem. Rev. **99**, 2071–2084 (1999)
14. P. Wasserscheid, W. Keim, Ionic liquids—new solutions for transition metal catalysis. Angew. Chem. Int. Ed. **39**, 3772–3789 (2000)
15. H. Ohno, Functional design of ionic liquids. Bull. Chem. Soc. Jpn. **79**, 1665–1680 (2006)
16. M.A.P. Martins, C.P. Frizzo, D.N. Moreira, N. Zanatta, H.G. Bonacorso, Ionic liquids in heterocyclic synthesis. Chem. Rev. **108**, 2015–2050 (2008)
17. P. Hapiot, C. Lagrost, Electrochemical reactivity in room-temperature ionic liquids. Chem. Rev. **108**, 2238–2264 (2008)
18. R. Giernoth, Task-specific ionic liquids. Angew. Chem. Int. Ed. **49**, 2834–2839 (2010)
19. K. Binnemans, Ionic liquid crystals. Chem. Rev. **105**, 4148–4204 (2005)
20. T.L. Greaves, F.J. Drummond, Ionic liquids as amphiphile self-assembly media. Chem. Soc. Rev. **37**, 1709–1726 (2008)
21. K.V. Axenov, S. Laschat, Thermotropic ionic liquid crystals. Materials **4**, 206–259 (2011)
22. A. Bianchi, K. Bowman-James, E. García-España (eds.), *Supramolecular Chemistry of Anion* (Wiley, New York, 1997)
23. R.P. Singh, B.A. Moyer (eds.), *Fundamentals and Applications of Anion Separation* (Kluwer, New York, 2004)
24. I. Stibor (ed.), *Anion Sensing*, *Topics in Current Chemistry*, vol. 255 (Springer, Berlin, 2005), pp. 1–238
25. J.L. Sessler, P.A. Gale, W.-S. Cho, *Anion Receptor Chemistry* (RSC, UK, 2006)
26. R. Vilar (ed.), *Recognition of Anions*, *Structure and Bonding*, vol. 129 (Springer, Berlin, 2008), pp. 1–252
27. P.A. Gale, W. Dehaen (eds.), *Anion Recognition by Supramolecular Chemistry, Topics in Heterocyclic Chemistry*, vol. 24 (Springer, Berlin, 2010), pp. 1–370
28. K. Bowman-James, A. Bianchi, E. García-España (eds.), *Anion Coordination Chemistry* (Wiley, Weinheim, 2011)
29. B. Dong, H. Maeda, Ion-based materials comprising planar charged species. Chem. Commun. **49**, 4085–4099 (2013)
30. H. Maeda, Y. Bando, Recent progress in research on anion-responsive pyrrole-based π-conjugated acyclic molecules. Chem. Commun. **49**, 4100–4113 (2013)
31. H. Maeda. Supramolecular chemistry of pyrrole-based π-conjugated molecules. Bull. Chem. Soc. Jpn. **86**, 1359–1399 (2013)
32. H. Maeda, Y. Kusunose, Dipyrrolyldiketone difluoroboron complexes: novel anion sensors with C-H···X^- interactions. Chem. Eur. J. **11**, 5661–5666 (2005)
33. H. Maeda, Y. Haketa, T. Nakanishi, Aryl-substituted C_3-bridged oligopyrroles as anion receptors for formation of supramolecular organogels. J. Am. Chem. Soc. **129**, 13661–13674 (2007)
34. H. Maeda, Y. Ito, Y. Haketa, N. Eifuku, E. Lee, M. Lee, T. Hashishin, K. Kaneko, Solvent-assisted organized structures based on amphiphilic anion-responsive π-conjugated systems. Chem. Eur. J. **15**, 3706–3719 (2009)

35. H. Maeda, Y. Terashima, Y. Haketa, A. Asano, Y. Honsho, S. Seki, M. Shimizu, H. Mukai, K. Ohta, Discotic columnar mesophases derived from 'rod-like' π-conjugated anion-responsive acyclic oligopyrroles. Chem. Commun. **46**, 4559–4561 (2010)
36. H. Maeda, Y. Bando, Y. Haketa, Y. Honsho, S. Seki, H. Nakajima, N. Tohnai, Electronic and optical properties in the solid-state molecular assemblies of anion-responsive pyrrole-based π-conjugated systems. Chem. Eur. J. **16**, 10994–11002 (2010)
37. Y. Haketa, S. Sasaki, N. Ohta, H. Masunaga, H. Ogawa, N. Mizuno, F. Araoka, H. Takezoe, H. Maeda, Oriented salts: dimension-controlled charge-by-charge assemblies from planar receptor–anion complexes. Angew. Chem. Int. Ed. **49**, 10079–10083 (2010)
38. H. Maeda, K. Naritani, Y. Honsho, S. Seki, Anion modules: building blocks of supramolecular assemblies by combination with π-conjugated anion receptors. J. Am. Chem. Soc. **133**, 8243–8896 (2011)
39. Y. Haketa, S. Sakamoto, K. Chigusa, T. Nakanishi, H. Maeda, Synthesis, crystal structures, and supramolecular assemblies of pyrrole-based anion receptors bearing modified pyrrole β-substituents. J. Org. Chem. **76**, 5177–5184 (2011)
40. B. Dong, Y. Terashima, Y. Haketa, H. Maeda, Charge-based assemblies comprising planar receptor–anion complexes with bulky alkylammonium cations. Chem. Eur. J. **18**, 3460–3463 (2012)
41. Y. Bando, S. Sakamoto, I. Yamada, Y. Haketa, H. Maeda, Charge-based and charge-free molecular assemblies comprising π-extended derivatives of anion-responsive acyclic oligopyrroles. Chem. Commun. **48**, 2301–2303 (2012)
42. Y. Haketa, Y. Honsho, S. Seki, H. Maeda, Ion materials comprising planar charged species. Chem. Eur. J. **18**, 7016–7020 (2012)
43. B. Dong, T. Sakurai, Y. Honsho, S. Seki, H. Maeda, Cation modules as building blocks forming supramolecular assemblies with planar receptor–anion complexes. J. Am. Chem. Soc. **135**, 1284–1287 (2013)
44. Y. Terashima, M. Takayama, K. Isozaki, H. Maeda, Ion-based materials of boron-modified dipyrrolyldiketones as anion receptors. Chem. Commun. **49**, 2506–2508 (2013)
45. Y. Bando, T. Sakurai, S. Seki, H. Maeda, Corannulene-fused anion-responsive π-conjugated molecules that form self-assemblies with unique electronic properties. Chem. Asian J. **8**, 2088–2095 (2013)
46. Y. Terashima, T. Sakurai, Y. Bando, S. Seki, H. Maeda, Assembled structures of anion-responsive π-systems tunable by alkyl/perfluoroalkyl segments in peripheral side chains. Chem. Mater. **25**, 2656–2662 (2013)
47. H. Maeda, W. Hane, Y. Bando, Y. Terashima, Y. Haketa, H. Shibaguchi, T. Kawai, M. Naito, K. Takaishi, M. Uchiyama, A. Muranaka, Chirality induction by formation of assembled structures based on anion-responsive π-conjugated molecules. Chem. Eur. J. **19**, 16263–16271 (2013)
48. B. Dong, T. Sakurai, Y. Bando, S. Seki, K. Takaishi, M. Uchiyama, A. Muranaka, H. Maeda, Ion-based materials derived from positively and negatively charged chloride complexes of π-conjugated molecules. J. Am. Chem. Soc. **135**, 14797–14805 (2013)
49. B.W. Laursen, F.C. Krebs, Synthesis of a triazatriangulenium salt. Angew. Chem. Int. Ed. **39**, 3432–3434 (2000)
50. B.W. Laursen, F.C. Krebs, Synthesis, structure, and properties of azatriangulenium salts. Chem. Eur. J. **7**, 1773–1783 (2001)
51. Y. Yamamoto, T. Fukushima, Y. Suna, N. Ishii, A. Saeki, S. Seki, S. Tagawa, M. Taniguchi, T. Kawai, T. Aida, Photoconductive coaxial nanotubes of molecularly connected electron donor and acceptor layers. Science **314**, 1761–1764 (2006)
52. A. Saeki, S. Seki, T. Sunagawa, K. Ushida, S. Tagawa, Charge-carrier dynamics in polythiophene films studied by in situ measurement of flash-photolysis time-resolved microwave conductivity (FP-TRMC) and transient optical spectroscopy (TOS). Philos. Mag. **86**, 1261–1276 (2006)

53. T. Umeyama, N. Tezuka, S. Seki, Y. Matano, M. Nishi, K. Hirao, H. Lehtivuori, V.N. Tkachenko, H. Lemmetyinen, Y. Nakao, S. Sakaki, H. Imahori, Selective formation and efficient photocurrent generation of [70]fullerene–single-walled carbon nanotube composites. Adv. Mater. **22**, 1767–1770 (2010)
54. Y. Yasutani, A. Saeki, T. Fukumatsu, T. Koizumi, S. Seki, Unprecedented high local charge-carrier mobility in P3HT revealed by direct and alternating current methods. Chem. Lett. **42**, 19–21 (2013)
55. K. Sato, S. Takeuchi, S. Arai, M. Yamaguchi, T. Yamagishi, 4, 6-Bis(imidazolio)pyrimidine as a new anion receptor. Heterocycles **73**, 209–215 (2007)
56. A. Rit, T. Pape, F.E. Hahn, Self-assembly of molecular cylinders from polycarbene ligands and Ag^{I} or Au^{I}. J. Am. Chem. Soc. **132**, 4572–4573 (2010)
57. H. Maeda, Y. Ito, Y. Kusunose, T. Nakanishi, Dipyrrolylpyrazoles: anion receptors in protonated form and efficient building blocks for organized structures. Chem. Commun. 1136–1138 (2007)
58. H. Maeda, K. Chigusa, T. Sakurai, K. Ohta, S. Uemura, S. Seki, Ion-pair-based assemblies comprising pyrrole–pyrazole hybrids. Chem. Eur. J. **19**, 9224–9233 (2013)

Chapter 10
Nanotechnology and Nanomaterials in Photodeformable Liquid Crystalline Polymers

Li Yu, Haifeng Yu and Quan Li

Abstract The combination of nanomaterials and ordered deformable soft materials is emerging as an enabling system in nanoscience and nanotechnology. In this context, nanomaterial functionalized photoresponsive liquid crystalline polymers are very promising and versatile systems due to their dynamic function. Moreover, the unique characteristic of nanomaterials combined with the mechanical, self-organizing and stimuli-responsive properties of deformable liquid crystalline polymers opens up new and exciting possibilities. In this chapter, we present recent developments of photodeformable behaviors of liquid crystalline polymers functionalized with nanomaterials. The main emphasis revolves around how the physicochemical properties of different nanomaterials modulate the reversible photomechanical behaviors of liquid crystalline polymers and their potential application in devices such as optically controlled switches and soft actuators.

10.1 Introduction

Stimuli-responsive materials have sparked enormous interest in recent years due to their potential applications in micro-machines, soft robots, biomedical systems, etc. [1–6]. A variety of intelligent polymeric materials such as shape memory polymers [7, 8], polymer gels [9, 10], conducting polymers [11, 12], and dielectric elastomers [13, 14] have been developed for these applications. Compared to other stimulus-driven methods including pressure [15], heat [16, 17], electric field

L. Yu · H. Yu (✉)
Department of Materials Science and Engineering, College of Engineering, Peking University, Beijing, China
e-mail: yuhaifeng@pku.edu.cn

Q. Li (✉)
Liquid Crystal Institute, Kent State University, Kent, OH, USA
e-mail: qli1@kent.edu

Q. Li (ed.), *Nanoscience with Liquid Crystals*, NanoScience and Technology,
DOI: 10.1007/978-3-319-04867-3_10,

[18, 19], magnetic field [20], and pH [21], light is particularly attractive because it can provide remote, instant and precise noncontact control. Thus, photoresponsive polymers that can deform in response to light have become a hot topic in both academic and industrial fields. Among all the photoresponsive polymers, liquid crystalline polymers (LCPs) with anisotropic mesogens in the side or main chains have been attracting more and more attentions because of their advantages: (1) They elegantly combine self-organization properties of liquid crystals (LCs) with the mechanical properties of polymers; (2) They show reversible photodeformable behaviours because of the photoisomerization of photo crosslinking moieties such as azobenzene and cinnamate groups; (3) They exhibit large shape deformations with relatively weak forces due to the amplification of the response by the cooperative motion of mesogenic moieties; and (4) Their preferential deformations can be precisely controlled by the orientation of mesogenic moieties.

In recent years, nanotechnology and nanomaterials have been widely used in designing advanced functional materials based on photoresponsive LCPs. Although several reviews have concentrated on the photodeformable effect of LCPs and their applications in soft actuators [22], to date the influence of nanostructures and nanomaterials on the photodeformable properties of LCPs has not been summarized. In this chapter, we mainly focus on the utilization of special nanostructures and amazing physicochemical properties of nanomaterials to manipulate the photomechanical behaviors of LCPs. Furthermore, their potential applications as light-driven devices and other future prospects are proposed.

10.2 Mechanism of Photoinduced Deformation in LCPs

Generally, azobenzene is one of the most extensively studied photochromic groups due to its unique properties such as photoisomerization, photoinduced phase transition and photo-triggered cooperative motion in LC systems [23–31]. Especially, the LCPs with azobenzene acting as both a mesogen and a photoresponsive chromophore in the side chain dominate the research field of photo-driven LCPs. In this system, "bimetal model" is reasonable to explain the photomechanical behaviors of LCPs [32]. As shown in the left of Fig. 10.1, when mesogens are oriented parallel to the surface of LCP films, defined as the homogeneous alignment, volume contraction is often generated along this pre-aligned direction leading to the anisotropic bending behavior towards a light source. According to the "bimetal model", *trans*-to-*cis* photoisomerization takes place only in the surface area of LCPs since the molar extinction coefficient (ε) of azobenzene moieties at around 360 nm is very large (on the order of 10^4 L/mol cm), causing incident photons to be absorbed just in the surface of LCP films (a thickness of several tens of microns). Thus, these LCP films can be considered as a bilayer structure in which the upper layer is active layer and the bottom layer is passive one. As a result, the volume contraction is produced only in the surface layer from the irradiated side, which causes LCPs to bend toward the light source. On the

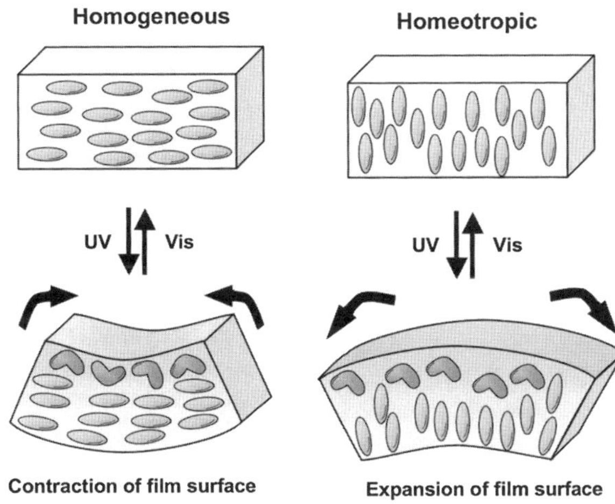

Fig. 10.1 Plausible mechanism of different photoinduced deformation behaviors of LCPs with their mesogens in homogeneous and homeotropic alignment states, respectively. Reproduced with permission from [33]. Copyright 2013

contrary, when mesogens are oriented perpendicular to the surface of LCPs, a different photodeformable behavior of bending away from the light source occurs due to the volume expansion of the upper layer (active layer) of LCP films [33]. Additionally, the photodeformable behavior of LCPs is reversible because of the reversible photoinduced phase transition and the change of mesogenic orientation.

10.3 Application of Nanotechnology and Nanomaterials in Photo-Driven Actuators of LCPs

Because of the unique photodeformable behavior of LCPs, a variety of soft actuators such as flexible inchworm walkers [34], plastic motors [35] and high-frequency oscillators [36] have been designed. For these soft photo-driven actuators based on LCPs, many factors determine the photo-driven behaviors of actuators, including the wavelength, polarization direction and intensity of the light source, and inherent parameters of actuators such as the concentration of azobenzene moieties, the crosslinking density and the molecular arrangement within actuators. Besides, their mechanical properties, conductivity and overall physical attributes should also be taken into consideration for practical applications (see in Fig. 10.2).

In order to meet these requirements for the application of LCPs in different areas, the research concerning nanotechnology and nanomaterials in photo-driven actuators made from LCPs has been focused on the following aspects: (1) Design of special nanostructures as the nanotemplate for the molecular arrangement of mesogens; (2) Utilization of nanomaterials as a photo-absorbing agent and thermal source to trigger the phase transition of azobenzene due to their photothermal

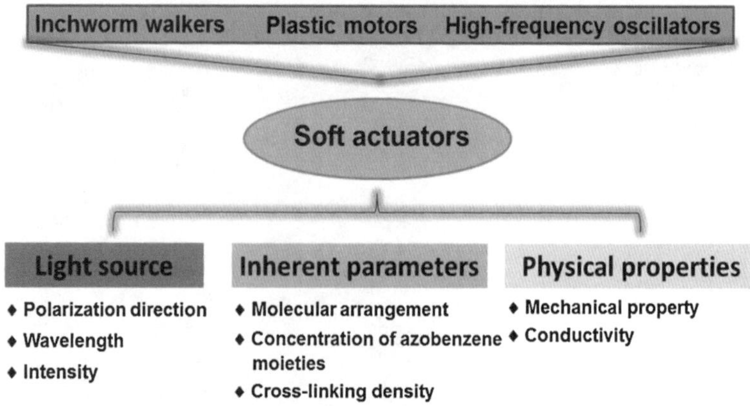

Fig. 10.2 Influencing factors for potential applications of soft actuators based on LCPs

effect; and (3) Using upconversion nanomaterials to adjust the wavelength of the light inducing the deformation of LCPs. Herein, the influence of nanostructures and nanomaterials on the photomechanical effect of LCPs will be discussed in detail based on these three aspects.

10.3.1 Template for Alignment of Mesogens

As mentioned in Sect. 10.2, the alignment of mesogens and the pre-treatment condition play a dominated role in the photodeformable behaviors of LCPs. Mechanically-treated methods utilizing stretching or rubbing have been typically applied to align mesogenic molecules. For instance, Finkelmann and coworkers first fabricated crosslinked LCP (CLCP) film with reversible photoinduced deformation by a two-step method [37]. In this method, well-defined networks were synthesized in the first step and then mesogens were well aligned along with the stretching direction and cross-linked to fix the ordered mesogens in the second step. Although a large contraction ratio of 20 % was achieved, harsh conditions such as the fast reaction rate of vinyl groups and fixation of ordered mesogens with a constant load were required. It is well known that mechanical rubbing of a polymer film (polyvinyl alcohol or polyimide) on a substrate has been widely adopted for the production of LC display devices. Undoubtedly, broken debris and structural damage often occur because of the dust and electrostatic charge generated in the fabrication process. To overcome these existing issues of mechanical method for aligning LCs, a self-assembled nanoporous structure has been fabricated as a nanotemplate for LC alignment [38]. As shown in Fig. 10.3a, the nanoporous structures can be generated by the extrusion and uniaxial stretching of the polymer film, and polyethylene (PE), polypropylene (PP) and polyamide (PA)

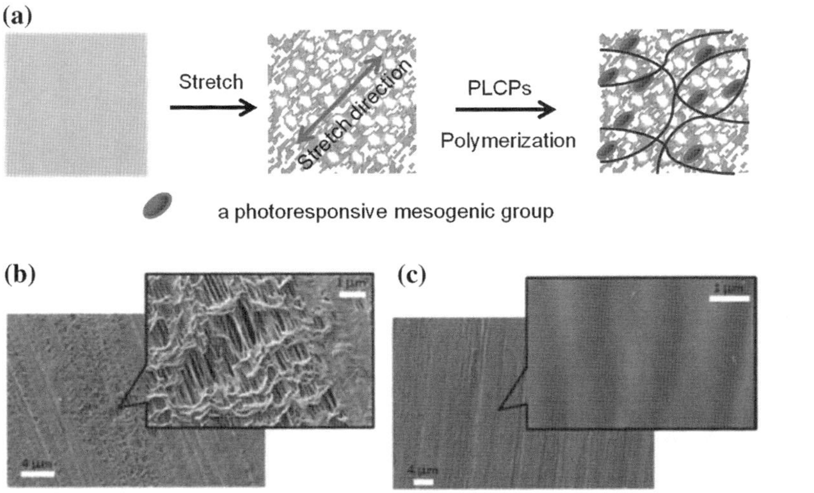

Fig. 10.3 Scheme of the preparation of nanoporous polymer template by stretch method and LCP composite film with mesogens highly aligning along with the stretch direction (**a**). SEM images of the initial porous polymer film (**b**) and the polymer composite film after filling with LC mixture and polymerization (**c**). Reproduced with permission from [38]. Copyright 2013

with a high crystallinity ($\sim$50 %) all can be used as such a polymer. Figure 10.3b shows a typical nanoporous structure of the polymer film, which just looks like composed of many parallel grooves. After immersing the nanoporous polymer film into the molten LC mixture and allowing thermal polymerization, the nanopores were filled with the CLCPs (see Fig. 10.3c). Upon irradiation with UV light, a strong bending behavior parallel to the stretching direction was observed in the nanocomposite film and it recovered to its initial state after the removal of the light source. Thus, the reversible photomechanical behavior can be attributed to the mesogens well aligned along the stretching direction induced by the self-assembled nanoporous structure. Moreover, the nanoporous polymer film functions as not only a template to orient mesogens but also an ideal matrix to provide excellent mechanical properties for these actuators. Although the excellent alignment behavior of LCs can be induced by these self-assembled nanoporous structures, the preparation process of the nanoporous structure is complicated and nanopore size distribution is not uniform. Additionally, the mechanism of LC alignment induced by the nanoporous structure is still unclear. Moreover, the morphology and size of nanopores are assumed to play an important role in inducing the LC alignment. Thus, the development of a convenient method to prepare nanoporous structures with uniform and controlled pore sizes is a promising way to align LCs and to explore the mechanism of LC alignment induced by the nanoporous structure.

For practical applications, the photoinduced deformation of LCPs should be accurately controlled in a designed way. As the photomechanical behavior of LCPs is closely related to the alignment direction of mesogens, multiple

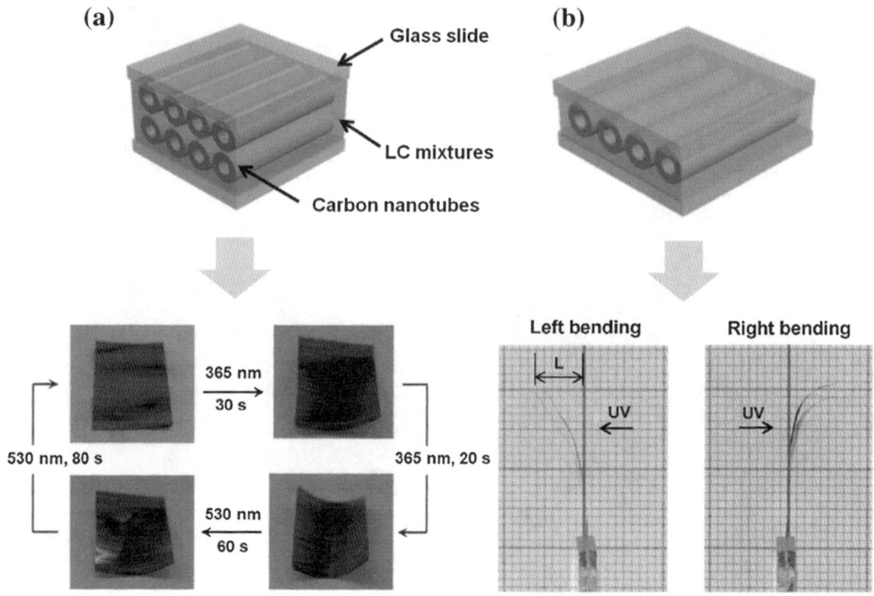

Fig. 10.4 Scheme of the preparation process of **a** CLCP nanocomposite film with mesogens oriented parallel to CNTs and **b** CLCP nanocomposite film with mesogens oriented perpendicularly to CNTs and their distinctive photomechanical behaviors. Reproduced with permission from [39, 40]. Copyright 2013

orientation directions are most desired, which can be induced by the self-assembled nanostructures. Recently, the CLCP, fabricated by utilization of ordered carbon nanotubes (CNTs) as a template to induce the LC alignment nanocomposites, were reported to exhibit excellent photodeformable behavior [39, 40]. In Fig. 10.4a, the CLCP nanocomposites showed distinctive photodeformable behaviors when they were fabricated with different preparation processes. Molten mixtures of the liquid crystalline monomer, initiator and crosslinker were injected into a LC cell that was made of two CNTs-sheet-covered glass slides with CNTs sheet inside. The freestanding CLCP nanocomposite film was obtained by photopolymerization at the liquid crystalline phase. A rapid and reversible photoinduced deformation was achieved by alternating UV and visible light irradiation, and the CLCP nanocomposite film curled along the aligned direction of CNTs indicating mesogens oriented parallel to the align direction of CNTs. This phenomenon might result from the structure of CNTs' sheet performing the same function as the surface of a rubbed polyimide film.

In Fig. 10.4b, the CLCP nanocomposite film was fabricated via a different process. The LC monomer, initiator and crosslinker were all dissolved in the solvent and doped onto the CNT strip. The CLCP nanocomposite film was achieved by the photopolymerization of CNT strip in the LC cell without CNT sheets. Interestingly, a unique photomechanical behavior of the CLCP nanocomposite film bending away from the light source was observed, indicating that mesogens would be oriented perpendicular to the alignment direction of CNTs. It was assumed that the mechanism of alignment of mesogens induced by CNTs was related to many facets such as the special nanostructures of CNTs, non-covalent interactions between the CNTs and mesogens, and the preparation method of the CLCP nanocomposite. These will be further elucidated through continued experiments and efforts in the future. Overall, the mechanical properties and conductivity of CLCP nanocomposites can be greatly improved by the incorporation of CNTs.

10.3.2 Nanomaterials for Wavelength Regulation

It is well known that the wavelength and intensity of the light source triggering the photodeformable behaviors of LCPs are greatly determined by their application areas. For example, near infrared (NIR) light is a more attractive source of optical stimulus in biological systems compared to UV or visible light because the relatively long wavelength light scatters less and penetrates deeper into tissues and also causes less damage to organisms. As a result, the control of the wavelength of the activating light source by nanomaterials has been a hot topic in recent years. Based on the mechanism that the material responds to light, two strategies to manipulate this wavelength have been developed. One is to utilize the photothermal effect of nanomaterials and the other is to use the frequency shifting of upconversion nanomaterials.

10.3.2.1 Photothermal Effect

The photothermal effect of nanomaterials such as CNTs, graphenes, graphene oxides, and gold nanoparticles has been used in biomedical domain for hyperthermia therapy [41]. The photothermal effect of nanomaterials is shown in Fig. 10.5, in which nanomaterials strongly absorb the light with a characteristic frequency and transfer the light energy into heat due to internal conversion or the losses in the surface plasmon resonance (SPR). Because the photodeformable behavior of LCPs is attributed to the phase transition from an LC to isotropic state, which can be induced by the light or heat, the photothermal effect of nanomaterials can be considered as a promising method for controlling the responsive wavelength for the photoinduced deformation.

Several groups have selected CNTs as the photo-absorbing agent to induce the photodeformable behavior of LCPs because of their strong absorption in the visible and NIR region [42–45]. However, sensitivity and stability are relatively low in the case of an indirect photo-actuation of CNTs randomly dispersed in LCPs. To achieve excellent photodeformable behavior, both excellent dispersion property and alignment of CNTs in LCPs are required. Chen and coworkers fabricated

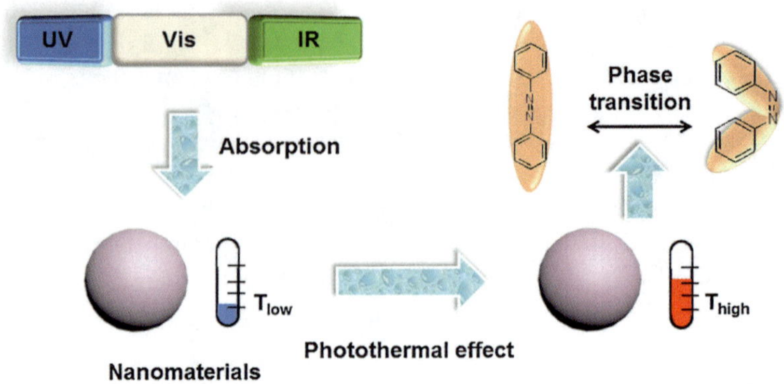

Fig. 10.5 Schematic illustrations of the photothermal effect of nanomaterials and the mechanism of the photomechanical behavior induced by the photothermal effect of nanomaterials

photo-driven deformable bilayer nanocomposite films based on the CNTs/LC elastomers and silicone, in which the CNT/LC elastomer acted as an active layer with silicone as the passive layer [46]. A large and stable deformation can be reversibly triggered by IR light due to the mechanical stretching induced by the photothermal effect of CNTs and their self-assembling behavior in the LCPs. The inchworm walker and artificial arm were realized by delicate structural design based on these bilayer nanocomposite films.

Similar to CNTs, gold-based nanomaterials have been explored for their excellent photothermal properties and are widely applied in the biomedical field due to their stability, biocompatibility and wavelength tunable SPR position. The SPR of gold nanoparticles is closely related to their shape, size and structure, which means that the characteristic wavelength of the SPR can be accurately prescribed by controlling the synthesis of gold nanoparticles [47–52]. Although few studies concerning the use of the photothermal effect of gold nanoparticles to induce photomechanical behavior of LCPs have been reported, gold nanoparticles are more convenient and suitable for regulating the wavelength sensitivity of nanocomposites when compared to CNTs.

10.3.2.2 Upconversion Nanomaterials

Upconversion nanomaterials are "frequency modulation" particles that can absorb multiple photons with the wavelength of λ_1 and then emit corresponding photons with the principal wavelength of λ_2 (Fig. 10.6), where the emission wavelength is shorter than the excitation wavelength. If the emission wavelength of upconversion nanomaterials λ_2 can match with that of the light inducing the *trans*-to-*cis* isomerization of azobenzene chromophores, the reversible photomechanical behavior of LCPs can be accomplished by light with a wavelength of λ_1. The above-mentioned process shows how upconversion nanomaterials can be used to regulate

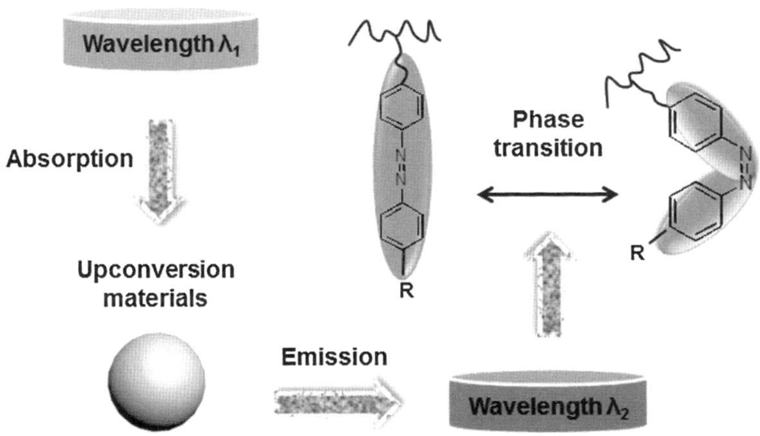

Fig. 10.6 Schematic illustration of the frequency modulating function of upconversion nanomaterials and the mechanism of the photomechanical behavior induced by upconversion nanomaterials

the wavelength of the stimulating light source. However, the emission wavelength of upconversion nanomaterials λ_2 is typically located in the visible spectrum and the wavelength of light triggering the *trans*-to-*cis* isomerization of azobenzene chromophores is located in the UV region. This mismatching of wavelengths presents a challenge for this strategy.

In order to overcome this issue, a solution is to adjust the wavelength of the light inducing the *trans*-to-*cis* isomerization to visible region by designing molecular structures containing azobenzene chromophores and to fabricate corresponding upconversion nanomaterials that have the appropriate wavelength, simultaneously. For example, a longer conjugated azotolane structure grafted to the side chain of LCPs followed by mixing the upcoversion nanoparticles. The resultant LCP films underwent photoinduced bending and unbending deformation by manipulating the actinic visible light [53]. For the purpose of matching the wavelength of the light inducing the *trans*-to-*cis* isomerization, corresponding upconversion nanoparticles of $NaYF_4$ were subsequently prepared [54]. As shown in Fig. 10.7, the main emission wavelengths of the upconversion nanomaterials were 450 and 475 nm that were in accordance with the wavelength of *trans*-to-*cis* photoisomerization of the azotolane moieties. Therefore, the photoinduced bending behavior of these LCPs can be triggered by NIR light with a wavelength of 980 nm. However, a relatively high excitation power density of 15 W/cm^2 of NIR light was required and the associated heating effect could restrain the potential applications of this scheme. Thus, a red-to-blue triplet-triplet annihilation (TTA) based upconversion system with a high absolute quantum yield of 9.3 $\pm$ 0.5 % was prepared. The generated composite film bent towards the light source when it was irradiated with 635-nm laser at relatively low power density of 200 mW/cm^2 [55].

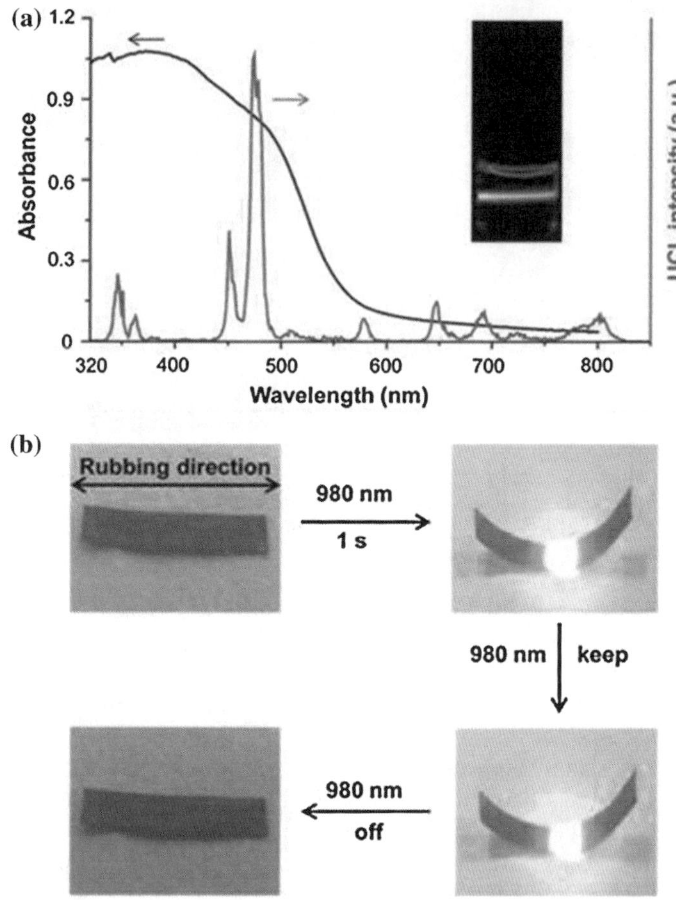

Fig. 10.7 **a** UCL emission spectrum (*blue line*) of a colloidal $CHCl_3$ solution of UCNPs excited with a 980 nm CW laser and the UV–vis absorption spectrum (*black line*) of the azotalane CLCP film. **b** Photographs of the azotolane CLCP/UCNP composite film bending toward the light source along the alignment direction of the mesogens, remaining bent in response to the CW NIR irradiation at 980 nm, and becoming flat again after the light source was removed. Reproduced with permission from [49]. Copyright 2013

10.4 Soft Actuators Based on Nanomaterials Functionalized LCPs

As mentioned in Sect. 10.3.2, nanomaterials can be utilized to allow design flexibility of the wavelength of the stimulating light source triggering the photomechanical behavior of LCPs. At the same time, the functionalization of LCPs has been achieved by the incorporation of nanomaterials, which greatly extends their applications in many fields. Compared to another important class of smart moving

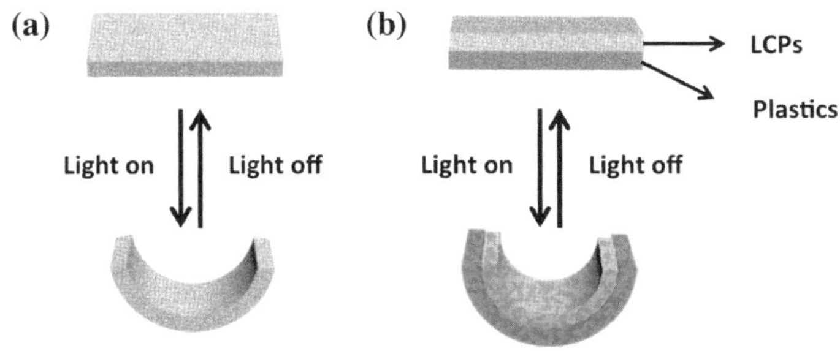

Fig. 10.8 **a** Reversible photomechanical behavior of LCPs. **b** Reversible photomechanical behavior of bilayer structure composed of LCPs acting as an active layer and plastics without photoresponsive behavior as a passive layer

polymers, namely shape memory polymers, LCPs exhibit reversible photomechanical behaviors (see Fig. 10.8a). However, the poor film-forming characteristics and mechanical properties of LCPs severely limit their practical applications. In order to overcome this issue, a strategy to use a bilayer structure composed of LCPs and normal polymer plastics, e.g. polyethylene (PE), polycarbonate (PC) and silicone, was developed [56]. In this bilayer structure, LCP is an active layer that provides reversible photomechanical response and the photo-inert plastics is a passive layer that supplies engineered excellent film-forming and mechanical properties (see in Fig. 10.8b). Adhesion between the active layer and passive layer is important for the reversible photomechanical behavior of the bilayer structure and ultimately for the reliability and stability of soft actuators based on this approach. Thus, adhesive interlayers and crosslinking between film layers are ways usually used to enhance the mechanical connection between the active layer and the passive layer. A variety of soft actuators with sophisticated structure have been designed based on the bilayer configuration and will be well described in the next section.

10.4.1 Optically Controlled Switching

In general, optically controlled switching circuits have been realized through photosensitive properties such as resistance that can translate light signals into electrical ones. Compared to the traditional light-sensitive properties, the reversible photodeformable behavior mechanism of LCPs is distinct and simple [40]. By incorporation of CNT, the CNT/LCP nanocomposite exhibits unique reversible photomechanical behavior and excellent overall conductivity due to the templating and high conductivity of CNTs respectively. When the nanocomposite strip is irradiated with UV light from the right side, it quickly bends to the left side, i.e. away from the UV source, to contact the left conductive wire making the left electric circuit path continuous. Removal of the UV source allows the

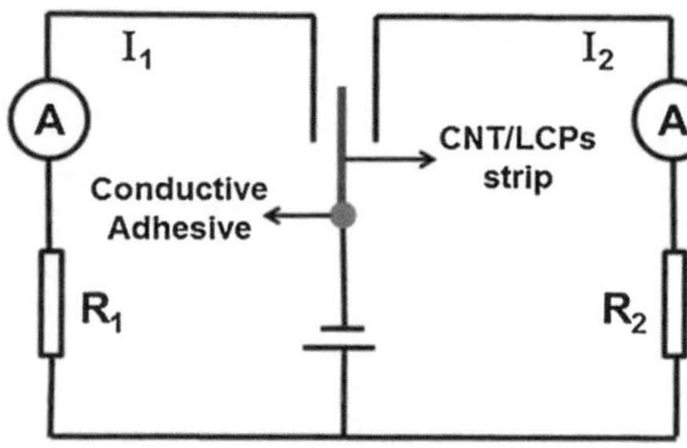

Fig. 10.9 Application of the electrically conductive CNT/LCP nanocomposite strip for a remote electric switch. Reproduced with permission from [40]. Copyright 2013

nanocomposite strip to return to its initial position and the electric circuit at the left is open. Similarly, when the nanocomposite strip is UV irradiated from the left side, it quickly bends to the right side to contact the right conductive wire making the right electric circuit path complete, which is similarly reversible. Based on this demonstration of a photo regulation mechanism through LCPs, it can be clearly seen that CNT/LCP nanocomposites can effectively act as optically controlled switches (Fig. 10.9). Moreover, reliability and stability of this optical controlled switching is promising because the photomechanical behavior of the nanocomposite exhibits no decay even after one hundred cycles.

10.4.2 IR-Triggered Artificial Arm

As mentioned above, simple and reversible photoinduced bending behavior has been achieved in LCPs by phase transition or change of mesogenic ordering in the LCP microstructures. However, active origami structures usually exhibit complex three dimensional (3D) shapes upon exposure to an external stimulus in environment. Thus, the realization of complex 3D shapes in LCPs could be a key step in enabling their practical application. From Fig. 10.10, it can be observed that distinctive 3D shapes occur upon IR irradiation when different macroscopic structures of soft actuators are designed based on the bilayer structure of CNT/LCP nanocomposite and polycarbonate (PC) [46]. The artificial arm, a most important part of robot, has been drawing extensive attention in the literature and new approaches are continuously being developed. An IR-triggered artificial arm has been successfully fabricated based on CNT/LCP nanocomposites/PC bilayer, for example. In comparison to other artificial arms developed by conducting polymers, dielectric elastomers and shape memory polymers, it needs no electric wire and can grip irregularly shaped objects in different conditions, even in water. This kind of artificial arm can be remotely and accurately controlled and has wide potential applications especially in the aeronautics and astronautics fields.

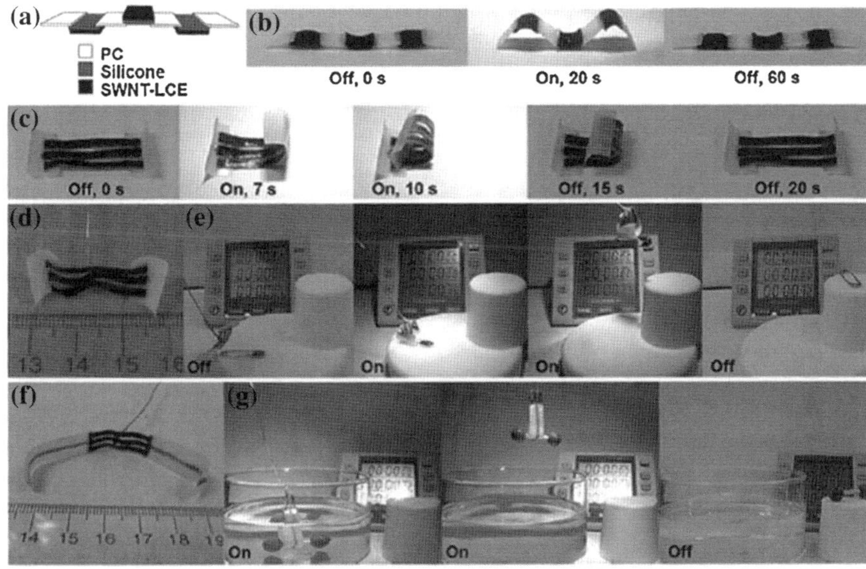

Fig. 10.10 **a** Macroscopic structure and **b** complex 3D folding behavior of the device composed of bilayer structure CNT/LCPs nanocomposite and PC. **c** Reversible folding and unfolding behavior of the artificial arm based on CNT/LCPs nanocomposite and PC. **d–g** Grip irregular shape objects with this artificial arm in the condition of air and water. Reproduced with permission from [46]. Copyright 2013

10.4.3 Inchworm Walker Devices

Inchworm, an interesting and typical movement mode of active origami, also can be reproduced based on the bilayer structure of CNT/LCP nanocomposite and PC [46]. In general, the structure of active origami exhibiting inchworm movement is complicated. In some cases, sophisticated structures composed of four different layers are required to imitate the inchworm movement in the inchworm walker device (Fig. 10.11b). The movement of the inchworm walker device is enabled by the reversible photomechanical behavior of CNT/LCPs combined with the macroscopic structure of the inchworm walker device. The device bends away from the light source exhibiting an arc shape upon IR irradiation. Due to the unique shape of the bottom PC layer and friction interaction between the bottom PC layer and the substrate, the device then apparently creeps forwards a distance as it completely recovers to its initial shape after the IR source is removed. Because of the reversible photomechanical behavior of CNT/LCPs, the inchworm walker device can creep forward continuously as long as it undergoes an on-off cycle of light irradiation.

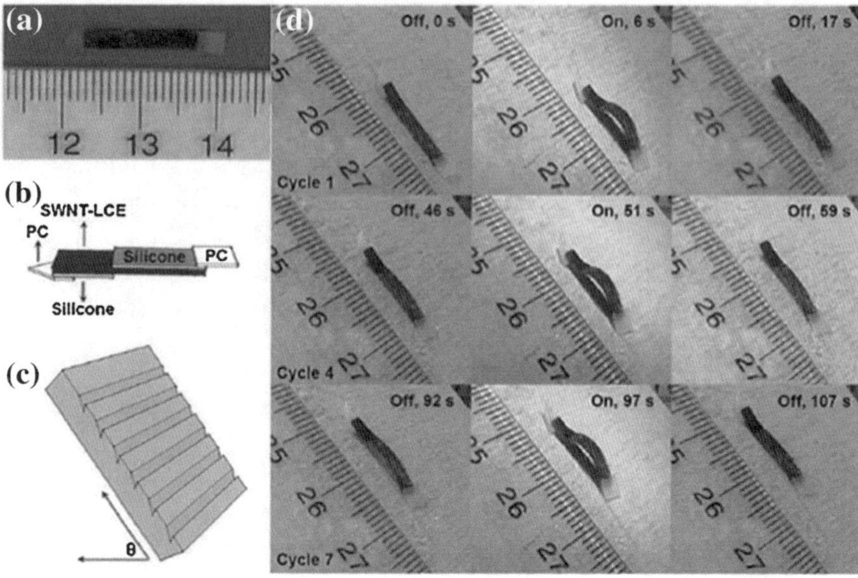

Fig. 10.11 **a** Photograph and **b** macroscopic structure of the inchworm walker device. **c** Scheme of a ratcheted wood substrate. **d** The process of creep behavior of the inchworm walker device during cycles of IR on and off. Reproduced with permission from [46]. Copyright 2013

10.5 Conclusions and Outlook

It is the elegant combination of self-organization properties of LCs with mechanical performance of organic polymers that enables photoresponsive LCPs to directly convert light energy into mechanical work. These kinds of LCPs with their unique characteristics show potential applications in light-driven sensors, actuators and micro-opto-mechanical systems (MOMs). From the viewpoint of practical applications, many characteristics of these devices based on LCPs including the triggering wavelength, complex 3D shape changes and various mechanical functions must be accurately controlled. In order to meet these requirements, self-assembled nanostructures and nanomaterials have been utilized as the template to induce the alignment of mesogens and brought the active components together to render LCPs with new functionality. A variety of excellent soft actuators including optically controlled switching, an IR triggered artificial arm and inchworm walker have been developed based on these LCP nanocomposites. Although the photodeformable behavior of LCPs can be greatly influenced by the incorporation of nanomaterials, the interactions between LCPs and nanomaterials, and the relationship of the self-assembled nanostructure with these interactions, the photomechanical behavior of these systems continues to be explored with considerable effort. Furthermore, a variety of actuators with complicated deformation behaviors based on LCPs nanocomposites are urgently required for diverse applications. The photoresponsive behavior in such systems can be activated quickly and with high energy transfer efficiency. Several of these new routes have been initiated only during the last few years and some may develop into highly successful research fields and applications. In most of the areas discussed here, the community is still at a relatively early stage of exploration. Undoubtedly, the technical impact of the various themes will be assessed in the coming years. Ultimately, these are very exciting research developments for photodeformable LCPs and nanomaterials, which shed light both on their utility and on the complex, fundamental interactions within these nanostructured and responsive nanocomposites.

References

1. Q. Li, *Intelligent Stimuli-Responsive Materials: From Well-Defined Nanostructures to Applications* (Wiley, Hoboken, 2012)
2. H.F. Yu, T. Ikeda, Photocontrollable liquid-crystalline actuators. Adv. Mater. **23**, 2149–2180 (2011)
3. Y. Wang, Q. Li, Light-driven chiral molecular switched of motors in liquid crystals. Adv. Mater. **24**, 1926–1945 (2012)
4. M.H. Li, P. Keller, J.Y. Yang, P.-A. Albouy, An artificial muscle with lamellar structure based on a nematic triblock copolymer. Adv. Mater. **16**, 1922–1925 (2004)
5. A. Lendlein, R. Langer, Biodegradable, elastic shape-memory polymers for potential biomedical applications. Science **296**, 1673–1676 (2002)
6. L. Ionov, Biomimetic 3D self-assembling biomicroconstructs by spontaneous deformation of thin polymer films. J. Mater. Chem. **22**, 19366–19375 (2012)
7. A. Lendlein, S. Kelth, Shape-memory polymers. Angew. Chem. Int. Ed. **41**, 2034–2057 (2002)
8. T. Xie, Tunable polymer multi-shape memory effect. Nature **464**, 267–270 (2010)
9. Y. Osada, H. Okuzaki, H. Hori, A polymer gel with electrically driven motility. Nature **355**, 242–244 (1992)
10. K. Kajiwara, S.B. Rossmurphy, Polymers-synthetic gels on the move. Nature **355**, 208–209 (1992)
11. E. Smela, Conjugated polymer actuators for biomedical applications. Adv. Mater. **15**, 481–494 (2003)
12. W. Lu, A.G. Fadeev, B. Qi, E. Smela, B.R. Mattes, J. Ding, G.M. Spinks, J. Mazukiewicz, D. Zhou, G.G. Wallace, D.R. MacFarlane, S.A. Forsyth, M. Forsyth, Use of ionic liquids for pi-conjugated polymer electrochemical devices. Science **297**, 983–987 (2002)
13. Q.M. Zhang, V. Bhatti, X. Zhao, Giant electrostriction and relaxor ferroelectric behavior in electron-irradiated poly(vinylidene fluoride-trifluoroethylene) copolymer. Science **280**, 2101–2104 (1998)
14. Q.M. Zhang, H.F. Li, M. Poh, F. Xia, Z.Y. Cheng, H.S. Xu, C. Huang, An all-organic composite actuator material with a high dielectric constant. Nature **419**, 284–287 (2002)
15. F. Ilievski, A.D. Mazzeo, R.F. Shepherd, X. Chen, G.M. Whitesides, Soft robotics for chemists. Angew. Chem. Int. Ed. **123**, 1930–1935 (2011)
16. J.S. Leng, X. Lan, Y.J. Liu, S.Y. Du, Shape-memory polymers and their composites: stimulus methods and applications. Prog. Mater. Sci. **56**, 1077–1135 (2011)
17. M. Behl, A. Lendlein, Triple-shape polymers. J. Mater. Chem. **20**, 3335–3345 (2010)
18. S. Daunert, E.A. Moschou, M.J. Madou, L.G. Bachas, Voltage-switchable artificial muscles actuating at near neutral pH. Sens. Actuators **115**, 379–383 (2006)

19. J.S. Leng, W.M. Huang, X. Lan, Y.J. Liu, S.Y. Du, Significantly reducing electrical resistivity by forming conductive Ni chains in a polyurethane shape-memory polymer/carbon-black composite. Appl. Phys. Lett. **92**, 204101 (2008)
20. R. Mohr, K. Kratz, T. Weigel, M. Lucka-Gabor, M. Moneke, A. Lendlein, Initiation of shape-memory effect by inductive heating of magnetic nanoparticles in thermoplastic polymers. Proc. Natl. Acad. Sci. U.S.A. **103**, 3540–3545 (2006)
21. K.D. Harris, C.W.M. Bastiaansen, J. Lub, D.J. Broer, Self-assembled polymer films for controlled agent-driven motion. Nano Lett. **5**, 1857–1860 (2005)
22. J. Wei, Y. Yu, Photodeformable polymer gels and crosslinked liquid-crystalline polymers. Soft Matter **8**, 8050–8059 (2012)
23. C. Yelamaggad, S.K. Prasad, Q. Li, in Chapter 4 in *Liquid Crystals Beyond Displays: Chemistry, Physics, and Applications,* ed. by Q. Li. Photo-Stimulated Phase Transformations in Liquid Crystals and Their Non-display Applications (Wiley, Hoboken, 2012)
24. H.F. Yu, A. Asaoka, A. Shishido, T. Iyoda, T. Ikeda, Photoinduced nanoscale cooperative motion in a novel well-defined triblock copolymer. Small **3**, 768–771 (2007)
25. H.F. Yu, H. Liu, T. Kobayashi, Fabrication and photoresponse of supramolecular liquid-crystalline microparticles. ACS Appl. Mater. Interfaces **3**, 1333–1340 (2011)
26. Y. Li, M. Wang, T.J. White, T.J. Bunning, Q. Li, Azoarenes bearing opposite chiral configurations: light-driven dynamic reversible handedness inversion in self-organized helical superstructure. Angew. Chem. Int. Ed. **52**, 8925–8929 (2013)
27. T.-H. Lin, Y. Li, C.-T. Wang, H.-C. Jau, C.-W. Chen, C–.C. Li, H.K. Bisoyi, T.J. Bunning, Q. Li, Red, green and blue reflections enabled in optically tunable self-organized 3D cubic nanostructured thin film. Adv. Mater. **25**, 5050–5054 (2013)
28. M. Mathews, R. Zola, S. Hurley, D.-K. Yang, T.J. White, T.J. Bunning, Q. Li, Light-driven reversible handedness inversion in self-organized helical superstructures. J. Am. Chem. Soc. **132**, 18361–18366 (2010)
29. H.F. Yu, T. Iyoda, T. Ikeda, Photoinduced alignment of nanocylinders by supramolecular cooperative motions. J. Am. Chem. Soc. **128**, 11010–11011 (2006)
30. R. Sun, X. Ma, M. Gao, H. Tian, Q. Li, Light-driven linear helical supramolecular polymer formed by molecular-recognition-directed self-assembly of bis-*p*-sulfonatocalix[4]arene and pseudorotaxane. J. Am. Chem. Soc. **135**, 5990–5993 (2013)
31. H.F. Yu, J. Li, T. Ikeda, T. Iyoda, Macroscopic parallel nanocylinder array fabrication using a simple rubbing technique. Adv. Mater. **18**, 2213–2215 (2006)
32. T. Ikeda, M. Nakano, Y. Yu, O. Tsutsumi, A. Kanazawa, Anisotropic bending and unbending behavior of azobenzene liquid-crystalline gels by light. Adv. Mater. **15**, 201–205 (2003)
33. M. Konda, Y.L. Yu, T. Ikeda, How does the initial alignment of mesogens affect the photoinduced bending behavior of liquid-crystalline elastomers? Angew. Chem. Int. Ed. **45**, 1378–1382 (2006)
34. M. Yamada, M. Kondo, R. Miyasato, Y. Naka, J. Mamiya, M. Kinoshita, A. Shishido, Y. Yu, C.J. Barrett, T. Ikeda, Photomobile polymer materials-various three-dimensional movements. J. Mater. Chem. **19**, 60–62 (2009)
35. M. Yamada, M. Kondo, J. Mamiya, Y. Yu, M. Kinoshita, C.J. Barrett, T. Ikeda, Photomobile polymer materials: towards light-driven plastic motors. Angew. Chem. Int. Ed. **47**, 4986–4988 (2008)
36. C.L. Oosten, C.W.M. Bastiaansen, D.J. Broer, Printed artificial cilia from liquid-crystal network actuators modularly driven by light. Nature Mater. **8**, 677–682 (2009)
37. H. Finkelmann, H.J. Kock, H. Rehage, Investigations on liquid crystalline polysiloxanes, 3.Liquid crystalline elastomers-a new type of liquid crystalline material. Makromol. Chem. Rapid Commun. **2**, 317–322 (1981)
38. A. Rvabchun, A. Bobrovskv, J. Stumpe, V. Shibaev, Novel generation of liquid crystalline photo-actuators based on stretched porous polyethylene films. Macromol. Rapid Commun. **33**, 991–997 (2012)

39. W. Wang, X. Sun, W. Wu, H. Peng, Y. Yu, Photoinduced deformation of crosslinked liquid-crystalline polymer film oriented by a highly aligned carbon nanotube sheet. Angew. Chem. Int. Ed. **51**, 4644–4647 (2012)
40. X.M. Sun, W. Wang, L.B. Qiu, W.H. Guo, Y.L. Yu, H.S. Peng, Unusual reversible photomechanical actuation in polymer/nanotube composites. Angew. Chem. Int. Ed. **51**, 8520–8524 (2012)
41. H.M. Chen, Z.P. Zhen, T. Todd, P.K. Chu, J. Xie, Nanoparticles for improving cancer diagnosis. Mater. Sci. Eng. R **74**, 35–69 (2013)
42. S.V. Ahir, E.M. Terentjev, Photomechanical actuation in polymer-nanotube composites. Nature Mater. **4**, 491–495 (2005)
43. L. Yang, K. Setyowati, A. Li, S. Gong, J. Chen, Reversible infrared actuation of carbon nanotubes-liquid crystalline elastomer nanocomposites. Adv. Mater. **20**, 2271–2275 (2008)
44. Y. Ji, Y.Y. Huang, R. Rungsawang, E.M. Terentjev, Dispersion and alignment of carbon nanotubes in liquid crystalline polymers and elastomers. Adv. Mater. **22**, 3436–3440 (2010)
45. C.S. Li, Y. Liu, C.-W. Lo, H.R. Jiang, Reversible white-light actuation of carbon nanotube incorporated liquid crystalline elastomer nanocomposites. Soft Matter **7**, 7511–7516 (2011)
46. R.R. Kohimeyer, J. Chen, Wavelength-selective, IR light-driven hinges based on liquid crystalline elastomer composites. Angew. Chem. Int. Ed. **52**, 9234–9237 (2013)
47. C. Xue, Y. Xue, L. Dai, A. Urbas, Q. Li, Size and shape dependent fluorescence quenching of gold nanoparticles on perylene dye. Adv. Opt. Mater. **1**, 581–587 (2013)
48. C. Xue, K. Gutierrez-Cuevas, M. Gao, A. Urbas, Q. Li, Photo-modulated self-assembly of hydrophobic thiol monolayer-protected gold nanorods and their alignment in thermotropic liquid crystal. J. Phys. Chem. C **117**, 21603–21608 (2013)
49. C. Xue, O. Birel, Y. Li, X. Ma, M. Gao, A. Urbas, Q. Li, Metal complex monolayer-protected gold nanorods: a parallel facile synthesis and self-assembly. J. Colloid Interface Sci. **398**, 1–6 (2013)
50. X. Ma, A. Urbas, Q. Li, Controllable self-assembling of gold nanorods via on and off supramolecular noncovalent interactions. Langmuir **28**, 16263–16267 (2012)
51. J.M. El Khoury, X. Zhou, L. Qu, L. Dai, A. Urbas, Q. Li, Organo-soluble photoresponsive azo hybrid gold nanorods. Chem. Commun. 2109–2111 (2009)
52. M. Rycenga, C.M. Cobley, J. Zeng, W.Y. Li, C.H. Moran, Q. Zhang, D. Qin, Y.N. Xia, Controlling the synthesis and assembly of silver nanostructures for plasmonic applications. Chem. Rev. **111**, 3669–3712 (2011)
53. R.Y. Yin, W.X. Xu, M. Kondo, C. Yen, J. Mamiya, T. Ikeda, Y.L. Yu, Can sunlight drive the photoinduced bending of polymer films? J. Mater. Chem. **19**, 3141–3143 (2009)
54. W. Wu, L.M. Yao, T.S. Yang, R.Y. Yin, F.Y. Li, Y.L. Yu, NIR-light-induced deformation of cross-linked liquid-crystal polymer using upconversion nanophosphors. J. Am. Chem. Soc. **133**, 15810–15813 (2011)
55. Z. Jiang, M. Xu, F.Y. Li, Y.L. Yu, Red-light controllable liquid-crystal soft actuators via low-power excited upconversion based on triplet-triplet annihilation. J. Am. Chem. Soc. **135**, 16446–16453 (2013)
56. F.T. Cheng, R.Y. Yin, Y.Y. Zhang, C.C. Yen, Y.L. Yu, Fully plastic microrobots which manipulate objects using only visible light. Soft Matter **6**, 3447–3449 (2010)

Chapter 11
Self-Assembled Liquid Crystalline Conjugated Polymers: Synthesis, Development, and Their Advanced Electro-Optical Properties

Benedict A. San Jose and Kazuo Akagi

Abstract In this chapter, recent developments in self-assembled liquid crystalline conjugated polymers leading to advanced functionalities are discussed. Liquid crystallinity in conjugated polymers enables the self-assembly of a hierarchical structure or facilitates the formation of higher-ordered structures through liquid crystalline phases. Combining liquid crystallinity with the inherent photoluminescence and electrical conductivity of these polymers, which arise from their conjugated main chain structure, advanced functionalities such as linearly and circularly polarized luminescence, ferroelectricity, and dynamic switching are achieved. Such developments may pave the way for the creation of next-generation materials for applications in organic optoelectronics.

11.1 Introduction

Conjugated polymers are attractive materials for research due to their unique structure, which consists of covalently bonded atoms with alternating single and multiple bonds. This conjugated system of atomic bonds results in a delocalization of the electrons across all of the adjacent parallel aligned π-orbitals of the atoms, which increases the molecular stability and thereby lowers the energy gap of the molecule. This distinct feature of conjugated polymers and the phenomena of electron delocalization in conjugated polymer systems lead to functionalities such as electrical conductivity, fluorescence, and photo-harvesting behavior, making them attractive materials for the development of next-generation optoelectronic devices [1–8].

The electronic and optoelectronic properties of conjugated polymers can be modified and tuned by controlling the conformation, arrangement, and alignment

B. A. San Jose · K. Akagi (✉)
Department of Polymer Chemistry, Kyoto University, Katsura, Kyoto 615-8510, Japan
e-mail: akagi@fps.polym.kyoto-u.ac.jp

Q. Li (ed.), *Nanoscience with Liquid Crystals*, NanoScience and Technology,
DOI: 10.1007/978-3-319-04867-3_11, © Springer International Publishing Switzerland 2014

of the main chains, for which self-organization [9, 10], spontaneous alignment [11, 12], externally forced alignment [13–20], and intra- and inter-molecular interactions [20–27] can be intentionally incorporated. This control means that not only the primary structure but also the higher-order structure of conjugated polymers must be considered in the pursuit of advanced optoelectronic properties and functions. Self-assembly is one of the most reliable methods to construct higher-ordered and even super-hierarchical structures in conjugated polymers. The forces of self-assembly are classified as π-electron overlap interactions, van der Waals interactions, hydrogen-bonding interactions, and electrostatic interactions. Liquid crystalline behavior in polymers facilitates self-assembly and can be introduced to generate advanced functionalities such as linearly (LPL) and circularly polarized luminescence (CPL), ferroelectricity, and photo-stimuli responsiveness.

Liquid crystals (LCs) are described as a fluid phase that flows like a liquid and is oriented in a crystalline manner. LCs are divided into two types: thermotropic LCs, where the LC phase transition is dependent on temperature; or lyotropic LCs, where the LC phase transition occurs as a function of solvent concentration. To introduce liquid crystallinity to conjugated polymers, LC moieties can be introduced to the polymer side chains for side chain-type liquid crystallinity. On the other hand, designing conjugated polymers with rigid main chain structures with flexible alkyl side chains for solubility enables main chain-type liquid crystallinity.

It is well-known that polymers with LC moieties spontaneously align to form multidomains by virtue of the orientation of the LC moieties. Further linear arrangement of these polymers can be achieved by applying an external force such as shear stress or an electric or magnetic field [13–20] to construct a monodomain structure on a macroscopic level. LC polymers in a monodomain structure exhibit excellent anisotropic properties in their electrical conductivity and LPL. Moreover, when LC polymers are helically arranged into higher-ordered structures by chiral moieties or chiral dopants, the LC polymers can exhibit circular polarized functionalities such as circular dichroism (CD) and CPL. The schematic in Fig. 11.1 shows the formation of multidomains, monodomains, and helical assemblies through spontaneous orientation and linear and helical arrangements of LC conjugated polymers.

In this chapter, we focus on π-electron overlap interactions and van der Waals interactions present in LC conjugated polymers that lead to self-assembly, the formation of higher-ordered structures, and advanced functionalities. In Part 1, we discuss the hierarchical self-assembly of poly(meta-phenylene)s exhibiting a unique π-stabilized helicene helical structure in the main chain that enables the formation of polymer whiskers. In Part 2, we introduce the formation of a higher-ordered chiral nematic LC phase in di-substituted polyacetylene and thiophene-based copolymers exhibiting high degrees of CPL. In Part 3, we review the development of poly(meta-phenylene)s and poly(bithienylene-phenylene)s showing dynamic switching functionalities such as ferroelectric liquid crystallinity and fluorescence-quenching switching of LPL and CPL.

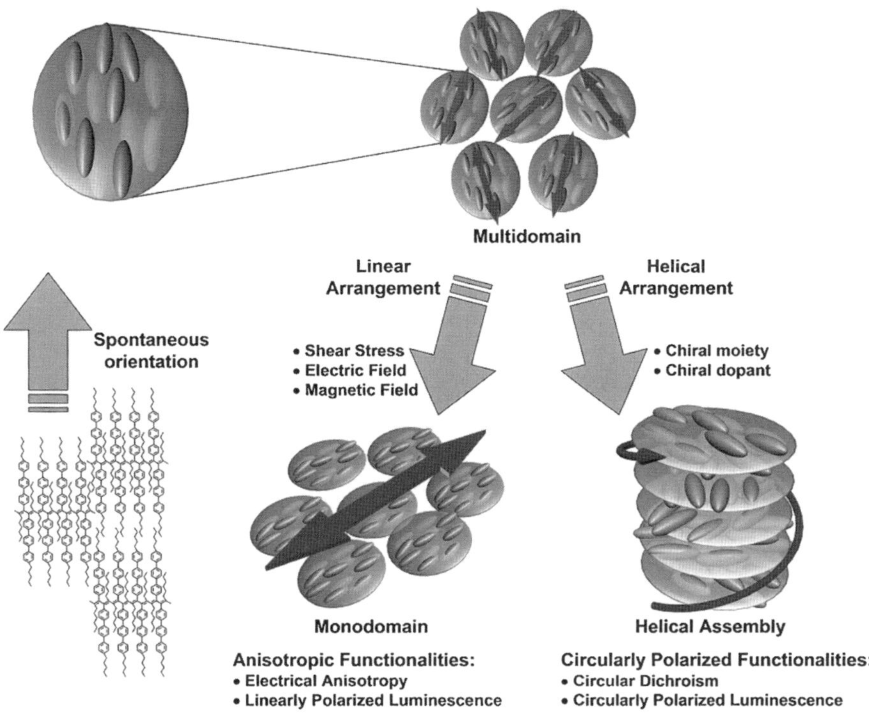

Fig. 11.1 Schematic representation of the spontaneous orientation, linear, and helical arrangement of LC conjugated polymers. Reprinted with permission from [17]. Copyright 2013, The royal chemical society

11.2 Hierarchical Self-assembly of Liquid Crystalline Conjugated Polymers

Self-assembly is one of the most promising ways to construct a higher-ordered and super-hierarchical structure of conjugated polymers. π-Electron overlap interactions or van der Waals interactions enable self-assembly of conjugated polymers. It is well-known that π-conjugated polymers with planar main chains tend to form π-stacked structures owing to interchain π-electron overlap interactions. It was elucidated that poly(meta-phenylene) (PMP) derivatives bearing a long alkyl group as a side chain has a helicene-like helical structure directed by intrachain π-electron overlaps based on a *cisoid* linkage between the 1- and 5- positions of the phenylene rings.

Self-assembling PMPs are able to form peculiar whiskers with a hierarchical structure consisting of intrachain helices owing to its rigidity and the high crystallinity of the directly meta-linked phenylene units in the conjugated backbone. The higher-ordered structure of PMP whiskers would be able to amplify the intrinsic functionalities of conjugated polymers, such as electrical conductivity and fluorescence.

The use of liquid crystalline functionalities in conjugated polymers facilitates their macroscopic alignment upon application of an external force such as shear stress or an electric or magnetic field [13–20]. Macroscopic alignment leads to an enhancement of the anisotropic functionalities of self-assembled conjugated polymers, which could pave the way for the development of advanced electroconductive or fluorescent materials [27].

11.2.1 Whisker Morphology Consisting of Hierarchically Self-assembled Polymer Helices

A polymer whisker is a spontaneously formed assembly with a higher-ordered structure and could be used as a single fiber for amplification of the intrinsic properties of the polymer. The whiskers of aromatic polyesters such as poly(*para*-oxybenzoyl) and poly(2-oxo-6-naphthoyl) have been intensively investigated [31, 32], while there have been only few reports on the whiskers of conjugated polymers such as poly(3-alkylthiophene) [33]. When a helical conjugated polymer with a whisker structure is designed, the intrachain and interchain interactions must be balanced by precisely controlling the van der Waals, π-electron overlap, and electrostatic interactions.

Helical conjugated polymers have been attracting much interest because they can provide us with unique optical functionalities such as circularly polarized luminescence [20–24] and unprecedented physicochemical properties such as induced solenoidal magnetism [34]. Various types of helical conjugated polymers, represented by helical polyacetylenes [35, 36], amino acid-containing substituted polyacetylene derivatives [37], and polythiophene derivatives [38], have been reported. These helical conjugated polymers are constructed with intrachain-spiral, intrachain-twisted ribbon, or interchain-helically π-stacked structures [39, 40]. However, there are few reports of conjugated polymers having a helicene-like helical structure [41], which is favorable for maintaining π-conjugation on the main chain despite the intrachain helical structure. The meta-phenylene fragment is likely to be a suitable building block for the construction of a helicene-like helical structure. This hypothesis is supported by previous studies showing that meta-phenylene ethynylene oligomers spontaneously form helical structures depending on the solvent and temperature [42] and that the meta-deciphenyl compound [43] and PMP [44] have helical structures where each turn of the helix has five aromatic rings.

In this study, we designed and synthesized a series of novel PMP derivatives with chiral groups as side chains, **(*R*)-/(*S*)-Poly-1** (Fig. 11.2) [10]. **Poly-1** features a chiral side chain with either **(*R*)-** or **(*S*)**-2-nonanol chiral alkyl groups linked via

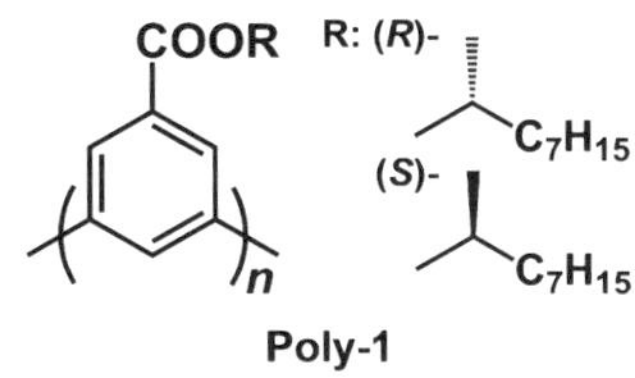

Fig. 11.2 Structures of the chiral PMP derivatives, ***(R)*-/*(S)*-Poly-1**

Table 11.1 Liquid crystal phase transition temperatures of **Poly-1**

Polymers	Heating/Cooling (°C)
(*R*)-poly-1	G 126 Col_h 210 I
	G 117 Col_h 205 I
(*S*)-poly-1	G 126 Col_h 210 I
	G 117 Col_h 205 I

G: glass state, Col_h: hexagonal columnar phase, I: isotropic phase

an ester moiety. The PMPs have helicene-like helical structures through intrachain π-electron overlap interactions. The PMPs self-assemble to form whiskers when slowly cooled from the isotropic phase. X-ray diffraction (XRD) analyses indicate that the whisker is composed of a hexagonal columnar packing of the helical main chains.

The PMPs with chiral alkyl groups in the side chains, ***(R)*-/*(S)*-Poly-1**, have number average molecular weights (M_n) of 3,500 and 3,700, respectively, and degrees of polymerizations (DPs) of 14 and 15, respectively. The thermal behavior of the polymers was evaluated by differential scanning calorimetry (DSC) analyses and polarizing optical microscopy (POM) observations, and the LC phases were assigned from the XRD measurements. ***(R)*-/*(S)*-Poly-1** show a hexagonal columnar phase (Col_h) from 117 to 205 °C during the cooling process. The LC phase transition values and assigned LC phases of the polymers are listed in Table 11.1.

It is found that ***(R)*-Poly-1** forms whiskers when slowly cooled from the isotropic phase, as depicted in Fig. 11.3. The observed whiskers of ***(R)*-Poly-1** are 5-7 μm in diameter and approximately 100–200 μm in length. The size of the whisker depends not only on the polymer structure including the side chains but also on the cooling rate from the isotropic phase. A large area of the POM micrograph is dark, but the circularly polarized micrograph indicates that the dark area contains many whiskers (Fig. 11.3b).

The structures of the polymers in the crystalline phases were investigated by XRD measurements. Figure 11.4 shows the XRD profile and pattern of ***(R)*-Poly-1** in the solid state. ***(R)*-Poly-1** shows an intense peak at 23.8 Å and six weak peaks at 13.7, 11.7, 8.7, 7.8, 6.5, and 5.8 Å in the small angle region. A diffuse halo is present at 4.4 Å, and a sharp but small halo is present at 3.7 Å in the wide-angle region. The peaks that represent the reciprocal Bragg spacings with ratios of $1{:}\sqrt{3}{:}2{:}\sqrt{7}{:}3{:}\sqrt{13}{:}4$ were assigned to the Miller indices of (100), (110), (200), (210), (300), (310), and (400), respectively. The reflections have a hexagonal arrangement with a lattice constant of $a = 27.5$ Å. The diffuse peak in the

Fig. 11.3 **a** Polarized optical micrograph and **b** circularly polarized optical micrograph of **(*R*)-Poly-1** in the solid state. Reprinted with permission from [10]. Copyright 2011, American Chemical Society

wide-angle region (4.4 Å) corresponds to the distance of the side chains. The periodicity of 3.7 Å corresponds to the distance between the π-stacks of the phenylene cores. These results show that the polymer has a helicene-like helical structure with a helical pitch of 3.7 Å. Similar XRD results were obtained for **(*S*)-Poly-1**. Hence, the whisker has a hexagonal columnar structure consisting of discotically packed helical conjugated polymers, as illustrated in Fig. 11.4.

In summary, **Poly-1** forms peculiar whiskers with a hierarchical structure consisting of intrachain helices owing to its rigidity and the high crystallinity of the directly meta-linked phenylene units in the conjugated backbone, as shown in Fig. 11.5.

11.2.2 Macroscopic Alignment of the Self-assembled Polymer Helix by a Magnetic Field

In the previous section, we reported PMP derivatives bearing chiral side chains having a helicene-like helical structure through intrachain π-electron overlap interactions, which lead to a self-assembled whisker morphology due to interchain van der Waals interactions. However, each whisker of the PMP derivative is still randomly oriented, even if it has a self-assembled structure. It is therefore desirable to align the PMP derivative to construct a higher-ordered hierarchical structure and also to evolve an anisotropic nature in its potentially profound optoelectrical properties.

Here, we synthesized liquid crystalline PMP derivatives by introducing phenylcyclohexyl (PCH) or cyanobiphenyl (CB) LC groups into the side chains and macroscopically aligned the polymers in LC phases using a magnetic field of 12 T [13]. **Poly-2** ($M_n = 8{,}400$) has PCH side chains linked with flexible dodecamethylene chains via ester moieties. **Poly-3** ($M_n = 8{,}500$), on the other hand, has a CB LC moiety in its side chain (Fig. 11.6).

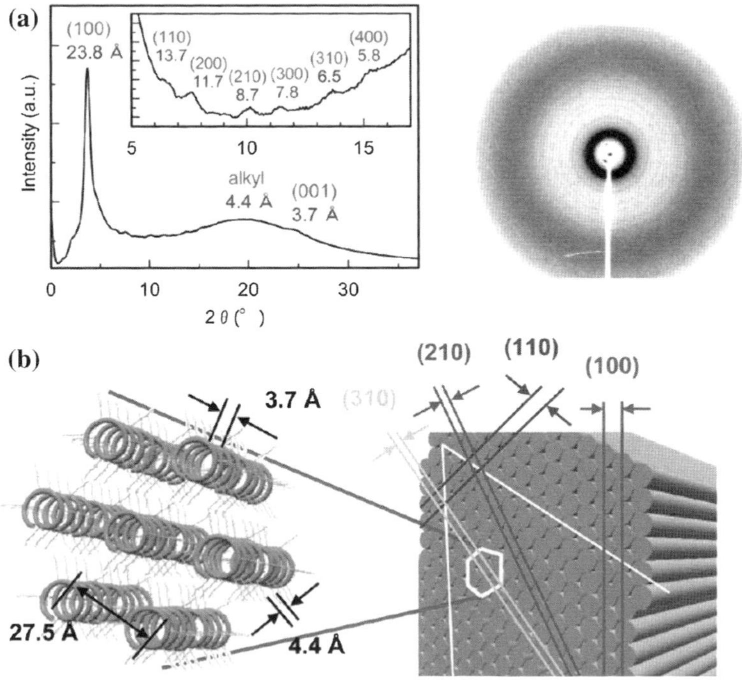

Fig. 11.4 **a** X-ray diffraction profile and pattern of **(*R*)-Poly-1** at the solid state. *Inset* shows the expanded XRD profiles in the region of 5–17° in 2θ. **b** The assignments for hexagonal columnar (Col_h) crystal are given in the schematic figure of self-assembled polymers. Reprinted with permission from [10]. Copyright 2011, American Chemical Society

The macroscopic alignment of the polymers was carried out under a magnetic field. The polymer sample supported on an aluminum plate was inserted into a mantle heater and placed in the middle of a superconducting magnet. The polymer was first heated to an isotropic temperature and then cooled to its LC temperature, at which a magnetic field of 12 T was applied for 30 min. Then, the polymer was slowly cooled to room temperature under a magnetic field.

POM characterization revealed that **Poly-2** has a batonnet-like texture from 125 to 94 °C and a fan-shaped texture from 94 to 80 °C during the cooling process. On the other hand, **Poly-3** has a sandy texture from 142 to 112 °C and a batonnet-like texture from 112 to 93 °C during the cooling process. XRD measurements revealed that **Poly-2** has a discotic nematic (N_D) phase at higher LC temperatures and a discotic lamellar (D_L) phase at lower LC temperatures. **Poly-3** showed a N_D phase and a hexagonal columnar (Col_h) phase at higher and lower LC temperatures, respectively. The phase transition temperatures of **Poly-2** and **Poly-3** are summarized in Table 11.2.

The macroscopic alignment of the polymers was carried out by applying a magnetic field of 12 T in their LC states. Figure 11.7 shows the XRD images and profiles of the aligned films of the polymers. **Poly-2** has an intense peak of 4.5 Å

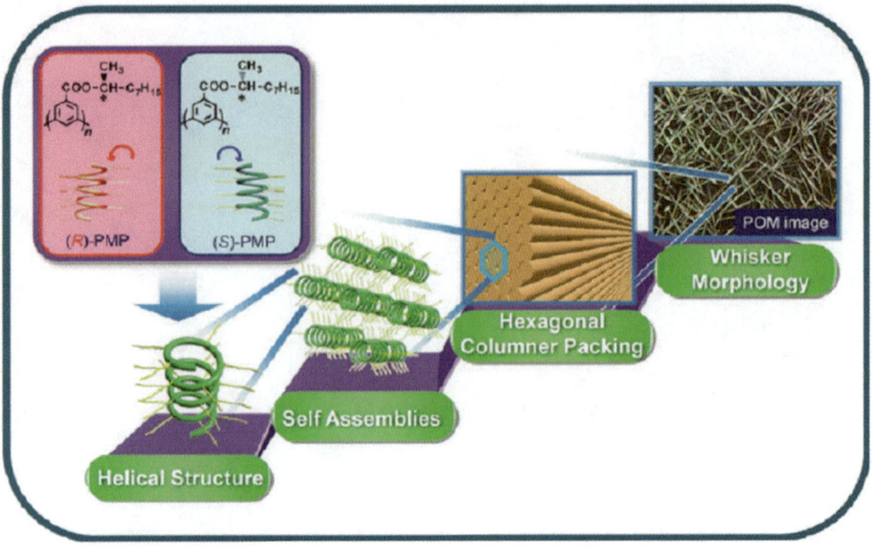

Fig. 11.5 **Poly-1** forms peculiar whiskers with a hierarchical structure consisting of intrachain helices owing to its rigidity and the high crystallinity. Reprinted with permission from [10]. Copyright 2011, American Chemical Society

Fig. 11.6 Structures of the LC PMP derivatives, **Poly-2** having a phenylcyclohexyl LC moiety and **Poly-3** having a cyanobiphenyl LC moiety

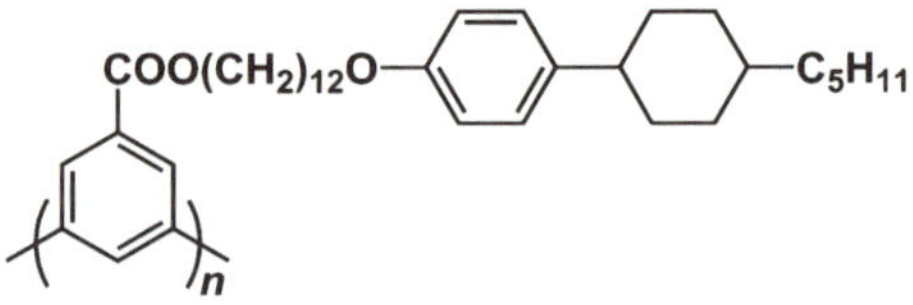

Poly-2

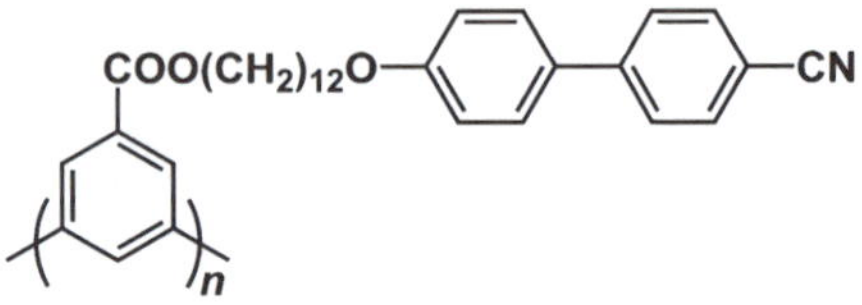

Poly-3

Table 11.2 Liquid crystal phase transition temperatures of **Poly-2** and **Poly-3**

Polymers	Heating/Cooling (°C)
Poly-2	G 84 D_L98 N_D 139 I
	G 80 D_L 94 N_D 125 I
Poly-3	G 98 Col_h 114 N_D 152 I
	G 93 Col_h 112 N_D 142 I

G: glassy state, D_L: discotic lamella phase, N_D: discotic nematic phase, Col_h: hexagonal columnar phase, I: isotropic phase

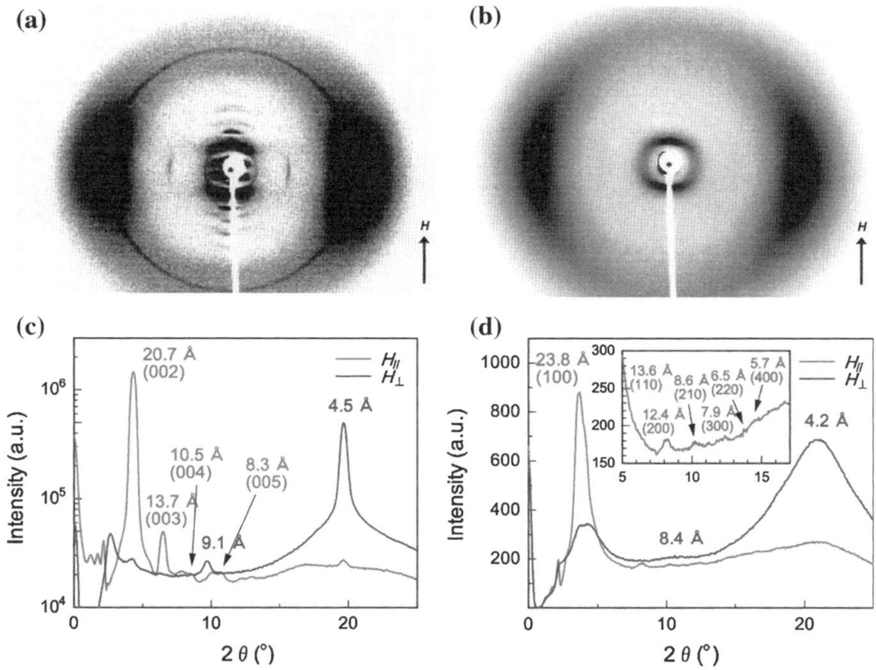

Fig. 11.7 X-ray diffraction pattern of the aligned films of **a Poly-2**, and **b Poly-3**. X-ray diffraction profiles of magnetically aligned polymers **c Poly-2**, and **d Poly-3** at room temperature. Reprinted with permission from [13]. Copyright 2009, Elsevier

and a small peak of 9.1 Å, which are assigned to the interdistance of the LC side chains and the secondary diffraction, respectively. They are observed perpendicular to the direction of the applied magnetic field. Several diffractions of 20.7, 13.7, 10.5 and 8.3 Å, where the reciprocals of these distances are in the ratio of 2:3:4:5, are observed parallel to the magnetic field. They correspond to the reflections of the (002), (003), (004), and (005) indices of a discotic lamellar phase with a lattice parameter of $a = 41.4$ Å. The columns constituting the helical main chains are aligned perpendicular to the magnetic field (Fig. 11.7c).

On the other hand, **Poly-3** showed a halo perpendicular to the magnetic field, corresponding to the interdistance between the LC side chains (Fig. 11.7b). Furthermore, an intense diffraction peak of 23.8 Å and weak diffraction peaks of 13.6, 12.4, 8.6, 7.9, 6.5, and 5.7 Å are observed parallel to the magnetic field. The reciprocals of the peaks are in the ratio of $1:\sqrt{3}:2:\sqrt{7}:3:2\sqrt{3}:4$, and the peaks correspond to the diffractions of the (100), (110), (200), (210), (300), (220), and (400) indices of a hexagonal columnar structure with a lattice parameter of $a = 27.5$ Å. Such a diffraction pattern comes from a uniaxially aligned hexagonal columnar structure in which the columns are aligned perpendicular to the magnetic field (Fig. 11.7d).

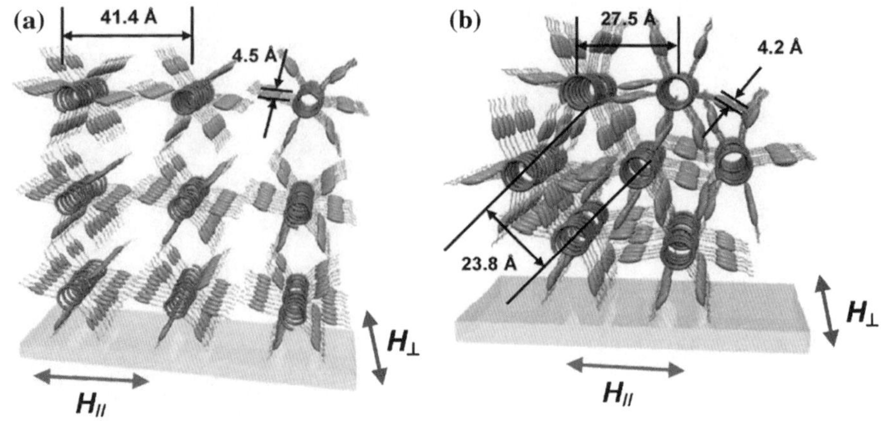

Fig. 11.8 Schematic representation of LC PMPs after magnetically forced alignment. The **a** discotic lamellar (D_L) structure of **Poly-2** and the **b** hexagonal columnar (Col_h) structure of **Poly-3** at room temperature. Reprinted with permission from [13]. Copyright 2009, Elsevier

Lastly, it is worthwhile to discuss the mechanism of the magnetically forced alignment of the PMP derivatives. First of all, one must consider the orientation directions of the main chain columns, i.e. the axes of the columns are parallel or perpendicular to the glass substrate. Actually, the latter (parallel alignment of the column axis) is favored because the terminal alkyl or cyano group of the LC side chain tends to be aligned perpendicular to the glass substrate rather than parallel to it. Next, the column, which is lying parallel to the glass substrate, has two choices for the magnetically forced alignment, i.e. parallel or perpendicular to the magnetic field. Taking into account that the diamagnetic susceptibility of the helicene-like π-conjugated disc consisting of six meta-phenylene rings could be much larger than those of the LC side chains consisting of PCH or CB, the plane of the π-conjugated disc should be parallel to the magnetic field, and therefore, the axis of the column should be aligned perpendicular to the magnetic field. This holds true in **Poly-2** with the discotic lamellar structure and **Poly-3** with the hexagonal columnar one, as described in Fig. 11.8.

We synthesized PMP derivatives with LC groups composed of PCH or CB moieties in the side chains. The polymers showed discotic liquid crystallinities based on the helical structure of the main chains, where the helical chains were self-assembled to form columns along the helical axis. The macroscopic alignment of the polymers was carried out using a magnetic field of 12 T in the liquid crystal phases. It was clarified from the XRD measurements that the main chains were aligned perpendicular to the magnetic field. As a consequence, the present study demonstrated the formation of macroscopically aligned hierarchical structures based on self-assembled helical conjugated polymers.

11.3 Helical π-Stacked Self-assemblies of Liquid Crystalline Conjugated Polymers Showing Circularly Polarized Luminescence with High Dissymmetry Factors

Polymeric materials exhibiting circularly polarized luminescence have been widely researched as photonic materials in optical amplifiers, light emitting diodes, and optical information storage [45–47]. In addition, LC displays (LCDs) using a CPL backlight have been shown to exhibit steeper transmission-voltage characteristics and higher brightness compared with LCDs using linearly polarized luminescence (LPL) illumination [45, 46]. CPL is defined as luminescent light having different intensities for the left (I_L) and right (I_R) circularly polarized components. The degree of circular polarization is described by the emission dissymmetry factor, $g_{em} = (I_L - I_R)/[(I_L + I_R)/2] = \Delta I/I$, where $-2 < g_{em} < 2$.

CPL from a π-conjugated polymer was first reported by Meijer et al. as a polythiophene (PT) having a chiral side chain with g_{em} values up to 5.0×10^{-3} [48]. Subsequently, circularly polarized electroluminescence (CPEL) [47] was reported by the same research group using an EL cell with poly(*para*-phenylene vinylene) (PPV) bearing a chiral side chain as an emissive layer. The CPEL device has reported values of g_{em} up to -1.7×10^{-3}. In chiral-substituted conjugated polymers, the chiroptical properties are assumed to evolve mainly from interchain interactions within chiral aggregates.

With the aim of further increasing CPL g_{em} values, research groups looked into chiral liquid crystalline conjugated polymers. The liquid crystallinity of the polymers enhances the self-assembly, while chiral moieties at the side chains or chiral dopants direct the polymer chains to form helical π-stacked assemblies. Oda et al. constructed a CPEL device using thermotropic liquid crystalline polyfluorene (PF) bearing chiral side chains with g_{em} values up to -2.5×10^{-1} [49]. The combination of well-defined chiral packing of the liquid crystalline PF and chiral interchain interactions is the key for the increase in the g_{em} values of the PF-based CPEL device. Furthermore, Chen et al. has reported thermotropic poly(*para*-phenylene) (PPP) and PT copolymers with monomer units bearing CB or (–)-cholesterol pendant groups forming a chiral nematic (N*-LC) phase and exhibiting g_{em} values up to -4.9×10^{-1} [50].

It has been shown that thermotropic N*-LC materials demonstrate supramolecular helical ordering that leads to CPL with high dissymmetry factors [49–51]. On the other hand, it would be appealing to investigate lyotropic N*-LC systems as alternative circularly polarized optical materials for use in optoelectronic devices and displays. The effects of the solvent, solution concentration, and chiral dopant employed in the lyotropic N*-LC system would be of particular interest, especially in relation to the helical structure of the polymer LC phase and its chiroptical properties. It was reported that di-substituted polyacetylene (di-PA) adopting a poly(diphenylacetylene) (PDPA) structure with alkyl side chains exhibits lyotropic LC behavior [18, 19]. The PDPA structure, with phenyl moieties

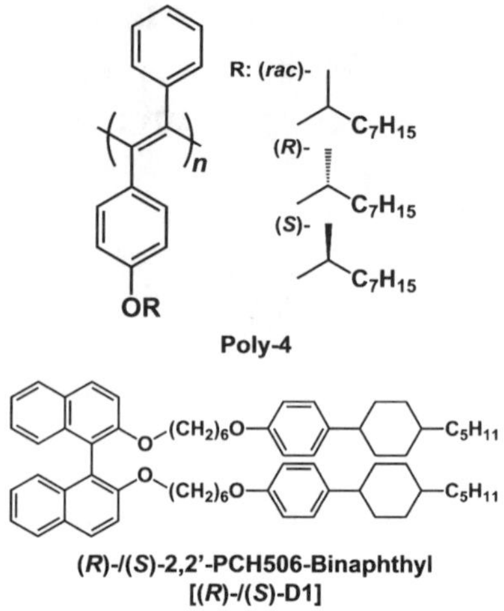

Fig. 11.9 Structures of the chiral di-substituted LCPA derivatives, **(*rac*)-/(*R*)-/(*S*)-Poly-4**, and chiral dopants, **(*R*)-/(*S*)-D1**

directly attached to the polymer backbone, has a stiff main chain that acts as an LC mesogen, which self-assembles into a lyotropic LC phase above a critical concentration.

11.3.1 Lyotropic Di-substituted Polyacetylenes that Exhibit High Dissymmetry Factors in Circularly Polarized Luminescence Through the Chiral Nematic Liquid Crystal Phase

Our group has synthesized chiral di-substituted liquid crystalline polyacetylene (di-LCPA) derivatives that exhibit main chain-type liquid crystallinity, where the LC mesogen consists of a stilbene structure along the polyene main chain [21]. **Poly-4** has a 4-nonyloxy phenyl group with either racemic (*rac*)-, (*R*)-, or (*S*)-handedness and a PDPA main chain structure that is responsible for both the green fluorescence and lyotropic liquid crystallinity (**(*rac*)-/(*R*)-/(*S*)-Poly-4**: $M_n = 22{,}000$, 23,000, 25,000, respectively) (Fig. 11.9). Axial chiral dopants were also synthesized as described in a previous report [35, 36]. **(*R*)-/(*S*)-2,2'-PCH506-Binaphthyl**, abbreviated as **(*R*)-/(*S*)-D1**, has PCH LC moieties that are substituted at the 2,2′ positions of the binaphthyl rings. The substituted PCH LC moieties of **(*R*)-/(*S*)-D1** improve the miscibility between the polymer and the chiral dopant and also give rise to the chiral induction between them.

The main chains of **Poly-4** have a stiff polymer structure because its side chains are composed of phenyl moieties directly attached to the main chain, forming a stilbene fragment enforced by π-conjugation. A polymer exhibiting this elevated level of stiffness should be infusible upon heating, resulting in an LC with no thermotropic behavior. Instead, by virtue of the nonyloxy group on the phenyl

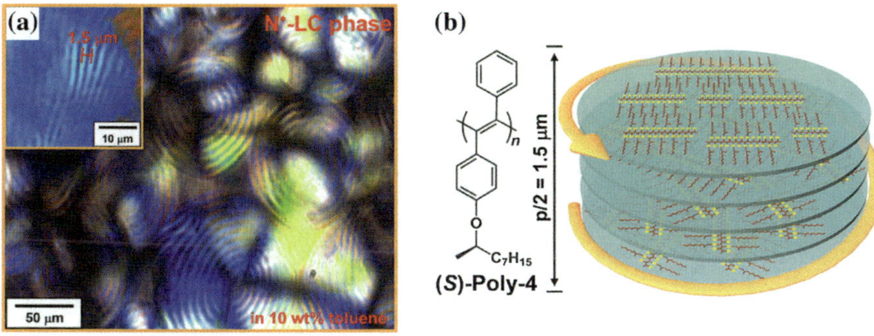

Fig. 11.10 **a** POM image of the chiral nematic (N*-LC) phase of **(*S*)-Poly-4** in 10 wt % lyotropic LC solution in toluene showing a double-spiraled texture. *Inset* shows a fingerprint texture with a helical half pitch of 1.5 μm. **b** Schematic representation of the N*-LC phase of **(*S*)-PA17**. Reprinted with permission from [18]. Copyright 2012, American Chemical Society

moiety, **Poly-4** is soluble in organic solvents, such as a toluene, and shows lyotropic LC behavior. In the case of the PDPA structure of **Poly-4**, the stilbene fragment might be suitable for the formation of interchain π-electron overlapped associations through van der Waals interactions, which enables the polymers to exhibit lyotropic LC.

In the lyotropic LC phase of **(*S*)-Poly-4**, the polymer main chains act as LC mesogens, which spontaneously align to form a helically twisted structure in an N*-LC phase. POM analysis reveals a lyotropic N*-LC phase prepared from 10 wt % solution using toluene as a solvent (Fig. 11.10a). **Poly-4** demonstrates lyotropic liquid crystallinity at a critical concentration range from 5 to 10 wt % in toluene. Analogous to thermotropic LCs, which show an isotropic phase above a critical temperature, lyotropic LCs become isotropic liquids below a concentration of 5 wt % in toluene, where the polymer main chains (which act as LC mesogens) are dispersed randomly without any ordering. At the lyotropic LC concentration range, there are weak π-interactions between the aromatic stilbene structure and toluene. In such a situation, the toluene enhances the attraction between the polymer main chains, inducing spontaneous self-assembly to occur; the toluene also provides enough fluidity in the system to allow liquid crystallinity.

Figure 11.10a shows a double-spiraled texture characteristic of the N*-LC phase [52]. The inset of Fig. 11.10a shows a fingerprint texture with a distance of 1.5 μm between striae, which corresponds to the half helical pitch of the N*-LC phase. Figure 11.10b shows a schematic representation of the N*-LC phase of **(*S*)-Poly-4**.

In Fig. 11.11, the chiral induction of **(*rac*)-Poly-4** forming an N*-LC phase is illustrated. The POM image of **(*rac*)-Poly-4** shows a Schlieren texture characteristic of a lyotropic N-LC phase [52], prepared from a 10 wt % toluene solution (Fig. 11.11a). Upon addition of chiral dopant **(*S*)-D1** at 10 wt %, the N-LC phase of **(*rac*)-Poly-4** changes into an N*-LC phase (Fig. 11.11b). The POM image of **(*S*)-D1/(*rac*)-Poly-4** reveals a fingerprint texture with a helical half-pitch of 2.0 μm.

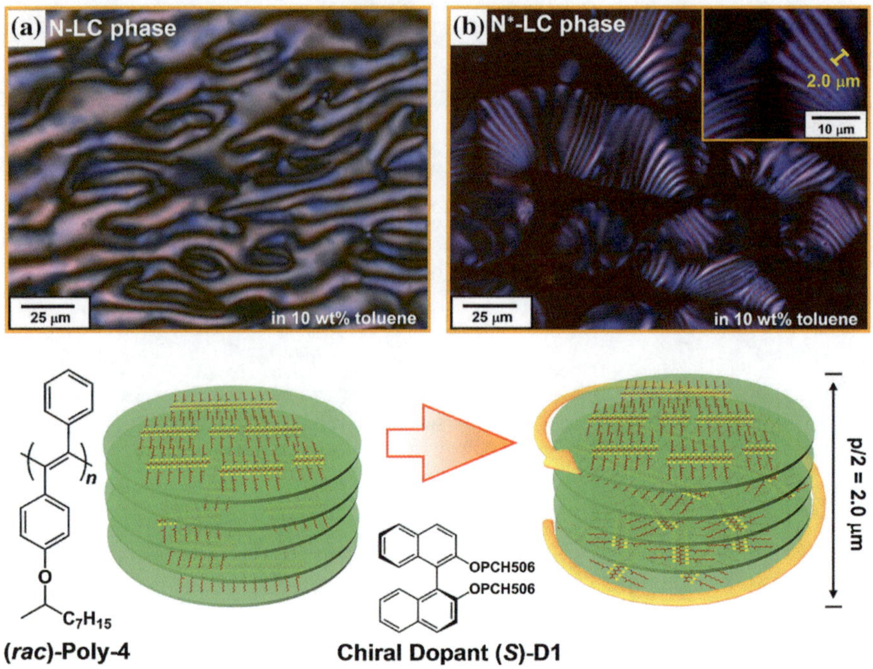

Fig. 11.11 Schematic model of the chiral induction of ***(rac)*-Poly-4** into a N*-LC phase. Upon the addition of the chiral dopant **(*S*)-D1** at 10 wt % into the N-LC phase of ***(rac)*-Poly-4** (*left*), a N*-LC phase is induced (*right*). **a** POM image of the N-LC phase of ***(rac)*-Poly-4** in 10 wt % lyotropic LC solution in toluene showing a Schlieren texture. **b** POM image of the N*-LC phase of **(*S*)-D1/(*rac*)-Poly-4** in 10 wt % lyotropic LC solution in toluene showing a fingerprint texture. *Inset* shows the helical half pitch of 2.0 μm. Reprinted with permission from [18]. Copyright 2012, American Chemical Society

The N*-LC film samples were prepared by dissolving chiral **Poly-4** or achiral **Poly-4** with chiral dopant (10 wt %) in toluene (5–10 wt %) to form a lyotropic LC state. The LC solution was subsequently cast onto a quartz plate, which was later exposed to toluene vapor for one hour and dried slowly at room temperature. Toluene vapor exposure of the N*-LC film was performed to increase the domain size of the N*-LC phase.

The di-LCPAs in solution and in cast film showed monosignate and bisignate Cotton effects in their CD spectra, respectively. These results suggest the formation of intrachain and interchain helical structures, respectively. In spite of this, the di-LCPAs in solution and in cast film exhibited no CPL [21]. However, **(*R*)-/(*S*)-Poly-4** in an N*-LC film showed bisignate CPL bands centered around the PL emission band (Fig. 11.12a). **(*R*)-Poly-4** exhibited positive and negative CPL bands at wavelengths shorter and longer than 530 nm, respectively. **(*S*)-Poly-4** exhibited CPL bands with handedness opposite to those **(*R*)-Poly-4**. The CPL spectra of **(*R*)-/(*S*)-Poly-4** in an N*-LC film showed g_{em} values of 1.8×10^{-1} and

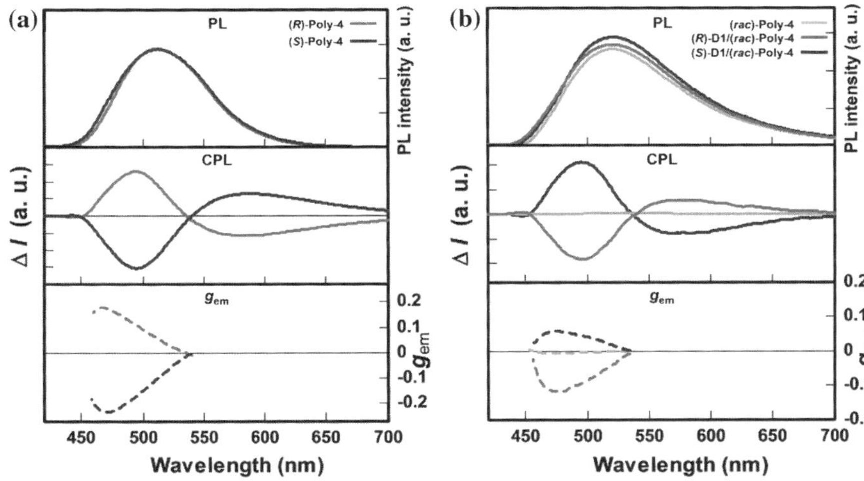

Fig. 11.12 PL (*upper*), CPL spectra (*middle*), and g_{em} (*lower*) spectra of **a (*R*)-/(*S*)-Poly-4** N*-LC film, and **b (*R*)-/(*S*)-D1** doped **(*rac*)-Poly-4** (10 wt %) N*-LC film. The measurements were performed with excitation using unpolarized light at 367 nm. Reprinted with permission from [18]. Copyright 2012, American Chemical Society

Table 11.3 Emission dissymmetry factors (g_{em}) of **Poly-4** and doped **Poly-4** in N*-LC films

Polymers	g_{em} [a] (wavelength/nm)
(*R*)-Poly-4	1.8×10^{-1} (466)
(*S*)-Poly-4	-2.3×10^{-1} (471)
(*R*)-D1/(*rac*)-Poly-4	-1.2×10^{-1} (478)
(*S*)-D1/(*rac*)-Poly-4	5.9×10^{-2} (473)

[a] Wavelength at which g_{em} was evaluated

-2.3×10^{-1}. These high g_{em} values suggest the formation of higher-ordered helical π-stacked structures in the N*-LC films of **(*R*)-/(*S*)-Poly-4**.

The N*-LC film of **D1D**oped **(*rac*)-Poly-4** exhibited CPL similar to that of **(*R*)-/(*S*)-Poly-4** (Fig. 11.12b). The CPL spectra of **D1D**oped **Poly-4** films showed handedness opposite to their counterparts in **(*R*)-/(*S*)-Poly-4**, with g_{em} values up to the order of 10^{-1}. The CPL spectrum of **(*rac*)-Poly-4** showed no CPL; therefore, the CPL of **D1D**oped **(*rac*)-Poly-4** originated from the chiral induction of **(*rac*)-Poly-4**. The g_{em} values of the chiral **Poly-4** and doped **Poly-4** N*-LC films are listed in Table 11.3.

The PDPA main chain structure of the di-LCPAs, along with the chirality induced with either chiral nonyloxy phenyl moieties or chiral dopants, allows the formation of a highly ordered lyotropic N*-LC phase. We found that the highly ordered lyotropic N*-LC phase is indispensable for the generation of CPL with high g_{em} values.

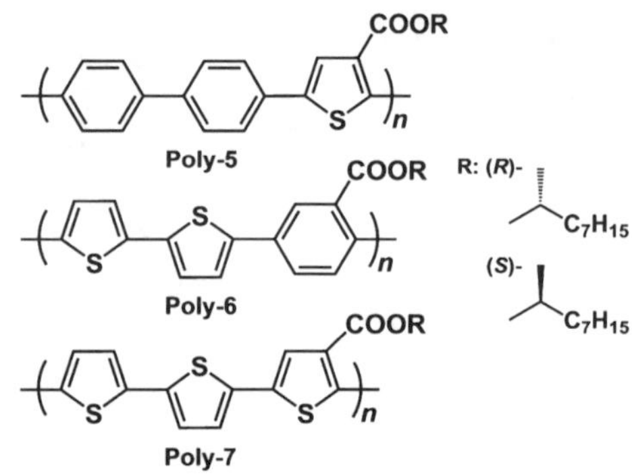

Fig. 11.13 Structures of the RGB conjugated polymers, ***(R)-/(S)*-Poly-5**, **Poly-6**, and **Poly-7**

Aside from investigating lyotropic liquid crystallinity in di-PA, our group also looked into thermotropic chiral liquid crystalline thiophene-based copolymers that exhibit red, green, and blue (RGB) CPL and white CPL [22].

11.3.2 Helically π-Stacked Thiophene-based Copolymers that Exhibit RGB and White Circularly Polarized Luminescence

We synthesized chiral liquid crystalline thiophene-based copolymers by incorporating a chiral nonyloxycarbonyl group into their side chains [22]. Generally, the fluorescence color of polymers depends on the structure of the repeating unit and the conjugation length of the polymer main chain. Increasing the number of thiophene rings in the repeating unit can create a red shift in the π-π^* transition band, whereas increasing the number of phenylene rings creates a blue shift of the π-π^* transition band. Copolymers with thiophene and phenylene repeating units were designed and synthesized to exhibit fluorescence colors ranging from blue to red. **Poly-5** has two phenylene units with a chiral thiophene unit [P(PPT*)] and exhibits blue fluorescence ((***R***)-/(***S***)-**Poly-5**: $M_n = 23{,}000$, 24,000, respectively). **Poly-6**, which has two thiophene units with a chiral phenylene unit [P(TTP*)], exhibits green fluorescence ((***R***)-/(***S***)-**Poly-6**: $M_n = 4{,}800$, 4,800, respectively). A homopolymer consisting of two thiophene units followed by a chiral thiophene unit [P(TTT*)], **Poly-7**, shows red fluorescence ((***R***)-/(***S***)-**Poly-7**: $M_n = 4{,}300$, 4,600, respectively). The structures of the copolymers are shown in Fig. 11.13.

In particular, **Poly-5**, **-6**, and **-7**, which have two *non*-substituted aromatic rings in their repeating units, partially form one-handed interchain helical structures, even in a good solvent such as chloroform, because the polymers tend to exhibit interchain π-stacking due to their highly crystalline main chains. As a result, each polymer exhibits CPL both in solution and in spin cast film. We found that self-ordering of the polymer aggregates in the thermotropic N*-LC phase during annealing at the LC temperature is essential for enhancing the CPL and achieving dissymmetry factors on the order of 10^{-1}.

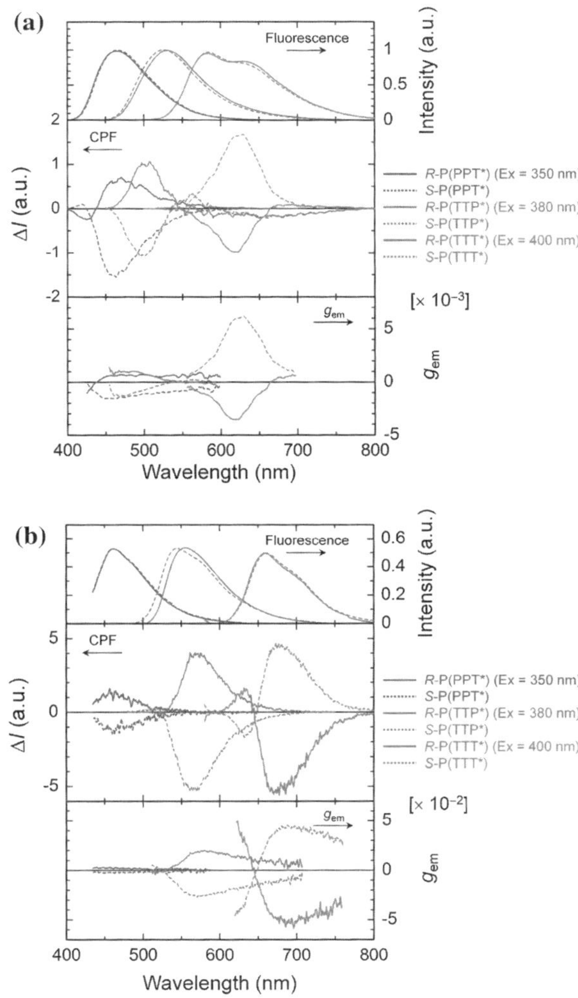

Fig. 11.14 **a** The PL (fluorescence) (*upper*), CPL (or CPF, circularly polarized fluorescence) (*middle*), and g_{em} (*lower*) spectra of **Poly-5** (P(PPT*), **Poly-6** (P(TTP*), and **Poly-7** (P(TTT*) in **a** chloroform ($c = 1.0 \times 10^{-4}$ M) and **b** as-cast films. Reprinted with permission from [19]. Copyright 2012, American Chemical Society

Figure 11.14a shows the CPL spectra of the **Poly-5**, **-6**, and **-7** in chloroform ($c = 1.0 \times 10^{-4}$ M), which exhibit circular polarization in their fluorescence bands. The CPL signals for the polymers with (*R*)- and (*S*)- configurations appear in opposite directions, resulting in almost mirror images in solution and cast films; for example, the **(*R*)-Poly-5** exhibits a positive signal, whereas the **(*S*)-Poly-5** exhibits a negative signal. The g_{em} values of the polymer solutions are approximately on the order of 10^{-3}. Figure 11.14b shows the CPL spectra of the polymer films. The g_{em} values of the polymer films for **Poly-6** and **Poly-7** increased relative to their corresponding solutions by approximately one order of magnitude; this increase may be attributed to the strong π-π interactions between the polymer backbones in the solid state.

Table 11.4 Emission dissymmetry factors (g_{em}) of **Poly-5, -6, -7** in chloroform ($CHCl_3$), cast films, and annealed cast films

Polymers	g_{em} [a] (wavelength/nm)		
	In $CHCl_3$	Cast film	Annealed film
(*R*)-Poly-5	0.72×10^{-3} (447)	3.1×10^{-3} (457)	-6.9×10^{-3} (456)
(*S*)-Poly-5	-1.5×10^{-3}(455)	-2.9×10^{-3} (463)	1.1×10^{-2} (454)
(*R*)-Poly-6	1.1×10^{-3} (500)	1.8×10^{-2} (568)	2.9×10^{-1} (558)
(*S*)-Poly-6	-2.2×10^{-3} (502)	-2.6×10^{-2} (565)	-1.1×10^{-2} (534)
(*R*)-Poly-7	-4.0×10^{-3} (623)	-4.8×10^{-2} (680)	0.96×10^{-1} (666)
(*S*)-Poly-7	3.0×10^{-3} (627)	4.1×10^{-2} (680)	-1.5×10^{-1} (666)

[a] Wavelength at which g_{em} was evaluated

The polymer films were annealed for 30 min at 150 °C, which falls in the LC temperature range, and then their CPL spectra were measured. After annealing, the absolute g_{em} values increased by at least one order of magnitude, producing g_{em} values as high as 10^{-1}. The g_{em} values of **Poly-5**, **-6**, and **-7** in solution, cast film, and annealed N*-LC films are listed in Table 11.4. These g_{em} values are very high, despite the regiorandom structure of the polymers. This value may be due to the helical π-stacking that occurs during annealing at the LC temperature. Annealing in the LC phase encourages the self-assembly of the polymer backbone and the formation of macroscopic π-π stacks, which leads to high g_{em} values [49]. The hierarchical arrangement of higher-ordered structures by chiral liquid crystallinity, such as the self-ordering of the polymer aggregates in the N*-LC phase by annealing at the LC temperature, is a promising method to enhance the dissymmetry factor of chiral conjugated polymers up to the order of 10^{-1}.

Additionally, the development of RGB-colored luminescence allows for white light emission by mixing the polymers in the appropriate ratios, which generates circularly polarized white luminescence. We mixed a solution of the **Poly-5**, **Poly-6**, and **Poly-7**, which exhibit blue, green, and orange fluorescence, respectively, in a ratio estimated from the inverse of their quantum yields to achieve circularly polarized white luminescence. The actual ratio of **Poly-5**, **Poly-6**, and **Poly-7** employed in the mixture was 1:1:5 (5.0×10^{-5} M/5.0×10^{-5} M/2.5×10^{-4} M) based on the inverse quantum yields of 1/0.26, 1/0.23, and 1/0.05, respectively. Figure 11.15 shows the CPL spectra of the polymer mixture. The inset shows a photo of the polymer mixture in chloroform, which emits white light. System 1 is the mixture of **(*R*)-Poly-5**, **(*R*)-Poly-6**, and **(*S*)-Poly-7**, which displays a positive g_{em}, and system 2 is the mixture of **(*S*)-Poly-5**, **(*S*)-Poly-6**, and **(*R*)-Poly-7**, which displays a negative g_{em}. These mixtures displayed mirrored CPL spectra from 400 to 700 nm and a g_{em} on the order of 10^{-3}, demonstrating circularly polarized white luminescence.

However, the cast films prepared using the above mixtures showed red, not white, fluorescence due to polymer aggregation, which allows either an energy transfer from the excited states of the **Poly-5** and **Poly-6** to that of **Poly-7** or the reabsorption of the fluorescent emissions of **Poly-5** and **Poly-6** by **Poly-7**. To

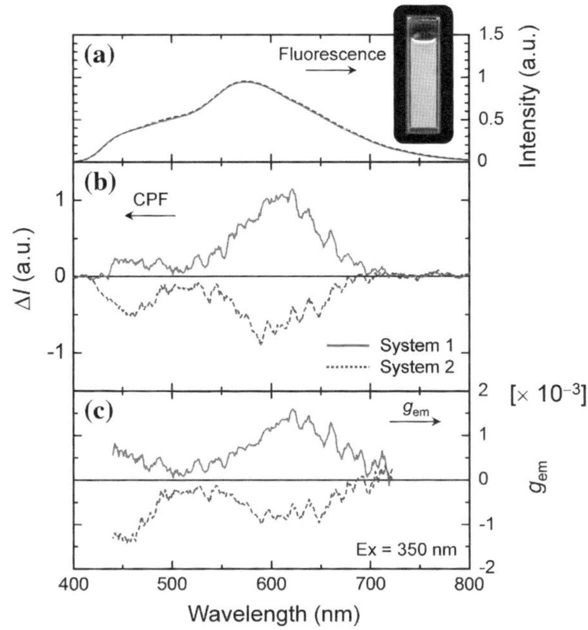

Fig. 11.15 **a** PL (fluorescence), **b** CPL (or CPF, circularly polarized fluorescence), and **c** g_{em} spectra of the polymer mixtures in chloroform, [**Poly-5**, 5.0 × 10^{-5} M; **Poly-6**, 5.0 × 10^{-5} M; **Poly-7**, 2.5 × 10^{-4} M], where system 1 is the mixture of **(*R*)-Poly-5**, **(*R*)-Poly-6**, and **(*S*)-Poly-7**, and system 2 is a mixture of **(*S*)-Poly-5**, **(*S*)-Poly-6**, and **(*R*)-Poly-7**. The *inset* shows a photo of system 1 in chloroform, irradiated with 365 nm UV light. Reprinted with permission from [19]. Copyright 2012, American Chemical Society

suppress both energy transfer and reabsorption, the interchain distances in the polymers need to be increased. Therefore, the polymer mixture in system 1 was dispersed in an excess of polystyrene by dissolving both in chloroform and then casting the solution onto a quartz substrate. This cast film sufficiently separated the polymers from each other in terms of their interchain distances to create a so-called "pseudodilute solution" state. Figure 11.16 depicts the RGB fluorescence of the individual polymers and the white fluorescence of the mixture, system 1, in chloroform and as films on a quartz substrate. The RGB and white fluorescence are circularly polarized emissions.

In this section, we described the emergence of CPL with high dissymmetry factors brought about by the higher-ordered helical arrangement of the N*-LC phase of liquid crystalline conjugated polymers. Furthermore, liquid crystalline conjugated polymers can also exhibit dynamic switching functionalities owing to the use of advanced functional moieties such as ferroelectric LC and photoresponsive moieties.

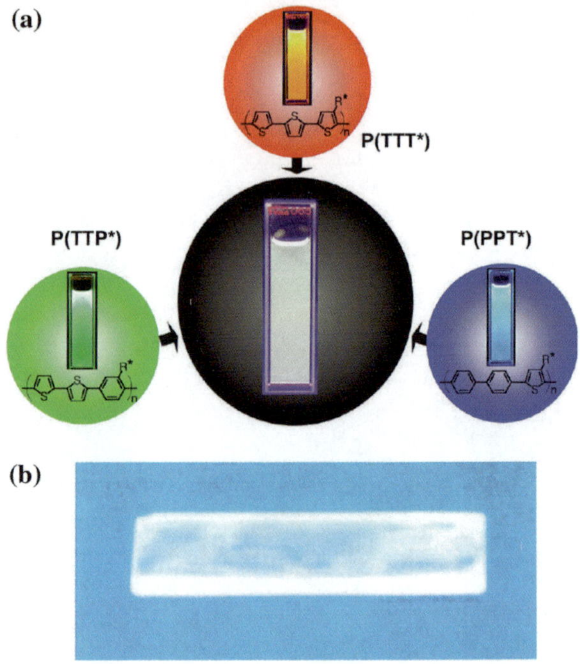

Fig. 11.16 **a** RGB and white fluorescence of system 1, the mixture of **(*R*)-Poly-5** (P(PPT*)), **(*R*)-Poly-6** (P(TTP*)), and **(*S*)-Poly-7** ((P(TTT*)), in chloroform and **b** white fluorescence of the cast film of system 1 on a quartz substrate. The film was prepared by dissolving the mixture and an excess of polystyrene in chloroform followed by casting onto the quartz substrate. UV light at 365 nm was used for excitation. Reprinted with permission from [19]. Copyright 2012, American Chemical Society

11.4 Dynamic Switching Functionalities of Liquid Crystalline Conjugated Polymers

Liquid crystalline conjugated polymers have attracted much research interest because they produce anisotropies in their electrical and optical properties when they are macroscopically aligned [28–30]. The macroscopic alignment of LC polymers can be achieved by external forces such as shear stress and electric and magnetic fields [13–20]. However, it takes a significant amount of time to complete the alignment owing to the high viscosity common to macromolecules and polymers. This makes it difficult to achieve rapid control over the alignment of LC conjugated polymers and a dynamic switching of the LPL and/or anisotropic conductivity through external forces. One of the promising approaches to overcome this obstacle is to add ferroelectricity to the LC conjugated polymers because ferroelectricity should enable a quick response to an electric field by virtue of the large interaction between the spontaneous polarization and the electric field. It has been found that the chiral smectic C (SmC*) phase is essential in generating ferroelectric liquid crystallinity (FLC) in structurally rigid π-conjugated polymers [14]. Meanwhile, it is known that ferroelectric materials can be used as memory devices by utilizing a residual polarization, which is defined as a polarization that remains even after the electric field is turned off. Therefore, FLC conjugated polymers could be candidates for organic memory devices with a quick switching function, as they may have large spontaneous and residual polarizations.

Furthermore, aside from the dynamic switching of the SmC* phase upon the application of an external electric field, the utilization of photoresponsive moieties will add light-stimuli responsive functionalities to LC conjugated polymers. Macroscopically aligned fluorescent polymers showing dynamic switching between fluorescence and quenching may show dynamic switching of the LPL, which could be practical for optical read-write memory functional materials [53]. To realize this, LC conjugated polymers having a novel structural feature such as a photoresponsive moiety in the side chain causing switchable fluorescence and quenching modes were prepared. Such a novel design will lead to dynamic switching of the LPL through photo-stimulation [15].

This novel structural design strategy was further extended to chiral LC conjugated polymers adopting a helical π-stacking structure to exhibit dynamic switching of chiroptical functionalities, such as CPL [23].

11.4.1 Ferroelectric Liquid Crystalline Poly(meta-phenylene)

It is worth noting that the orientation force, $F(\varepsilon_\alpha)$, due to the interaction between an ordinary LC molecule and an electric field, E, is expressed by the following equation:

$$F(\varepsilon_\alpha) = -\varepsilon_\alpha E^2/2,$$

where $\varepsilon_\alpha = \varepsilon_{//} - \varepsilon_\perp$, an anisotropic dielectric constant [54]. On the other hand, the orientation force in an FLC molecule, F_{FLC}, is given as follows:

$$F_{\mathrm{FLC}} = F(\varepsilon_\alpha) + F(P_{\mathrm{S}} \cdot E),$$

where P_{S} is the spontaneous polarization. The spontaneous polarization is generated by the uniaxial alignment of the dipole moments in the mesophase. The second term of the orientation force, $F(P_{\mathrm{S}} \cdot E)$, due to an interaction between the spontaneous polarization and the electric field, is much larger than the first term, $F(\varepsilon_\alpha)$. This means that the FLC molecule is more quickly aligned by an electric field. Namely, the response time (τ) for the orientation in an FLC molecule is a few microseconds, which is much less than in an ordinary LC molecule.

$$\tau = \eta/P_{\mathrm{S}} \cdot E(E > E_{\mathrm{C}}),$$

where η is a viscosity coefficient and E_{C} is a coercive electric field. This situation is not limited to LC conjugated polymers, even if the relatively higher viscosity of the polymer must be taken into consideration. This motivates us to incorporate an FLC group into the side chain of a conjugated polymer to develop an advanced LC conjugated polymer with dynamic switching functionalities. LC molecules are

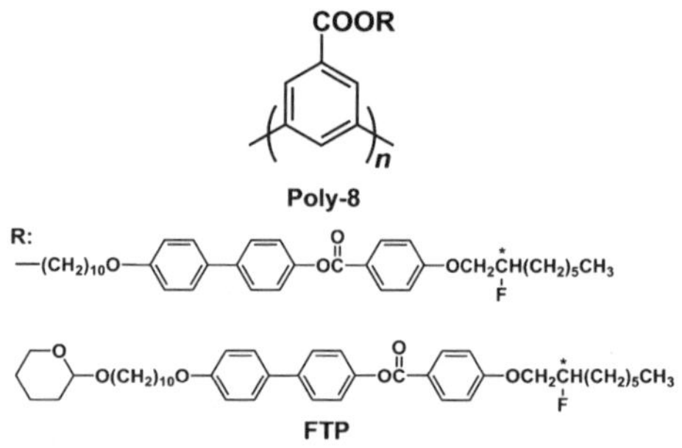

Fig. 11.17 Structures of the FLC PMP, **Poly-8**, and FLC molecule, **FTP**

generally composed of a rigid moiety (a mesogenic core), such as PCH or CB groups, and flexible alkyl chain moieties attached to both sides of the mesogenic core. For ferroelectricity to occur, a spontaneous polarization generated by an assembly of dipole moments is required. In the case of LC molecules, spontaneous polarization is realized only in an SmC* phase [55–57]. In other words, the SmC* phase is responsible for ferroelectricity. Thus, FLC molecules must satisfy the following conditions: (i) be optically active (chiral) and (ii) have a large dipole moment whose component perpendicular to the molecular axis is nonzero. Because the fluorine atom has a large electronegativity [58] and a small van der Waals radius [59], it affords a large dipole moment and thermal stability [60, 61], giving rise to a favorable FLC molecule. This also holds for side chain-type FLC polymers. In this work, with the goal of a quick electroresponsive conjugated polymer, we designed and synthesized a PMP to meet the aforementioned requirements.

We synthesized a ferroelectric liquid crystalline PMP by incorporating a fluorine-containing chiral LC group into the side chain [14]. Figure 11.17 shows the structure of the FLC-PMP, **Poly-8**, (M_n = 4,200) bearing a chiral FLC side chain and the FLC molecule: a fluorine-containing triphenyl compound, **FTP**.

Upon heating **Poly-8** to its isotropic phase and subsequent cooling to 144 °C, a striped fan texture characteristic of the SmC* phase was observed [52]. To measure the electroresponsive behavior of **Poly-8**, sample cells used for POM observation were prepared by injecting the melted polymer into two indium tin oxide (ITO) glasses separated with a 30 μm Teflon spacer. Note that the ITO glasses have no alignment treatment such as a coating of polyimide film. Figure 11.18 shows the changes in the optical textures of **Poly-8** under an electric field. When an electric field of 10 V was applied to the sample cell, the stripes of the fan-shaped texture in the SmC* phase (Fig. 12.18a) instantaneously disappeared to give an ordinary fan-shaped texture consistent with a SmC phase (Fig. 11.18b). The direction of polarization in the fan-shaped texture rapidly reversed, indicating a transition to another SmC phase, that is, a polarization inversion occurred when the opposite electric field was applied (Fig. 11.18c). The switching between the bistable SmC phases occurred quickly, that is, in less than 1 s.

The response time of 1 s in **Poly-8** is much smaller than those observed for ordinary LC conjugated polymers (approximately 20–30 min for complete alignment) in N-LC or smectic (Sm LC) phases under an electric field [62]. Note that

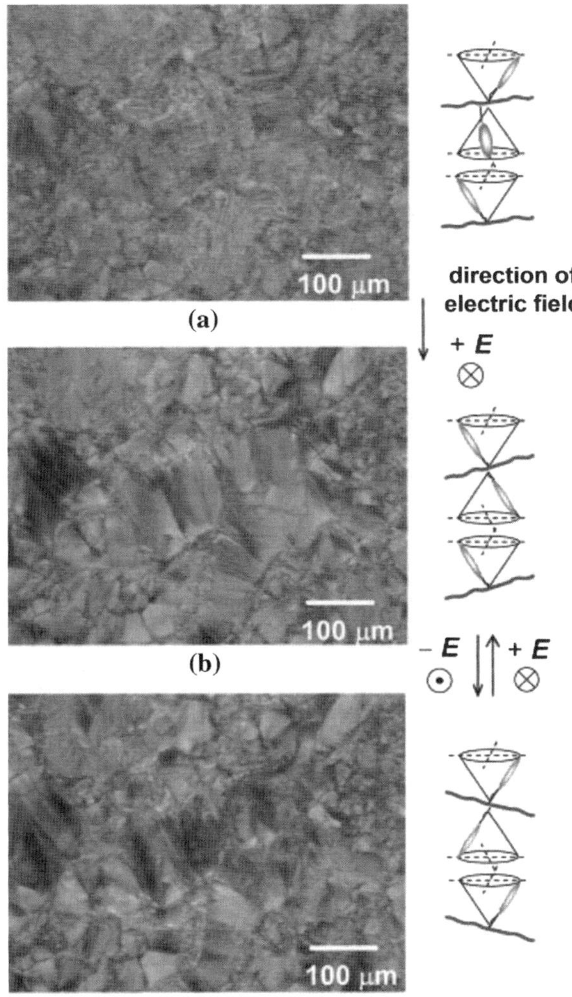

Fig. 11.18 Polarizing optical micrographs under crossed polarizers of **Poly-8** at 140 °C in cooling **a** before applying an electric field (SmC*), **b** with an applied field of +10 V (SmC) and **c** with an applied field of −10 V (SmC). The stripes of the fan-shape texture in SmC* phase disappeared by applying an electric field to give an ordinary fan-shape texture of SmC phase. The direction of the polarization in the fan-shape texture became inverse when opposite electric current was applied. Reprinted with permission from [14]. Copyright 2008, Wiley

although there are very fast switching (<1 ms) FLC displays using FLC polymers [63], these polymers are based on non-conjugated main chains. In this sense, **Poly-8** is the first example to realize such a smooth electroresponse through a FLC functionality, so far as conjugated polymers are concerned. The quick electroresponse of **Poly-8** is owing to the spontaneous polarization characteristics of the FLC polymers. This is because the response time (τ) is proportional to the inverse of the product of the spontaneous polarization (P_S) and the electric field (E); $\tau = \eta / P_S \cdot E$. The huge spontaneous polarization is generated by the macroscopic alignment of the polymer in the SmC phase. It is worthwhile to stress that the quick response time of less than one second in **Poly-8** is mainly due to the large spontaneous polarization and partly due to the oligomeric nature, which is evident

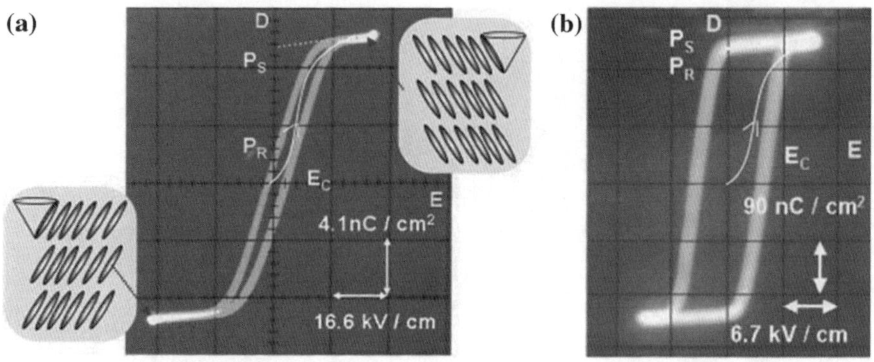

Fig. 11.19 Ferroelectric hysteresis loops of **a** FLC molecule (**FTP**), and **b Poly-8** under electric field at temperatures of SmC* phases. Vertical axis (D): Electric displacement (nC/cm^2). Horizontal axis (E): Electric field (kV/cm). Electric displacement is related with electric polarization (P) as follows; $D = \varepsilon_0 E + P$. P_S: spontaneous polarization; P_R: residual polarization; E_C: coercive electric field. Reprinted with permission from [14]. Copyright 2008, Wiley

from its low molecular weight ($M_n = 4{,}200$). That is, the latter should contribute to a decrease in the rotational viscosity of the side chains in the ferroelectric switching motion along the Goldstone mode. It is of interest that the texture of the SmC phase remained unchanged even after the electric field was turned off. This is attributed to the stable layered structures of the SmC phase of **Poly-8**; this behavior leads to a large residual polarization, as will be discussed below.

The dependence of the polarization of **Poly-8** on the electric field was investigated using a Sawyer–Tower circuit [64]. The measurements were carried out at particular temperatures, where the polymers showed SmC* phases during the cooling processes. A sinusoidal wave of 50 Hz was applied to the samples. The sample cell used was the same as that for the POM observations mentioned above, and the electrode area for the cell was 0.22 cm^2 (0.47 cm × 0.47 cm). Figure 11.19 shows the so-called D-E curves for **Poly-8**, together with those for the FLC molecule, **FTP**. Note that the electric displacement (D) is related to the polarization (P), that is, $D = \varepsilon_0 E + P$, where ε_0 is the vacuum permittivity (dielectric constant). In the figure, the coercive field (E_C) is the electric field necessary for polarization inversion, and it is nonzero to give a hysteresis loop in the D-E curve. In fact, the polarization inversion was observed in **Poly-8**. As the strength of the electric field increased from zero voltage, corresponding to SmC* phase, the polarization of the polymer first increased from zero value to attain the saturated value, that is, the spontaneous polarization, P_S. Next, as the electric field decreased, the polarization also decreased to pass the point represented by the residual polarization, P_R, where the electric field is zero. Subsequently, as the electric field of opposite sign increased, the polarization further decreased and then had an opposite value at the point corresponding to a coercive electrical field, E_C. The opposite polarization then increased to attain the value of P_S in another SmC state. In this state, the LC side chains of the polymers are tilted towards the opposite direction.

Table 11.5 Spontaneous Polarization (P_S), Residual Polarization (P_R), and Coercive Electric Field (E_C) of FLC Molecule, **FTP**, and **Poly-8**

	P_s (nC/cm^2)	P_R (nC/cm^2)	E_c (kV/cm)
FTP[a]	10	3	3
Poly-8[b]	200	200	5

[a] Measurement carried out at SmC* phase at 123 °C
[b] Measurement carried out at SmC* phase at 134 °C

The measured electroresponsive behaviors of the FLC molecule and **Poly-8** are given in Table 11.5. The FLC molecule, **FTP**, which can be used as a substituent of **Poly-8** after deprotection of the tetrahydropyranyl group, has a P_S value of 10 nC/cm^2, while the value of P_S of **Poly-8** is 200 nC/cm^2. The value is much larger, by at least one order of magnitude, than that of the **FTP** (10 nC/cm^2). The increase in P_S is due to the polymerization, which accumulates the FLC moieties having spontaneous polarizations into a polymer. Meanwhile, the E_C value of **Poly-8** is 5 kV/cm, which is not so large compared with that of **FTP** (3 kV/cm). This means that the polarization inversions of **Poly-8** can proceed under electric fields comparable to or smaller than those used for FLC molecules, in spite of the large spontaneous polarizations of the polymers. The observed high P_S and P_R values of **Poly-8** are due to the high stereoregularity of the main chain of **Poly-8**, which stabilizes the tilted layered structure of the SmC phase. It should be noted that the residual polarization, P_R, is the same as that of the spontaneous polarization, P_S, in **Poly-8**. This means that the macroscopically aligned polarization generated by the electric field remains unchanged even after the electric field is turned off. It is therefore suggested that **Poly-8** might be feasible as a polymer-based ferroelectric memory material, as well as a linearly dichroic blue fluorescent material with quickly rapid electroresponsive switching functionality.

11.4.2 Dynamic Switching of Linearly and Circularly Polarized Luminescence of Liquid Crystalline Photoresponsive Conjugated Polymers

In addition to investigating ferroelectric liquid crystalline conjugated polymers with dynamic switching functionalities under an electric field, our group has also developed photoresponsive liquid crystalline conjugated polymers with dynamic switching of linearly and circularly polarized luminescence. This was accomplished through the use of a photoisomerizable moiety in the polymer side chain.

Among various photoresponsive compounds, dithienylethene (DE) derivatives are the most feasible photochromic materials due to their outstanding fatigue resistance, thermal stability, and reversible properties [65–67]. Addition of this photoresponsive moiety into an aromatic conjugated polymer is attractive as a method of dynamic control of luminescence through photoisomerization-driven switching [68–76]. If the substituent of the main chain-type LC conjugated polymer is a photoresponsive moiety, the polymer should not only be macroscopically aligned by application of an external force but also photochemically switched by irradiation with external light. Particularly, the LPL of the macroscopically aligned LC conjugated polymer can be dynamically controlled by virtue of the reversible photoswitching of the photoresponsive moiety upon irradiation with ultraviolet (UV) and visible (vis) light.

When the polymer-based LPL device is fabricated, the relative arrangement between the LPL polymer film and the linear polarizer should be rigorously controlled. Namely, the aligned LPL polymer film must be arranged exactly parallel or perpendicular to the linear polarizer. Otherwise, the light passing through the LPL film and the polarizer is insufficiently polarized, which gives rise to incompletely polarized light. This type of fabrication becomes much more difficult and time consuming when the size of the device becomes smaller (micro- or nano-size). One way to avoid such difficulties in the device fabrication is to address CPL because CPL is not influenced by the macroscopic alignment of the polymer, and the degree of circular polarization of the polymer film depends on the one-handed screwing in the helical polymer and/or the helical π-stacking of the polymer. In addition, CPL is easily transformed into LPL by passing the light of CPL through a 1/4 (quarter) wave plate.

Here, we have designed and synthesized multifunctional poly(bithienylene-phenylene)s with either racemic (**Poly-9**) (M_n = 14,000) or chiral moieties ((***R***)-/(***S***)-**Poly-10** (M_n = 10,000, 8,000, respectively), which exhibit fluorescence, liquid crystallinity, and photoresponsive properties (Fig. 11.20). The polymers are composed of a π-conjugated main chain, poly(bithienylene-phenylene), which acts as a fluorescence moiety and mesogen core, and photochromic DE moieties [71, 72] are linked with racemic or chiral alkyl groups in the side chains. The DE photoresponsive moiety isomerizes between its closed and open forms upon irradiation of UV and visible light, respectively.

The phase transition of **Poly-9** in its open form was elucidated through POM. *Schlieren* textures from temperatures of 60 °C to over 200 °C were observed, which are characteristic of the N-LC phase. Liquid crystallinity was observed during both heating and cooling processes, indicating an enantiotropic nature. This suggests that **Poly-9** has a main chain-type liquid crystallinity and might be macroscopically aligned to form a monodomain structure in its LC phase.

The photoisomerization of the DE moiety upon irradiation with UV and visible light was examined by ^{1}H NMR measurements in chloroform-*d* ($CDCl_3$). Two broad signals observed at 5.30 and 5.34 ppm were assigned to the methylene protons neighboring the DE moiety in its open form. Irradiation with UV light (λ = 254 nm, 16 W) caused photoisomerization, resulting in new broad signals at 5.05 and 5.08 ppm, corresponding to the methylene protons neighboring the DE moiety in its closed form. The integrated intensities of the proton signals indicated that the open forms of 42 and 30 % were converted into the closed forms in the photostationary states (PSS) of **Poly-9** and **Poly-10**, respectively. Therein, the PSS is defined as {[closed form]/([open form] + [closed form])} × 100 [65, 77]. In

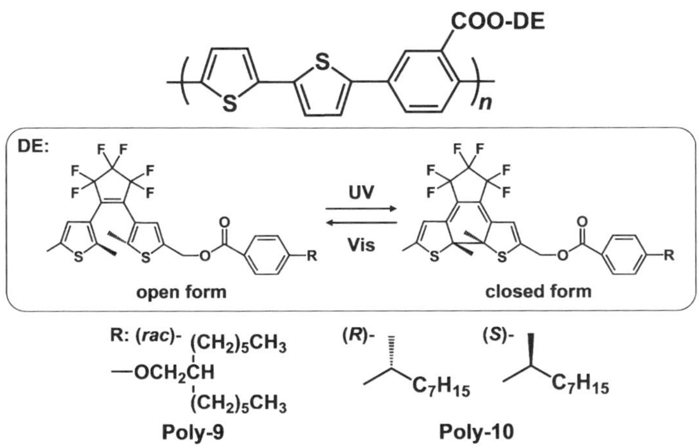

Fig. 11.20 Structures of the poly(bithenylene-phenylene)s, **Poly-9**, and **(*R*)-/(*S*)-Poly-10** having dithienylethene (DE) moieties in open form and closed form upon exposure to Vis and UV light, respectively

other words, 58 and 70 % of the DE moieties of **Poly-9** and **Poly-10**, respectively, remained unchanged in their open forms, even after UV light irradiation. After the subsequent irradiation of visible light ($\lambda > 400$ nm, 100 W), all signals of the open form reappeared. Note that because the closed form is not completely formed even after UV light irradiation, the term PSS will be used hereafter instead of the closed form.

The fluorescence band of **Poly-9** at 524 nm (open form) in chloroform drastically decreased in intensity upon being irradiated with UV light for 1 min. However, the quenched fluorescence of the open form was regenerated by irradiation with visible light for 5 min, as shown in Fig. 12.21a. The ratio of fluorescence intensity (I_{open}/I_{PSS}) between the open form and the PSS was 94 for **Poly-9**. Fluorescence quantum yields of the open form and the PSS of **Poly-9**, evaluated using quinine sulfate in 1.0 M sulfuric acid as a standard, were 0.23 and 0.002, respectively. The quenching of the fluorescence may be due to an energy transfer through reabsorption, where the photon being emitted from the excited main chain is absorbed by the DE moiety in its closed form. The excited DE moiety, however, releases the photon through a non-radiative transition. This causes the apparent disappearance of the fluorescence in the closed form. The above argument can be supported by the fact that the fluorescence band of the main chain at 524 nm is very close to the absorption band approximately 550 nm of the DE moiety in the closed form. A similar quenching in fluorescence was also observed in the cast film, as seen in Fig. 11.21b. The ratio of fluorescence intensity (I_{open}/I_{PSS}) at 575 nm between the open form and the PSS in the **Poly-9** film was 323. The values of I_{open}/I_{PSS}, i.e. the degree of quenching, were three times larger in the cast films than in solution. The reversible changes between the quenching and emission were repeated over more than 20 cycles in the cast films of **Poly-9**, as shown in Fig. 12.21c.

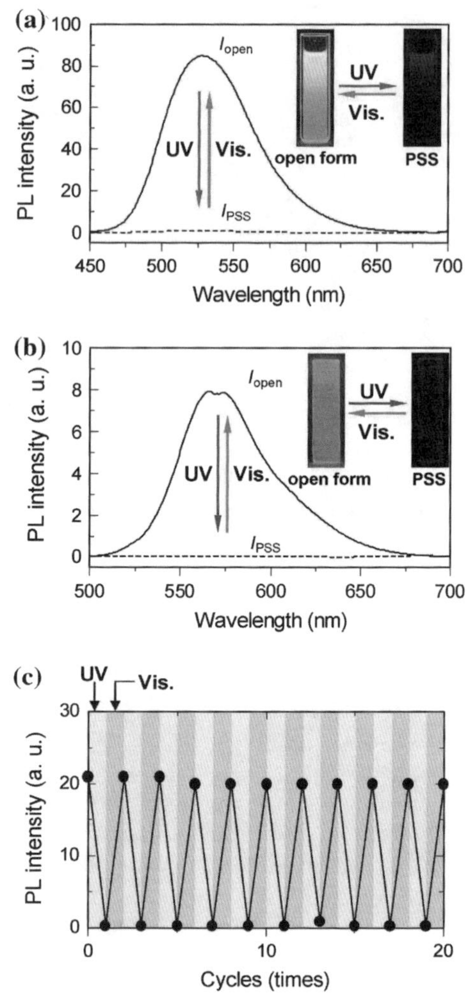

Fig. 11.21 PL (fluorescence) spectra (excitation light of $\lambda = 350$ nm, 100 W) of **Poly-9** **a** in chloroform solution (4.0×10^{-5} M) and **b** in cast film. *Inset* shows photographs of emitting and quenching between open form and PSS in solution (excitation light of $\lambda = 330$–390 nm, 16 W, handy lamp). Reversible changes of fluorescence intensities of **Poly-9** at 575 nm **c** in cast films upon alternative irradiations of UV (*blue area*) and vis (*yellow area*) light. Reprinted with permission from [15]. Copyright 2008, American Chemical Society

Next, we carried out the macroscopic alignment of **Poly-9** (open form) films by the rubbing technique [78]. The polymers on the quartz substrates were heated to the LC temperature region, and they were rubbed with a glass rod. The polymers were then cooled to room temperature. **Poly-9** (open form) was well aligned parallel to the rubbing direction to give a monodomain structure. It is understood that the main chain of **Poly-9** consists of bithienylene-phenylene units, which can be regarded as a mesogenic moiety, and hence it is aligned parallel to the rubbing direction.

Figure 11.22 shows the LPL of the aligned film of **Poly-9** (open form), in which the polarizer was placed parallel or perpendicular to the rubbing direction. The LPL intensity at 575 nm under nonpolarized light excitation ($\lambda = 380$ nm) in the parallel direction ($I_{//,open}$) was larger than that in the perpendicular one ($I_{\perp,open}$).

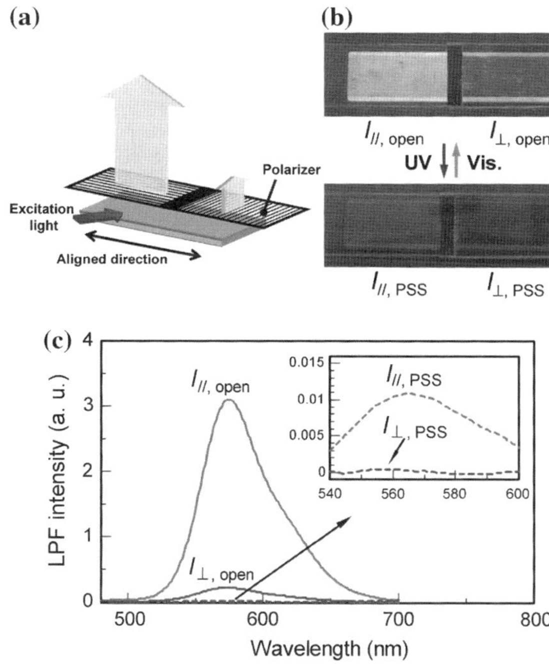

Fig. 11.22 **a** Schematic illustration of linearly polarized luminescence (LPL). The aligned **Poly-9** film on the quartz substrate (size: 10 × 35 mm) shows LPL upon an irradiation of excitation light ($\lambda = 330{-}390$ nm, 16 W, handy lamp). Two polarizers are arranged parallel and perpendicular to the aligned direction of the film (*right*). **b** Both the parallel and perpendicular components of LPLs are reversibly switched in intensity between the open and PSS forms, by the irradiations of UV ($\lambda = 254$ nm, 16 W, handy lamp) and visible ($\lambda > 400$ nm, 100 W) light (*left*). **c** Spectroscopic changes of LPLs of the aligned **Poly-9** film. *Inset* shows the expanded spectra of LPLs in the PSS form. Reprinted with permission from [15]. Copyright 2008, American Chemical Society

The dichroic ratio (R) and order parameter (S) of the aligned film (open form) were evaluated to be 16.0 and 0.83, respectively, where R and S are defined as $R = (I_{//,\mathrm{open}})/(I_{\perp,\mathrm{open}})$ and $S = (R - 1)/(R + 2)$, respectively. It is evident that the aligned **Poly-9** film has a high anisotropy in LPL and a high order parameter comparable to the highest reported values of N-LC polymers. It is of keen interest that the parallel fluorescence of **Poly-9** at 575 nm for the open form ($I_{//,\mathrm{open}}$) was quenched in the PSS and recovered by visible light irradiation. The ratio of the LPL intensity between the open form and the PSS in the parallel direction, $(I_{//,\mathrm{open}})/(I_{//,\mathrm{PSS}})$ was more than 310, which is comparable to that of the nonoriented film (323). Similar emission and quenching phenomena were observed in the perpendicular intensities between the open form ($I_{\perp,\mathrm{open}}$) and the PSS ($I_{\perp,\mathrm{PSS}}$), where the ratio was more than 480. It is found that the LPL of the aligned **Poly-9** film is photochemically switched by irradiations of UV and visible light. Taking into account the LPL intensities in terms of the open form and the PSS, and also

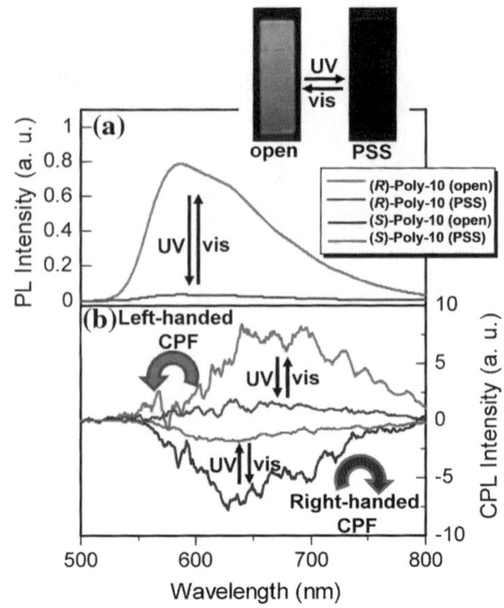

Fig. 11.23 **a** PL (fluorescence) spectra and **b** CPL (or CPF, circularly polarized fluorescence) spectra of **(*R*)-Poly-10** cast films. The wavelength of excitation light for fluorescence was 380 nm. The *inset* shows photographs of the emitting film of the open form and the quenching film of the PSS of **(*R*)-Poly-10** (excitation wavelength $\lambda = 370$ nm, 4 W, handheld lamp). Reprinted with permission from [20]. Copyright 2010, Wiley

the parallel and perpendicular components in the aligned film, the relative intensities at 575 nm of ($I_{//,\mathrm{open}}$), ($I_{\perp,\mathrm{open}}$), ($I_{//,\mathrm{PSS}}$), and ($I_{\perp,\mathrm{PSS}}$) were evaluated to be 7,800, 480, 25, and 1, respectively. This implies that four types of fluorescence intensities are realized in the same polymer and are even controllable using both LC-based macroscopic alignment and photochemical switching.

The cast films of **(*R*)-/(*S*)-Poly-10** showed CD signals with bisignate Cotton effects in the π-π transition region of the polymer main chain, which implies that the polymers in the film are each assembled to form a one-handed helical interchain π-stacking structure. Interestingly, there is no difference in the CD spectra between the open form and the PSS, irrespective of the (*R*)- and (*S*)- configuration. This indicates that the helical π-stacked structure of the **Poly-10** film remains unchanged even after the reversible photoisomerization between the open form and the PSS in the side chain. This may allow us to exploit the chiroptical properties such as CPL during the photochemical isomerization, as discussed below.

It is of interest that the polymers show CPL as well as CD in cast films, although they show neither CPL nor CD in solution, except the weak CD signal that corresponds to the overlapped absorption bands of the dithienylethene and phenylene linked with the chiral alkyl moiety. The formation of helical π-stacking structures of **Poly-10** is supported by the observation of its CPL in cast films, as shown in Fig. 11.23. The open forms of **(*R*)-/(*S*)-Poly-10** show negative and positive signs in the CPL spectra, respectively. Figure 11.23 shows that the CPL intensity of the **Poly-10** film drastically decreases, irrespective of the configuration, upon photoisomerization from the open form to the PSS. Namely, CPL quenching occurs in the PSS, which is associated with the fluorescence quenching itself. The reverse photoisomerization from the PSS to the open form causes a reappearance of the CPL band. Such a photoswitching behavior was observed over 10 cycles. The ratio of CPL intensities ($\Delta I_{open}/\Delta I_{PSS}$) between the open form and the PSS of the **(*R*)-Poly-10** film at 650 nm was 7, which is smaller than the corresponding ratios of the fluorescence intensities ($I_{open}/I_{PSS} = 80$) for **(*R*)-Poly-10**. The degree of circular polarization in the fluorescence was evaluated using g_{em}. The g_{em} values for the **Poly-10** films were estimated to be on the order of 10^{-2} at 650 nm. Finally, it should be emphasized that the helical π-stacked structure formed in the films of the presented conjugated polymers is so rigid that it is not affected by the photochemical opening and closing isomerization reaction of the DE moiety in the solid state. This allowed us to control the reversible switching of the CPL between emitting and quenched states through photochemical irradiation while maintaining the chirality of the polymers.

The linearly polarized luminescence in the aligned **Poly-9** film, which was prepared by rubbing in the LC phase, showed reversible quenching and emitting behaviors owing to the photochemical isomerization upon irradiation with ultraviolet and visible light. The cast films of **(*R*)-/(*S*)-Poly-10** exhibited right- or left-handed CPL with relatively large g_{em} factors, on the order of 10^{-2}. The emission and quenching of the CPL of the cast films are reversibly switched through the photoisomerization of the DE moiety introduced in the side chain. The presented polymers could be useful for advanced functional optical materials with photochemically switchable LPL and CPL properties.

11.5 Summary

In this chapter, a review of recent developments of self-assembled liquid crystalline conjugated polymers with advanced functionalities was discussed. In Part 1, it was shown that the unique π-stabilized helicene-type intrachain helical main chain structure of PMP has a high degree of crystallinity. This enables self-assembly to higher-ordered whisker structures. In addition, self-assembled LC-substituted PMP was macroscopically aligned using an external magnetic field. In Part 2, the higher-ordered N*-LC phase in di-substituted polyacetylenes and thiophene-based copolymers enables the emergence of CPL with high dissymmetry factors. In Part 3, the use of dynamic switching systems such as ferroelectric liquid crystalline moieties in PMP and photoresponsive dithienylethene moieties in poly(bithienylene-phenylene)s leads to advanced switchable functionalities in response to external stimuli.

Furthermore, the recent developments in liquid crystalline conjugated polymers described in this chapter may accelerate the further development of other self-assembled polymers with novel functionalities. Research into liquid crystalline conjugated polymers exhibiting innovative functionalities such as helical inversion switching through external stimuli, photoswitching between RGB fluorescence, and the thermal-responsive selective reflection of circularly polarized luminescence is anticipated in the near future. The synergistic use of the self-assembly exhibited by liquid crystalline conjugated polymers with other advanced functionalities may lead to the emergence of next-generation optoelectronic organic materials.

Acknowledgments The authors are grateful to Dr. Kiyoshi Suda, Dr. Hiroyuki Hayasaka, and Mr. Kazuyoshi Watanabe (Department of Polymer Chemistry, Kyoto University) for their valuable contributions in the development and synthesis of the liquid crystalline conjugated polymers discussed in this chapter. This work was supported by Grants-in-Aid for Science Research (S) (No. 20225007), (A) (No. 25246002), and (No. 25620098) from the Ministry of Education, Culture, Sports, Science and Technology, Japan.

References

1. D.T. McQuade, A.E. Pullen, T.M. Swager, Conjugated polymer-based chemical sensors. Chem. Rev. **100**, 2537–2574 (2000)
2. J. Liu, J.W.Y. Lam, B.Z. Tang, Acetylenic polymers: syntheses, structures, and functions. Chem. Rev. **109**, 5799–5867 (2009)
3. T. Masuda, Substituted polyacetylenes. J. Polym. Sci. Part A Polym. Chem. **45**, 165–180 (2007)
4. C.V. Hoven, A. Garcia, G.C. Bazan, T.Q. Nguyen, Recent applications of conjugated polyelectrolytes in optoelectronic devices. Adv. Mater. **20**, 3793–3810 (2008)
5. J.H. Burroughes, D.D.C. Bradley, A.R. Brown, R.N. Marks, K. Mackay, R.H. Friend, P.L. Burns, A.B. Holmes, Light-emitting diodes based on conjugated polymers. Nature **347**, 539–541 (1990)
6. R.H. Friend, R.W. Gymer, A.B. Holmes, J.H. Burroughes, R.N. Marks, C. Taliani, D.D.C. Bradley, D.A. Dos Santos, J.L. Bredas, M. Logdlund, W.R. Salaneck, Electroluminescence in conjugated polymers. Nature **397**, 121–128 (1999)
7. S. Guenes, H. Neugebauer, N.S. Sariciftci, Conjugated polymer-based organic solar cells. Chem. Rev. **107**, 1324–1338 (2007)
8. S.W. Thomas III, G.D. Joly, T.M. Swager, Chemical sensors based on amplifying fluorescent conjugated polymers. Chem. Rev. **107**, 1339–1386 (2007)
9. F.J.M. Hoeben, P. Jonkheijm, E.W. Meijer, A.P.H.J. Schenning, About Supramolecular assemblies of π-conjugated systems. Chem. Rev. **105**, 1491–1546 (2005)
10. K. Suda, K. Akagi, Self-assembled Helical conjugated poly(*meta*-phenylene) derivatives that afford whiskers with Discotic Hexagonal columnar packed structure. Macromolecules **44**, 9473–9488 (2011)
11. M. Grell, W. Knoll, D. Lupo, A. Meisel, T. Miteva, D. Neher, H.G. Nothofer, U. Scherf, A. Yasuda, Blue polarized electroluminescence from a liquid crystalline polyfluorene. Adv. Mater. **11**, 671–675 (1999)
12. H. Sirringhaus, R.J. Wilson, R.H. Friend, M. Inbasekaran, W. Wu, E.P. Woo, M. Grell, D.D.C. Bradley, Mobility enhancement in conjugated polymer field-effect transistors through chain alignment in a liquid-crystalline phase. Appl. Phys. Lett. **77**, 406–408 (2000)
13. K. Suda, J. Yoshida, S. Nimori, K. Akagi, Hierarchical structures of self-assembled helical conjugated polymers - magnetically forced alignment of liquid crystalline poly(*meta*-phenylene) derivatives. Synth. Met. **159**, 943–948 (2009)

14. K. Suda, K. Akagi, Electro-optical behavior of ferroelectric liquid crystalline polyphenylene derivatives. J. Polym. Sci. Part A Polym. Chem. **46**, 3591–3610 (2008)
15. H. Hayasaka, K. Tamura, K. Akagi, Dynamic switching of linearly polarized emission in liquid crystallinity-embedded photoresponsive conjugated polymers. Macromolecules **41**, 2341–2346 (2008)
16. K. Akagi, Synthesis and properties of liquid-crystalline-conjugated polymers. Bull. Chem. Soc. Jpn. **80**, 649–661 (2007)
17. J. Oguma, R. Kawamoto, H. Goto, K. Itoh, K. Akagi, Magnetically forced alignment of liquid crystalline aromatic conjugated polymers and their electrical and optical anisotropies. Synth. Met. **119**, 537–538 (2001)
18. B.A. San Jose, S. Matsushita, Y. Moroishi, K. Akagi, Disubstituted liquid crystalline polyacetylene derivatives that exhibit linearly polarized blue and green emissions. Macromolecules **44**, 6288–6302 (2011)
19. G. Kwak, M. Minakuchi, T. Sakaguchi, T. Masuda, M. Fujiki, Poly(diphenylacetylene) bearing long alkyl side chain via silylene linkage: its lyotropic liquid crystallinity and optical anisotropy. Chem. Mater. **19**, 3654–3661 (2007)
20. B.A. San Jose, K. Akagi, Liquid crystalline polyacetylene derivatives with advanced electrical and optical properties. Polym. Chem. **4**, 5144–5161 (2013)
21. B.A. San Jose, S. Matsushita, K. Akagi, Lyotropic chiral nematic liquid crystalline aliphatic conjugated polymers based on di-substituted polyacetylene derivatives that exhibit high dissymmetry factors in circularly polarized luminescence. J. Am. Chem. Soc. **134**, 19795–19807 (2012)
22. K. Watanabe, I. Osaka, S. Yorozuya, K. Akagi, Helically π-stacked thiophene-based copolymers with circularly polarized fluorescence: high dissymmetry factors enhanced by self-ordering in chiral nematic liquid crystal phase. Chem. Mater. **24**, 1011–1024 (2012)
23. H. Hayasaka, T. Miyashita, K. Tamura, K. Akagi, Helically π-stacked conjugated polymers bearing photoresponsive and chiral moieties in side chains: reversible photoisomerization-enforced switching between emission and quenching of circularly polarized fluorescence. Adv. Funct. Mater. **20**, 1243–1250 (2010)
24. K. Watanabe, H. Iida, K. Akagi, Circularly polarized blue luminescent spherulites consisting of hierarchically assembled ionic conjugated polymers with a helically π-stacked structure. Adv. Mater. **24**, 6451–6456 (2012)
25. E. Yashima, K. Maeda, H. Iida, Y. Furusho, K. Nagai, Helical polymers: synthesis, structures, and functions. Chem. Rev. **109**, 6102–6211 (2009)
26. E. Yashima, K. Maeda, T. Nishimura, Detection and amplification of chirality by helical polymers. Chem. Eur. J. **10**, 42–51 (2004)
27. C. Li, M. Numata, A.H. Bae, K. Sakurai, S. Shinkai, Self-assembly of supramolecular chiral insulated molecular wire. J. Am. Chem. Soc. **127**, 4548–4549 (2005)
28. K. Akagi, H. Shirakawa, Morphological alignment of liquid crystalline. Conducting polyacetylene derivatives. Macromol. Symp. **104**, 137–158 (1996)
29. K. Akagi, H. Shirakawa, *Polymeric Materials Encyclopedia: Synthesis, Properties and Applications* (CRC Press, Boca Raton, 1996)
30. K. Akagi, H. Shirakawa, in *Electrical and Optical Polymer Systems: Fundamentals, Methods, and Applications*, ed. by D.L. Wise, G.E. Wnek, D.J. Trantolo, T.M. Cooper, J.D. Gresser (Marcel Dekker, New York, 1998)
31. Y. Yamashita, Y. Kato, S. Endo, K. Kimura, Preparation and formation mechanism of Poly(oxycarbonyl-1,4-phenylene) whiskers from 4-Acetoxybenzoic acid. Makromol. Chem. Rapid Commun. **9**, 687–691 (1988)
32. K. Kimura, Y. Kato, T. Inabe, Y. Yamashita, Formation mechanism of Poly(4'-oxy-4-biphenylcarbonyl) crystals obtained by solution polymerization. Macromolecules **28**, 255–259 (1995)
33. K.J. Ihn, J. Moulton, P. Smith, Whiskers of Poly(3-alkylthiophene)s. J. Polym. Sci. Part B Polym. Phys. **31**, 735–742 (1993)

34. D.S. Suh, T.J. Kim, A.N. Aleshin, Y.W. Park, G. Piao, K. Akagi, H. Shirakawa, J.S. Qualls, S.Y. Han, J.S.J. Brooks, Helical polyacetylene heavily doped with iodine: magnetotransport. Chem. Phys. **114**, 7222–7227 (2001)
35. K. Akagi, G. Piao, S. Kaneko, K. Sakamaki, H. Shirakawa, M. Kyotani, Helical polyacetylene synthesized with a chiral nematic reaction field. Science **282**, 1683–1686 (1998)
36. K. Akagi, Helical polyacetylene: asymmetric polymerization in a chiral liquid-crystal field. Chem. Rev. **109**, 5354–5401 (2009)
37. B.S. Li, K.K.L. Cheuk, F. Salhi, J.W.Y. Lam, J.A.K. Cha, X. Xiao, C. Bai, B.Z. Tang, Tuning the chain helicity and organizational morphology of an L-valine-containing polyacetylene by pH change. Nano Lett. **6**, 323–328 (2001)
38. M.M. Bouman, E.W. Meijer, Stereomutation in optically active regioregular polythiophenes. Adv. Mater. **7**, 385–387 (1995)
39. A. Khan, S. Muller, S. Hecht, Practical synthesis of an amphiphilic, non-ionic poly(para-phenyleneethynylene) derivative with a remarkable quantum yield in water. Chem. Commun. 584–586 (2005)
40. W. Vanormelingen, L. Pandey, M. Van der Auweraer, T. Verbiest, G. Koeckelberghs, Steering the conformation and chiroptical properties of poly(dithienopyrrole)s substituted with chiral OPV side chains. Macromolecules **43**, 2157–2168 (2010)
41. M.M. Green, N.C. Peterson, T. Sato, A. Teramoto, R. Cook, S. Lifson, A Helical polymer with a cooperative response to chiral information. Science **268**, 1860–1866 (1995)
42. J.C. Nelson, J.G. Saven, J.S. Moore, P.G. Wolynes, Solvophobically driven folding of nonbiological oligomers. Science **277**, 1793–1796 (1997)
43. D.J. Williams, H.M. Colquhoun, C.A.O. Mahoney, The structure of poly(m-phenylene): a prediction from single-crystal X-ray studies of m-Deciphenyl and m-Undeciphenyl. Chem. Commun. 1643–1644 (1994)
44. N. Kobayashi, S. Sasaki, M. Abe, S. Watanabe, H. Fukumoto, T. Yamamoto, Chain distortion of m-Linked aromatic polymers: poly(m-phenylene) and poly(m-pyridine). Macromolecules **37**, 7986–7991 (2004)
45. M. Grell, D.D.C. Bradley, Polarized luminescence from oriented molecular materials. Adv. Mater. **11**, 895–905 (1999)
46. M. Schadt, Liquid crystal materials and liquid crystal displays. Annu. Rev. Mater. Sci. **27**, 305–379 (1997)
47. E. Peeters, M.P.T. Christiaans, R.A.J. Janssen, H.F.M. Schoo, H.P.J.M. Dekkers, E.W. Meijer, Circularly polarized electroluminescence from a polymer light-emitting diode. J. Am. Chem. Soc. **119**, 9909–9910 (1997)
48. B.M.W. Langeveld-Voss, R.A.J. Janssen, M.P.T. Christiaans, S.C.J. Meskers, H.P.J. Dekkers, E.W. Meijer, Circular dichroism and circular polarization of photoluminescence of highly ordered Poly{3,4-di[(S)-2-methylbutoxy]thiophene}. J. Am. Chem. Soc. **118**, 4908–4909 (1996)
49. M. Oda, H.G. Nothofer, G.L.U. Scherf, S.C.J. Meskers, D. Neher, Circularly polarized electroluminescence from liquid-crystalline chiral polyfluorenes. Adv. Mater. **12**, 362–365 (2000)
50. S.H. Chen, B.M. Conger, J.C. Mastrangelo, A.S. Kende, D.U. Kim, Synthesis and optical properties of thermotropic polythiophene and poly(p-phenylene) derivatives. Macromolecules **31**, 8051–8057 (1998)
51. J.N. Wilson, W. Steffen, T.G. McKenzie, G. Lieser, M. Oda, D. Neher, U.H.F. Bunz, Chiroptical properties of poly(p-phenyleneethynylene) copolymers in thin films: large g-Values. J. Am. Chem. Soc. **124**, 6830–6831 (2002)
52. I. Dierking (ed.), *Textures of Liquid Crystals* (Wiley, Weinheim, 2003)
53. S. Kawata, Y. Kawata, Three-dimensional optical data storage using photochromic materials. Chem. Rev. **100**, 1777–1788 (2000)
54. D.M. Ivory, G.G. Miller, J.M. Sowa, L.W. Shacklette, R.R. Chance, R.H. Baughman, Highly conducting charge-transfer complexes of poly(p-phenylene). J. Chem. Phys. **71**, 1506–1507 (1979)

55. M. Mitsuishi, S. Ito, M. Yamamoto, H. Endo, S. Hachiya, T. Fischer, W. Knoll, Optical characterization of a ferroelectric liquid crystalline polymer studied by time-resolved optical waveguide spectroscopy. Macromolecules **31**, 1565–1574 (1998)
56. N. Leister, D. Geschke, Pyroelectric investigations of polarization and charge distributions in sandwich cells containing a ferroelectric liquid crystalline polymer. Liq. Cryst. **24**, 441–449 (1998)
57. A. Fukuda, Y. Takanishi, T. Isozaki, K. Ishikawa, H. Takezoe, Antiferroelectric chiral smectic liquid crystals. J. Mater. Chem. **4**, 997–1016 (1994)
58. V. Reiffenrath, J. Krause, H.J. Plach, G. Weber, New liquid-crystalline compounds with negative dielectric anisotropy. Liq. Cryst. **5**, 159–170 (1989)
59. P. Balkwill, D. Bishop, A. Pearson, I. Sage, Fluorination in nematic systems. Mol. Cryst. Liq. Cryst. **123**, 1–13 (1985)
60. M.A. Osman, Molecular structure and mesomorphic properties of thermotropic liquid crystals. III. Lateral substituents. Mol. Cryst. Liq. Cryst. **128**, 45–63 (1985)
61. C.J. Booth, J.W. Goodby, J.P. Hardy, O.C. Lettington, K.J. Toyne, Fluorosubstituted chiral liquid crystals derived from (R)-2-(4-Hydroxyphenoxy)propan-1-ol. J. Mater. Chem. **3**, 935–941 (1993)
62. H. Goto, X. Dai, H. Narihiro, K. Akagi, Synthesis of polythiophene derivatives bearing ferroelectric liquid crystalline substituents. Macromolecules **37**, 2353–2362 (2004)
63. H. Endo, S. Hachiya, T. Sekiya, K. Kawasaki, Rotational viscosity of ferroelectric liquid-crystalline polysiloxanes. Liq. Cryst. **12**, 147–155 (1992)
64. A.K. Gathania, B. Singh, K.K. Raina, Switching Dynamics in Ferroelectric Liquid Crystal Mixture. Jpn. J. Appl. Phys. **43**, 8168–8172 (2004)
65. M. Irie, Diarylethenes for memories and switches. Chem. Rev. **100**, 1685–1716 (2000)
66. B.L. Feringa (ed.), *Molecular Switches* (Wiley, Weinheim, 2001)
67. H. Tian, S. Yang, Recent progresses on diarylethene based photochromic switches. Chem. Soc. Rev. **33**, 85–97 (2004)
68. T. Koshido, T. Kawai, K. Yoshino, Novel photomemory effects in photochromic dye-doped conducting polymer and amorphous photochromic dye layer. Synth. Met. **73**, 257–260 (1995)
69. S. Murase, M. Teramoto, H. Furukawa, Y. Miyashita, K. Horie, Photochemically induced fluorescence control with intermolecular energy transfer from a fluorescent dye to a photochromic diarylethene in a polymer film. Macromolecules **36**, 964–966 (2003)
70. D. Medvedeva, A. Bobrovsky, N. Boiko, V. Shibaev, I. Zavarzin, M. Kalik, M. Krayushkin, A combination of selective light reflection and fluorescence modulation in a cholesteric polymer matrix. Macromol. Rapid Commun. **26**, 177–182 (2005)
71. T. Kawai, T. Sasakia, M. Irie, A photoresponsive laser dye containing photochromic dithienylethene units. Chem. Commun. 711–712 (2001)
72. T.B. Norsten, N.R. Branda, Axially coordinated porphyrinic photochromes for non-destructive information processing. Adv. Mater. **13**, 347–349 (2001)
73. T. Kawai, T. Kunitake, M. Irie, Novel photochromic conducting polymer having diarylethene derivative in the main chain. Chem. Lett. **9**, 905–906 (1999)
74. H. Cho, E. Kim, Highly fluorescent and photochromic diarylethene oligomer bridged by p-phenylenevinylene. Macromolecules **35**, 8684–8687 (2002)
75. S. Wang, X. Li, B. Chen, Q. Luo, H. Tian, Photochromic copolymers containing bisthienylethene units. Macromol. Chem. Phys. **205**, 1497–1507 (2004)
76. Y.C. Jeong, S.I. Yang, E. Kim, K.H. Ahn, A high-content diarylethene photochromic polymer for an efficient fluorescence modulation. Macromol. Rapid Commun. **27**, 1769–1773 (2006)
77. M. Irie, K. Sakemura, M. Okinaka, K. Uchida, Photochromism of dithienylethenes with electron-donating substituents. J. Org. Chem. **60**, 8305–8309 (1995)
78. D. Demus, J. Goodby, G.W. Gray, H.W. Spiess, V.I. Vill, *Handbook of Liquid Crystals* (Wiley, Weinheim, 1998)

Chapter 12
Solubilization and Delivery of Drugs from GMO-Based Lyotropic Liquid Crystals

Nissim Garti, Dima Libster and Abraham Aserin

Abstract Self-assembled lyotropic liquid crystals (LLCs) of lipids and water are interesting mesophases that can be considered among other applications as delivery vehicles due to their remarkable structural complexity. The phase behavior of mixtures of glycerol monooleate (monoolein, GMO) and phytantriol were particularly well studied in drug delivery systems, food products, and encapsulation and crystallization of proteins. We are summarizing in this chapter our contribution to modified LLCs and their structural features and recent systematic efforts to utilize them for solubilization, and potential release of drugs and biomacromolecules. Some of the most interesting applications are the implementation of peptides and proteins in the reversed cubic and hexagonal mesophases, to enhance penetrating pattern of a model drugs. LLC vehicles were shown to allow "on demand" targeted release, based on controlling the polymorphism of lyotropic liquid crystalline mesophases. A new pathway to responsive LLCs, enabling the controlled release of hydrophilic drugs diffusing through the water channels of the mesophases was also investigated. The system is capable of self-assembling into a reverse bicontinuous cubic phase of *Im*3*m* symmetry and transform into a reverse columnar hexagonal as a function of pH, pressure or enzymes. LLCs were shown to entrap several peptides enzymes, proteins and nucleotides into cubic and lamellar monoolein-based mesophases in order to protect them and enable their release. The main outcomes of the described research demonstrated that control of the physical properties of hexagonal LLC on different length scales is key for rational design of these systems as delivery vehicles for both low molecular weight therapeutics and biomacromolecules.

N. Garti (✉) · D. Libster · A. Aserin
Casali Institute of Applied Chemistry, The Institute of Chemistry,
The Hebrew University of Jerusalem, 91904 Jerusalem, Israel
e-mail: garti@vms.huji.ac.il

Q. Li (ed.), *Nanoscience with Liquid Crystals*, NanoScience and Technology,
DOI: 10.1007/978-3-319-04867-3_12,

12.1 Characterization and Structural Considerations

12.1.1 Introduction

Lyotropic liquid crystals (LLCs) are interesting mesophases that can be considered among other applications as delivery vehicles due to their remarkable structural complexity and high solubilization capacity. The remarkable ability along other very unique characteristic is to release the encapsulated bioactives 'on demand' and to enhance their transport across membranes.

Liquid crystals (LCs) are self-assembled organized mesophases with properties intermediate to those of crystalline solids and isotropic liquids [1]. In LC phases, long-range periodicity exists, although the molecules exhibit a dynamical disorder at atomic distances, as in liquids. Accordingly, these materials can also be considered ordered fluids [2]. LLC are composed of at least two components: an amphiphilic molecule and its solvent. A hydrophilic solvent, such as water, hydrates the polar moieties of the amphiphiles via hydrogen bonding, while the flexible aliphatic tails of the amphiphiles aggregate into fused hydrophobic regions based on van der Waals interactions. In addition to morphologic dependence on the chemical composition, LLC are also sensitive to external stimuli such as temperature and pressure [1–3]. As a function of the molecular shape of the surfactants, packing parameters, and interfacial curvature energy considerations, LLC can be formed with aqueous domains ranging from planar bilayer lamellae to extended, cylindrical channels, to three-dimensional interconnected channels and manifolds [4]. These mesophases are defined as lamellar (L_α), hexagonal (H), bicontinuous cubic [Q (or V)], and discontinuous cubic (I) phases, based on their symmetry [5]. In addition, most lyotropic mesophases exist as symmetric pairs, a "normal" (type I) oil-in-water system, consisting of lipid aggregates in a continuous water matrix, and a topologically "inverted" (type II) water-in-oil version. The headgroups hydrated by water are arranged within a continuous non-polar matrix, which is composed of the fluid hydrocarbon chains [5]. In addition to its biological significance, inverse lipid phases could be useful as host systems for incorporation of food additives [6, 7], the crystallization of membrane proteins for drug delivery [6–8], and for inorganic synthesis [9] and nanoparticles of pharmaceutically active metals such as gold.

The lamellar structure does not possess any intrinsic curvature and is considered as the midpoint of an ideal, symmetrical LLC phase progression (Fig. 12.1) [4, 5]. The current chapter is mainly focused on the inverted (W/O) mesophases (cubic and hexagonal), representing an important class of nanostructures for potential pharmaceutical applications.

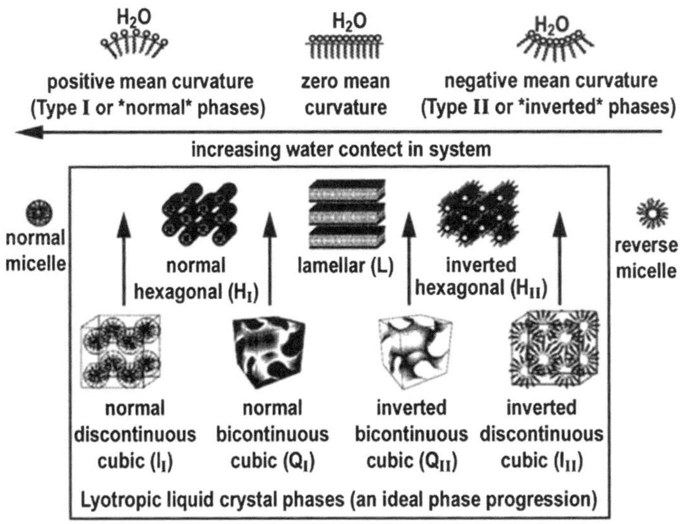

Fig. 12.1 Schematic representations of common LLC-phases formed by amphiphiles in water (Taken from [1])

12.1.2 The Amphiphiles

Only a few amphiphiles can mimic the behavior of biological lipids and form inverted mesophases. The unsaturated monoglycerides (monoolein and monolinolein) belong to this category. In less recent studies monolinolein was considered as amphiphile of choice because it is very fluid at room temperature while monoolein (GMO) is partially solid. However, GMO has less tendency than monolinolein to oxidize and hydrolyze and is commercially available in high purity. Therefore, GMO is the polar lipid that is commonly used as a food emulsifier and was selected in our studies as well in most others as the major compound for structuring mesophases. Presently, GMO is the preferred amphiphile for formulating LC phases for scientific research and drug delivery. This is a non-toxic, biocompatible, and biodegradable lipid, which possesses low water solubility, but swells and forms several LC phases in excess water [10].

Another lipid, phytantriol, was shown to form cubic LC. The advantage of this lipid, compared to GMO, is that it is not susceptible to esterase catalyzed hydrolysis and therefore provides additional stability to the mesophases. However, in many cases, there is an advantage of partial hydrolysis of the mesophase in the guts helping to release the drugs from the vehicles. Phytantriol is commonly used as an ingredient in the cosmetics industry for improving moisture retention. Its phase behavior as a function of water concentration and temperature is very similar to that of GMO, although structurally they are very different (Fig. 12.2) [11]. The health authorities did not yet cleared phytantriol for use in oral or other pharmaceutical application.

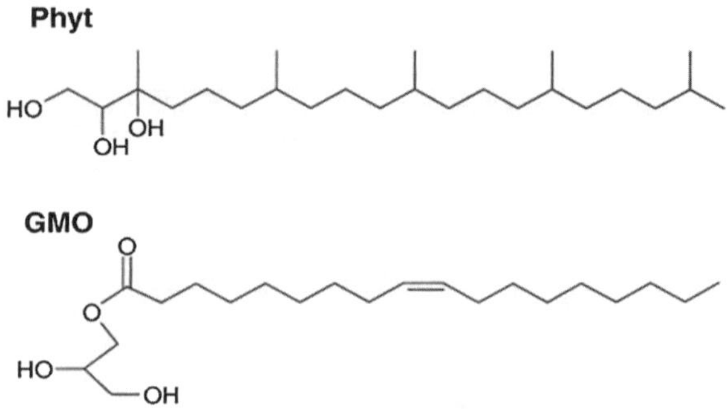

Fig. 12.2 Chemical structures for phytantriol (Phyt) and glyceryl monooleate (GMO) (Taken from [11])

Synthetic phospholipids mixtures were also shown to form lyotropic liquid crystals with interesting properties. For example ternary lipid mixtures comprising of two lipids, dioleoylphosphatidylcholine (DOPC), dioleoylglycerol (DOG), and cholesterol, of molar ratios 1:2:1 and 1:2:2, in excess water induced formation of three dimensional (3D) hexagonal mesophases [12]. Other synthetic such as polyoxyethylene-10-oleyl ether [13] also can form LLC but since these structures were less studied and have less potential to serve as delivery vehicles they will not be discussed in this chapter.

12.1.3 Cubic Phases

The bicontinuous cubic phase is the most complicated among all the LLC. Here the lipids are located in a complex, optically isotropic 3D lattice. Their topology is differentiated as bicontinuous or micellar- the first one more ordered than the second. Cubic structures can be either normal (type I, O/W) or inverse (type II, W/O) and resulting in seven cubic space groups. Only three space groups are both inverse and bicontinuous structures (Fig. 12.3) and very extensively studied: the gyroid (G) type (*Ia*3*d*, denoted Q^{230}), the diamond (D) type (primitive lattice *Pn*3*m*, denoted Q^{229}), and the primitive (P) type (body-centered lattice *Im*3*m*, denoted Q^{224}) [14]. The aqueous channels of G-surface consist of two separate, left-handed and right-handed helical channels. The aqueous channels can extend through the matrix, but the centers of the water channels never intersect. Hence they are connected and the structure adopts a helical arrangement. The diamond structure, D, is characterized by a bilayer, which separates two interpenetrating aqueous channel systems forming a diamond lattice. In this configuration four aqueous channels of the D-surface meet at a tetrahedral angle of 109.5°. Structure P contains two aqueous channel systems that are separated by a bilayer. The unit cell possesses three mutually perpendicular aqueous channels, connected to contiguous unit cells, taking shape of a cubic array. The structures can be identified by small angle x-rays scattering (SAXS), each of them shows typical scattering diffraction patterns (Fig. 12.3).

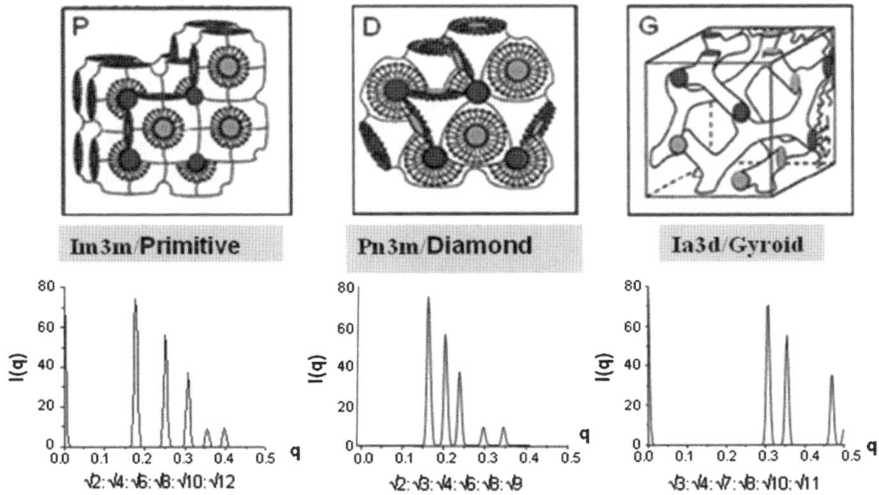

Fig. 12.3 Schematic illustrations of the inverse bicontinuous cubic phases Ia3d (*gyroid* type), Pn3m (*diamond* type), and Im3m (*primitive* type). (Taken from [24])

12.1.4 The Reverse Hexagonal Mesophase

The reverse hexagonal mesophase is of a primitive type symmetry (*P6mm*) and characterized by one cylinder per unit cell corner. These densely packed, straight, water-filled cylinders exhibit two-dimensional ordering. Each cylinder is surrounded by a layer of surfactant molecules that are perpendicular to the cylinder interface such that their hydrophobic moieties point outward from the water rods (Fig. 12.4). There is a growing indication that inverse hexagonal mesophases play structural and dynamic roles in biological systems [5, 15, 16]. These systems are assumed to be active as transient intermediates in biological phenomena that require topological rearrangements of lipid bilayers, such as membrane fusion/fission and the trans-bilayer transport of lipids and polar solutes [5, 15, 16]. H_{II} mesophases have recently been considered promising drug delivery vehicles, mainly owing to their unique structural features [9, 17, 18].

Another LC structure, based on a close packing of inverse micelles, is a 3D hexagonal inverse micellar phase, of space group *P63/mmc*, discovered by Shearman et al. [12]. Ternary lipid mixtures comprising of two lipids, dioleoylphosphatidylcholine (DOPC), dioleoylglycerol (DOG), and cholesterol, of molar ratios 1:2:1 and 1:2:2, in excess water induced formation of 3D hexagonal mesophase over a temperature range of 16–52 °C and a wide range of pressures (1–3000 bar). The new inverse micellar phase consists of an "hcp" packing of identical, spherical inverse micelles, with two lattice parameters, a and c, that were found to be $a = 71.5$ Å and $c = 116.5$ Å. This is the first new inverse LLC phase, which was reported for two decades and it is the only known phase whose structure consists of a close packing of identical inverse micelles.

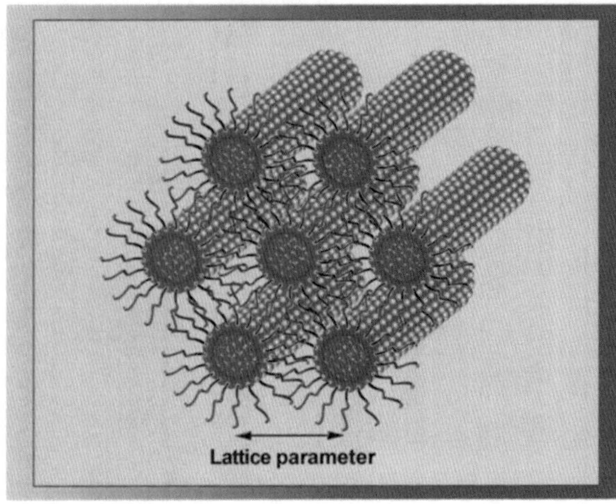

Fig. 12.4 Schematic presentation of H_{II} mesophase showing the packing of water-filled rods surrounded by lipid layers

In a very recent study we examined theoretically the structure of water inside the water cylinder of inverse hexagonal mesophase (H_{II}) of monoolein (GMO), using methods of Molecular Dynamics. Due to the complex geometry of the system, a novel method for obtaining the distribution of water at the initial moment of calculation process is developed and applied. The initial density of water was obtained by fitting the final results within predictor–corrector scheme.

Distribution of hydrogen bonds, density as well as radial distribution of water molecules, across the water cylinder were calculated. They showed the presence of water structure deep in the cylinder (about 6 Å below the GMO heads). The obtained results may help to understand the role of water structure in processes of insertion of external molecules inside GMO/water structure.

12.1.5 The Lamellar Mesophase

Lamellar LLC systems are thermodynamically stable, optically isotropic systems, which are formed with low energy input the lamellar phase is periodic in one dimension. At low temperatures, the acyl chains adopt a close-packed hexagonal arrangement (the phase is called L_β). The L_β may represent metastable polymorphs which give rise to more stable phase (Lc) wherein the acyl chains are packed in a very highly ordered almost crystalline manner [19]. Above the chain melting transition temperature, the acyl chains undergo extensive trans/gauche isomerization reminiscing of a fluid hydrocarbon. This is the lamellar liquid crystalline phase (L_α) that is characterized by a rapid translational motion in the plane and by considerable although incomplete, short-range disorder [20]. Much work was done on this membrane-like mesophases including delivery aspects but they will not be covered in this chapter.

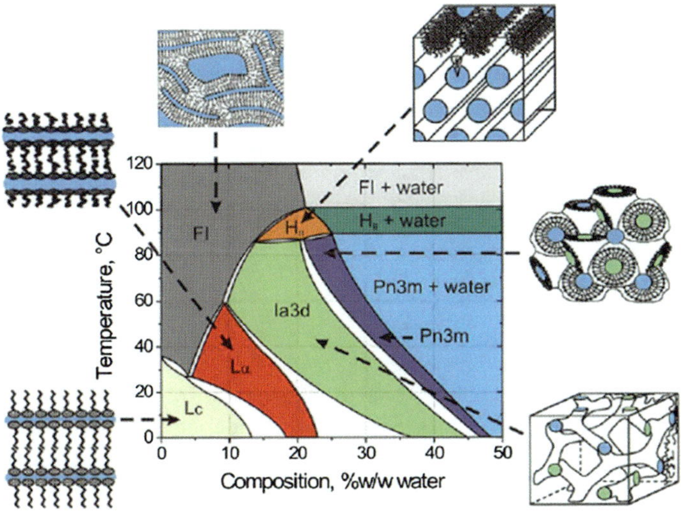

Fig. 12.5 Temperature-composition phase diagram of the monoolein/water system (up to 50 wt% water). A cartoon representation of the various phase states is included in which colored zones represent water. The mesophases are as follows: Lc-crystalline lamellar, Lα-lamellar, Ia3d-gyroid inverted bicontinuous cubic, Pn3m-primitive inverted bicontinuous cubic, H_{II}-inverted hexagonal, and F1-reverse micelles fluid phases (Taken from [8])

12.1.6 Phase Behavior

Most of the common surfactants form direct (type I) phases, where the interface bends away from the polar solvent, toward the tail region. However, most biological amphiphiles (such as phospholipids) form "type II" LLC, where the interface curves toward the polar region. Only a few synthetic amphiphiles can mimic the behavior of biological lipids and form inverted mesophases. The unsaturated monoglycerides (monoolein and monolinolein) belong to this category [10]. GMO is the preferred amphiphile for formulating LC phases for scientific research and drug delivery. This is a non-toxic, biocompatible, and biodegradable lipid, which possesses low water solubility, but in excess water it swells and forms several LC phases [21].

The phase diagram of monoolein/water as shown in Fig. 12.5 revealed complex structural behavior. At room temperature the following phase sequence existed upon increasing hydration: lamellar crystalline phase (L_c) in coexistence with a L_2 phase, lamellar mesophase (L_α), and the inverted bicontinuous cubic mesophases-gyroid *Ia3d* and diamond *Pn3m*. Upon heating, at about 85 °C, the cubic phase is transformed into the H_{II} mesophase, followed by the micellar phase.

These concentration and temperature-dependent structural transitions can be qualitatively explained in terms of effective critical packing parameter (CPP), as developed by Israelachvili et al. [22]. According to this theory, amphiphiles possess geometric parameters characterized by the critical packing parameter, CPP = $V_s/a_0 l$, where V_s is the hydrophobic chain volume, a_0 is the polar headgroup area, and l is the length of the chain in its molten state. The packing parameter is useful in predicting which phases can be preferentially formed by a given lipid, since it connects the molecular shape and properties to the favored curvature of the polar–apolar interface, and therefore the topology and shape of the aggregate. The main factors responsible for alterations of the mentioned parameters are the molecular shape of the surfactant, the chemical composition, and the temperature. According to the theory developed by Israelachvili and coworkers, inverse mesophases are formed by amphiphiles with CPP >1, which adopt inversed cone shaped geometry. Such lipids should possess a small headgroup area as compared to that of the tail region. In contrast, direct mesophases are preferred when the surfactant head cross-section is larger than that of the tail, resulting in CPP <1. In the case of lamellar mesophase these parameters are equal and hence result in CPP = 1. The size of the polar headgroup area is dictated by both molecular shape and the degree of hydration.

Within the boundaries of a given mesophase the headgroups' area normally increases with increasing hydration and decreases upon temperature rise. The length and the volume of the lipophilic tails are also greatly affected by a temperature increase. Stronger thermal motion of the tails eventually decreases the length of the tails but increases their volume.

It was shown for monoglyceride-based LLC that H_{II} mesophases are formed from amphiphiles with CPP ~ 1.7, cubic phases with CPP ~ 1.3, and lamellar structures with CPP = 1 [2]. The lamellar phase is characterized by zero curvature since the cross sections of the polar heads and the lipophilic tails are similar. Upon increased hydration, cubic phases (*Ia3d* and *Pn3m*) with higher curvature and consequently higher CPP are preferred. The effective CPP theory can supply a reasonable explanation for the temperature-induced structural shifts from lamellar through cubic to reverse hexagonal phases, requiring higher curvature than in the lamellar phase. Increasing the thermal motion of both the hydrocarbon chains and the water molecules would increase the CPP values via expanding the volume of the lipophilic moiety, but decreasing the chain length and the headgroup area. This leads to an increase in curvature and therefore induces the formation of cubic and hexagonal mesophases.

The CPP concept may only be used as a "rule of thumb" to predict transitions between various forms of LLC. For example, it was shown that it fails to predict the appearance of the inverse bicontinuous phases and an intermediate phase [12, 23]. This is the reason that a thermodynamic approach considering the total free energy of the LLC systems was applied by Seddon and coworkers [12, 24, 25], however, this is beyond the scope of this chapter.

12.1.7 H_{II} Mesophase Composed of GMO/Triglyceride/Water as Drug Delivery Systems

The cubic phase in the GMO-water system is transformed into an H_{II} mesophase upon heating at ca. 85 °C [10]. CPP theory can supply a reasonable explanation the temperature-induced structural shifts from lamellar through cubic to reverse hexagonal phases, requiring greater curvature than in the lamellar phase. Increasing the thermal motion of both the hydrocarbon chains and the water molecules would increase the CPP values via expanding the volume of the lipophilic moiety, but decreases the chain length and the headgroup area. This leads to an increase in curvature and therefore induces the formation of cubic and hexagonal mesophases [1, 22].

Systematic research was conducted to decrease the cubic to hexagonal temperature transition and stabilize the GMO-based H_{II} mesophase at room temperature [26]. To achieve this goal, experiments relative to incorporation of triglycerides (TAGs) with various chain lengths to the binary GMO/water system were conducted. Amar-Yuli and Garti surmised that immobilization of a TAG between GMO tails would lead to a change in the geometry of monoolein molecules from cylindrical to wedge-shaped and thereby an increase in the CPP value of the system [26]. This should encourage transition from lamellar or cubic phases to hexagonal structures. In addition, the immobilization of TAG in the system was expected to reduce the packing frustration, stabilizing the hexagonal LLC at room temperature. These experimental results showed that a critical and optimal chain length of the triglyceride is required to induce the formation of H_{II} at room temperature. Among the examined TAGs, tricaprylin (C_8) was the most successful flexibly accommodating between the tails of the GMO.

The structural properties of ternary hexagonal mesophases composed of GMO, tricaprylin, and water were extensively and systematically studied in our group as was shown in numerous publications by Garti and his students [27–31]. Several additives, including dermal penetration enhancers, were solubilized to control the physical properties of these carriers [28–30], such as viscosity and thermal stability. For instance, the synergistic solubilization of two major hydrophilic (vitamin C, ascorbic acid, AA) and lipophilic (vitamin E, D-alpha-tocopherol, VE) antioxidants within H_{II} mesophases was reported by Bitan-Cherbakovsky et al. [32, 33]. This enabled expanding conditions to obtain stable H_{II} mesophase at room temperature. In addition, it was shown that phosphatidylcholine (PC) can be embedded into the ternary GMO/TAG/water system [34–37]. PC was incorporated into the ternary mesophases, as it is known that phospholipid-based structures possess relatively high thermal stability, and enhance transdermal drug permeation [38–40] and transmembrane transport across the digestive tract [41, 42].

Incorporation of PC to the ternary system caused competition for water binding between the hydroxyl groups of GMO and the phosphate groups of the phosphatidylcholine (PC), leading to dehydration of the GMO hydroxyls in favor of the phospholipid hydration [36]. On the macroscopic level, this was correlated with improvement of elastic properties and thermal stability of the H_{II} mesophase [34]. Structural flexibility and stability is essential for a drug delivery system, especially for tuning its physical properties (such as viscosity) and composition.

Libster et al. also explored and demonstrated correlations between the microstructural and mesostructural properties of the reverse hexagonal LLC, using ESEM technique [35]. It was shown that the mesoscopic organization of these systems is based on an alignment of polycrystalline domains. The topography of H_{II} mesophases, imaged directly in their hydrated state, as a function of aqueous phase concentration, was found to possess fractal characteristics, indicating a discontinuous and disordered alignment of the corresponding internal water rods on the mesoscale. Fractal analysis indicated that the mesoscale topography of the H_{II} phase was likely to be influenced by the microstructural parameters and the water content of the samples [35]. Garti and coworkers also made considerable progress elucidating the solubilization of therapeutic peptides into the H_{II} mesophase [34, 36, 43–46].

Lately, Cohen-Avrahami et al. used for the first time cell penetrating peptides (CPPs) as enhancers to overcome the stratum corneum barrier [47]. Na-DFC is a common NSAID (non-steroid anti-inflammatory drug) used to treat mild to moderate pain, particularly when inflammation is also present. This study utilized the advantages of the H_{II} mesophase as a transdermal vehicle, in addition to those of the CPPs as skin penetration enhancers, for the development of improved drug delivery vehicles in which the drug diffusion rate might be carefully controlled.

The selected CPPs, representing prominent members of these families, were chosen for their great efficiency and short length; their amino acid sequences are the following: "RALA", RALARALARALRALAR, is a 16-amino-acid peptide belonging to a synthetic family of CPPs, based on GALA, amphipatic peptides named after their alanine-leucine-alanine repeats that exhibit improved membrane permeability. "PEN", penetratin (RQIKIWFQNRRMKWKK), is a 16-amino-acid sequence based on the active penetrating peptide produced by the homeodomain of the antennapedia homeoprotein. "NONA", nona-arginine, RRRRRRRRR. Arginine was shown to play a key role in CPP attachment to membranes; thus, much research was done concerning the penetrating activity of polyarginine through membranes. The maximal solubilization load of Na-DFC, RALA and PEN into the H_{II} structures were found to be 5 wt%, while NONA solubilization capacity was 6 wt%.

SAXS data of the Na-DFC-loaded systems revealed that the drug caused shrinkage of the hexagonal channels diameter. The lattice parameter sharply decreases from 57.55 ± 0.5 Å in the blank system to 51.55 ± 0.5 Å in the 5 wt% loaded system (Fig. 12.6), leading to denser packing of the system. The investigators assumed that such a "kosmotropic effect" of the Na-DFC might originate from the drug's location between the GMO molecules, causing an enhancement of the interactions between them. This assumption was supported by the slight increase in the domain lengths of the hexagonal clusters in the presence of Na-DFC from 392 to 493 ± 50 Å (Fig. 12.6), hinting a stabilizing effect. The main effect appeared in the low concentration regime, whereas the additional

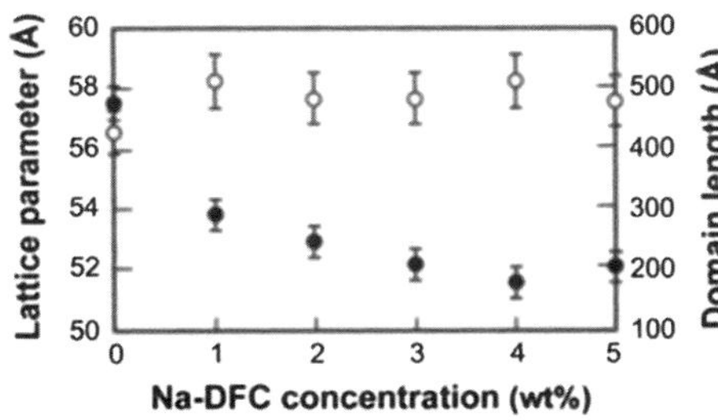

Fig. 12.6 The lattice parameter (●) and the domain length (o) obtained from the SAXS measurements, versus solubilized Na-DFC concentration at the H_{II} mesophases

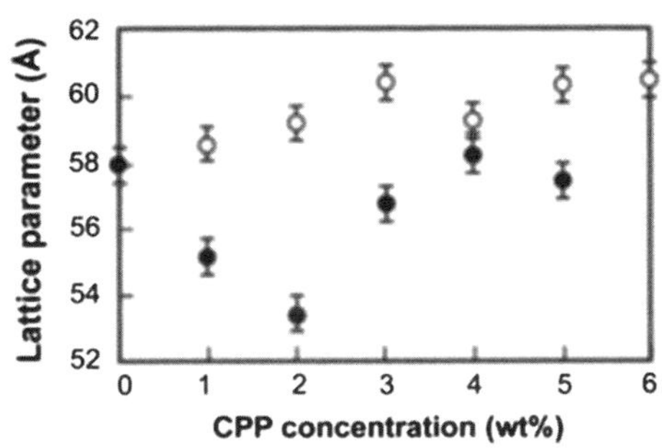

Fig. 12.7 The lattice parameter (α), calculated from the SAXS measurements, versus NONA (o) and PEN (●) concentration in the mesophase

Na-DFC quantities had a weaker influence. These effects were noticed until the system transformed into a micellar structure.

In addition, DSC results suggested enhanced water binding in the presence of the drug. Shrinkage of the aqueous columnar cylinders (as detected also by SAXS) caused a change in the structure curvature (curvature more concave toward the inner water phase) and an increase in the fraction of the bound water at the expense of the free water that populated the inner channels. It also indicated that Na-DFC interacts with the surfactant tails, reducing the number of molecules that are free to participate in the melting process. Combined DSC measurements and ATR-FTIR analysis led the authors to conclude that Na-DFC solubilization effect was primarily manifested at the interfacial region, when the drug intercalated between the GMO molecules. Na-DFC influenced the intermolecular bonding at this molecular region. It strengthened the interactions between GMO tails, loosened the repulsion tension between the carbon atoms at the headgroups, and enabled the formation of a tighter structure with higher curvature.

Each CPP was separately solubilized within the H_{II} for identification of its specific location and chemical interactions within the mesophase. The SAXS measurements indicated two trends in the effect of the peptides. Up to 2 wt% of PEN solubilization and initial shrinkage of the aqueous cylinders to a minimum of 53.4 ± 0.5 Å was detected. With increasing solubilization loads of PEN to 5 wt%, the shrinkage changed, and a swelling effect of the cylinders was detected until their diameter reached their original radius of 57.9 ± 0.5 Å (Fig. 12.7).

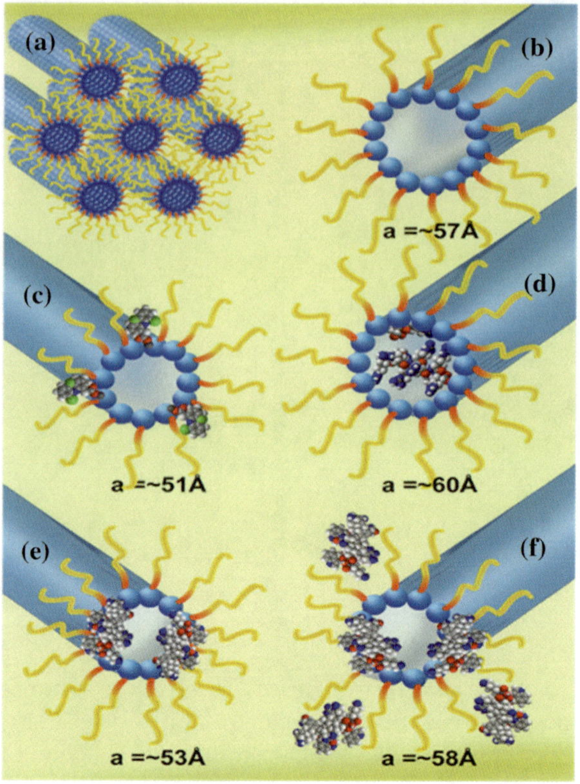

Fig. 12.8 **a** Schematic illustration of the H_{II} mesophase general structure. The water populates the inner region of the cylinders; the GMO tails as well as the TAG point outward. **b** A single "blank", empty system of the H_{II} mesophase, **c** Na-DFC-loaded system. The drug populates the interfacial area and causes channel shrinkage. **d** NONA-loaded H_{II} mesophase. The peptide populates the aqueous channels and swells them. **e**, **f** PEN-loaded systems. At low PEN concentration (**e**), it is adsorbed on the GMO headgroups and causes channel shrinkage. With increasing PEN concentration (**f**), it populates an additional hosting region at the interface

NONA solubilization had a minor, yet consistent swelling effect on the cylinders. The initial lattice parameter in the blank system was 57.9 ± 0.5 Å, and it slightly increased with NONA loading to a maximal value of 60.5 ± 0.5 Å at 6 wt% NONA. Its effect was found to be focused mostly on the water and the GMO carbonyls. The stretching mode of the water was gradually shifted from 3361 to 3346 cm^{-1} with increasing NONA concentration in the mesophase, indicating stronger hydrogen bonding of the water with the peptide.

It was summarized that structural investigation revealed that the solubilization sites of the different guest molecules depend on their molecular structure and differ significantly. Na-DFC populated the interfacial region, enhances the interactions between GMO tails, and shrinks and stabilizes the H_{II} mesophase (Fig. 12.8). PEN solubilization is concentration-dependent. The initial PEN loads populate the hydrophilic headgroup area, whereas the higher PEN loads pack closer to the GMO tails (Fig. 12.8e, f). The hydrophilic NONA populated the inner channels region and swelled the mesophase (Fig. 12.8d). RALA acted as a chaotropic agent at the H_{II} mesophase, interacting mostly with the water within the channels and enhancing the hydration of the GMO headgroups.

12.1.8 Molecular Interactions of LLCs with Proteins and Nucleotides

Murgia et al. entrapped several nucleotides into cubic and lamellar monoolein-based mesophases in order to protect them and enable their release [48]. Nucleoside and nucleotide analogues, represent a class of innovative therapeutics for the treatment of viral diseases and cancer. Simultaneously, nucleotides can constitute a promising choice to promote molecular recognition. However, these substances need to be protected since they can be easily recognized and degraded by different extracellular nucleases, resulting in poor in vivo pharmacokinetic properties. The usually severe cytotoxic side effects of nucleoside and nucleotide analogues, particularly myelotoxicity, can be partially relieved by lipid derivatization of the phosphate group of the active nucleotide. The resulting drug is amphiphilic and can be delivered to the therapeutic target by incorporation in lipid-based nanovectors.

In this regard these investigators used as simple models for both hydrophilic and hydrophobic modified/functionalized nucleotide-based drugs, AMP, GMP, UMP, and CMP (XMPs), along with two hydrophobically functionalized nucleotides (nucleolipids), i.e. the 1-palmitoyl-2-oleoyl-*sn*-glycerol-3-phosphoadenosine (POPA) and the hexadecylphosphoadenosine (HPA) (Fig. 12.9). In these nucleolipids a nucleic base is enzymatically exchanged with the choline headgroup of a lipid precursor. The resulting systems were investigated mainly through SAXS and ^{31}P-NMR techniques.

The GMO-based LC phases here used can be easily regarded as membrane models as a result of lipid type and bilayer occurrence. These lipid LC systems have been shown to constitute suitable matrices to entrap either hydrophilic nucleotides or amphiphilic nucleolipids and are therefore promising lipid-based nanovectors to protect and deliver nucleotide-analogue drugs. However, a drastically different behavior regarding the long-term stability of the nanodevice is displayed. All these molecules contain a phosphate moiety. It has been ascertained that in the case of the amphiphilic additives (POPA and HPA), which locate at the GMO polar–apolar interface, the LC systems were found endowed of high stability, more than 2 years. Differently, the hydrophilic XMP molecules, which are located in the aqueous domain of the GMO LC phases, undergo a hydrolysis process at the phosphoester bond, due to a preferential orientation with respect to the MO interface. The XMPs degradation, within about 4 months, causes an

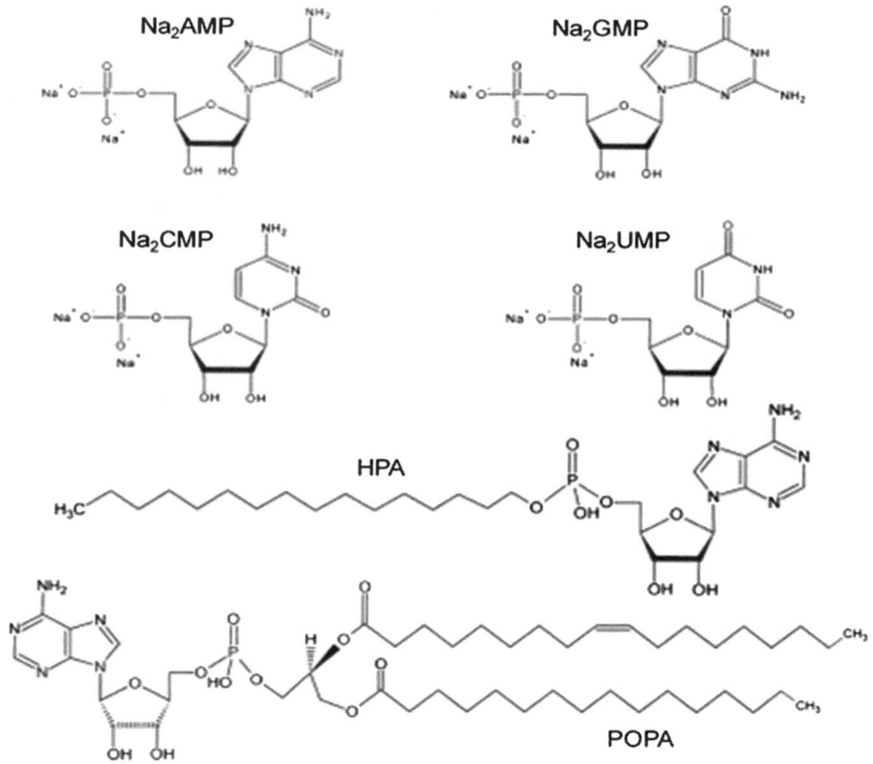

Fig. 12.9 Chemical structures of mononucleotides disodium salts and nucleolipids. (Taken from [48])

impressive alteration of the interface arrangement, resulting in structural transition from cubic to hexagonal mesophase. It was demonstrated that the relatively small amount of HPO_4^{2-} anion resulting from the XMPs hydrolysis were sufficient to induce a cubic-to-hexagonal phase transition as a result of different, and specific interactions at the polar-apolar interface that favored a reverse curvature. The authors concluded that the presence of unesterified phosphate groups in fully hydrophilic drug molecules (for instance XMPs) may have dramatic effects on the stability of the lipid bilayer of a biological membrane.

Amar-Yuli et al. combined the potential of both liquid crystalline structure as well as glycerol as cosolvent to enhance insulin thermal stability and moderate the aggregation progress [49]. Insulin was incorporated into several modified reverse hexagonal systems based on friendly surfactant and polyols to explore the impact of the protein confinement on its stability, unfolding behavior, and morphology with severe external conditions, low pH, and higher temperatures (up to 70 °C).

The investigators focused on solubilizing insulin in GMO/decane/water, GMO/decane/glycerol/water, and GMO/phosphatidylcholine (PC)/decane/glycerol/water

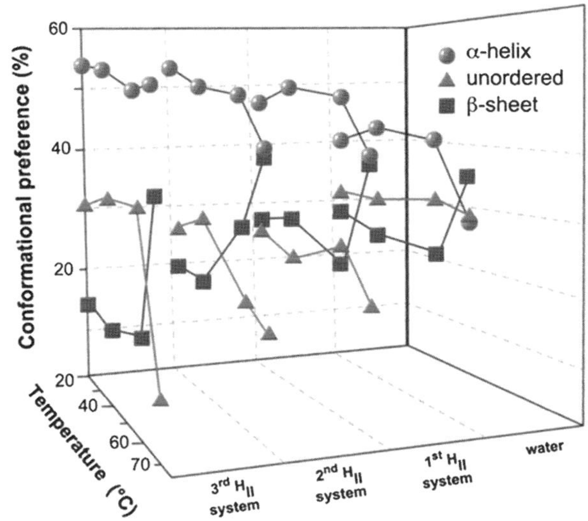

Fig. 12.10 Insulin secondary structure (α-helix, β-sheets, and unordered conformations, turn together with random coils) during heating from 25 to 70 °C in different hosting media: water, water-filled H_{II} phase (first H_{II} system), water/glycerol-filled H_{II} phase (second system), and water/glycerol-filled H_{II} phase containing PC (third system). Each system contains 4 wt% insulin in the polar phase

mixtures and compared them to bulk solution (water). The incorporation of insulin within the water and especially in the glycerol/water cylinders compels the protein to be assembled in a restricted, confined configuration compared to its rearrangement in solution. The phenomenon of a greater degree of compactness of the tertiary fold of insulin in the presence of glycerol is known and here is overstressed upon its incorporation within the water and glycerol filled cylinders, which limits its free volume. The ability to obtain more confined protein structure, suggesting a more compact configuration of solvated dimeric insulin, is expected to be mirrored by an enhancement of the protein stability.

Comparison of insulin secondary structure during heating from 25 to 70 °C in different hosting media: water, water-filled H_{II} phase (first H_{II} system), glycerol/water-filled H_{II} phase (second system), and glycerol/water-filled H_{II} phase containing PC (third system), is shown in Fig. 12.10. Each system depicts the corresponding content of α-helix, β-sheets, and unordered conformations (turns together with random coils) of insulin. The dependence of the insulin secondary structure on prolonged heating (for 60 min) at 60 and 70 °C was tested. The pronounced modifications were observed only at 70 °C. These temperatures were selected on the basis of previous reports demonstrating that the insulin aggregation process begins at 55 °C (but is surmised to be too slow to be detected); however, at 60 °C, immediately after the dimer dissociation, the aggregation process was detectable and accelerated with increasing temperature (to 70 °C) or upon prolonged heating.

All fourth matrices did not reveal any significant changes in content of the conformational elements up to 60 °C (Fig. 12.10). However, prolonged incubation of insulin at 60 °C for 60 min caused an increase in the β-sheet configurations from 25 to 36 % only when insulin was confined in the water-filled H_{II} phase (first H_{II} system, data not shown). Once the temperature reached 70 °C, marked increase in the parallel and anti-parallel β-sheet bands (Fig. 12.10) were detected in all

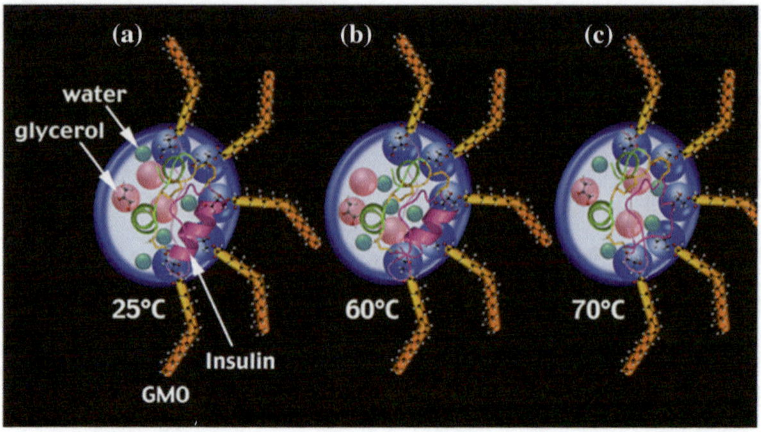

Fig. 12.11 Schematic illustration of the interactions between the components, representing the second and third H_{II} systems (water/glycerol-filled H_{II} phases, in the absence and presence of PC, respectively) at three temperatures; **a** 25, **b** 60, and **c** 70 °C. For simplification PC and tricaprylin are not shown in the image

systems, suggesting an initiation of insulin aggregation, with noticeable variations between the hosting systems. The total β-sheets of insulin in water increased from ~25 to 37 % at 70 °C, and after 60 min treatment further increased to 56 %. Similar quantities of β-sheets were obtained at 70 °C when insulin was confined within the H_{II} cylinders (39, 40, and 36 % in the water-, glycerol/water-H_{II} phase, and glycerol/water-filled H_{II} phase containing PC, respectively). On the other hand, heat treatment of insulin for 60 min revealed the differences in system efficiency in intermolecular β-sheet growth. While an increase of ~20 % in β-sheet content was recorded in the bulk solution, only a slight increase of 5 % was observed when insulin was incorporated in the water-filled H_{II} channels. Moreover, when glycerol was present in these narrow channels (in the second and third systems) insulin retained its β-sheet content although severe conditions (70 °C for 60 min) were applied. The addition of PC contributed to the stability of the secondary structure of insulin.

Figure 12.11 illustrates the interactions between the components, representing the second and third H_{II} systems. It should be stressed that in these new systems, the dehydration process ended at 60 °C, whereas further increase in the temperature to 70 °C caused the formation of stronger hydrogen bonds between the GMO hydroxyls and water. Above 60 °C, owing to protein-solvation effect, the water and glycerol molecules were probably excluded from the protein surface and redistributed themselves in the vicinity of the protein molecule with a preferential binding of the solvent components. The increase in the strength of hydrogen bonds as detected by FTIR is considered to be responsible for the more stable insulin secondary structure, which is probably the reason for hampering the formation of fibrillar aggregates, associated with several neurodegenerative diseases.

The results presented in this study provide valuable insights into the protection abilities and mechanism of the used colloid structures, containing cosolvent as a medium for long-term protein stability. Moreover, these findings yield valuable information regarding the effect of these colloid structures confinement on insulin fibrillation and/or aggregation.

12.1.9 Solubilization and Delivery of Biomacromolecules

Solubilization of biomacromolecules into LLCs is of great practical interest. Mezzenga et al. investigated the effects of confining polymeric forms of glucose in the water domains of liquid crystals composed of monoglyceride and water solutions [50]. Using rheological and SAXS methods the authors studied the effect of sugar concentration and molecular weight on cubic-to-H_{II} transitions by further solubilization of maltose and dextran I in the system. It was found that higher concentrations of sugars resulted in a decrease of the *Pn*3*m*–H_{II} transition temperature, as a consequence of the Hofmeister effect. However, no dependence on molecular weight of the sugar was observed. In the case of a polymeric sugar, as soon as the end-to-end distance of the polymer approached the diameter of water channels, its molecular weight made it more difficult for the guest molecule to fit within narrow water channels. Dextrans of larger molecular weights than dextran I did not fit into the water channels of the cubic phases, inducing a phase separation. Moreover, it was noticed that dextran I, solubilized into the *Ia*3*d* phase channels, induced the appearance of a *Pn*3*m* phase, which possesses larger water channels. Hence, when the chain end-to-end distance of the polysaccharide approached the diameter of the water channels, it induced phase transitions toward structures with different topologies, thus enhancing the solubilization capacity of large sugars.

Peptides and proteins are increasingly considered for the development of new therapeutic compounds [51, 52]. Peptides of various sizes are currently tested for a broad spectrum of diseases, including skin cancer, acne, psoriasis, hypertension, hepatitis, and rheumatism. These biomacromolecules are usually found to be very effective, and in most cases low doses of them are needed for good medical treatment. Nevertheless, when delivered orally these peptides and proteins are susceptible to cleavage by various enzymes, mainly the human digestive proteases, leading to practically poor bioavailability and poor pharmaceutical efficacy. As a result of these critical limitations, excessively large doses of the peptide-based drugs are usually required to obtain therapeutic effects in vivo, which in turn often cause a wide range of hazardous side effects [51, 52]. This is the main reason that the medical application of these otherwise high potential therapeutics is currently very limited and even those few that are used at the present time are far from being efficient. In this respect, LLCs seem to be promising candidates as alternative delivery means for various pharmaceuticals. Such vehicles can provide enhanced drug solubility, relative protection of the solubilized drugs, and controlled release of drugs, while avoiding substantial side effects [53–55]. The efficiency of the LLC delivery system can be further enhanced in transdermal delivery, where peptide degradation and low absorption rates are much less relevant.

Significant progress has been made during the last decade in the characterization of the interactions of peptides and proteins with LLCs, mainly for crystallographic and drug delivery purposes. LLCs have been shown to provide sustained release of drug molecules with a range of physicochemical properties [10, 23, 54]. The cubic phase was shown to host a range of water-soluble biomacromolecules for use in controlled release applications. Clogston and Caffrey systematically examined biomacromolecules ranging from a single amino acid (tryptophan), to complex proteins and nuclear DNA [56]. These investigators found that for a given cubic phase, the rate of diffusion depends on the molecular size of the specific diffusing molecule. Shah and Paradkar prepared in situ a cubic phase system of GMO [23]. This system provided protection to the metalloenzyme seratiopeptidase (STP) in the gastric environment and gave delayed and controlled release with no initial burst after oral administration.

Although the cubic phase has been proposed and studied as a drug delivery vehicle, there is relatively little information about the interactions of peptides and proteins with the H_{II} mesophases and the therapeutic potential of these structures. Libster et al. explored and controlled the physical properties of H_{II} mesophases to use these systems as drug delivery vehicles for biologically active peptides and proteins. The results of this structural research enabled significant expansion of the application spectrum of hexagonal LLC, utilizing them for the solubilization of peptides and proteins [34, 36, 43, 57], into this mesophase and its utilization as sustained drug delivery vehicle. Two model cyclic peptides, cyclosporin A (11 amino acids) and desmopressin (nine amino acids), of similar molecular weight but with very different hydrophilic and lipophilic properties, were chosen to demonstrate the feasibility of using the H_{II} mesophase [57]. In addition, a larger peptide, RALA [44] (16 amino acids), was solubilized into the H_{II} structures as a model skin penetration enhancer. Finally, a larger macromolecule, LSZ protein (129 amino acids), was directly incorporated into a GMO-based H_{II} mesophase [45, 46].

With the aim of designing a biologically inspired carrier in which the encapsulation and the delivery of DNA can be efficiently controlled, Amar-Yuli et al. have designed two lipid-based columnar hexagonal LLCs [58], which can accomplish two opposite roles while maintaining the same liquid crystalline symmetry. The first system was based on a nonionic lipid, such as monoolein, while the second system was modified by a low additional amount of the oleyl amine cationic surfactant. DNA was enzymatically treated to generate a broad distribution of contour lengths and diffusion characteristic times [58].

The impact of DNA confinement on these two columnar hexagonal structures was investigated by SAXS. Both the neat neutral and cationic LLC systems had a columnar hexagonal symmetry, the main difference being the reduced lattice of 49.8 Å in the cationic system compared to the 55.5 Å of the neutral formulation. This difference was interpreted to be the result of the kosmotropic effect of the oleyl amine, which, due to its charged nature, dehydrates the surfactant polar heads and reduces the LLC lattice [58].

A very different effect on the H_{II} lattice parameter was found when the DNA (1.4 wt% from the aqueous phase) was incorporated into the two systems. In the nonionic LLC system, the lattice parameter decreased from 55.5 to 50.8 Å (±0.5 Å) in the presence of DNA. In contrast, in the cationic columnar phase the lattice parameter increased from 49.8 to 59.2 Å. These effects can be rationalized by considering the negative charge of the DNA: when added to the neutral formulation, DNA leads to a negative effective charge which dehydrates the lipid polar heads and reduces the lattice parameter; on the other hand, when added to the cationic formulation, DNA partially neutralizes the overall positive charge, moderates the dehydration effect caused by the cationic surfactants and induces, at least partially, swelling of the lattice [58].

The authors [58] postulated two different possibilities of arrangements for the DNA within the nonionic and cationic columnar hexagonal phases. Being strongly hydrophilic, the DNA double-strands must be either segregated outside the water channels, within the grain boundaries, or confined within the relative narrow aqueous cylinders of the H_{II} structure. The strong impact of DNA on the lattice of the mesophases provides a robust indication that the DNA molecules were indeed confined within the water channels of the reverse hexagonal phases, although the different trends observed for the lattice evolution in the nonionic and cationic case suggest two different types of interactions among the lipids and the DNA [58].

This issue was directly addressed by performing ATR-FTIR analysis on the two formulations and indirectly by following the release of the DNA from each LLC system by UV spectrophotometry. In the nonionic-based H_{II} mesophase the results provided an evidence for a breakage of GMO–GMO and GMO–D_2O hydrogen bonds due to incorporation of DNA molecules. Furthermore, the carbonyl absorption revealed a slight modification in the area of their peaks, when the DNA was present. In the absence of DNA, the number of hydrogen-bonded carbonyls was calculated to be 72 % and it slightly increased to 78 % in the presence of DNA. It was concluded that the DNA molecules interfere with the interfacial region of the cylinders, with tangible effects on the water (D_2O) and GMO molecules, including the hydroxyl and carbonyl groups and the ester moiety. The decrease in lattice parameter upon the addition of DNA, as detected by the SAXS measurements, is consistent with a dehydration of the GMO hydroxyl groups, i.e., a breakage of hydrogen bonds between GMO polar groups caused by the presence of the highly polar DNA. Additionally, the modifications within the vicinity of the GMO–water interface imply a less ordered state with increased content of *gauche* conformations, which is again coherent with the shrinkage of the H_{II} cylinders.

According to Amar-Yuli et al., these results indicated that the DNA is confined within the aqueous domains of the H_{II} cylinders [58], stabilized by hydrogen bonds with the water and/or GMO headgroups, and leading to an observable rupture of the H-bonds in the H-bond donor part of the GMO polar heads (OH) and a decrease in the frequency of the H-bond acceptor bands of the GMO polar heads (CO–O). The role of DNA in this case is thus consistent with moving away bound water from the interfacial region of the water channels.

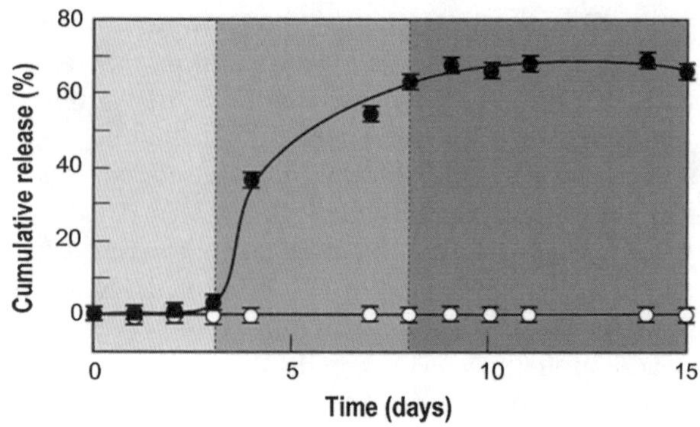

Fig. 12.12 The release profile of DNA from the two columnar H_{II} mesophases considered: release from the nonionic system (*filled circles*, GMO/tricaprylin/water/ascorbic acid) and release from the cationic-based columnar hexagonal phase (*open circles*, GMO/oleyl amine/tricaprylin/water/ascorbic acid) at room temperature

Examination of the second H_{II} system which contains the cationic surfactant oleyl amine shows very moderate changes upon incorporation of DNA in all the bands considered when compared to the nonionic mesophase. These data clearly suggested that in the presence of a cationic surfactant decorating the interface of the water channels, the interactions between the lipid polar heads and DNA change in nature, as the hydrogen bonding of D_2O and the GMO polar heads is mostly unaffected by the presence of DNA. It was inferred that the DNA is confined in the water channels also in the cationic H_{II} mesophases, and electrostatically bound to their surfaces [58].

The DNA cumulative release profile from the two mesophases is depicted in Fig. 12.12, which illustrates the difference in release rate from the nonionic and cationic H_{II} host mesophases. While practically no release of DNA molecules was found in the presence of oleyl amine even after 15 days, three main steps on the release profile can be identified when DNA is confined within the pure nonionic columnar system. During the first 3 days, no release is detected, and this can be attributed to the lag time needed for DNA molecules to diffuse out of the H_{II} cylinders. In the range of 4–9 days the release of DNA from the LLC system into the excess water begins, progressing at relatively high rate, as indicated by the sharp slope observed in Fig. 12.12. After 9 days the release of DNA slowed down significantly, and reached a plateau. At this point a significant amount of DNA, approximately 65 % of the initial content, has already been released by the nonionic H_{II} mesophase. These results are consistent with the conclusions based on SAXS and ATR-FTIR analyses stating that hydrogen bonds between DNA, water and GMO headgroups are responsible for the DNA confinement within the water channels, in the case of the nonionic hexagonal mesophase, which allows transport of DNA into the excess water. In contrast, the presence of cationic oleyl amine in the ionic hexagonal mesophase leads to such a strong electrostatic confinement of the DNA at the water–lipid interface that release into the excess water is prevented: this formulation is then more reliable for permanent encapsulation and protection of the DNA rather than a controlled release [58].

In order to gain insight on the correlations between the release loads and the molecular conformation of the DNA molecules released, single molecule AFM was used to provide the needed information on DNA contour length distributions (Fig. 12.13a). Figure 12.13a depicts the DNA contour length distribution prior to the release experiment. As can be seen, several DNA fragments are generated upon the enzymatic treatment and can be divided into three main lengths: relatively small fragments ($\leq$100 nm contour length), medium size DNA strands (130–160 nm) and large macromolecules ($\sim$190–200 nm).The contour length distribution of DNA after its release from the nonionic system was examined on the third, forth and the fifteenth days of release, corresponding to the lag time, the sharp slope and plateau regions as identified in the release profile (Fig. 12.13a–d).

During the lag time, on the third day of the release experiment, only very short length DNA fragments with contour length below 40 nm could be observed by AFM analysis (Fig. 12.13b). These very short strands were too small and in too low concentration to be detected by the spectrophotometer resulting in the flat absorption signal. The following day (day 4), in correspondence of the sharp increase in the diffusion as measured by the UV–vis absorption, the medium size DNA strands, mainly 40–100 nm in contour length, started to appear coexisting with the very short DNA fragments (Fig. 12.13c). Finally, after 15 days, when the release profile has reached a plateau, the longest fragments, 190–200 nm in contour length, also appeared in the AFM analysis of the excess water (Fig. 12.13d). Interestingly, after 15 days, but only after this time, the contour length distribution of the DNA released matches closely the initial distribution, indicating that release is nearly entirely completed (Fig. 12.13d).

Quite in contrast, AFM examination of the excess water in the case of the cationic hexagonal phase reveals no traces of DNA including the smallest fragments detectable by high resolution AFM (Fig. 12.13e), confirming the strong binding of the DNA to the cationic lipids [58].

Figure 12.14 summarizes the main findings of this work. When confined within nonionic columnar hexagonal phases, DNA interacts with the polar heads of the lipids via hydrogen bonding (see figure), but this enables a controlled release of the DNA in excess water following three main release stages. When positive charges belonging to cationic lipids are decorating the water channels, binding between the lipid and DNA becomes strong, and any release of DNA is completely suppressed at the charge ratio used. It is reasonable to anticipate that further adjustment of the positive: negative charge ratio or use of variable ionic strengths to partially screen electrostatic attraction can be used as additional means to fine tune the DNA release kinetics [58].

12.2 LLC as Drug Delivery Vehicles

Liquid crystalline phases, such as lamellar, reverse hexagonal, and cubic, present interesting properties for a topical delivery system, and hence were considered and have been studied as delivery vehicles of pharmaceuticals via the skin and mucosa

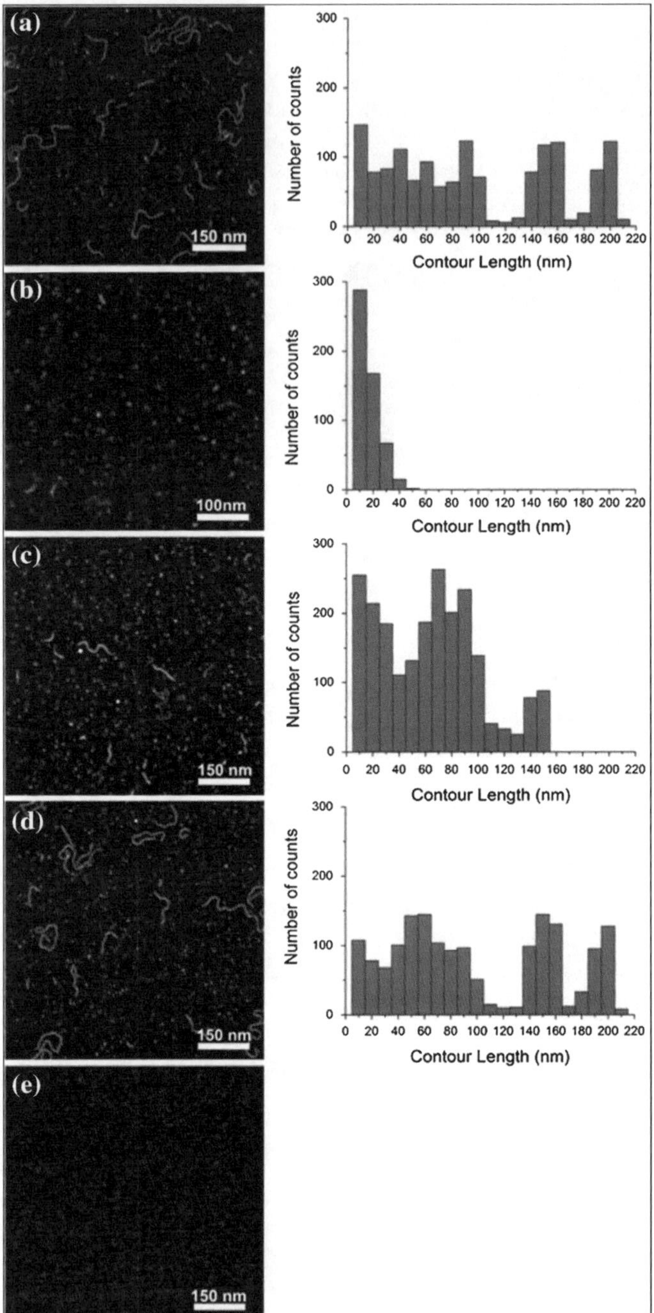

Fig. 12.13 AFM image and the corresponding contour length distribution of DNA fragments: **a** before incorporating DNA into the H_{II} nonionic LLC system; **b** after 3 days of release in the excess water environment; **c** after 4 days of release and **d** after 15 days of release at room temperature. **e** AFM image after 15 days of DNA release from the cation-based H_{II} LLC system

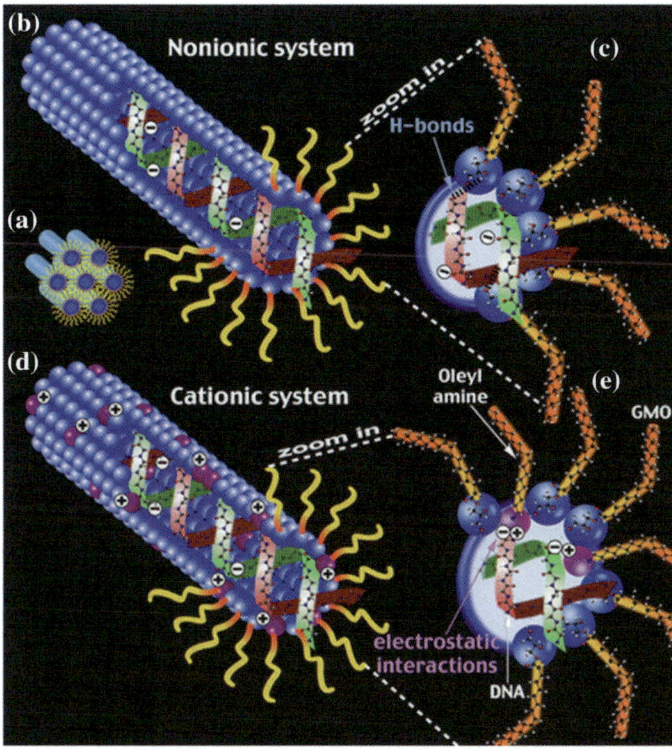

Fig. 12.14 Schematic summary illustrating a typical columnar hexagonal phase (**a**), the DNA entrapment and interactions with the surfactants forming the nonionic (**b** and **c**) and cationic (**d** and **e**) H_{II} systems. The molecular ratio between all the components is not in scale in the image

[59, 60]. These phases are: (a) thermodynamically stable; (b) consist of nano-scaled hydrophilic and hydrophobic domains, which are separated by the surfactant self-assembled layers; (c) contain an extremely large surface and present the ability to incorporate compounds independent of their solubility, to protect them from physical and enzymatic degradation, and to sustain their delivery; and (d) may consist of permeation enhancers as the structure-forming lipid/solvent [54, 61].

The success of an innovative delivery system for bioactives and specially for proteins or peptides will depend on its ability to fulfill some or all of the following parameters and perquisites:

i. In vitro and in vivo stability
ii. Improved systemic availability
iii. Prolonged biological half-life ($T_{1/2}$)
iv. Patient convenience and compliance
v. Reduced dosing frequency
vi. Local toxicity and safety concerns
vii. Product life cycle management.

The primary progress achieved in the field of transdermal delivery uses cubic and hexagonal liquid crystalline phases [43, 62, 63]. This chapter is focusing on the current knowledge of topical, mainly transdermal routes of liquid crystal-based delivery systems.

12.2.1 Monoolein and Phytantriol: Main Building Blocks of Lipids Mesophases

Only a few synthetic amphiphiles can mimic the behavior of biological lipids and form inverted mesophases. The unsaturated monoglycerides (monoolein and monolinolein) belong to this category [10, 64]. GMO is the preferred amphiphile for formulating LC phases for scientific research and drug delivery. This is a non-toxic, biocompatible, and biodegradable lipid, which possesses low water solubility, but swells and forms several LC phases in excess water.

Another lipid, phytantriol, was also recently shown to form cubic LC. Phytantriol is commonly used as an ingredient in the cosmetics industry for improving moisture retention [11, 65]. Its phase behavior as a function of water concentration and temperature is very similar to that of GMO, although structurally they are very different [11, 66] (Fig. 12.2).

The cubic and hexagonal phases based on GMO/water and GMO/oleic acid/water, respectively, are well studied and were shown to have the ability to sustain the release of incorporated compounds [59, 67, 68]. Moreover, each component constructing either the cubic or the hexagonal phase is a penetration enhancer by itself. GMO is known to promote ceramide extraction and enhancement of lipid fluidity in the *stratum corneum*, and oleic acid is considered to increase epidermal permeability via a mechanism involving perturbation of the *stratum corneum* lipid bilayers and lacunae formation [69].

12.2.2 Hexagonal and Cubic Mesophases and Their Dispersions as Carriers of Hydrophilic Drugs Cubosomes, Hexosomes and Micellosomes

Exceptional structural properties allow the lyotropic mesophases to entrap water-soluble compounds within the aqueous domains, and hydrophobic molecules via direct interactions within their lipid hydrophobic compartments. LLCs have been shown to provide sustained release of drug molecules with a range of physico-chemical properties [18, 53, 55]. In addition to its biological significance, inverse lipid phases could be useful as host systems for the crystallization of membrane proteins [8] and for inorganic synthesis [70].

One of the main factors that hamper the success of LLC technology in pharmaceutical applications, especially in injectable formulations, is the high viscosity of these mesophases. To solve this problem, aqueous dispersions of bulk lamellar (called vesicles or liposomes), cubic (cubosomes), and reverse hexagonal (hexosomes) LLC mesophases have been developed. When the bulk mesophases are dispersed in aqueous phase containing a stabilizer (usually an amphiphilic polymer Pluronic F-127), nano-soft particles of 100–500 nm are formed. These dispersed soft particles were prepared and studied during the last decades as potential drug delivery vehicles [62, 71, 72] and as carriers of food supplements and additives [7]. Cubosomes and hexosomes (Fig. 12.15) offer the benefits of their unique internal structures, dispersion ability in water, large surface area, and high solubilization capacities combined with low viscosity [27]. It is assumed that F-127 provides steric stabilization by having its propylene oxide (PO) moieties adsorbed on the surface of the stabilized particle, while its more hydrophilic ethylene oxide (EO) ends extend into the external continuous phase [73]. The choice of the proper dispersing agent for these mesophases is essential and critical to the stabilization of the final dispersions. Some other surfactants also were shown to facilitate the dispersion of monoolein into stabilized nanoparticles, including ethyl hydroxyethyl cellulose ether (EHEC) [74], Tween 80 [75], and ethoxylated phytosterol [76]. The penetration properties of the dispersion of the hexagonal phase discussed earlier was studied by dispersing this liquid crystalline phase in excess water in the presence of a dispersing agent (i.e. Poloxamer) [27]. Dispersions of cubic phase have been suggested as suitable systems for percutaneous delivery of small molecules such as indomethacin [71].

12.2.3 Oral Administration Using LLC

Oral administration of various pharmaceuticals, and pharmaceuticals is especially macromolecular drugs that are insoluble in water or oil, is extremely challenging task. A specially difficult goal is to force large molecular weight protein and other bioactives a result of their large molecular weight, hydrophilicity, and susceptibility to enzymatic inactivation in the gastrointestinal tract. Parenteral routes of administration enable total systemic availability and fastest onset of action. The most commonly used parenteral routes are intravenous (IV), intramuscular (IM), and subcutaneous (SC). However, administration with needles causes pain and can lead to the transmission of blood-borne pathogens induced by accidental needle pricks or intentional needle reuse. The consequence is reduced patient compliance [77]. To raise therapeutic efficacy, research has focused on finding alternative means of administration, particularly noninvasive routes, such as pulmonary, intranasal, or transdermal.

The oral mucosal route offers a number of advantages over parenteral and other non-invasive routes for the systemic delivery of biologically active peptides and proteins as well as conventional drugs [78]. However, most drugs absorbed via the

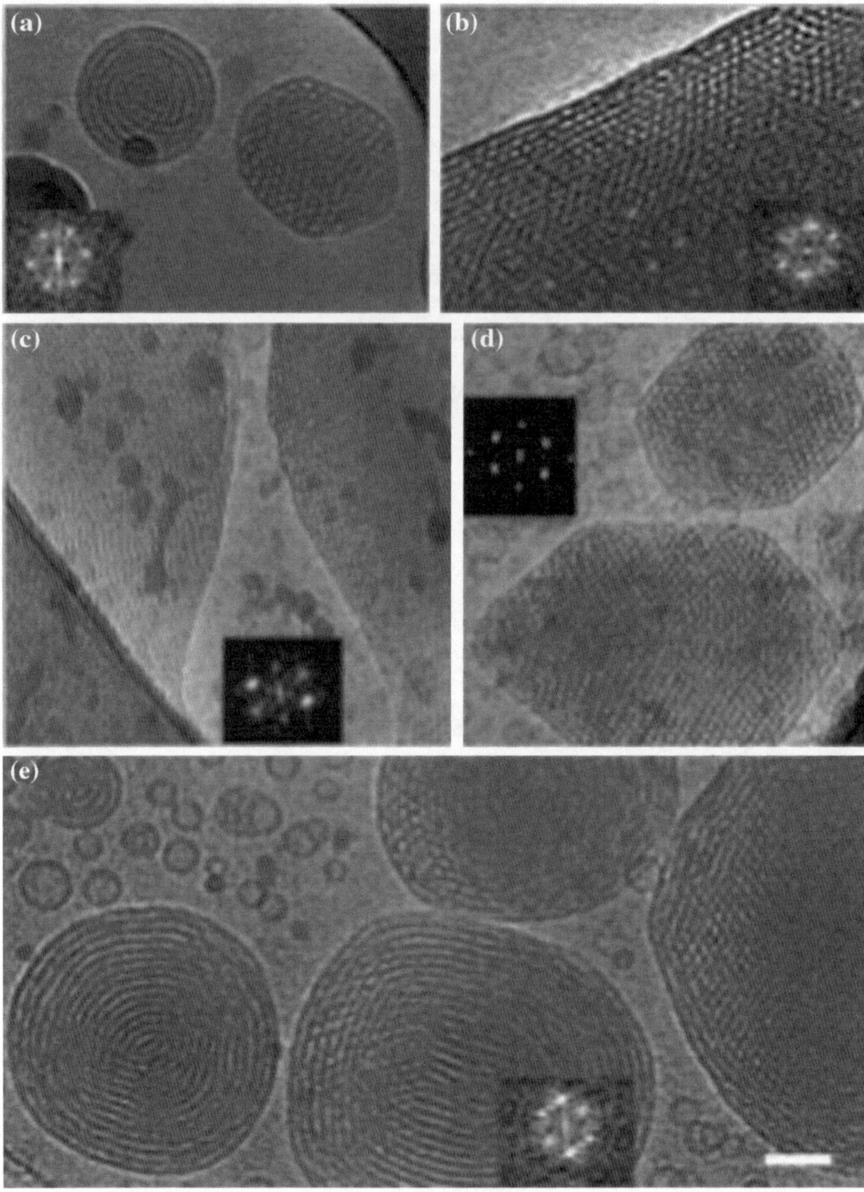

Fig. 12.15 Cryo-TEM images of hexosome dispersions with 20 wt% inner water content stabilized with **a–b** 0, **c** 0.4, **d** 1, and **e** 1.6 wt% F127 in the external aqueous phase. Images **b** and **c** depict only the very large particles that were seen in the sample. The small particles are similar to those seen in image (**d**). The bar corresponds to 50 nm and is appropriate to images (**a–e**)

mucosal membrane of the oral cavity have exhibited low bioavailability due largely to the low mucosal membrane permeability, relatively small surface area available for absorption and poor retention of the drug and/or delivery systems at the site of absorption. These problems can be often be overcome by utilizing mucoadhesive dosage forms that provide increased residence time leading to an increased total permeability of macromolecular drugs such as peptides and proteins. LLC phases of glyceryl monooleate (GMO) are considered as attractive buccal delivery carriers for peptide and protein drugs since they have bioadhesive properties and the ability to enhance the ex vivo buccal membrane permeability of a peptide drug [59, 79, 80].

Attempts are made in various labs around the world including in our lab but no good and conclusive results are available. Therefore this chapter will not deal with LLC as possible potential alternative to IV treatment.

12.2.4 Transdermal Delivery from LLC

Transdermal delivery has lately emerged as a valuable alternative method to improve bioavailability and increase pharmaceutical efficacy of various drugs. Transdermal administration can potentially minimize side effects as well as first-pass metabolism. It has been used successfully for hormone replacement therapy, smoking cessation, and pain management. However, there have been challenges in expanding use of the technology to numerous types of pharmaceuticals, including extremely hydrophilic or lipophilic drugs, peptides, proteins, and other macromolecules. In these LLCs, seem to be promising candidates as unconventional delivery means for various drugs. Such vehicles can provide enhanced drug solubility, relative protection of the solubilized drugs, and controlled release of drugs, avoiding substantial side effects. LLCs exhibit interesting properties for a topical delivery system, and hence were considered and have been studied as delivery vehicles of pharmaceuticals via the skin and mucosa.

We will focus on lyotropic hexagonal and cubic and their nano-dispersions as topical delivery vehicles. Recent advances in transdermal and mucosal drug delivery via LLC carriers are demonstrated and discussed.

The major benefits of transdermal delivery are its relatively easier accessibility to the skin, assisting in high patient compliance, avoidance of the gastrointestinal tract, and the ability to achieve sustained release. These advantages have led to considerable advancement in the development of transdermal administration over the last decade [81]. The transdermal route is considered a promising approach to advance the delivery of peptides and to minimize side effects and first-pass metabolism [82, 83]. However, the skin has evolved into a highly effective barrier around the human body. As a result, despite the significant promise of transdermal administration, it is still not trivial to deliver large molecular weight drugs through the skin. Most of the relevant drugs do not cross skin at therapeutic rates and it is not surprising, therefore, that fewer than 20 drugs have been approved by the FDA for transdermal delivery since the first (transdermal) patch was introduced more than 25 years ago [83]. The properties of the skin barrier are derived from the outermost layer of the skin, called *stratum corneum* (SC). The SC consists of 10–15 layers of corneocytes and varies in thickness from approximately 10–15 μm in the dry state to about 40 μm when hydrated. The "brick and mortar model" is usually applied to describe the structure of *stratum corneum*. It comprises keratin-rich corneocytes (bricks) in an intercellular matrix (mortar) composed primarily of long chain ceramides, free fatty acids, triglycerides, cholesterol, cholesterol sulfate, and sterol/wax esters [69, 84].

Owing to the SC composition, its lipid phase behavior is different from that of other biological membranes. The hydrocarbon chains are arranged into regions of crystalline, lamellar gel, and lamellar LC phases, thereby creating various domains within the lipid bilayers [84]. Intrinsic and extrinsic proteins, such as enzymes, may also affect the lamellar structure of the SC. Water is a vital component of the structure, acting as a plasticizer to prevent cracking of the SC.

There are two major mechanisms for penetration enhancement by these dedicated molecules. The first mechanism is the interaction of the enhancers with the *polar headgroups* of the lipids. Consequently, the lipid–lipid headgroup interactions and the packing order of the lipids are disturbed, leading to the increased diffusion of hydrophilic drugs. The hydrophobic tails of the lipids can also be affected as a result of their headgroup disturbance by hydrophilic enhancers. Accordingly, structural rearrangements in the membrane occur, which can improve the penetration of lipophilic drugs.

The second possible enhancing mechanism is the interaction of the enhancers with the *non-polar tails* of the lipids. Lipophilic penetration enhancers can interact with the hydrocarbon tails of the membrane lipids, affecting their arrangement and leading to liquid-like behavior. In that case, the penetration of lipophilic drugs is improved due to the increased fluidization of the hydrocarbon chains. These structural modifications can also affect the order of the polar headgroups and also enhance the penetration of hydrophilic drugs [69]. A variety of amphiphilic molecules was found to act as penetration enhancers. For instance, phospholipids, such as phosphatidylcholine (PC), are known as membrane recognition agents and transport enhancer agents, and facilitate the bioavailability of the guest molecules transdermally [39–42]. Oleic acid is a powerful buccal and transdermal permeation enhancer since it has been shown to effectively increase the percutaneous and transmucosal absorption rates and can be easily incorporated into the cubic phase of GMO. Oleic acid was shown to alter the intercellular lipid fluidity within the *stratum corneum* and in buccal membranes, and to disrupt lipid bilayers, thereby affecting penetration [59].

Currently, various technologies are used to increase the permeability of macromolecules by exploring iontophoresis, electroporation, ultrasound, and microporation using electrical current/voltage, radio frequency, and microneedles to open the skin. Although mechanical abrasion and chemical enhancers may increase drug permeation, their effects on the skin's inherent rate-controlling properties are difficult to control and they may irritate the skin [85].

The limitations and the difficulty in crossing the barrier of the SC by various molecules can be addressed and partially overcome via the skin application of penetration enhancers, compounds that increase the diffusion coefficient of the drug through the SC.

12.2.5 Delivery of Lipophilic Drugs from LLC Systems

Most lipophilic drugs show extremely poor skin penetration, unless a special chemical or physical treatment is performed [82]. Keeping such behavior in mind, the main objective of using LLC is to increase the skin penetration of these drugs, when incorporated in the proposed carriers.

Lopes et al. investigated cubic and hexagonal phases containing chemical enhancers as delivery vehicles for cutaneous penetration of cyclosporin A in vitro (using porcine ear skin) and in vivo (using hairless mice) [63]. Cyclosporin A (CysA) was selected as a model peptide since it is a cyclic and highly lipophilic peptide that presents extremely poor skin penetration, unless a chemical or physical strategy is used [86, 87]. Additionally, this peptide is an immune-suppressant agent and has therapeutic potential in the treatment of skin inflammatory disorders [86].

The penetration of CysA in the skin and its percutaneous delivery were assessed in an in vitro model of porcine ear skin using a Franz diffusion cell. The quantity of CysA detected in SC and in the epidermis and dermis layers [E + D] was indicative of drug penetration into the skin, whereas the amount of drug in the receptor phase was indicative of its percutaneous delivery. The in vitro skin penetration and percutaneous delivery of the cubic and hexagonal liquid crystalline phases compared to the control formulation (olive oil) containing CysA are depicted in Fig. 12.16.

Incorporation of CysA in both liquid crystalline formulations enhanced the in vitro penetration of the peptide in the skin, probably due to the action of monoolein (in combination with oleic acid in the case of hexagonal phase) as a penetration enhancer [60]. Several differences between the two liquid crystalline phases were detected (Fig. 12.16). The cubic phase increased the penetration of CysA in the SC and [E + D], but did not influence the percutaneous delivery of the peptide compared with the control formulation. The hexagonal phase increased the penetration of CysA in the [E + D] as well as its percutaneous delivery. No enhancement of the peptide concentration in the SC was observed when the hexagonal phase was used, since the increased penetration of CysA was into deeper skin layers and through the skin. Whereas the cubic phase formulation favored retention of CysA in the skin in vitro, the hexagonal phase favored its penetration into deeper skin layers and its percutaneous delivery.

The cubic and hexagonal phases used in this study differ in their internal structure and composition, since oleic acid was added (5 %) to the hexagonal phase. Thus, the authors attributed the differences observed in the in vitro skin

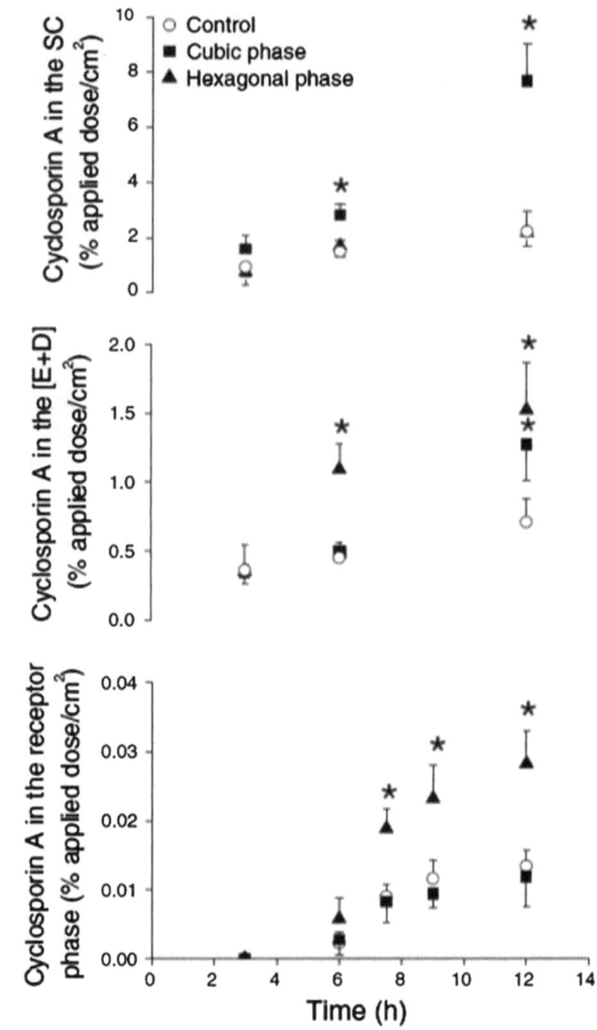

Fig. 12.16 Time-course of the in vitro skin penetration and percutaneous delivery of CysA incorporated in different formulations. The number of replicates is 6–8 per experimental group. *$p < 0.05$ compared to control formulation

penetration of CysA with the difference of constituents of both phases. Addition of oleic acid, a penetration enhancer, resulted in greater CysA penetration into deeper skin layers and through this tissue.

According to the researchers, the structure of the delivery system can also influence the skin penetration of CysA based on tests of non-peptidic compounds. In this matter, the release of non-peptidic compounds was shown to be influenced by the structure of the delivery system, drug/system interactions, and system physical–chemical characteristics [88]. On the basis of FT-IR measurements (data not shown), similar interactions between CysA and monoolein were observed in both cubic and hexagonal phases; thus, they should not contribute to the differences in skin penetration. Alternatively, cubic and hexagonal phases present different internal structures, viscosities, and drug release properties (including for CysA) that might influence their interactions with SC and thus the delivery of CysA to the skin [63].

The penetration of CysA, incorporated into the same three formulations, into the SC and [E + D] was further evaluated using an in vivo mouse model [63]. Once more, the entrapment of CysA in the liquid crystalline phases led to increased skin penetration of the peptide in vivo at 6 h post-application (Fig. 12.17). Even though the skin models used are different (mouse skin used in the in vivo experiment is more permeable than porcine skin), the results obtained in vivo confirmed the in vitro observation that monoolein-based delivery systems enhanced skin penetration of CysA.

Taking the in vitro and in vivo models together, Lopes et al. concluded that the penetration enhancement that was caused by structural alteration of the SC can further induce changes in the deeper skin layers depending on the penetration enhancer used and its concentration [63].

To obtain liquid crystalline systems that are expected to cause less skin irritation, with a larger surface area to interact with the skin, greater fluidity, and ability to be incorporated into other product formulations, Lopes et al. also tested the hexagonal dispersion [62]. Lopes and coworkers probed whether the hexagonal phase (GMO/oleic acid/water) nanodispersion had the ability to increase skin penetration of the relatively large peptide CysA. Correspondingly, after 6–12 h, the skin was separated to three sections and the quantity of CysA detected in SC and [E + D] was indicative of drug penetration in the skin, whereas the quantity of drug in the receptor phase was indicative of its percutaneous delivery (Fig. 12.18). Compared to the control formulation (olive oil), the hexagonal nanodispersion significantly enhanced skin penetration of CysA in the SC and [E + D] at 6 and 12 h post-application. Using the hexagonal phase nanodispersion revealed delivery of 7.12 and 3.85 % of the applied dose/cm^2 of CysA to the SC and [E + D], respectively, after 6 h, and 13.10 and 5.06 % of the applied dose/cm^2 to the SC and [E + D], respectively, 12 h after administration. The maximum concentrations were ~2-fold greater than those obtained using the control formulation. It should be noted that in vivo skin penetration experiments were carried out as well, and confirmed the in vitro observations. It was concluded that the hexagonal dispersion provided an efficient delivery of CysA, especially to deeper skin layers than that delivered by other methods.

Another successful application of LLC was topical application of vitamin K (vitK) [17]. VitK is important for the prevention of vascular manifestations of aging, for the suppression of pigmentation, and for the resolution of bruising [89, 90]. However, transdermal administration of vitK is hampered by its high lipophilicity. In this regard, Lopes et al. incorporated vitamin K in the reversed hexagonal phase gel and in an MO-based dispersion, and evaluated the efficacy of such systems to increase the topical delivery of this bioactive [17]. The MO/water systems allowed incredible structural flexibility, maintaining the hexagonal symmetry in the range of vitK concentration from 2 to 20 % at room temperature, MO content ranging from 58 to 87 %, and water from 8 to 40 % (Fig. 12.19).

Fig. 12.17 In vivo skin penetration of CysA incorporated in different formulations. SC: stratum corneum, [E + D]: epidermis without stratum corneum + dermis. The number of replicates is 4–6 per experimental group. *p < 0.05 compared to control formulation

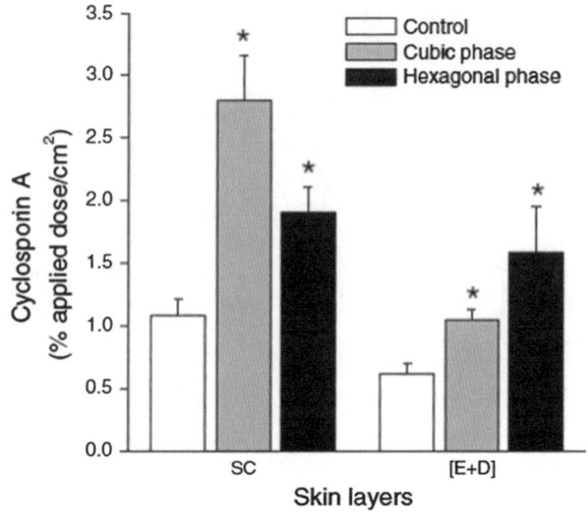

Fig. 12.18 In vitro penetration of CysA in the SC and [E + D] at 6 and 12 h following its topical application using the hexagonal phase nanodispersion or the control formulation. SC: stratum corneum, [E + D]: epidermis (without SC) plus dermis, * p < 0.05. Five to six batches of each formulation were tested, and the results are shown as mean T SD

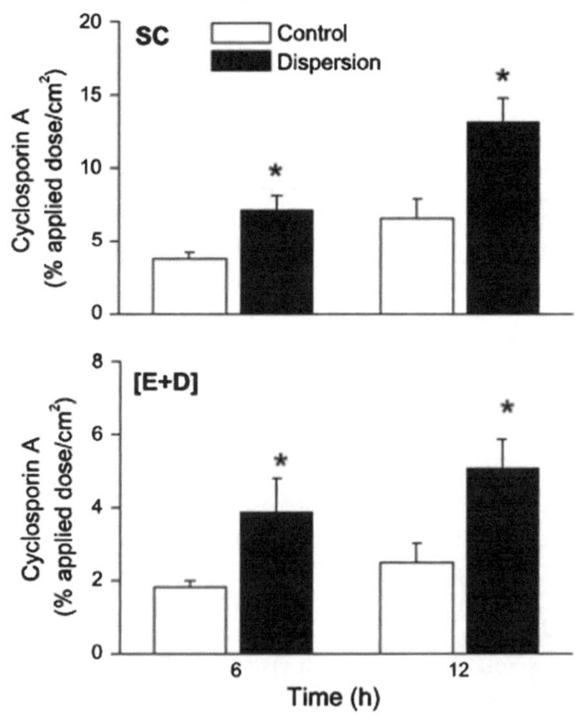

These investigators used bulk hexagonal mesophase and the corresponding hexagonal nano-dispersions (diameter = 315 ± 12 nm), containing 2.5 % vitK for transdermal delivery experiments. The skin penetration of vitK from the

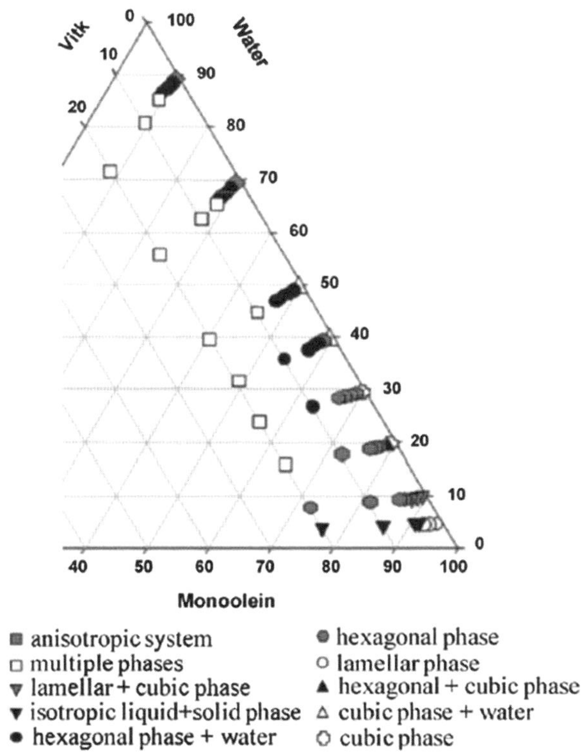

Fig. 12.19 Partial phase diagram of MO/water/vitK at 25 °C. (Taken from [17])

nano-dispersions and vaseline solution (as control system) is demonstrated in Fig. 12.20. It was observed that a small quantity of vitK incorporated in vaseline crossed the SC and reached viable layers of the skin; a greater quantity of vitK was retained in the SC. Specifically, 6.52 ± 0.48 and 9.50 ± 0.97 μg/cm^2 of vitK was found in the SC at 3 and 12 h post-application, respectively, when the vaseline solution was used; whereas 1.35 ± 0.28 and 4.90 ± 1.28 μg/cm^2 of vitK was delivered to [E + D] at the same time points. Lipophilic drugs such as vitK generally present a large partition in the SC. However, their penetration in viable epidermis is low, presumably due to their accumulation in the SC, which might result in limited effectiveness or even lack of effect. The quantity of vitK delivered to the SC was significantly increased ($p < 0.05$) at 3 h post-application when the hexagonal phase gel was used (14.72 ± 1.95 μg/cm^2, 2.2-fold increase) and at 9 h post-application when the nanodispersion was used (17.83 ± 1.75 μg/cm^2, 2.7-fold increase) (Fig. 12.20a–c). At 12 h post-application (the last time point studied), the gel and the nanodispersion delivered ~2–2.2 times more vitK to the SC than the vaseline solution. As to the penetration of vitK in the [E + D], a significant augment (~3-fold, $p < 0.05$) was observed at 3 h post-application using either the gel or the nanodispersion (Fig. 12.20d–f). After 12 h, vitK delivery to [E + D] was 2 and 3.7 times greater using the gel and the nanodispersion,

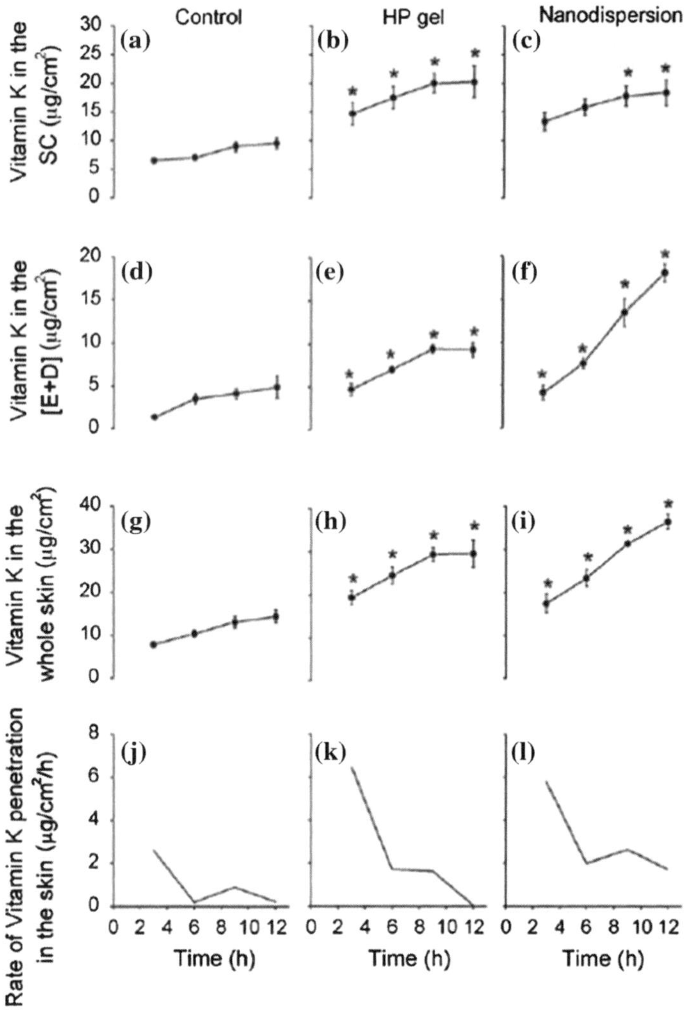

Fig. 12.20 Time-course of in vitro vitK penetration in the SC (**a–c**), [E + D] (**d–f**) and whole skin (SC + [E + D]) (**g–i**) at 3, 6, 9 and 12 h. The figure also shows the rate of skin penetration, calculated using the penetration of vitK in the whole skin (**j–l**). The number of replicates is 6–8 per experimental group. $^{*}p < 0.05$ compared to control. SC, stratum corneum; [E + D], epidermis without stratum corneum plus dermis; HP gel: hexagonal phase gel. (Taken from [17])

respectively. Based on the time-course of vitK content in the whole skin (Fig. 12.20g–i), the rate of vitK penetration was analyzed, revealing that the rate of penetration of vitK in the whole skin was maximal at 3 h post-application, and then gradually decreased (Fig. 12.20j–l). This was so regardless of the formulation used, although higher rates were attained with the gel and nanodispersion. However, at later stages (6–12 h), the rate of vitK penetration in the skin was greater

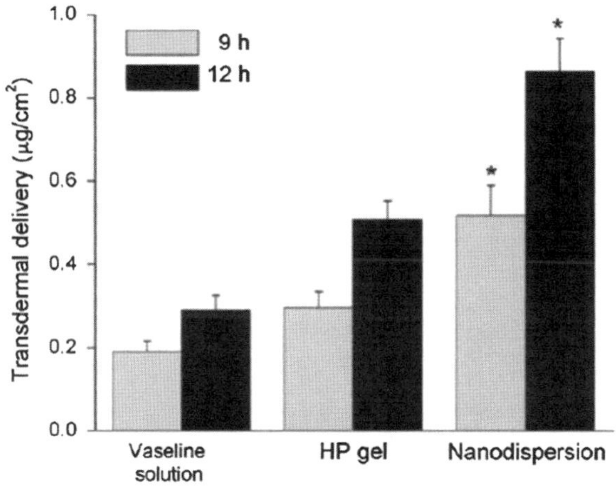

Fig. 12.21 Transdermal delivery of vitK at 9 and 12 h post-application using vaseline solution (control formulation, 2.5 % Vaseline, w/w), hexagonal phase gel (HP gel) and nanodispersion. The number of replicates is 6–8 per experimental group. $^{*}p < 0.05$ compared to the vaseline solution. (Taken from [17])

when the nanodispersion was used. Certainly, such 2.0–3.7-fold increase in the skin penetration of vitK may have a positive influence on therapeutic efficacy of this bioactive.

Penetration of vitK across the skin was also examined. It was detected that a very small fraction of the applied vitK was delivered across the skin (Fig. 12.21): only 0.29 ± 0.04 μg/cm^2 reached the receptor phase at 12 h. The use of the dispersion, but not of the gel, led to a significant augment of vitK transdermal delivery ($\sim$3 times, $p < 0.05$) at 12 h. Therefore it was concluded that it was unlikely that topical vitK would be absorbed in sufficient quantities to alter the systemic levels and affect the synthesis of coagulation factors.

A targeted delivery to selected skin layers might be obtained depending on the system used (bulk mesophase or hexosomes), as summarized in Fig. 12.22. After long periods of contact of the formulations with the skin (9–12 h), the hexagonal phase gel increased vitK retention in more superficial skin layers (SC), whereas an improved delivery of vitK into deeper skin layers ([E + D]) was observed when the nano-dispersion was used (Fig. 12.22a). As a result, the hexagonal phase gel provided vitK localization within the skin, and a minimal transdermal delivery (similar to the vaseline solution) was observed. On the other hand, greater quantities of vitK in the skin associated with an increased transdermal delivery were observed when the nanodispersion was used (Fig. 12.22b).

The observed differences in delivery of vitK between the hexagonal phase gel and the MO-based nanodispersion were attributed to several structural factors. First, since the fluid and particulate nanodispersion possessed a larger area of contact with the skin, this might improve system/skin interaction. The second important aspect was the concentration of monoglyceride in the formulations. It has already been shown that monoglyceride concentration exhibited dual effect in formulations containing cyclosporin A: increased MO concentrations (more than 20 %, w/w) led to increased retention of cyclosporin in the skin, whereas smaller

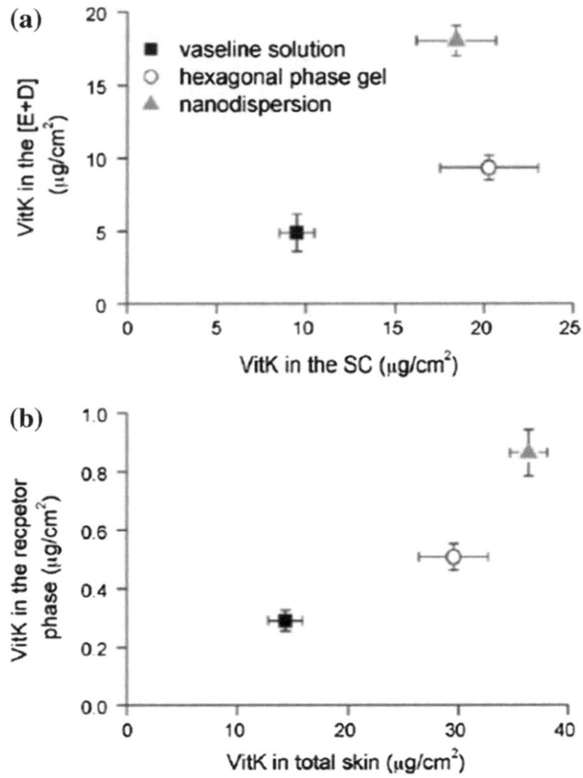

Fig. 12.22 Relationship between skin penetration and transdermal delivery of vitK at 12 h post-application using different formulations. **a** Relationship between vitK concentration in the stratum corneum (SC) and epidermis without SC + dermis ([E + D]). **b** Relationship between vitK concentration in the receptor phase (an index of transdermal delivery) and total skin (an index of skin penetration). The number of replicates is 6–8 per experimental group. (Taken from [17])

quantities influenced cyclosporin delivery across the skin [87]. MO content in the nanodispersion is less than in the hexagonal phase gel (MO at 15 % is present in the nanodispersion and at 77.5 % in the gel), which might account for the increased delivery of vitamin K to deeper skin layers and across this tissue when the nanodispersion was used.

The authors concluded that MO-based systems enhanced topical delivery of vitK in a time-dependent manner. An increased delivery of compounds to selected skin layers might be obtained depending on the system used and on the MO concentration. The nanodispersion enabled greater delivery of vitK to the [E + D]. The results obtained in this study can be useful to increase the effectiveness of topical vitK for resolution and prevention of bruising and other vascular manifestations of aging.

Mucosal delivery of lipophilic compounds from LLC was also tested, as was shown in case of progesterone [72]. Progesterone, named for its progestational role in maintaining pregnancy, is traditionally regarded as a "female hormone", playing a key role in generating female-typical behaviors such as lordosis and maternal behavior. The major problem with its therapeutic use is selection of a proper route of administration, since conventional delivery methods, including oral, intramuscular, vaginal, and topical suffer from serious drawbacks [72].

In this respect, Swarnakar et al. tested a newer method for mucosal delivery of progesterone, using hexosomes, and studied the possible mechanism of penetration of these carriers [72]. These investigators used GMO/oleic acid hexagonal system with ethanol (for decreasing viscosity), stabilized with Pluronic F-68, which is more hydrophobic (compared to more commonly used F-127), for increasing the critical packaging parameter. The drug had no influence of its own on the lattice parameter (64.2 Å) as well as particle size of hexosomes ($\sim$257 nm). IR technique was used to study the correlation between the barrier function of buccal epithelial lipids, thus being able to relate the transbuccal drug flux to epithelial lipid structure and organization. It was revealed that utilization of hexosomes induces lipid disorder in buccal epithelium and strongly alters lipid packing of the bilayer. The authors hypothesized that incorporation of oleic acid and hence formation of hexosomes disrupted the epithelial lipids and decreased the diffusional resistance to permeants. This hypothesis was further confirmed by confocal laser microscopy (CLSM) of mucosa treated with hexosomes. CLSM was utilized to explore probable the penetration pathway through mucosa. Since intact mucosa does not contain a significant number of pathways of sufficient width to allow nanoparticles to pass, either transcellular or paracellular pathways can be considered. Pores, which were detected in epithelium of mucosa, suggested that these regions may contain structural irregularities within the intercellular lipids. These irregularities were assumed to function as "virtual channels" through which the nanodispersion could penetrate.

Furthermore, ex vivo release studies of progesterone showed that permeation through mucosa has been increased to a greater extent and large flux values were obtained using hexosomes. The nano-dispersion exhibited increased flux (4.67 ± 0.14 μg cm^{-2} h^{-1}) and decreased lag time (1.54 h) of the drug compared to progesterone-loaded gel (0.93 ± 0.14 μg cm^{-2} h^{-1}, lag time 2.27 h) and plain progesterone suspension (1.13 ± 0.13 μg cm^{-2} h^{-1}, 2.24 h; $p < 0.05$). The observed flux was five-fold greater than that of progesterone-loaded gel and nearly four-fold greater than plain progesterone suspension. The very low permeability of plain drug suspension is due to its extreme hydrophobicity and low solubility in PBS (11.4 μg/ml), which in turn leads to a very low concentration gradient across the mucosa. Although gel has high viscosity it did not greatly alter the flux with respect to plain drug suspension. At the same time, gel-loaded formulation containing oleic acid increased the mucosal permeability progressively, i.e., a time-dependent change in lipid integrity of the barrier [59], due to which the system probably entered the mucosa through virtual channels.

Im et al. formulated monoolein-based cubic crystalline phase and its nanoparticle system of oregonin and hirsutanonol (both diarylheptanoid derivatives) to treat atopic dermatitis [91]. Diarylheptanoids have a therapeutic effect on the dermal inflammation as selective 5-lipoxygenase inhibitor and thus are expected to be applicable for atopic dermatitis [91]. For the effective treatment of atopic dermatitis with topical administration of therapeutics, penetration and accumulation of the agent into the relevant layers of the skin is required. Therefore, Im et al. explored the localization factors for local skin delivery of oregonin and hirsutanonol [91]. The effects of lipophilicity of drug, vehicle type, and skin condition on skin permeation and deposition of the drugs were evaluated using the rat skin mounted in Franz diffusion cells. The average size of oregonin and hirsutanonol-loaded cubic nanodispersion were in the range of 131.77 ± 3.80 and 94.20 ± 1.67 nm, respectively.

The investigators presumed that the greater aqueous solubility of oregonin hindered its deposition in the skin. Greater skin deposition of hirsutanonol was attributed to its greater lipophilicity compared to oregonin. The enhanced delivery of the drugs to the skin by monoolein-based systems was assigned to the action of monoolein as penetration enhancer. Another interesting finding was that the quantity of hirsutanonol delivered to the skin layer (i.e., permeation and deposition) was significantly increased at 12 h post-application when the nanodispersion was used as the vehicle. The reason for this may be that the nano-dispersion presents a larger surface area for interaction with the skin, and the cubic nanodispersion itself can act as a penetrating enhancer owing to its minimal size. Another important finding was that 2–3 times greater quantities of oregonin and hirsutanonol were deposited in the skin when administered to delipidized skin. This was attributed to the lowered barrier property of the delipidized skin, compared to the regular one. On the whole, this study demonstrated that topical delivery of oregonin and hirsutanonol can be significantly enhanced via monoolein-based systems.

12.2.6 Administration of an Amphiphilic Drug

Propranolol hydrochloride (PHCl) is an important water soluble cationic amphiphilic drug. PHCl is a nonselective β-adrenoreceptor blocker that is widely used in the treatment of many cardiovascular diseases, such as hypertension, angina pectoralis, cardiac arrhythmia, and myocardial infarction. However, propranolol is subject to an extensive and highly variable hepatic first pass metabolism following oral administration, with oral bioavailability between 15 and 23 % [92, 93]. Consequently, transdermal delivery would be a valuable alternative method. Lamellar and cubic LLC systems based on monoolein and phytantriol were developed by Costa-Balogh et al. for delivery of this amphiphilic drug [93]. Diolein (DO) and cineol were also incorporated to selected formulations. DO was added into the cubic mesophases as a viscosity modifier. DO together with PHCl stabilized the cubic structures, leading to an increase in viscosity. Cineol, an essential oil (monoterpene) also known as eucalyptol, is a penetration enhancer.

The *Pn*3*m* cubic phase was not stable to the addition of the amphiphilic drug, and the successively increasing drug loads leads to the phase changes to the *Ia*3*d* cubic phase and to the lamellar (Lα) phase. Phytantriol (PHY) formed bicontinuous cubic phases at lower water contents than MO. In addition, PHY-based formulations generally presented greater viscosity and smaller lattice parameters than the GMO-based mesophases, but in both systems the drug load induced a decrease in viscosity that was related with phase changes. The changes in phase behavior caused by the addition of DO and/or PHCl were explained in terms of the packing concept of amphiphilic molecules. The addition of Diolein (DO) to the MO or to the PHY, led to the formation of H_{II} mesophase upon hydration. DO promoted structures with negative curvature, such as the reversed hexagonal phase, and PHCl promoted normal, lamellar, structures. When both DO and PHCl were added to the MO/water cubic phase, these opposing effects compensated for each other, and the cubic phase appeared to be the most stable structure, possessing increased viscosity [93].

The phase behavior and the viscosity of the LLC formulations affected both the release of the drug and skin permeation. In general, formulations with greater viscosity presented a slower drug release, especially for a specific lipid and drug load. DO prevented the formation of the less viscous phases and, for the smaller drug loads, it stabilized the MO cubic phases and retarded release, resulting in substantial control over the respective release rate. PHY formulations exhibited a slower drug release than MO formulations. PHCl was not completely released from the carriers. This effect was more dominant for the systems that contained DO. This was explained by the drug partitioning into the lipid bilayers due to its amphiphilic nature. The systems with cineol demonstrated the highest permeability, much above the permeability of the other lamellar formulations. Less viscous lamellar phases, which favored faster release, seemed to be more adapted to transdermal delivery. Consequently, with this work it was shown that systems involving lamellar phases of monoolein and cineol were good candidates to be used as skin permeation enhancers for propranolol hydrochloride, and could possibly be used in transdermal delivery systems of this drug. These findings supported earlier results of Namdeo and Jane [94], who developed a new system for the enhanced transdermal delivery of propranolol hydrochloride (PH). The synthesized prodrugs, propranolol palmitate hydrochloride (PPH) and propranolol stearate hydrochloride (PSH), self-assembled to form gel based on lamellar liquid crystals, upon mixing alcoholic solution of prodrug with an aqueous solution in a specified ratio. This pharmacogel provided high transdermal permeation with many variables to regulate the delivery and presented a great potential in percutaneous delivery.

12.2.7 Release of Proteins as Drugs

Libster et al. systematically investigated solubilization of delivery of desmopressin as a model hydrophilic peptide, using H_{II} mesophases based on GMO tricaprylin and water [43]. Desmopressin is a synthetic analogue of the antidiuretic hormone vasopressin, which is mainly used for treatment of enuresis in young children, central diabetes insipidus, hemophilia A, von Willebrand disease, and trauma-induced injuries. The very high hydrophilicity of desmopressin and its enzymatic degradation in the gastrointestinal tract seem to be the major reasons for the poor bioavailability of the drug [95–97]. In this respect, transdermal administration of desmopressin can therefore be a valuable alternative. The concentration-dependent SAXS measurements revealed a significant increase (up to 6 Å) in the lattice parameter of the hexagonal mesophases as a result of the peptide incorporation. This chaotropic effect of the peptide was thought to be related to its interactions with hydroxyls of GMO in the outer interface region, as was inferred from ATR-FTIR examination. It was found that neither the inner shell of the interface nor the acyl chains were affected by the peptide. Desmopressin molecules probably did not penetrate deeply enough into the interface region of the mesophase. As a result, it could not interact with the inner parts of the host GMO molecules, including the carbonyl groups, the ester moiety, and the lipophilic parts. The desmopressin mobility in the water cylinders of the hexagonal structure was compared to its diffusion in pure water. Inclusion of the peptide into the mesophase domains radically hindered its mobility, decreasing the diffusion coefficients by orders of magnitude. Rapid increase in peptide mobility was observed above 45 °C, similar to the process noticed in the water solution. It was realized that the diffusion of the peptide in the hexagonal phase, compared to water solution, was slowed down by two factors (Fig. 12.23).

The first was the physical restriction of the peptide motion, more specifically its diffusion within the water cylinders, owing to the geometric constrain of hexagonal architecture. The second factor is the chemical interactions of the peptide with the polar heads of monoolein. These two effects were separated and quantified by SD-NMR analysis (SD-NMR). Rankin's model [43] was used to calculate the theoretical diffusion coefficients of desmopressin within the channels of the H_{II} mesophase, assuming no interactions of the peptide with GMO and TAG. Using the theoretical diffusion coefficient of the drug and the measured diffusion values, the observed decrease in diffusion coefficients of the peptide in the H_{II} mesophase was clarified (Fig. 12.23), defining two obstruction factors. The obstruction factors enabled quantification of the effects of both physical restriction (β) and the chemical interactions (γ).

With this approach, it turned out that between 25 and 45 °C both β and γ factors hold similar values, indicating that the physical restriction and the chemical interactions contributed almost equally to the decrease in the mobility of the peptide within the H_{II} mesophase, compared to the water solution. However, above 45 °C increased values of γ-factor suggested considerable weakening of the GMO and desmopressin interactions and even repulsive interactions between the surfactant and the peptide at 55 °C. This might trigger the observed decreased thermal stability of the loaded mesophase. Thus, it was concluded that the peptide dictated the phase behavior of the hexagonal structure, exhibiting the onset of the critical behavior at 45 °C.

It would be interesting to compare these results with the findings of Ericsson et al. who measured diffusion of this peptide in the aqueous channels of the cubic phase [53]. According to these investigators, the self-diffusion data indicated that desmopressin interacted significantly with the monoolein-water interface. For example, the desmopressin diffusion coefficients in the cubic phase at 40 °C was about *a factor 9 smaller* than in ^2H_2O solution, a difference that is larger than what

(a)

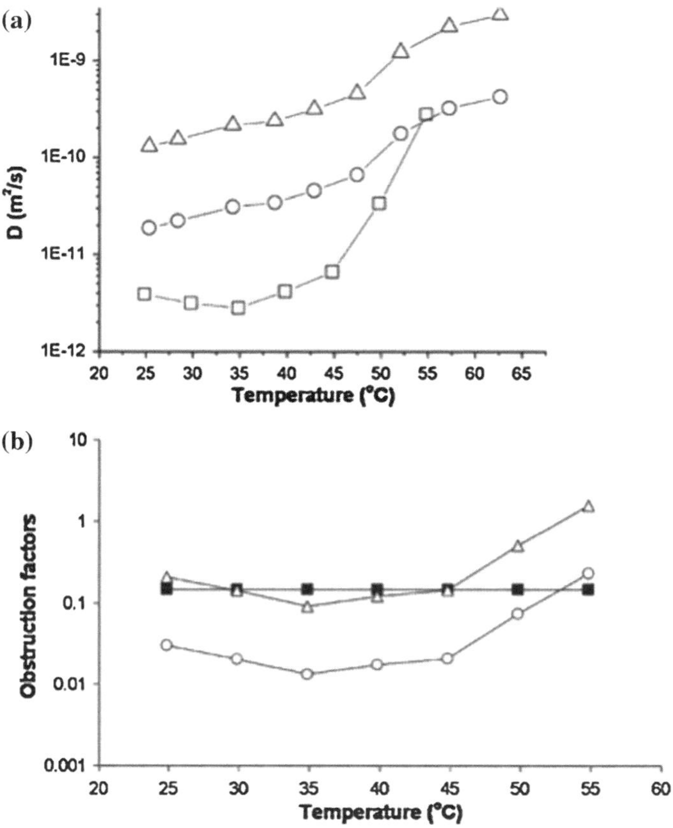

Fig. 12.23 **a** Desmopressin diffusion coefficients as a function of temperature in (△) 8 wt% in D_2O solution (○) according to the Renkin model, (□) 8 wt% in H_{II} mesophase as a function of temperature. **b** Obstruction factors calculated from the SD-NMR analysis as a function of temperature (○). The overall obstruction factor (λ); (■) β is the obstruction factor responsible for the physical restriction of the peptide in the hexagonal structure; (△) γ is the obstruction factor associated with the chemical interactions between the peptide and GMO

is expected from pure obstruction effects; *a reduction of 3* was expected from an inclusion of a solute in the water channel of a cubic phase. In addition, it was found that at 40 °C the desmopressin diffusion coefficient in the hexagonal mesophase was about *a factor 57 smaller* than in ^2H_2O solution, while a *reduction of 6.9* (Renkin model) was expected from the peptide physical restriction within the H_{II} mesophase. Comparing these data to those obtained for the cubic phase, it is clearly shown that desmopressin is retarded much more significantly within the hexagonal phase, compared to cubic phase. As shown, one source of this diffusion impediment is the physical restriction of the peptide within the H_{II} mesophase, which is more significant, compared with that within the cubic phase (*a reduction of 6.9 vs. 3*). This is due to greater curvature and narrower aqueous channels of the

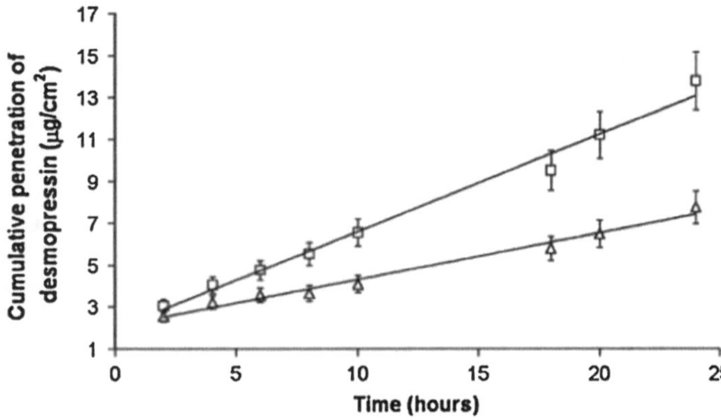

Fig. 12.24 Cumulative transdermal penetration of 1 wt% desmopressin (□) via water solution and (Δ) via H_{II} mesophase

hexagonal phase, containing less water. Second, a more important and dominant reason is that the increased physical restriction of the peptide within the hexagonal structure imposed drastically stronger chemical interactions (via hydrogen bonding) of the peptide with monoolein. This eventually led to a significant hindering of its diffusion within the H_{II} mesophase, compared to cubic structure (a *reduction of 57 vs. 9*). Such difference may have remarkable implications for drug release. One direct conclusion from these studies is that a carrier with hexagonal architecture should normally be a superior matrix for *sustained release* of hydrophilic drugs than one with cubic structure. In particular, it has been demonstrated that the release of hydrophilic drugs from the phytantriol-based Q_2 structure is significantly faster than from the H_2 phase (prepared by addition of 10 % vitamin E acetate to phytantriol) at 37 °C [98].

The applicability and efficiency of the H_{II} mesophase [43] and the corresponding hexosomes [76] as carriers for a transdermal delivery of desmopressin was also tested. In the case of bulk hexagonal mesophase, reduced values of steady-state flux of the drug and its permeability coefficients through the skin were obtained, and compared to water solution (Fig. 12.24). In other words, the peptide molecules penetrated more easily and freely through the skin from water solution than from hexagonal phase vehicles. Hence, *a sustained release of desmopressin* was obtained via the H_{II} mesophase.

Hexosomes were also examined as carriers for transdermal release of desmopressin [76]. In this case relatively low release rates were obtained for the first 10 h and drastically increased rates during succeeding hours (Fig. 12.25).

When comparing the release from hexagonal phase in its bulk form and its dispersion form, it should be considered that hexosomes are *discontinuous* particles, imposing a much longer diffusion pathway for the drug, and consequently requiring more time, compared to the *continuous* bulk H_{II} mesophase. In addition, the H_{II} mesophase is only 5 wt% of the whole dispersion. Thus, the discontinuity

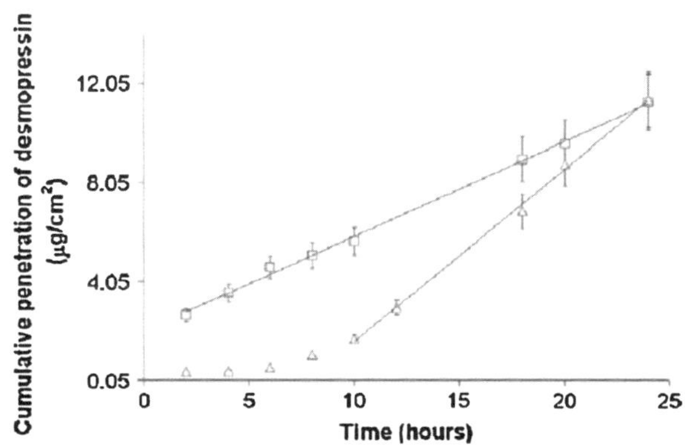

Fig. 12.25 Cumulative transdermal penetration of 0.5 wt% desmopressin (□) via water solution, (Δ) via hexagonal dispersion, stabilized with 0.1 wt% PhEO (Taken from [76])

of the dispersion and the relatively low overall concentration of hexagonal phase in the dispersion could be the two major factors that led to the observed slow permeation of the drug in the first 10 h after the application of the drug-containing dispersion. On the other hand, desmopressin is likely to diffuse into the external water. Accordingly, after the end of the lag time (the subsequent 14 h in our experiment), the release of the drug will be more significant via the *intercellular matrix* from the external water of the dispersions, although still accompanied with the release of the drug from the hexagonal particles via *inter-cluster penetration*. Such combined release of desmopressin via two different routes simultaneously dramatically increased the penetration of the drug in the 10–24 h range, compared to the first 10 h. Hence, hexosomes present a very flexible delivery system for desmopressin, allowing relative control over its exact diffusion rate (sustained or enhanced) as a function of time.

Based on the idea of faster release from cubic phase, phytantriol-based and monoolein-based cubic (Q_2) and (H_{II}) mesophases were developed Boyd and coworkers to allow change to the nanostructure in response to external change in temperature. This was done to control drug release rates in vitro and in vivo [55]. These investigators optimized the phytantriol:vitamin E acetate and GMO:oleic acid (OA) ratios to obtain a "switch" temperature for the Q_2 to H_2 phase transition of approximately 37 °C. A model hydrophilic drug, glucose, was used for in vitro release and in vivo absorption studies. The temperature responsiveness was investigated in vivo in rats by determination of drug absorption before and after application of a heat and cool pack at the site of subcutaneous injection to control the subcutaneous temperature and hence the liquid crystalline matrix and drug release and absorption rates. Experiments were conducted to investigate the effect of changing phase structure on drug release in situ. This was accomplished by "switching" the temperature from 30 to 40 °C to stimulate the phase changes between the Q_2 and H_2 phases, respectively. The reversibility of the transition was also tested by switching back to the starting temperature to revert to the original phase structure, and hence original drug release rate. The release data on switching

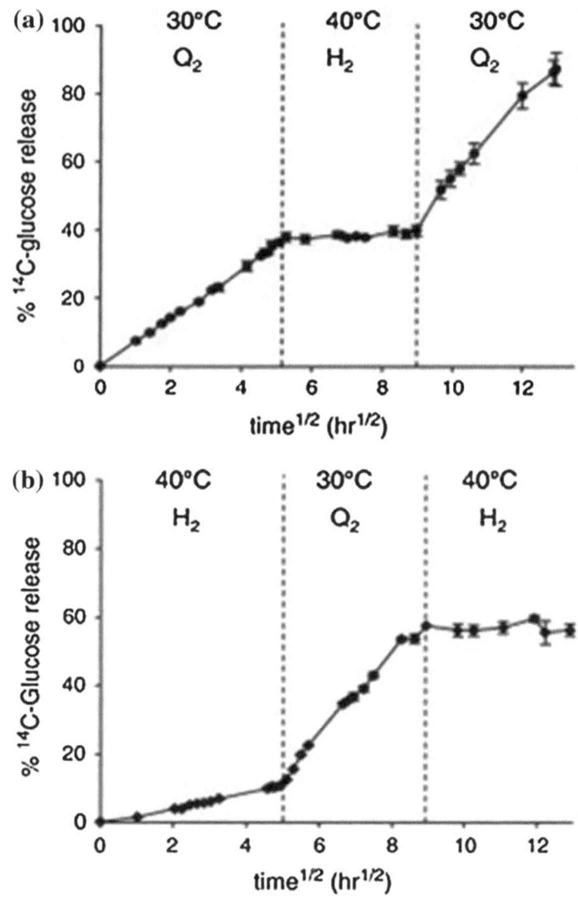

Fig. 12.26 Dynamic release profiles for glucose into phosphate buffered saline from phytantriol + 3 % vitamin E acetate with changing temperature, plotted against the square root of time (data are mean ±S.D., $n = 3$). Temperature was switched from 30 °C → 40 °C → 30 °C at the times indicated by the *dashed lines* (**a**) and from 40 °C → 30 °C → 40 °C (**b**). (Taken from [55])

from 30 °C → 40 °C → 30 °C, and hence from Q_2 → H_2 → Q_2 are illustrated in Fig. 12.26a. A clear suppression of drug release on switching the phase structure from Q_2 to H_2 was detected. On switching the phase structure back to Q_2, the system appeared to return to close to the original release rate. The linear profile when plotted against $t^{1/2}$ again demonstrates that the diffusion of drug within the matrix is controlling the release rate, and that the diffusion behavior is reversible on changing back to the original Q_2 phase. The calculated diffusion coefficients did not support a strictly reversible transition between structures—the diffusion coefficient in the Q_2 phase, whilst initially maintained at 30 °C, agreed well with the "static" diffusion coefficient for the Q_2 phase; however, after switching back from the H_2 phase to the Q_2 phase, the diffusion coefficient had increased significantly to more than 160×10^{-8} cm^2 s^{-1}. The profiles were linear, indicating that an effectively equilibrium structure had been achieved in both cases. The diffusion coefficient for glucose in the H_2 phase in the interim period at 40 °C was also significantly different from that in the static experiment, i.e., less than

1×10^{-8} cm^2 s^{-1}. This study provided initial proof of concept for the use of liquid crystalline matrices as on-demand stimuli-responsive systems for hydrophilic drugs.

12.2.8 Specific Utilization of Penetration Enhancers for Delivery of Hydrophilic Drugs from LLC

As mentioned earlier, penetration enhancers are extremely important for specific and rationally tailored delivery of pharmaceuticals. One novel class of such compounds are cell penetration enhancers (CPEs), which are unique peptides, first recognized in biological systems such as HIV, herpes enzymes, some fly [99] and frog species [100], and more [101]. These peptides have the ability to cause a perforation in the membranes of living cells thereby enhancing the penetration of molecules into the cells and target specific molecules to specific cells or nuclei [102, 103].

Cohen-Avrahami et al. examined the feasibility of a system based on H_{II} mesophase loaded with a skin penetration enhancer and sodium diclofenac (Na-DFC) [44]. They used a penetration enhancer (RALA, a 16 amino-acid long peptide) belonging to a synthetic family of CPEs, based on GALA, amphipathic peptides. These peptides contain an alanine-leucine-alanine repeating unit that exhibited improved membrane permeability [104].

The penetration mechanism of these peptides into the cell and into the nucleus was proposed to be based on peptide aggregation within the bilayer surface. When a peptide aggregate reaches a critical size, a transitional destabilization of the membrane bilayers occurs [105] and a pore is formed. The size of the pore depends on the number of peptides forming it, which limits the size of the molecules that can penetrate the pore [106]. Combining several experimental techniques (SAXS, ATR-FTIR, and DSC) these authors revealed that RALA was located in the aqueous cylinders and slightly swelling them by enhanced hydration of the GMO headgroup. Both the GMO and water exhibited a stronger bound state in the presence of RALA, indicating its chaotropic effect on the bulk water, which caused their migration towards the GMO headgroup. Moreover, RALA incorporation into the H_{II} mesophase retained its native random coil structure, preserving its skin penetrating activity with the solubilization in the mesophase.

The effect of these peptides on skin permeation efficiency was determined by diffusion experiments of Na-DFC through porcine skin using Franz diffusion cells. All systems were composed of H_{II} mesophases loaded with 1 wt% Na-DFC and 1 wt% of each CPP (the control was a system with Na-DFC and no CPP). These experiments indicated that all three peptides increased significantly the diffusion of Na-DFC through the skin.

NONA was found to be the most efficient CPP, significantly enhancing the transdermal penetration (a 2.2-fold increase in the total amount of Na-DFC that diffused through the skin). PEN and RALA caused 1.9- and 1.5-fold increases, respectively, compared with the control system. In all tested systems, there was a gradual and linear increase in the cumulative penetration of Na-DFC with time.

A calculation of Na-DFC percentage that permeated through the skin showed that at the blank systems, the total amount that diffused through the skin during the 24 h experiment was 0.9 % from the initial applied dose. The amount released at the NONA, PEN, and RALA systems was 1.9, 1.8 and 1.3 %, respectively. The permeability coefficients (K_p, calculated as $cm \cdot h^{-1}$), derived from the steady state flux of Na-DFC, revealed the same tendency, a 1.5-fold increase in the presence of RALA and 2.3- and 2.2-fold increases with PEN and NONA, respectively. Since the skin permeation studies revealed different profiles for each of the three peptides, a more detailed analysis using the well established technique of "emptying experiments" was conducted by the authors. These experiments aimed to determine whether the rate-determining step of Na-DFC skin permeation from the mesophase is migration out of the mesophase or permeation through the skin. In addition, this analysis provides information on the rate-determining step in the CPP enhancement.

The "emptying" experiments revealed a surprising result. The control system, without CPP, released the largest quantities of Na-DFC with time. The release from the PEN systems was the second in order, whereas NONA and RALA systems released the lowest quantities of Na-DFC (Fig. 12.27).

Higuchi diffusion equation was applied to analyze Na-DFC release from the examined mesophases. Higuchi model implies that drug release is primarily controlled by diffusion through the matrix and can be described by the following equation:

$$Q = [D_m(2A - C_d)\, C_d\, t]^{1/2}$$

Q is the mass of drug released at time t and is proportional to the apparent diffusion coefficient of the drug in the matrix, D_m; the initial amount of drug in the matrix, A; and the solubility of the drug in the matrix, C_d. The slope of a linear fit of the data from this plot is proportional to the "apparent diffusion coefficient" for the drug in the matrix and permits preliminary assessment of diffusion as the primary means of drug release from the correlation coefficient for the linear fit, and second, a means to compare the diffusion of a drug from the different matrices into the release medium. The linearity of the plots was found to be >0.95 for all CPPs, indicating existence of a diffusion-controlled transport mechanism (Fig. 12.27). The slope was the greatest for the blank system, 2.8 $h^{-1/2}$, and it decreased to 2.0, 1.3, and 1.2 in the PEN, NONA, and RALA systems, respectively.

The "blank" systems released the highest amounts of Na-DFC, and the release from the CPP loaded systems was much lower. The order of Higuchi slopes was exactly the same; "blank" > PEN > NONA > RALA, and their values were 3.0, 0.9, 0.7, and 0.6 $h^{-1/2}$, respectively

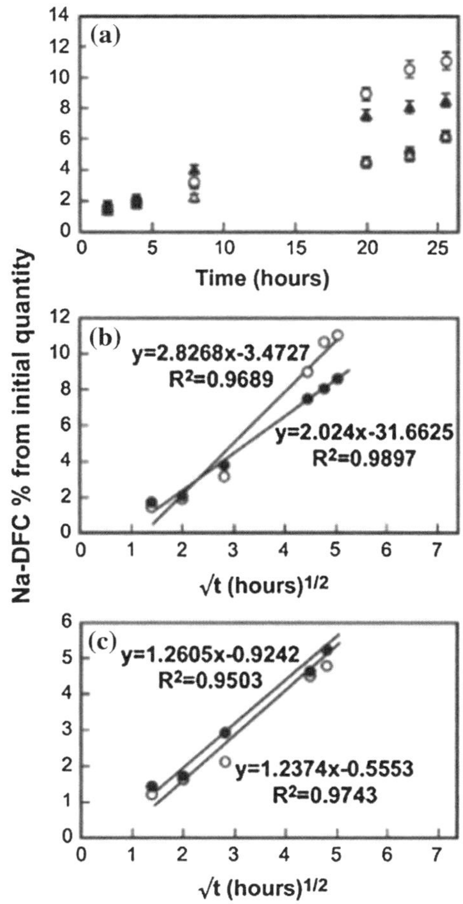

Fig. 12.27 **a** Na-DFC release to water from the HII mesophase containing: no CPP (o), RALA (●), NONA (Δ), and PEN (▲). Na-DFC release vs. the square root of time from **b** the "blank" (no CPP), (o) and PEN (●) systems and **c** RALA (●) and NONA (o) systems

Considering the Na-DFC's amphiphilic nature and the H_{II} mesophase structure, in which the aqueous phase is tightly packed within the lipid structure, with the limited accessibility to the surrounding media, one can assume that the drug diffusion out of the mesophase occurs through the lipophilic oily regions. The CPPs incorporated within the mesophase slowed the drug diffusion rate by decreasing its mobility through the mesophase by any specific molecular interfacial interactions. Nevertheless, the emptying experiments showed that the main enhancing effect of the CPPs on skin penetration does not take place in the step of drug migration out of the mesophase. Apparently, to be due to the enhanced permeation capability of the skin membrane and via any kinetic diffusion controlled process. It should be stressed that RALA and NONA, which populate the inner aqueous channels, revealed a similar diffusion profile with the same slope, $1.2 \pm 0.1\ h^{-1/2}$. PEN that populates the outer interfacial region is less disturbed in its diffusion, causing a release profile with a slope of $2\ h^{-1/2}$. Nevertheless, this interesting mechanism of the drug migration from the H_{II} mesophase in the presence of CPPs molecules cosolubilized within the mesophase should be further investigated.

Based on the principle that the H_{II} phases released model hydrophilic and hydrophobic drugs more slowly than the GMO cubic phase matrix, Fong et al. designed phytantriol and GMO-based bicontinuous cubic (Q_2) and H_{II} nanostructures, designed to allow change to the nanostructure in response to an external change in temperature, with an intention of eventual control of drug release rates in vivo [55]. Using glucose as a model hydrophilic drug, drug diffusion was shown to be reversible on switching between the H_{II} (very low release) and Q_2 nanostructures (high release), at temperatures above and below the physiological temperature, respectively (Fig. 12.26).

In this context, toward the design of an advanced drug delivery system based on light-triggered phase transition of liquid crystalline phases, Fong et al. reported the design of novel liquid crystalline matrix-gold nanorod hybrid materials [107]. Hydrophobized gold nanorods (GNRs) have been incorporated within the liquid crystalline matrix, composed of phytantriol and water, to provide remote heating, and trigger the phase transitions on irradiation at close to their resonant wavelength. The surface of plasmonic metal nanoparticles delivers heat into surrounding material on exposure to an appropriate light source at the plasmon resonance. The application of near-infrared sensitive nanorods has significance in designing systems for ophthalmic, subcutaneous, or deeper tissue applications. These investigators used plasmonic nanoparticles to achieve reversible phase transitions, hence offering a novel practical solution. The presence of nanorods at concentrations up to 3 nM did not change the lattice parameter of cubic (v_2), hexagonal (H_{II}) and micellar (L_2) systems in the absence of laser irradiation.

Irradiation of the system (at 808 nm NIR diode laser), on inclusion of a low concentration of GNRs (0.3 nM) in the matrix, did not induce a change in phase structure away from the *Pn*3*m* bicontinuous cubic phase. However, a small decrease in lattice dimension occurred, indicating that heating of the matrix has occurred. The structure relaxed back to the original position when the laser was off. On repeated application of the 5 s laser pulse, the system again displayed the heating effect and relaxed back to the starting position when irradiation was complete (Fig. 12.28). This contraction and expansion of the lattice on heating and cooling (the "breathing mode") was accompanied by concurrent expulsion and uptake of water from the matrix to satisfy the changes in lattice dimension.

In the case of higher concentrations of GNR, the 5 s on pulse for both the 1.5 and 3 nM systems did induce a phase change to the H_2 and L_2 phase structures. At 3 nM, complete transformation to the L_2 phase occurred by the end of the 5 s irradiation, while the lower 1.5 nM concentration resulted in a mixed $H_2 + L_2$ phase, indicating a reduced heating effect (Fig. 12.28). Again, when the laser was switched off, the system ultimately returned to the initial v_2 (*Pn*3*m*) phase structure. Interestingly, on conversion from the L_2 or $L_2 + H_2$ state back to the v_2

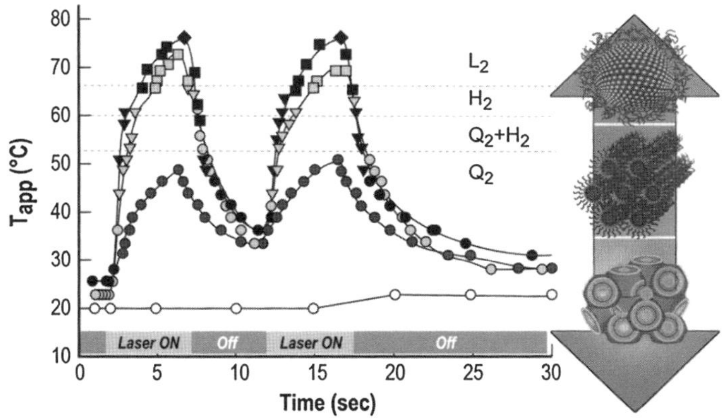

Fig. 12.28 Effect of laser irradiation on apparent temperature of the phytantriol + water matrix (T_{app}) with change in GNR concentration. [GNR] = 0 nM are *white* symbols, 0.3 nM *blue* symbols, 1.5 nM *yellow* symbols, and 3 nM *black* symbols. *Circles* indicate v2 phase, *triangles* indicate v2 + H2, *squares* indicate H2 + L2, and *diamonds* indicate L2. The cartoon on the *right* indicates the type of phase structure present with increasing temperature. (Taken from [107])

structure, the "gyroid" bicontinuous cubic phase with *Ia*3*d* space group was encountered. The gyroid phase coexisted with the H_2 phase initially and then with the v_2 (*Pn*3*m*) phase for approximately 5–6 s after the laser was switched off.

In addition, Fig. 12.28 clearly demonstrated that in the case where no GNRs were added to the matrix, there was no significant change in apparent sample temperature (T_{app}) upon laser irradiation. The reversibility of the heating effect in the presence of the GNR is evident from the T_{app} profiles for the three samples containing the nanorods. The sample containing 0.3 nM GNR heated to approximately 50 °C, just below the transition above which coexisting H_2 phase occurs. Increasing the nanorod concentration to 1.5 nM induces heating to approx 70 °C, while 3 nM GNR heated the matrix to an apparent temperature of 75 °C. The repeat irradiation provided the same peak temperature within 1–2 °C and reproducible kinetics of heating and cooling. The heating effect observed in these experiments is clearly a function of nanorod concentration, although the relationship between concentration and maximum temperature at differing irradiation conditions requires further investigation and is likely complicated by concurrent cooling. The "breaking wave" shape of the profiles, indicating nonlinear heating/cooling gradients in the material.

The authors [107] concluded that GNRs embedded in liquid crystalline matrices produce localized plasmonic heating of the hybrid matrix, enabling fine control over nanostructure. The phase transitions resulting from photothermal heating were fully reversible and specific to the GNR/laser wavelength combination. Localized plasmonic heating of the liquid crystal did not compromise the integrity of the lipid molecules in any of the mesophases. Undoubtedly, this research represented a significant advance toward effective, light-activated drug delivery systems with potential to solve unmet medical needs.

The most straightforward stimulus, which can be exploited for in vivo controlled release is certainly pH, because of the large pH changes occurring spontaneously within the mouth–stomach–intestine tract.

In this context, Negrini et al. presented a pH-responsive lipid-based LLC, which has a number of significant advantages for real oral-administration-controlled delivery [108]. First, the system is based on a simple formulation of monolinolein and linoleic acid, which maintains it entirely food-grade. Second, it offers a general, tunable release and diffusion strategy that is not specific to the particular drug ingredients. Finally, the release can be controlled in a fully reversible way and makes it suitable for a targeted delivery to specific points (pH) of the gastrointestinal tract. The lipid (monolinolein), the pH-responsive molecule (linoleic acid), and the model hydrophilic polyphenol drug (phloroglucinol). Because of the presence of the linoleic acid, which can be in either the deprotonated or protonated state when changing the pH from neutral to acidic conditions, respectively, the LLC undergoes a structural change from reverse bicontinuous cubic phase to reverse columnar hexagonal phase, which is accompanied by a change in the release rate by a factor of four, thus preventing the release in the stomach and making this system an ideal candidate for the targeted delivery of active ingredients in the basic environment typical of the intestinal tract.

This order-order transition can be explained by the presence of the ionizable carboxylic group of the linoleic acid (intrinsic $pK_a \approx 5$). The linoleic acid is negatively charged at pH 7; the electrostatic repulsions between the negatively charged headgroups stabilize the *Im*3*m* bicontinuous cubic phase. When the pH decreases below the pK_a value, the carboxyl group reprotonates to a large extent, the surface charge density on the water channels at the water–lipid interface decreases, and the linoleic acid became highly hydrophobic, acting as an oil and stabilizing the hexagonal phase at 37 °C.

This order-order transition is well-rationalized by the concept of the CPP expressed as v/Al, which is the ratio between the volume of the hydrophobic lipid tail, v, and the product of the cross-sectional lipid head area, A, and the lipid chain length, l. When linoleic acid is deprotonated (pH 7), the effective area A is large because of the electrostatic repulsive interactions among different lipid heads. When, however, the linoleic acid is mostly neutral (pH 2), A decreases and the CPP increases, promoting the transition from flat to reverse (water-in-oil) interfaces and inducing a bicontinuous cubic → reverse columnar hexagonal transition.

In vitro release studies were carried out first to establish the influence of the liquid crystalline symmetry on the release behavior. Figure 12.29a illustrates the drug release profiles from the liquid crystalline matrices, plotted as a percentage of the released drug against time. As can be observed, the release from *Im*3*m* is much more rapid and is nearly completed after 20 h; at this time, H_{II} has not yet released half of the initially loaded drug.

In Fig. 12.29b the profiles of drug released are plotted against the square root of time; the linear behavior confirms the Fickian diffusion release. Using the Higuchi equation, the diffusion coefficients are calculated to be $D_{pH\ 2} = 2.2 \times 10^{-8}\ cm^2/s$ and $D_{pH\ 7} = 30.6 \times 10^{-8}\ cm^2/s$ (slopes $= 1.34 \times 10^{-8}$ at pH 2 and 5×10^{-8} at

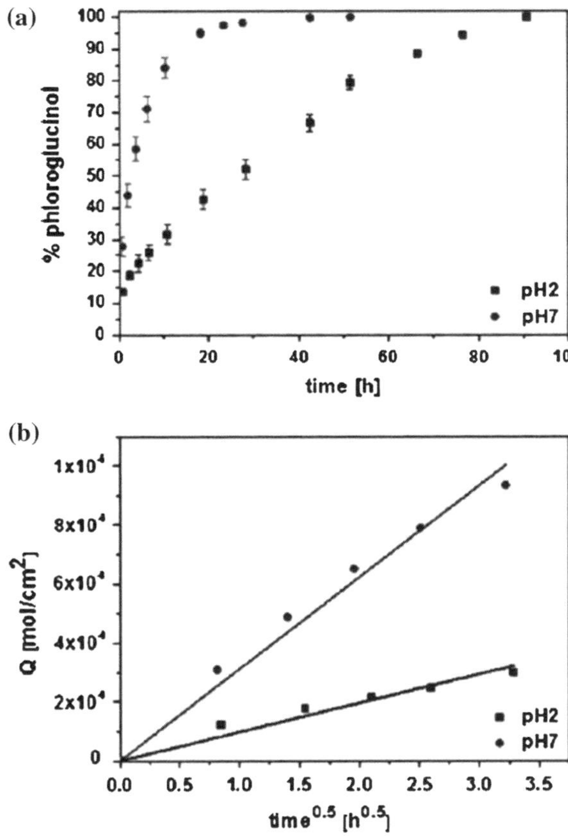

Fig. 12.29 pH-induced changes in the release of phloroglucinol from the reverse bicontinuous cubic and hexagonal phases at 37 °C. **a** Phloroglucinol released from the bicontinuous cubic phase at pH 7 (●) and reverse hexagonal phase at pH 2 (■) plotted against time. **b** Moles released per unit area plotted against the square root of time, illustrating the different diffusion-controlled behaviors in the two mesophases. (Taken from [108])

pH 7), respectively. It is therefore possible to conclude that the bicontinuous cubic phase releases almost 4 times faster than the reverse hexagonal phase, consistent with previous reported work. To demonstrate a pH-triggered on–off release behavior and tunable release profiles in dynamic conditions, the release study was also carried out by switching the pH between the values of 7 and 2 using the same setup, and the resulting release data are given in Fig. 12.30. The values of the slopes obtained in these conditions are 5.7×10^{-8} in the initial pH 7 conditions, 2.58×10^{-8} after switching the solution at pH 2, and 5.3×10^{-8} upon reverting the pH to 7 again. The values of the slopes at pH 7 agree well with those obtained by the static release experiment, whereas that measured at pH 2 shows a slightly faster release; this results from the relatively long time needed to induce a complete change of *Im*3*m* into H_{II} in bulk mesophases, leading to a long-living coexistence of

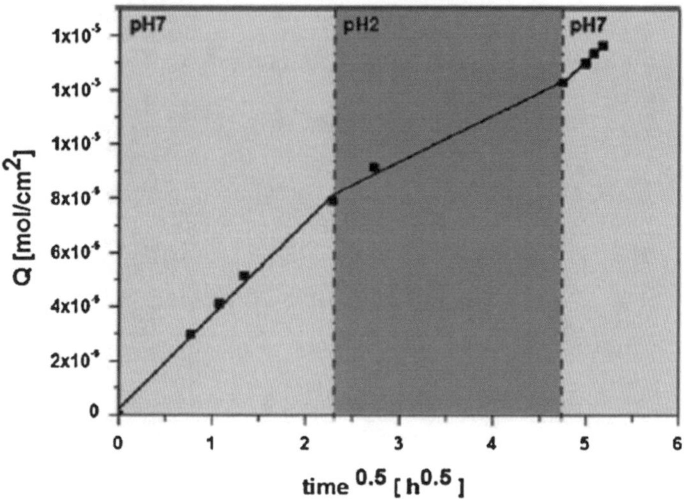

Fig. 12.30 Release profiles of phloroglucinol plotted against the square root of time, upon sequential switch in the pH of the excess solution from pH 7 → pH 2 → pH 7. (Taken from [108])

*Im*3*m* and H_{II} upon the switch of the pH from 7 to 2. It can be easily anticipated that, when applying these concepts to cubosomes and hexosomes of a few hundred nanometers in diameter, the order–order transitions will occur much faster, leading to a sharper pH-induced on–off–on release along the gastrointestinal tract.

Lee and Kellaway investigated the effect of polyethylene glycol 200 (PEG 200) on increasing the release of a lipophilic permeation enhancer oleic acid incorporated into the cubic phase of GMO and water [59]. They have also explored the combined effect of oleic acid and PEG 200 as a permeation enhancer and so-called co-enhancer, respectively, regarding their ability to enhance ex vivo buccal permeability of a model peptide from the cubic phase [D-Ala^2, D-Leu^5] enkephalin (DADLE, $Mw = 569.7$). The experiments concerning the in vitro release of oleic acid from the cubic phase of GMO and the permeation enhancing effect of oleic acid on the ex vivo buccal transport of DADLE from the cubic phase have revealed that most of the incorporated oleic acid was retained within the cubic phase, resulting in a poor permeability enhancing effect. Lee and Kellaway [59] hypothesized that increased release of oleic acid from the cubic phase would cause a greater permeation enhancing effect. In order to increase the amount of oleic acid released from the cubic phase, PEG 200 was co-administered with oleic acid due to the ability of PEGs to solubilize lipophilic compounds.

The in vitro release of oleic acid from the cubic phase increased as the PEG 200 content increased. This was explained by the impact of PEG 200 on solubilizing oleic acid. Due to the lipophilic nature of oleic acid its release from the cubic phase into aqueous PBS medium was not readily achieved. The solubilizing effect of PEG 200 on oleic acid incorporated into the cubic phase was not linear, i.e.

enhancement ratios, calculated from the cumulative quantity of oleic acid released with PEG 200 per cumulative quantity of oleic acid released without PEG 200, were 1.60 (1 % PEG 200), 6.05 (5 % PEG 200), and 9.93 (10 % PEG 200). However, the partitioning of oleic acid between the lipid domains and the aqueous regions of the cubic phase of glyceryl monooleate can be controlled by the PEG 200 concentration. Another reason for the increased release of oleic acid by PEG 200 might be that the polymer enhanced the penetration of water into the cubic phase matrix thereby leading to phase changes, affecting the release of oleic acid.

The addition of PEG 200 to the cubic phase containing 1.0 % oleic acid also facilitated the ex vivo permeability of DADLE across porcine buccal mucosa. On the other hand, 1 % oleic acid alone and the addition of 1 % PEG 200 to 1 % oleic acid did not show a statistically significant enhancing effect even though their mean fluxes slightly increased. Thus, the steady-state flux improvement may be due to increased oleic acid solubility brought about by the PEG 200 and the concomitant increase in oleic acid concentration in the buccal tissue. In addition, it was shown that PEG 200 alone in varying concentrations (1–10 % w/w) did not increase the flux of DADLE. Therefore, the appreciable enhancement effect of buccal permeation of DADLE from the cubic phase containing 1 % oleic acid and PEG 200 (5 and 10 %) were attributed to the solubilization of the lipophilic permeation enhancer by PEG 200. The results of this study demonstrated that the combination of lipophilic enhancer together with a cosolvent in the drug carrier may lead to optimization of the enhancing capability.

12.2.9 Probable Mechanisms of Delivery

The transdermal studies conducted to date are still lacking information concerning the mode of absorption and distribution in the skin. The following study by Bender et al. aimed to investigate distribution of a hydrophilic fluorescent model drug sulphorhodamine B (SRB) by visualizing its uptake in full-thickness human skin using four delivery systems [109]. The delivery vehicles applied were two bi-continuous lipid cubic systems consisting of either GMO or phytantriol (PT) and water. Water and a commercial ointment have been used as reference vehicles for comparison formulations with cubic internal structures to an emulsion with discrete micrometer-sized aggregates (ointment) and lack internal order (water). The formulations were applied on full thickness human skin (during 24 h) and thereafter investigated using two-photon microscopy (TPM). Two-photon excitation is an outcome of a nonlinear excitation process through simultaneous absorption of two photons, implying that near infrared light (NIR) can be used for excitation of fluorophores with fluorescence in the visible range of light. Since the NIR wavelengths lie in the so-called "optical window" of biological tissue, TPM enables imaging of fluorophores much deeper into highly light scattering and light absorbing tissue compared to confocal laser scanning microscopy, with minimal photo-bleaching and phototoxic effects.

The commercial ointment and cubic phases were found to reside in skin wrinkles due to the high viscosity and water insolubility of the formulations. Cubic phases, known to co-exist with excess water can, therefore, resist washing with water or PBS. Moreover, their rheological characterization has revealed elastic rather than viscous behavior. The micro-fissures reported in the present study were approximately 5 μm wide, and had an irregular and entangled structure. The intercellular pathway seems to be predominant when using the water vehicle and ointment, while the inter-cluster pathway seems to dominate the skin absorption from the cubic phases. The micro-fissures were not visible when performing TPM of skin auto-fluorescence [110]; however, these structures were expected to be present as a microscopic clustering of keratinocytes in normal skin. The elastic lipid cubic phases seemed to be able to penetrate deep into these pre-existing micro-fissures, giving rise to the observed fluorescence pattern. Even if the micro-fissures become narrower further into the skin, SRB fluorescence was detectable deeper than 30 μm in about 50 % of the GMO cubic and approximately 70 % of the phytantriol cubic phases. Thus, this route contributed to a more efficient delivery of hydrophilic substances using the bicontinuous cubic samples as drug delivery vehicles.

Another obvious difference that was obtained by the cubic formulations, compared to water, was the thin threadlike structures that presented down to a depth of approximately 10 μm in the former. The authors interpreted this pattern as arising from lateral diffusion of the formulation into the lipid matrix between the cell layers, which can be explained by an interaction between the lipid bilayers of the formulation and the cellular lipid matrix. The GMO cubic phase, compared to the PT cubic phase, penetrated the lipid matrix in a finer and more homogeneously distributed fluorescence pattern, implying different lipid bilayers-lipid matrix interactions between the cubic phase and the skin. This may be due to the different chain mobility of the oleoyl chains (GMO) and the phytanyl chains (PT).

12.3 Conclusions

Various liquid crystalline-based delivery systems have been developed for the delivery of pharmaceuticals. These delivery systems were found to be efficient in incorporating large quantities of drugs and could contain permeation enhancers for achieving improved bioavailability and were able to sustain and control their delivery.

It was shown that the properties of the liquid crystalline structures can be tuned to regulate the rate of transdermal and mucosal drug release. Based on the concept that the cubic phase provides a much faster release rate than the H_{II} mesophase, stimuli-responsive LLC systems were developed. Such polymorphism of lyotropic lipid-based mesophases in physiological conditions allows change to the nanostructure in response to external change in temperature, enabling control of drug release rates. On the whole, sustained release of hydrophilic drugs can be achieved using hydrophilic drugs, while in the case of lipophilic and an amphiphilic drugs increased penetration was demonstrated.

A novel approach to increase skin permeability for transdermal delivery utilizing pore-forming peptides incorporated within hexagonal mesophase was introduced. A recent investigation regarding distribution of hydrophilic fluorescent model drugs in full-thickness human skin was demonstrated. It was suggested that the intercluster penetration pathway is preferable for delivery of hydrophilic compounds via elastic cubic LLC, in contrast to the intercellular pathway when using the water solution.

The results of the demonstrated studies, although very interesting and promising, definitely do not cover all aspects and do not answer all the questions addressed by the fascinating field of lyotropic liquid crystals and their applications as drug delivery systems. However, recent achievements showed the great potential of these systems, providing new thoughts and interest in this rapidly growing field of research.

This chapter demonstrated that LC mesophases with complex architectures were rationally designed and prepared for the solubilization and potential administration of bioactive molecules. It was shown that comprehensive understanding of the structural properties of the carriers is imperative for rational and successful tailoring of delivery vehicles. The principal strategy was to first characterize the different levels of organization of these special materials and then to explore the detailed relationships and possible correlations between them, their structures, and the macroscopic properties.

The results lately obtained, although very interesting and promising, definitely do not cover all the aspects and do not answer all the questions addressed by the fascinating field of lyotropic liquid crystals and their applications as drug delivery systems. Clearly, many additional experiments need to be carried out in order to clarify the detailed structure, the exact properties, and specific potential of these systems. We hope that we at least opened a new window and provided new thoughts and interest into this rapidly growing field of research.

References

1. D.L. Gin, C.S. Pecinovsky, J.E. Bara, L. Kerr, Functional lyotropic liquid crystal materials. Struct. Bond. **128**, 181–222 (2008)
2. K. Larsson, Cubic lipid–water phases: structures and biomembrane aspects. J. Phys. Chem. **93**, 7304–7314 (1989)
3. I. Amar-Yuli, Hexagonal liquid crystals and hexosomes structural modifications and solubilization. Ph.D. dissertation, The Hebrew University of Jerusalem, Israel, 2008
4. S.M. Gruner, Stability of lyotropic phases with curved interfaces. J. Phys. Chem. **93**, 7562–7570 (1989)
5. J.M. Seddon, Structure of the inverted hexagonal (H_{ii}) phase, and non-lamellar phase-transitions of lipids. Biochim. Biophys. Acta **1031**, 1–69 (1990)
6. L. Sagalowicz, R. Mezzenga, M.E. Leser, Investigating reversed liquid crystalline mesophases. Curr. Opin. Colloid Interface Sci. **11**, 224–229 (2006)

7. L. Sagalowicz, M.E. Leser, H.J. Watzke, M. Michel, Monoglyceride self-assembly structures as delivery vehicles. Trends Food Sci. Tech. **17**, 204–214 (2006)
8. V. Cherezov, J. Clogston, M.Z. Papiz, M. Caffrey, Room to move: crystallizing membrane proteins in swollen lipidic mesophases. J. Mol. Biol. **357**, 1605–1618 (2006)
9. B.J. Boyd, S.-M. Khoo, D.V. Whittaker, G. Davey, C.J.H. Porter, A lipid-based liquid crystalline matrix that provides sustained release and enhanced oral bioavailability for a model poorly water soluble drug in rats. Int. J. Pharm. **340**, 52–60 (2007)
10. C.J. Drummond, C. Fong, Surfactant self-assembly objects as novel drug delivery vehicles. Curr. Opin. Colloid Interface Sci. **4**, 449–456 (2000)
11. S.B. Rizwan, T. Hanley, B.J. Boyd, T. Rades, S. Hook, Liquid crystalline systems of phytantriol and glyceryl monooleate containing a hydrophilic protein: characterisation, swelling and release kinetics. J. Pharm. Sci. **98**, 4191–4204 (2009)
12. G.C. Shearman, O. Ces, R.H. Templer, J.M. Seddon, Inverse lyotropic phases of lipids and membrane curvature. J. Phys. Condens. Matter **18**, S1105–S1124 (2006)
13. M. Makai, E. Csányi, Zs Németh, J. Pálinkás, I. Erős, Structure and drug release of lamellar liquid crystals containing glycerol. Int. J. Pharm. **256**, 95–107 (2003)
14. S.M. Gruner, M.W. Tate, G.L. Kirk, P.T.C. So, D.C. Turner, D.T. Keane, C.P.S. Tilcock, P.R. Cullis, X-ray-diffraction study of the polymorphic behavior of N-Methylated Dioleoylphosphatidylethanolamine. Biochemistry **27**, 2853–2866 (1988)
15. M. Gradzielski, Investigations of the dynamics of morphological transitions in amphiphilic systems. Curr. Opin. Colloid Interface Sci. **9**, 256–263 (2004)
16. L. Yang, L. Ding, H.W. Huang, New phases of phospholipids and implications to the membrane fusion problem. Biochemistry **42**, 6631–6635 (2003)
17. L.B. Lopes, F.F.F. Speretta, M.V.L.B. Bentley, Enhancement of skin penetration of vitamin k using monoolein-based liquid crystalline systems. Eur. J. Pharm. Sci. **32**, 209–215 (2007)
18. B.J. Boyd, D.V. Whittaker, S.-M. Khoo, G. Davey, Lyotropic liquid crystalline phases formed from glycerate surfactants as sustained release drug delivery systems. Int. J. Pharm. **309**, 218–226 (2006)
19. M. Caffrey, Kinetics and mechanism of the lamellar gel lamellar liquid-crystal and lamellar inverted hexagonal phase-transition in phosphatidylethanolamine: a real x-ray diffraction study using synchrotron radiation. Biochemistry **24**, 4826–4844 (1985)
20. R.P. Rand, V. Luzzati, X-ray diffraction study in water of lipids extracted from human erythrocytes - position of cholesterol in lipid lamellae. Biophys. J. **8**, 125–137 (1968)
21. S. Hyde, S. Andersson, K. Larsson, Z. Blum, T. Landh, S. Lidin, B.W. Ninham, *The Language of Shape. The Role of Curvature in Condensed Matter: Physics, Chemistry and Biology* (Elsevier, Amsterdam, 1997)
22. J.N. Israelachvili, D.J. Mitchell, B.W. Ninhan, Theory of self-assembly of hydrocarbon amphiphiles into micelles and bilayers. J. Chem. Soc. Faraday Trans. 2 **72**, 1525–1568 (1976)
23. M.H. Shah, A. Paradkar, Cubic liquid crystalline glyceryl monooleate matrices for oral delivery of enzyme. Int. J. Pharm. **294**, 161–171 (2005)
24. J.M. Seddon, J. Robins, T. Gulik-Krzywicki, H. Delacroix, Inverse micellar phases of phospholipids and glycolipids. Phys. Chem. Chem. Phys. **2**, 4485–4493 (2000)
25. J.M. Seddon, N. Zeb, R.H. Templer, R.N. McElhaney, D.A. Mannock, An Fd3m lyotropic cubic phase in a binary glycolipid/water system. Langmuir **12**, 5250–5253 (1996)
26. I. Amar-Yuli, N. Garti, Transitions induced by solubilized fat into reverse hexagonal mesophases. Colloids Surf. B **43**, 72–82 (2005)
27. I. Amar-Yuli, E. Wachtel, E. Ben Shoshan, D. Danino, A. Aserin, N. Garti, Hexosome and hexagonal phases mediated by hydration and polymeric stabilizer. Langmuir **23**, 3637–3645 (2007)
28. I. Amar-Yuli, E. Wachtel, D. Shalev, H. Moshe, A. Aserin, N. Garti, Thermally induced fluid reversed hexagonal (H_{II}) mesophase. J. Phys. Chem. B **111**, 13544–13553 (2007)
29. I. Amar-Yuli, E. Wachtel, D. Shalev, A. Aserin, N. Garti, Low viscosity reversed hexagonal mesophases induced by hydrophilic additives. J. Phys. Chem. B **112**, 3971–3982 (2008)

30. I. Amar-Yuli, A. Aserin, N. Garti, Solubilization of nutraceuticals into reverse hexagonal mesophases. J. Phys. Chem. B **112**, 10171–10180 (2008)
31. I. Amar-Yuli, D. Libster, A. Aserin, N. Garti, Solubilization of food bioactives within lyotropic liquid crystalline mesophases. Curr. Opin. Colloid Interface Sci. **14**, 21–32 (2009)
32. L. Bitan-Cherbakovsky, I. Amar-Yuli, A. Aserin, N. Garti, Structural rearrangements and interaction within H_{II} mesophase induced by cosolubilization of vitamin E and ascorbic acid. Langmuir **25**, 13106–13113 (2009)
33. L. Bitan-Cherbakovsky, I. Amar-Yuli, A. Aserin, N. Garti, Solubilization of vitamin E into H_{II} LLC mesophase in the presence and in the absence of vitamin C. Langmuir **26**, 3648–3653 (2010)
34. D. Libster, A. Aserin, E. Wachtel, G. Shoham, N. Garti, An H_{II} liquid crystal-based delivery system for cyclosporin a: physical characterization. J. Colloid Interf. Sci. **308**, 514–524 (2007)
35. D. Libster, P. Ben Ishai, A. Aserin, G. Shoham, N. Garti, From the microscopic to the mesoscopic properties of lyotropic reverse hexagonal liquid crystals. Langmuir **24**, 2118–2127 (2008)
36. D. Libster, P. Ben Ishai, A. Aserin, G. Shoham, N. Garti, molecular interactions in reverse hexagonal mesophase in the presence of cyclosporin A. Int. J. Pharm. **367**, 115–126 (2009)
37. P. Ben Ishai, D. Libster, A. Aserin, N. Garti, Y. Feldman, Molecular interactions in lyotropic reverse hexagonal liquid crystals: a dielectric spectroscopy study. J. Phys. Chem. B **113**, 12639–12647 (2009)
38. Y. Kurosaki, N. Nagahara, T. Tanizawa, H. Nishimura, T. Nakayama, T. Kimura, Use of lipid disperse systems in transdermal drug delivery: comparative-study of flufenamic acid permeation among rat abdominal skin, silicon rubber membrane and stratum-corneum sheet isolated from hamster-cheek pouch. Int. J. Pharm. **114**, 151–157 (1991)
39. E. Touitou, F. Levischaffer, N. Dayan, F. Akhaique, F. Riccieri, Modulation of caffeine skin delivery by carrier design: liposomes versus permeation enhancers. Int. J. Pharm. **103**, 131–136 (1994)
40. A. Spernath, A. Aserin, N. Garti, Fully dilutable microemulsions embedded with phospholipids and stabilized by short-chain organic acids and polyols. J. Colloid Interf. Sci. **299**, 900–909 (2006)
41. H. Brondsted, H.M. Nielsen, L. Hovgaard, Drug-delivery studies in caco-2 monolayers. 3. intestinal transport of various vasopressin analogs in the presence of lysophosphatidylcholine. Int. J. Pharm. **114**, 151–157 (1995)
42. D.-Z. Liu, E.L. LeCluyse, D.R. Thakker, Dodecylphosphocholine-mediated enhancement of paracellular permeability and cytotoxicity in caco-2 cell monolayers. J. Pharm. Sci. **88**, 1161–1168 (1999)
43. D. Libster, A. Aserin, D. Yariv, G. Shoham, N. Garti, Concentration- and temperature-induced effects of incorporated desmopressin on the properties of reverse hexagonal mesophase. J. Phys. Chem. B **113**, 6336–6346 (2009)
44. M. Cohen-Avrahami, A. Aserin, N. Garti, H_{II} mesophase and peptide cell-penetrating enhancers for improved transdermal delivery of sodium diclofenac. Colloids Surf. B **77**, 131–138 (2010)
45. T. Mishraki, D. Libster, A. Aserin, N. Garti, Lysozyme entrapped within reverse hexagonal mesophases: physical properties and structural behavior. Colloids Surf. B **75**, 47–56 (2010)
46. T. Mishraki, D. Libster, A. Aserin, N. Garti, Temperature-dependent behavior of lysozyme within the reverse hexagonal mesophases (H_{II}). Colloids Surf. B **75**, 391–397 (2010)
47. M. Cohen-Avrahami, D. Libster, A. Aserin, N. Garti, Sodium diclofenac and cell-penetrating peptides embedded in h_{ii} mesophases: physical characterization and delivery. J. Phys. Chem. B **115**, 10189–10197 (2011)
48. S. Murgia, S. Lampis, R. Angius, D. Berti, M. Monduzzi, Orientation and specific interactions of nucleotides and nucleolipids inside monoolein-based liquid crystals. J. Phys. Chem. B **113**, 9205–9215 (2009)

49. I. Amar-Yuli, D. Azulay, T. Mishraki, A. Aserin, N. Garti, The role of glycerol and phosphatidylcholine in solubilizing and enhancing insulin stability in reverse hexagonal mesophases. J. Colloid Interf. Sci. **364**, 379–387 (2011)
50. R. Mezzenga, M. Grigorov, Z. Zhang, C. Servais, L. Sagalowicz, A.I. Romoscanu, Polysaccharide-induced order-to-order transitions in lyotropic liquid crystals. Langmuir **21**, 6165–6169 (2005)
51. C.R. Daas, P.F.M. Choong, Biophysical delivery of peptides: applicability for cancer therapy. Peptides **27**, 3479–3488 (2006)
52. T.R.S. Kumar, K. Soppimath, S.K. Nachaegari, Novel delivery technologies for protein and peptide therapeutics. Curr. Pharm. Biotech. **7**, 261–276 (2006)
53. B. Ericsson, P.O. Eriksson, J.E. Löfroth, S. Engström, *Cubic Phases as Delivery Systems for Peptide Drugs, Polymeric Drugs and Drug Delivery System* (American Chemical Society, Washington, DC, 1991), pp. 251–265
54. J.C. Shah, Y. Sadhale, M.C. Dakshina, Cubic phase gels as drug delivery systems. Adv. Drug Deliv. Rev. **47**, 229–250 (2001)
55. W.K. Fong, T. Hanley, B.J. Boyd, Stimuli responsive liquid crystals provide 'on-demand' drug delivery in vitro and in vivo. J. Control. Rel. **135**, 218–226 (2009)
56. J. Clogston, M. Caffrey, Controlling release from the lipidic cubic phase. amino acids, peptides, proteins and nucleic acids. J. Control. Rel. **107**, 97–111 (2005)
57. D. Libster, A. Aserin, N. Garti, Interactions of biomacromolecules with reverse hexagonal liquid crystals: drug delivery and crystallization applications. J. Colloid Interf. Sci. **356**, 375–386 (2011)
58. I. Amar-Yuli, J. Adamcik, S. Blau, A. Aserin, R. Mezzenga, Controlled embedment and release of DNA from lipidic reverse columnar hexagonal mesophases. Soft Matter **7**, 8162–8168 (2011)
59. J. Lee, I.W. Kellaway, Combined effect of oleic acid and polyethylene glycol 200 on buccal permeation of [D-Ala(2), D-Leu(5)]enkephalin from a cubic phase of glyceryl monooleate. Int. J. Pharm. **204**, 137–144 (2000)
60. M.G. Carr, J. Corish, O.I. Corrigan, Drug delivery from a liquid crystalline base across visking and human stratum corneum. Int. J. Pharm. **157**, 35–42 (1997)
61. M.G. Lara, M.V.L.B. Bentley, J.H. Collet, In vitro drug release mechanism and drug loading studies of cubic phase gels. Int. J. Pharm. **293**, 241–250 (2005)
62. L.B. Lopes, D.A. Ferreira, D. de Paula, M.T.J. Garcia, J.A. Thomazini, M.C.A. Fantini, M.V.L.B. Bentley, Reverse hexagonal phase nanodispersion of monoolein and oleic acid for topical delivery of peptides: in vitro and in vivo skin penetration of cyclosporin A. Pharm. Res. **23**, 1332–1342 (2006)
63. L.B. Lopes, J.L.C. Lopes, D.C.R. Oliveira, J.A. Thomazini, M.T.J. Garcia, M.C.A. Fantini, J.H. Collett, M.V.L.B. Bentely, Liquid crystalline phases of monoolein and water for topical delivery of cyclosporin a: characterization and study of in vitro and in vivo delivery. Eur. J. Pharm. Biopharm. **63**, 146–155 (2006)
64. J. Barauskas, I. Švedaitė, E. Butkus, V. Razumas, K. Larsson, F. Tiberg, Synthesis and aqueous phase behavior of 1-glyceryl monooleyl ether. Colloids Surf. B **41**, 49–53 (2005)
65. J. Barauskas, T. Landh, Phase behavior of the phytantriol/water system. Langmuir **19**, 9562–9565 (2003)
66. Y.D. Dong, I. Larson, T. Hanley, B.J. Boyd, Bulk and dispersed aqueous phase behavior of phytantriol: effect of vitamin E acetate and F127 polymer on liquid crystal nanostructure. Langmuir **22**, 9512–9518 (2006)
67. J. Bender, M.B. Ericson, N. Merclin, V. Iani, A. Rosen, S. Engström, J. Moan, Lipid cubic phases for improved topical drug delivery in photodynamic therapy. J. Control. Rel. **106**, 350–360 (2005)
68. J. Borné, T. Nylander, A. Khan, Phase behavior and aggregate formation for the aqueous monoolein system mixed with sodium oleate and oleic acid. Langmuir **17**, 7742–7751 (2001)

69. H. Trommer, R.H.H. Neubert, Overcoming the stratum corneum: the modulation of skin penetration. A review. Skin Pharmacol. Physiol. **19**, 106–121 (2006)
70. T.M. Dellinger, P.V. Braun, Lyotropic liquid crystals as nanoreactors for nanoparticle synthesis. Chem. Mater. **16**, 2201–2207 (2004)
71. E. Esposito, R. Cortesi, M. Drechsler, L. Paccamiccio, P. Mariani, C. Contado, E. Stellin, E. Menegatti, F. Bonina, C. Puglia, Cubosome dispersions as delivery systems for percutaneous administration of indomethacin. Pharm. Res. **22**, 2163–2173 (2005)
72. N.K. Swarnakar, V. Jain, V. Dubey, D. Mishra, N.K. Jain, Enhanced oromucosal delivery of progesterone via hexosomes. Pharm. Res. **24**, 2223–2230 (2007)
73. J. Gustafsson, H. Ljusberg-Wahren, M. Almgren, K. Larsson, Submicron particles of reversed lipid phases in water stabilized by a nonionic amphiphilic polymer. Langmuir **13**, 6964–6971 (1997)
74. M. Almgren, J. Borné, E. Feitosa, A. Khan, B. Lindman, Dispersed lipid liquid crystalline phases stabilized by a hydrophobically modified cellulose. Langmuir **23**, 2768–2777 (2007)
75. J. Barauskas, A. Misiunas, T. Gunnarsson, F. Tiberg, M. Johnsson, "Sponge" nanoparticle dispersions in aqueous mixtures of diglycerol monooleate, glycerol dioleate, and polysorbate 80. Langmuir **22**, 6328–6334 (2006)
76. D. Libster, A. Aserin, D. Yariv, G. Shoham, N. Garti, Soft matter dispersions with ordered inner structures, stabilized by ethoxylated phytosterols. Colloids Surf. B **74**, 202–215 (2009)
77. L. Simonsen, A. Kane, J. Lloyd, M. Zaffran, M. Kane, Unsafe injections in the developing world and transmission of bloodborne pathogens: a Review. Bull. World Health Organ. **77**, 789–800 (1999)
78. M.E. de Vries, H.E. Bodd, J.C. Verhoef, H.E. Junginger, Developments in buccal drug delivery. Crit. Rev. Ther. Drug Carr. Syst. **8**, 271–303 (1991)
79. C.-M. Lehr, Bioadhesion technologies for the delivery of peptide and protein drugs to the gastrointestinal-tract. Crit. Rev. Ther. Drug Carr. Syst. **11**, 119–160 (1994)
80. A. Ahuja, R.K. Khar, J. Ali, Mucoadhesive drug delivery systems. Drug Dev. Ind. Pharm. **23**, 489–515 (1997)
81. A. Arora, M.R. Prausnitz, S. Mitragotri, Micro-scale devices for transdermal drug delivery. Int. J. Pharm. **364**, 227–236 (2008)
82. A. Goebel, R.H.H. Neubert, Dermal peptide delivery using colloidal carrier systems. Skin Pharmacol. Physiol. **21**, 3–9 (2008)
83. J.-H. Park, J.-W. Lee, Y.-C. Kim, M.R. Prausnitz, The effect of heat on skin permeability. Int. J. Pharm. **359**, 94–103 (2008)
84. H.A.E. Benson, Transdermal drug delivery: penetration enhancement techniques. Curr. Drug Delivery **2**, 23–33 (2005)
85. J. Hadgraft, Passive enhancement strategies in topical and transdermal drug delivery. Int. J. Pharm. **184**, 1–6 (1999)
86. J.I. Duncan, S.N. Payne, A.J. Winfield, A.D. Ormerod, A.W. Thomson, Enhanced percutaneous-absorption of a novel topical cyclosporine-a formulation and assessment of its immunosuppressive activity. Br. J. Dermatol. **123**, 631–640 (1990)
87. L.B. Lopes, J.H. Collet, M.V.L.B. Bentley, Topical delivery of cyclosporin A: an in vitro study using monoolein as a penetration enhancer. Eur. J. Pharm. Biopharm. **60**, 25–30 (2005)
88. N.H. Gabboun, N.M. Najib, H.G. Ibrahim, S. Assaf, Release of salicylic acid, diclofenac acid and diclofenac acid salts from isotropic and anisotropic nonionic surfactant systems across rat skin. Int. J. Pharm. **212**, 73–80 (2001)
89. W. Lou, A.T. Quintana, R.G. Geronemus, M.C. Grossman, Effects of topical vitamin K and retinol on laser-induced purpura on nonlesional skin. Dermatol. Surg. **25**, 942–944 (1999)
90. M.L. Elson, S. Nacht, Treatment of periorbital hyperpigmentation with topical vitamin K/ vitamin A. Cosmet. Dermatol. **10**, 32–34 (1999)
91. T.J. Im, M.J. Kang, D.W. Seo, J. Lee, Effect of cubic liquid crystalline systems on skin localization of oregonin and hirsutanonol. Biomolecules Ther. **16**, 226–230 (2008)

92. M. Guyot, F. Fawaz, Design and in vitro evaluation of adhesive matrix for transdermal delivery of propranolol. Int. J. Pharm. **204**, 171–182 (2000)
93. F.O. Costa-Balogh, E. Sparr, J.J.S. Sousa, A.A.C.C. Pais, Drug release and skin permeation from lipid liquid crystalline phases. Progr. Colloid Polym. Sci. **135**, 119–129 (2008)
94. A. Namdeo, N.K. Jain, Liquid crystalline pharmacogel based enhanced transdermal delivery of propranolol hydrochloride. J. Control. Rel. **82**, 223–236 (2002)
95. S. Lundin, P. Artursson, Absorption of a vasopressin analog, 1deamino-8-D-arginine-vasopressin (DDAVP), in a human intestinal epithelial-cell line, CaCO-2. Int. J. Pharm. **64**, 181–186 (1990)
96. A. Fjellestad-Paulsen, C. Söderberg-Ahlm, S. Lundin, Metabolism of vasopressin, oxytocin, and their analogs in the human gastrointestunal-tract. Peptides **16**, 1141–1147 (1995)
97. K. Fredholt, J. Østergaard, J. Savolainen, G.J. Friis, Alpha-chymotrypsin-catalized degradation of desmopressin (dDAVP): influence of pH, concentration and various cyclodextrins. Int. J. Pharm. **178**, 223–229 (1999)
98. K.W.Y. Lee, T.-H. Nguyen, T. Hanley, B.J. Boyd, Nanostructure of liquid crystalline matrix determines in vitro sustained release and in vivo oral absorption kinetics for hydrophilic model drugs. Int. J. Pharm. **365**, 190–199 (2009)
99. A. Joliot, C. Pernelle, H. Deagostini-Bazin, A. Prochiantz, Antennapedia homeobox peptide regulates neural morphogenesis. Proc. Natl. Acad. Sci. U.S.A. **88**, 1864–1868 (1991)
100. V. Molle, S. Campagna, Y. Bessin, N. Ebran, N. Saint, G. Molle, First evidence of the pore-forming properties of a keratin from skin mucus of rainbow trout (oncorhynchus mykiss, formerly salmo gairdneri). Biochem. J. **411**, 33–40 (2008)
101. Y.C. Kim, S. Late, A.K. Banga, P.J. Ludovice, M.R. Prausnitz, Biochemical enhancement of transdermal delivery with magainin peptide: modification of electrostatic interactions by changing pH. Int. J. Pharm. **362**, 20–28 (2008)
102. E. Vives, J. Schmidt, A. Pelegrin, Cell-penetrating and cell-targeting peptides in drug delivery. BBA-Rev. Cancer **1786**, 126–138 (2008)
103. M. Mae, U. Langel, Cell-penetrating peptides as vectors for peptide, protein and oligonucleotide delivery. Curr. Opin. Pharmacol. **6**, 509–514 (2006)
104. W. Li, F. Nicol, F.C. Szoka, GALA: a designed synthetic pH-responsive amphipathic peptide with applications in drug and gene delivery. Adv. Drug Deliv. Rev. **56**, 967–985 (2004)
105. S. Nir, J.L. Nieva, Interactions of peptides with liposomes: pore formation and fusion. Progr. Lipid Res. **39**, 181–206 (2000)
106. S. Nir, F. Nicol, F.C. Szoka, Surface aggregation and membrane penetration by peptides: relation to pore formation and fusion. Mol. Membr. Biol. **16**, 95–101 (1999)
107. W.K. Fong, T. Hanley, B. Thierry, N. Kirby, B.J. Boyd, Plasmonic nanorods provide reversible control over nanostructure of self-assembled drug delivery materials. Langmuir **26**, 6136–6139 (2010)
108. R. Negrini, R. Mezzenga, pH-responsive lyotropic liquid crystals for controlled drug delivery. Langmuir **27**, 5296–5303 (2011)
109. J. Bender, C. Simonsson, M. Smedh, S. Engström, M.B. Ericson, Lipid cubic phases in topical drug delivery: visualization of skin distribution using two-photon microscopy. J. Control. Rel. **129**, 163–169 (2008)
110. J. Paoli, M. Smedh, A.M. Wennberg, M.B. Ericson, Multiphoton laser scanning microscopy on non-melanoma skin cancer: morphologic features for future non-invasive diagnostics. J. Invest. Dermatol. **128**, 1248–1255 (2008)

Index

Q. Li (ed.), *Nanoscience with Liquid Crystals*, NanoScience and Technology,
DOI: 10.1007/978-3-319-04867-3,

Printed by Printforce, the Netherlands